DYNAMICAL EVOLUTION OF DENSE STELLAR SYSTEMS

IAU SYMPOSIUM No. 246

COVER ILLUSTRATION:

47 Tuc, one of the most massive Galactic globular clusters. Image based on data obtained with FORS1 on ESO's Very Large Telescope.

Credit ESO; see http://www.eso.org/public/outreach/press-rel/pr-2006/pr-20-06.html

INTERNATIONAL ASTRONOMICAL UNION

UNION ASTRONOMIQUE INTERNATIONALE

International Astronomical Union

DYNAMICAL EVOLUTION OF DENSE STELLAR SYSTEMS

PROCEEDINGS OF THE 246th SYMPOSIUM OF THE INTERNATIONAL ASTRONOMICAL UNION HELD IN CAPRI, ITALY SEPTEMBER 5–9, 2007

Edited by

ENRICO VESPERINI
Department of Physics, Drexel University, Philadelphia, PA, USA

MIREK GIERSZ
Nicolaus Copernicus Astronomical Center, Polish Academy of Sciences, Warsaw, Poland

and

ALISON SILLS
Department of Physics and Astronomy, McMaster University, Hamilton, ON, Canada

CAMBRIDGE
UNIVERSITY PRESS

CAMBRIDGE UNIVERSITY PRESS
The Edinburgh Building, Cambridge CB2 2RU, United Kingdom
32 Avenue of the Americas, New York, NY 10013–2473, USA
477 Williamstown Road, Port Melbourne, VIC 3207, Australia
Ruiz de Alarcón 13, 28014 Madrid, Spain
Dock House, The Waterfront, Cape Town 8001, South Africa

First published 2008

Printed in the United Kingdom at the University Press, Cambridge

Typeset in System LaTeX 2_ε

A catalogue record for this book is available from the British Library

Library of Congress Cataloguing in Publication data

ISBN 9780521874687 hardback
ISSN 1743-9213

Table of Contents

Preface xiii

Organizing committee xiv

Conference participants xvii

Part 1. CLUSTER FORMATION AND EARLY EVOLUTION

Formation of Stellar Clusters and the Importance of Thermodynamics for Fragmentation. 3
    R. S. Klessen, P. C. Clark & S. C. O. Glover

The Formation, Disruption and Properties of Pressure-Supported Stellar Systems and Implications for the Astrophysics of Galaxies. 13
    P. Kroupa

The Early Evolution of Dense Stellar Systems 23
    C. J. Clarke

Dynamical Masses of Young Star Clusters: Constraints on the Stellar IMF and Star-Formation Efficiency 32
    N. Bastian

The Influence of Gas Expulsion on the Evolution of Star Clusters 36
    H. Baumgardt & P. Kroupa

A Dynamical Origin for Early Mass Segregation in Young Star Clusters 41
    S. McMillan, E. Vesperini & S. Portegies Zwart

A Near-infrared Survey of the Rosette Complex: Clues of Early Cluster Evolution 46
    C. G. Román-Zúñiga, E. A. Lada & B. Ferreira

Changing Structures in Galactic Star Clusters 50
    S. Schmeja, M. S. N. Kumar, D. Froebrick & R. S. Klessen

The Formation and Dynamics of the SMC Cluster NGC 346 55
    L. J. Smith

Clustered Star Formation in the Magellanic Clouds 61
    D. A. Gouliermis

The Fraction of Runaway OB Stars in the SMC Field 63
    J. B. Lamb & M. S. Oey

The Relation Between Field Massive Stars and Clusters 65
    M. S. Oey, N. L. King, J. Wm. Parker & J. B. Lamb

Imprints of Stellar Encounters in the ONC 67
    C. Olczak & S. Pfalzner

Capture-Induced Binarity of Massive Stars in Young Dense Clusters 69
    S. Pfalzner & C. Olczak

On the Origin of Complex Stellar Populations in Star Clusters. 71
    J. Pflamm-Altenburg & P. Kroupa

Star Formation in Young Cluster NGC 1893. 73
    S. Sharma, A. K. Pandey, D. K. Ojha, W. P. Chen, S. K. Ghosh, B. C.
    Bhatt, G. Maheswar & R. Sagar

On the Origin of the Orion Trapezium System 75
    H. Zinnecker

Part 2. OPEN CLUSTERS

Open Clusters: Open Windows on Stellar Dynamics 79
    R. D. Mathieu

N-body Models of Open Clusters 89
    J. R. Hurley

Monte Carlo Simulations of Star Clusters with Primordial Binaries. Comparison
    with N-body Simulations and Observations. 99
    M. Giersz & D. C. Heggie

Defining the Binary Star Population in the Young Open Cluster M35 (NGC 2168) 105
    E. K. Braden, R. D. Mathieu & S. Meibom

Tidal Tails of the Nearest Open Clusters. 107
    Y. Chumak & A. Rastorguev

The WIYN Open Cluster Study Photometric Binary Survey: Initial Findings for
    NGC 188. 109
    P. M. Frinchaboy & D. Nielsen

Dynamics of the Open Cluster NGC 188: A Comparison to an N-body simulation
    of M67 111
    A. M. Geller, R. D. Mathieu, H. C. Harris & R. D. McClure

NIR Spectroscopy of the Most Massive Open Cluster in the Galaxy: Westerlund 1 113
    S. Mengel & L. E. Tacconi-Garman

The Population of Open Clusters in the Nearest kpc from the Sun 115
    S. Röser, N. V. Kharchenko, A. E. Piskunov, E. Schilbach & R. -D. Scholz

Tidal Radii and Masses of Galactic Open Clusters. 117
    E. Schilbach, N. V. Kharchenko, A. E. Piskunov, S. Röser & R. -D. Scholz

Part 3. GLOBULAR CLUSTERS

Modelling Individual Globular Clusters 121
    D. C. Heggie & M. Giersz

The Simple Underlying Dynamics of Globular Clusters 131
    I. R. King

Observational Evidence of Multiple Stellar Populations in Globular Clusters . . . 141
    G. Piotto

Effects of Stellar Collisions on Star Cluster Evolution and Core Collapse 151
  S. Chatterjee, J. M. Fregeau & F. Rasio

Multiple Stellar Populations in Globular Clusters: Connection of Information
  from the Horizontal Branch. 156
  F. D'Antona & V. Caloi

Why Haven't Loose Globular Clusters Collapsed yet? 161
  G. De Marchi, F. Paresce & L. Pulone

Dynamical Evolution of Rotating Globular Clusters with Embedded Black Holes 166
  J. Fiestas, O. Porth & R. Spurzem

Star Cluster Life-times: Dependence on Mass, Radius and Environment 171
  M. Gieles, H. J. G. L. M. Lamers & H. Baumgardt

Black Holes and Core Expansion in Massive Star Clusters 176
  A. D. Mackey, M. I. Wilkinson, M. B. Davies & G. F. Gilmore

Dynamical Evolution of Mass-Segregated Clusters 181
  E. Vesperini, S. McMillan & S. Portegies Zwart

N-body Simulations of Star Clusters 187
  P. Anders, H. J. G. L. M. Lamers & H. Baumgardt

Numerical Modelling of the Tidal Tails of NGC 5466. 189
  M. Fellhauer, N. W. Evans, V. Belokurov, M. I. Wilkinson & G. Gilmore

Mass-Loss Timescale of Star Cluster in External Tidal Field 191
  T. Fukushige & A. Tanikawa

Integrated Properties of Mass Segregated Star Clusters. 193
  E. Gaburov & M. Gieles

On the Efficiency of Field Star Capture by Star Clusters. 195
  S. Mieske & H. Baumgardt

Part 4. FEW-BODY SYSTEMS

Resonance, Chaos and Stability in the General Three-Body Problem. 199
  R. A. Mardling

The Problem of Three Stars: Stability Limit. 209
  M. Valtonen, A. Mylläri, V. Orlov & A. Rubinov

A Brief History of Regularisation 218
  S. Mikkola

Numerical Evolution of Single, Binary and Triple Stars. 228
  P. P. Eggleton

Full Ionisation in Binary-Binary Scattering. 233
  W. L. Sweatman

On the Calculation of Average Lifetimes for the 3-body problem 235
    D. Urminsky

Part 5. BINARY STAR DYNAMICS AND ITS INTERPLAY WITH CLUSTER DYNAMICAL EVOLUTION

Binary Stars and Globular Cluster Dynamics 239
    J. M. Fregeau

Evolution of Compact Binary Populations in Globular Clusters: a Boltzmann Study 246
    S. Banerjee & P. Ghosh

Effects of Hardness of Primordial Binaries on Evolution of Star Clusters 251
    A. Tanikawa & T. Fukushige

Dynamical Evolution of Star Clusters with Intermediate Mass Black Holes and
    Primordial Binaries 256
    M. Trenti

The Influence of Binary Stars on Post-Collapse Evolution 261
    R. Apple

The Binary Fraction of NGC 6397 263
    D. S. Davis, H. B. Richer, J. Anderson & J. Brewer

A Post-Newtonian Treatment of Relativistic Compact Object Binaries in Star
    Clusters 265
    J. M. B. Downing & R. Spurzem

The Formation of Contact and Very Close Binaries 267
    P. P. Eggleton & L. Kisseleva-Eggleton

Binaries and the Dynamical Mass of Star Clusters 269
    M. B. N. Kouwenhoven & R. de Grijs

Mass Transfer in Binary Systems: A Numerical Approach 271
    C. -P. Lajoie & A. Sills

Is our Sun a Singleton? 273
    D. Malmberg, M. B. Davies, J. E. Chambers, F. De Angeli, R. P. Church,
    D. Mackey & M. I. Wilkinson

Getting a Kick out of the Stellar Disk(s) in the Galactic Center 275
    H. B. Perets, G. Kupi & T. Alexander

A Search for Spectroscopic Binaries in the Globular Cluster M4 277
    V. Sommariva, G. Piotto, M. Rejkuba, L. R. Bedin, D. C. Heggie, A.
    Milone, R. D. Mathieu & A. Moretti

Part 6. EXOTIC STELLAR POPULATIONS

Blue Straggler Stars in Galactic Globular Clusters: Tracing the Effect of Dynamics
on Stellar Evolution 281
F. Ferraro & B. Lanzoni

Pulsars in Globular Clusters 291
S. M. Ransom

Observational Evidence for the Origin of X-ray Sources in Globular Clusters . . . 301
F. Verbunt, D. Pooley & C. Bassa

Black Hole Motion as Catalyst of Orbital Resonances 311
C. M. Boily, T. Padmanabhan & A. Paiement

Neutron Stars in Globular Clusters 316
N. Ivanova, C. O. Heinke & F. Rasio

Stellar Exotica in 47 Tucanae 321
C. Knigge, A. Dieball, J. Maíz-Apellániz, K. S. Long, D. R. Zurek &
M. M. Shara

Observations and Simulations of the Blue Straggler Star Radial Distribution: Clues
on the Formation Mechanisms. 326
B. Lanzoni

Where the Blue Stragglers Roam: Searching for a Link betwen Formation and
Environment 331
N. Leigh, A. Sills & C. Knigge

An X-ray Emitting Black Hole in a Globular Cluster 336
T. Maccarone, G. Bergond, A. Kundu, K. L. Rhode, J. J. Salzer, I. C. Shih
& S. E. Zepf

Central Dynamics of Globular Clusters: the Case for a Black Hole in ω Centauri 341
E. Noyola, K. Gebhardt & M. Bergmann

Formation and Evolution of Black Holes in Galactic Nuclei and Star Clusters . . 346
R. Spurzem, P. Berczik, I. Berentzen, D. Merritt, M. Preto &
P. Amaro-Seoane

The Imprints of IMBHs on the Structure of Globular Clusters: Monte-Carlo Simulations 351
S. Umbreit, J. M. Fregeau & F. A. Rasio

The Formation and Evolution of Very Massive Stars in Dense Stellar Systems . . 357
H. Belkus, J. Van Bever & D. Vanbeveren

On the Dynamical Capture of a MSP by an IMBH in a Globular Cluster 359
B. Devecchi, M. Colpi, M. Mapelli & A. Possenti

Unveiling the Core of M15 in the Far-Ultraviolet 361
A. Dieball, C. Knigge, D. R. Zurek, M. M. Shara, K. S. Long, P. A. Charles
& D. Hannikainen

Building Blue Stragglers with Stellar Collisions 363
E. Glebbeek & O. R. Pols

On the Origin of Hyperfast Neutron Stars.   365
    V. V. Gvaramadze, A. Gualandris & S. Portegies Zwart

Tracing Intermediate-Mass Black Holes in the Galactic Centre.   367
    U. Löckmann & H. Baumgardt

Environmental Effects on the Globular Cluster Blue Straggler Population: a
    Statistical Approach.   369
    A. Moretti, F. De Angeli & G. Piotto

Paucity of Dwarf Novae in Globular Clusters   371
    P. Pietrukowicz & J. Kaluzny

XMM-*Newton* and *Chandra* Observations of Neutron Stars and Cataclysmic
    Variables in the Globular Cluster NGC 2808.   373
    *M. Servillat, N. A. Webb, D. Barret, R. Cornelisse, A. Dieball, C. Knigge,
    K. S. Long, M. M. Shara & D. R. Zurek*

Part 7. GLOBULAR CLUSTER SYSTEMS

An Update on the ACS Virgo and Fornax Cluster Surveys   377
    *P. Côté, L. Ferrarese, A. Jordán, J. P. Blakeslee, C. -W. Chen, L. Infante,
    S. Mei, E. W. Peng, J. L. Tonry & M. J. West*

Giant Elliptical Galaxies: Globular Clusters and UCDs.   387
    W. E. Harris

Observational Constraints on the Formation and Evolution of Globular Cluster
    Systems   394
    S. E. Zepf

Dynamical Evolution of Globular Clusters in Hierarchical Cosmology   403
    O. Y. Gnedin & J. L. Prieto

Clues to Globular Cluster Evolution from Multiwavelength Observations of
    Extragalactic Systems   408
    A. Kundu, T. J. Maccarone & S. E. Zepf

The Origin of the Universal Globular Cluster Mass Function   413
    G. Parmentier & G. Gilmore

Masses and M/L Ratios of Bright Globular Clusters in NGC 5128.   418
    M. Rejkuba, P. Dubath, D. Minniti & G. Meylan

Slow Evolution of a System of Satellites Induced by Dynamical Friction   423
    S. E. Arena & G. Bertin

Sizes of Confirmed NGC 5128 Globular Clusters   425
    D. Geisler, M. Gomez, K. A. Woodley, W. E. Harris & G. L. H. Harris

Ultra-Compact Dwarf Galaxies – More Massive Than Allowed?   427
    M. Hilker, S. Mieske, H. Baumgardt & J. Dabringhausen

GMOS Spectroscopy of Globular Clusters in Dwarf Elliptical Galaxies   429
    B. W. Miller, J. Lotz, M. Hilker, M. Kissler-Patig & T. Puzia

Formation of Galactic Nuclei by Globular Cluster Merging   431
    P. Miocchi & R. Capuzzo Dolcetta

Dynamical Evolution of the Mass Function of the Galactic Globular Cluster System   433
    J. Shin, S. S. Kim & K. Takahashi

Part 8. COMPUTATIONAL ASPECTS OF SIMULATIONS OF DENSE STELLAR SYSTEMS

Dancing with Black Holes   437
    S. J. Aarseth

Virtual Laboratories and Virtual Worlds   447
    P. Hut

Current Status of GRAPE Project   457
    J. Makino

Fully Self-Consistent *N*-body Simulation of Star Cluster in the Galactic Center.   467
    M. Fujii, M. Iwasawa, Y. Funato & J. Makino

Test of the Accuracy of Approximate Methods to Handle Distant Binary-Single
Star Encounters   469
    Y. Funato, D. C. Heggie, P. Hut & J. Makino

TKira – a Hybrid N-body code   471
    *E. N. Mamikonyan, S. L. W. McMillan, S. F. Portegies Zwart &
    E. Vesperini*

6th and 8th Order Hermite Integrator Using Snap and Crackle   473
    K. Nitadori, M. Iwasawa & J. Makino

Embryo to Ashes *Complete* Evolutionary Tracks, *Hands-off*   475
    O. Yaron, A. Kovetz & D. Prialnik

Author index   477

Douglas Heggie.

This Symposium and this volume are dedicated to Douglas Heggie on the occasion of his 60th birthday.

Preface

Dense stellar systems are an interface between dynamics, stellar evolution, and formation of galaxies and provide us with an ideal laboratory to understand many different aspects of these important fields as well as to explore the interplay between them.

A wealth of observational data has now provided firm observational evidence showing that the dynamical evolution of a globular cluster, its structural and kinematical properties, the properties of its stellar population, the abundance of exotic objects such as pulsars, X-rays sources, and blue stragglers are closely related to each other: a full understanding of the evolution of star clusters can not be reached without properly considering the interplay between stellar dynamics and stellar evolution. An equally large amount of observational data from studies focusing on globular cluster systems in the Galaxy and in external galaxies have allowed us to explore the dependence of a number of properties of globular cluster systems on the type and the properties of the host galaxy. These studies have convincingly shown that the role played by the host galaxy in the formation and evolution of star clusters is an important additional element along with the effects of stellar dynamics, stellar evolution and their complex interplay.

The complete study of the formation and evolution of star clusters is a very challenging task which requires the collaboration and the exchange of ideas of astronomers and physicists with observational and theoretical expertise in Galactic and extra-galactic astronomy, stellar dynamics, hydrodynamics and stellar evolution. Expertise on the development of special-purpose hardware and software and, in general, on many aspects of computational physics also plays a key role in this endeavor.

IAU Symposium 246 "Dynamical Evolution of Dense Stellar Systems" brought together experts in all these areas and covered all the aspects of the study of star clusters with particular emphasis on the interplay between them and on the comparison between observations and simulations.

This Symposium was in honor of Douglas Heggie on the occasion of his 60th birthday. It was a great pleasure to celebrate Douglas and to dedicate the Symposium and these proceedings to him. Douglas has given and is still giving fundamental contributions to this field with his own research and with his always insightful, careful and kind advice, training and mentoring of many students and researchers working in this field.

Finally we would like to thank the financial support of the sponsors, all the participants for the excellent oral and poster presentations, the Scientific and the Local Organizing Committees for all the work done to make this Symposium a success.

Enrico Vesperini, Mirek Giersz and Alison Sills
Philadelphia, January 1, 2008

THE ORGANIZING COMMITTEE

Scientific

S. Aarseth (UK)

H. Baumgardt (Germany)

C. Boily (France)

M. Giersz (Poland)

D. Heggie (UK)

P. Hut (USA)

V. Kalogera (USA)

J. Makino (Japan)

R. Mardling (Australia)

S. McMillan (USA)

G. Meylan (Switzerland)

S. Mikkola (Finland)

S. Portegies Zwart (Netherlands)

F. Rasio (USA)

A. Sills (Canada)

R. Spurzem (Germany)

M. Trenti (USA)

E. Vesperini (chair-USA)

Local

E. Ferraro

M. Trenti (co-chair)

A. Pecoraro (co-chair)

Acknowledgements

The symposium is sponsored and supported by the IAU Division VII (Galactic System); and by the IAU Commissions No. 8 (Astrometry), No. 27 (Variable Stars), No. 28 (Galaxies), No. 33 (Structure and Dynamics of the Galactic System), No. 35 (Stellar Constitution), No. 37 (Star Clusters and Associations) and No. 40 (Radio Astronomy), No. 42 (Close Binary Stars); and by Working Groups on Star Formation and Massive Stars.

Funding by the
International Astronomical Union,
European Space Agency,
AstroSim Program, European Science Foundation,
School of Mathematics, University of Edinburgh,
Department of Physics, Drexel University
is gratefully acknowledged.

Participants

Sverre **Aarseth**, Cambridge University, UK,   sverre@ast.cam.ac.uk
Peter **Anders** University of Utrecht, Netherlands,   anders@astro.uu.nl
Rosemary **Apple** University of Edinburgh, UK,   r.apple@sms.ed.ac.uk
Serena **Arena** University of Milan, Italy,   serena.arena@unimi.it
Sambaran **Banerjee** Tata Institute of Fundamental Research, India,   sambaran@tifr.res.in
Nate **Bastian** University College London, UK,   bastian@star.ucl.ac.uk
Holger **Baumgardt** University of Bonn, Germany,   holger@astro.uni-bonn.de
Houria **Belkus** University of Brussel, Belgium,   hbelkus@vub.ac.be
Christian **Boily** Observatoire astronomique Strasbourg, France,   cmb@astro.u-strasbg.fr
Ella **Braden** University of Wisconsin, USA,   braden@astro.wisc.edu
Vittoria **Caloi** INAF, Italy,   vittoria.caloi@iasf-roma.inaf.it
Sourav **Chatterjee** Northwestern University, USA,   s-chatterjee@northwestern.edu
Yaroslav **Chumak** Sternberg Astronomical Institute, Russia,   chyo@mail.ru
Cathie **Clarke** Cambridge University, UK,   cclarke@ast.cam.ac.uk
Patrick **Côté** Herzberg Institute of Astrophysics, Canada,   patrick.cote@nrc-cnrc.gc.ca
Francesca **D'Antona** INAF-Observatory of Rome, Italy,   dantona@oa-roma.inaf.it
Saul **Davis** University of British Columbia, Canada,   sdavis@astro.ubc.ca
Guido **De Marchi** ESA, Netherlands,   gdemarchi@rssd.esa.int
Bernadetta **Devecchi** University of Milan, Italy,   bernadetta.devecchi@mib.infn.it
Andrea **Dieball** University of Southampton, UK,   andrea@astro.soton.ac.uk
Jonathan **Downing** ARI Heidelberg, Germany,   downin@ari.uni-heidelberg.de
Peter **Eggleton** LLNL, USA,   ppe@igpp.ucllnl.org
Michael **Fall** Space Telescope Science Inst., USA,   fall@stsci.edu
Michael **Fellhauer** Cambridge University, UK,   madf@ast.cam.ac.uk
Laura **Ferrarese** Herzberg Institute of Astrophysics, Canada,   laura.ferrarese@nrc-cnrc.gc.ca
Francesco **Ferraro** University of Bologna, Italy,   francesco.ferraro3@unibo.it
Jose **Fiestas** Heidelberg University, Germany,   fiestas@ari.uni-heidelberg.de
John **Fregeau** Northwestern University, USA,   fregeau@northwestern.edu
Marc **Freitag** Cambridge University, UK,   marc.freitag@gmail.com
Peter **Frinchaboy** University of Wisconsin, USA,   pmf@astro.wisc.edu
Michiko **Fujii** Tokyo University, Japan,   fujii@astron.s.u-tokyo.ac.jp
Toshiyuki **Fukushige** K&F Computing Research Co, Japan,   fukushig@kfcr.jp
Yoko **Funato** Tokyo University, Japan,   funato@artcompsci.org
Eughenii **Gaburov** University of Amsterdam, Netherlands,   egaburov@science.uva.nl
Karl **Gebhardt** University of Texas, USA,   gebhardt@astro.as.utexas.edu
Doug **Geisler** University of Concepcion, Chile,   dgeisler@astro-udec.cl
Gianpiero **Gervino** INFN, Italy,   gervino@to.infn.it
Mark **Gieles** ESO, Chile,   mgieles@eso.org
Mirek **Giersz** Copernicus Astronomical Center, Poland,   mig@camk.edu.pl
Evert **Glebbeek** University of Utrecht, Netherlands,   glebbeek@phys.uu.nl
Oleg **Gnedin** University of Michigan, USA,   ognedin@umich.edu
Dimitrios **Gouliermis** Max Planck Inst. for Astronomy-Heidelberg, Germany,   dgoulier@mpia-hd.mpg.de
Eva **Grebel** Universitat Basel, Switzerland,   grebel@ari.uni-heidelberg.de
Atakan **Gurkan** University of Amsterdam, Netherlands,   ato.gurkan@gmail.com
Vasilii **Gvaramdze** Sternberg Astronomical Institute, Russia,   vgvaram@yahoo.com
Bill **Harris** McMaster University, Canada,   harris@physics.mcmaster.ca
Douglas **Heggie** University of Edinburgh, UK,   d.c.heggie@ed.ac.uk
Michael **Hilker** ESO, Germany,   mhilker@eso.org
Jarrod **Hurley** Swinburne University, Australia,   jhurley@swin.edu.au
Piet **Hut** Institute for Advanced Study, USA,   piet@ias.edu
Natalia **Ivanova** CITA, Canada,   nata@cita.utoronto.ca
Vicky **Kalogera** Northwestern University, USA,   vicky@northwestern.edu
Sungsoo **Kim** Kyung Hee University, Korea,   sungsoo.kim@khu.ac.kr
Ivan **King** University of Washington, USA,   king@astro.washington.edu
Ludmila **Kisseleva-Eggleton** Expression College, USA,   lkisseleva@expression.edu
Ralf **Klessen** University of Heidelberg, Germany,   rklessen@ita.uni-heidelberg.de
Christian **Knigge** University of Southampton, UK,   christian@astro.soton.ac.uk
Thijs **Kouwenhoven** University of Sheffield, UK,   t.kouwenhoven@sheffield.ac.uk
Pavel **Kroupa** University of Bonn, Germany,   pavel@astro.uni-bonn.de
Arunav **Kundu** Michigan State University, USA,   akundu@pa.msu.edu
Charles **Lajoie** McMaster University, Canada,   lajoiec@mcmaster.ca
Joel **Lamb** University of Michigan, USA,   joellamb@umich.edu
Henny **Lamers** Utrecht University, Netherlands,   lamers@astro.uu.nl
Andrea **Lavagno** INFN, Italy,   alavagno@polito.it
Barbara **Lanzoni** University of Bologna, Italy,   barbara.lanzoni@bo.astro.it
Nathan **Leigh** McMaster University, Canada,   leighn@mcmaster.ca
Ulf **Löckmann** University of Bonn, Germany,   uloeck@astro.uni-bonn.de
Thomas **Maccarone** University of Southampton, UK,   tjm@phys.soton.ac.uk
Alasdair **Mackey** University of Edinburgh, UK,   dmy@roe.ac.uk
Jun **Makino** National Astronomical Observ., Japan,   makino@cfca.jp
Daniel **Malmberg** Lund Observatory, Sweden,   danielm@astro.lu.se
Ernest **Mamikonyan** Drexel University, USA,   ernie314@drexel.edu
Otonyo **Mangete** Drexel University, USA   otonyo@physics.drexel.edu
Rosemary **Mardling** Monash University, Australia,   Rosemary.Mardling@sci.monash.edu.au
Robert **Mathieu** University of Wisconsin, USA,   mathieu@astro.wisc.edu
Steve **McMillan** Drexel University, USA,   steve@physics.drexel.edu
Sabine **Mengel** ESO, Germany,   smengel@eso.org
Georges **Meylan** EPFL, Switzerland,   Georges.Meylan@epfl.ch
Steffen **Mieske** ESO, Germany,   smieske@eso.org
Seppo **Mikkola** Tuorla Observatory, Finland,   mikkola@utu.fi
Bryan **Miller** Gemini Observatory, Chile,   bmiller@gemini.edu
Paolo **Miocchi** University of Rome, Italy,   miocchi@uniroma1.it
Estelle **Moraux** Laboratoire d'Astrophysique de Grenoble, France,   Estelle.Moraux@obs.ujf-grenoble.fr
Keigo **Nitadori** University of Tokyo, Japan,   nitadori@cfca.jp
Eva **Noyola** Max Planck Institute, Germany,   noyola@mpe.mpg.de
Sally **Oey** University of Michigan, USA,   msoey@umich.edu
Cristoph **Olczak** University of Koeln, Germany,   olczak@ph1.uni-koeln.de

Francesco **Paresce** Centro Studi Spaziali, Italy, paresce@iasfbo.inaf.it
Genevieve **Parmentier** University of Liege, Belgium, gparm@astro.uni-bonn.de
Mario **Pasquato** University of Pisa, Italy, pasquato@df.unipi.it
Hagai **Perets** Weizmann Inst. of Science, Israel, hagai.perets@weizmann.ac.il
Susanne **Pfalzner** University of Cologne, Germany, pfalzner@ph1.uni-koeln.de
Jan **Pflamm-Altenburg** University of Bonn, Germany, jpflamm@astro.uni-bonn.de
Pawel **Pietrukowicz** Copernicus Astronomical Center, Poland, pietruk@camk.edu.pl
Giampaolo **Piotto** University of Padova, Italy, giampaolo.piotto@unipd.it
David **Pooley** University of Wisconsin-Madison, USA, dave@astro.wisc.edu
Simon **Portegies Zwart** University of Amsterdam, Netherlands, spz@science.uva.nl
Scott **Ransom** NRAO, USA, sransom@nrao.edu
Fred **Rasio** Northwestern University, USA, rasio@northwestern.edu
Marina **Rejkuba** ESO, Germany, mrejkuba@eso.org
Sigfried **Roeser** University of Heidelberg, Germany, roeser@ari.uni-heidelberg.de
Robert **Rood** University of Virginia, USA, rtr@mail.astro.virginia.edu
Carlos **Roman-Zuniga** CfA-Harvard, USA, cromanzu@cfa.harvard.edu
Mathieu **Servillat** CESR, France, mathieu.servillat@cesr.fr
Stefan **Schmeja** University of Porto, Portugal, sschmeja@astro.up.pt
Elena **Schilbach** University of Heidelberg, Germany, elena@ari.uni-heidelberg.de
Saurabh **Sharma** ARIES, India, saurabh@aries.ernet.in
Jihye **Shin** Kyung Hee University, Korea, jhshin@ap1.khu.ac.kr
Linda **Smith** Space Telescope Science Inst., USA, lsmith@stsci.edu
Veronica **Sommariva** ESO, Germany, vsommari@eso.org
Rainer **Spurzem** ARI Heidelberg, Germany, spurzem@ari.uni-heidelberg.de
Winston **Sweatman** Massey University, New Zealand, W.Sweatman@massey.ac.nz
Ataru **Tanikawa** Tokyo University, Japan, tanikawa@ea.c.u-tokyo.ac.jp
Michele **Trenti** Space Telescope Science Inst., USA, trenti@stsci.edu
Isaac **Tuchman** Hebrew University, Israel, tuchma@vms.huji.ac.il
Stefan **Umbreit** Northwestern University, USA, s-umbreit@northwestern.edu
David **Urminsky** University of Edinburgh, UK, s0450476@sms.ed.ac.uk
Sander **Valcke** University of Gent, Belgium, sander.valcke@ugent.be
Mauri **Valtonen** Tuorla Observatory, Finland, mavalto@utu.fi
Joris **Van Bever** St.Mary's University, Canada, vanbever@ap.smu.ca
Maureen **van den Berg** Harvard University, USA, maureen@head.cfa.harvard.edu
Anna Lisa **Varri** University of Milan, Italy, anna.varri@tiscali.it
Frank **Verbunt** University of Utrecht, Netherlands, verbunt@astro.uu.nl
Enrico **Vesperini** Drexel University, USA, vesperin@physics.drexel.edu
Ofer **Yaron** Tel Aviv University, Israel, oferya@post.tau.ac.il
Stephen **Zepf** Michigan State University, USA, zepf@pa.msu.edu
Hans **Zinnecker**, Astrophysikalisches Institut, Potsdam, Germany hzinnecker@aip.de

Part 1

Cluster Formation and Early Evolution

Dynamical Evolution of Dense Stellar Systems
Proceedings IAU Symposium No. 246, 2007
E. Vesperini, M. Giersz & A. Sills, eds.

Formation of Stellar Clusters and the Importance of Thermodynamics for Fragmentation

Ralf S. Klessen[1], Paul C. Clark[1] and Simon C. O. Glover[2]

[1]Zentrum für Astronomie der Universität Heidelberg, Institut für Theoretische Astrophysik,
Albert-Ueberle-Str. 2, 69120 Heidelberg, Germany

[2]Astrophysikalisches Institut Potsdam, An der Sternwarte 16, 14482 Potsdam, Germany

Abstract. We discuss results from numerical simulations of star cluster formation in the turbulent interstellar medium (ISM). The thermodynamic behavior of the star-forming gas plays a crucial role in fragmentation and determines the stellar mass function as well as the dynamic properties of the nascent stellar cluster. This holds for star formation in molecular clouds in the solar neighborhood as well as for the formation of the very first stars in the early universe. The thermodynamic state of the ISM is a result of the balance between heating and cooling processes, which in turn are determined by atomic and molecular physics and by chemical abundances. Features in the effective equation of state of the gas, such as a transition from a cooling to a heating regime, define a characteristic mass scale for fragmentation and so set the peak of the initial mass function of stars (IMF). As it is based on fundamental physical quantities and constants, this is an attractive approach to explain the apparent universality of the IMF in the solar neighborhood as well as the transition from purely primordial high-mass star formation to the more normal low-mass mode observed today.

Keywords. stars: formation – stars: mass function – early universe – hydrodynamics – equation of state – methods: numerical

1. Introduction

Identifying the physical processes that determine the masses of stars and their statistical distribution, the initial mass function (IMF), is a fundamental problem in star-formation research. It is central to much of modern astrophysics, with implications ranging from cosmic re-ionisation and the formation of the first galaxies, over the evolution and structure of our own Milky Way, down to the build-up of planets and planetary systems.

Near the Sun the number density of stars as a function of mass has a peak at a characteristic stellar mass of a few tenths of a solar mass, below which it declines steeply, and for masses above one solar mass it follows a power-law with an exponent $dN/d\log m \propto m^{-1.3}$. Within a radius of several kpc this distribution shows surprisingly little variation (Salpeter 1955; Scalo 1998; Kroupa 2001; Kroupa 2002; Chabrier 2003). This has prompted the suggestion that the distribution of stellar masses at birth is a truly universal function, which often is referred to as the Salpeter IMF, although note that the original Salpeter (1955) estimate was a pure power-law fit without characteristic mass scale.

The initial conditions in star forming regions can vary considerably, even in the solar vicinity. If the IMF were to depend on the initial conditions, there would be no reason for it to be universal. Therefore a derivation of the characteristic stellar mass that is based on fundamental atomic and molecular physics would be highly desirable. In this proceedings contribution we argue that indeed the thermodynamic properties of the star-forming cloud material determine the characteristic mass scale for fragmentation and subsequent stellar birth. The thermodynamic state of interstellar gas is a result of the balance between heating and cooling processes, which in turn are determined by fundamental atomic and molecular physics and by chemical abundances. The derivation of a characteristic stellar mass can thus be based on quantities and constants that depend solely on the chemical abundances in a molecular cloud. It also explains why deviations from the "standard" mode of star formation are likely to occur under extreme environmental conditions such as those occurring in the early universe or in circum-nuclear starburst regions.

2. Gravoturbulent Star Cluster Formation

Stars and star clusters form through the interplay between self-gravity on the one hand and turbulence, magnetic fields, and thermal pressure on the other (for recent reviews see Larson 2003; Mac Low & Klessen 2004; Ballesteros-Paredes *et al.* 2006). Supersonic turbulence, even if it is strong enough to counterbalance gravity on global scales, will usually *provoke* local collapse. Turbulence establishes a complex network of interacting shocks, where converging shock fronts generate clumps of high density. These density enhancements can be large enough for the fluctuations to become gravitationally unstable and collapse, which can occur when the local Jeans length becomes smaller than the size of the fluctuation. However, the fluctuations in turbulent velocity fields are highly transient. The random flow that creates local density enhancements can disperse them again. For local collapse to actually result in the formation of stars, Jeans-unstable shock-generated density fluctuations must collapse to sufficiently high densities on time scales shorter than the typical time interval between two successive shock passages. Only then are they able to 'decouple' from the ambient flow and survive subsequent shock interactions. The shorter the time between shock passages, the less likely these fluctuations are to survive. Hence, the timescale and efficiency of protostellar core formation depend strongly on the wavelength and strength of the driving source as well as on the dynamic response of the gas, as defined by the equation of state, i.e. the balance between heating and cooling processes.

The velocity field of long-wavelength turbulence is found to be dominated by large-scale shocks which are very efficient in sweeping up molecular cloud material, thus creating massive coherent structures. When a coherent region reaches the critical density for gravitational collapse, its mass typically exceeds the local Jeans limit by far. Inside the shock compressed region, the velocity dispersion is much smaller than in the ambient turbulent flow and the situation is similar to localized turbulent decay. These are the conditions for the formation of star clusters. The efficiency of turbulent fragmentation is reduced if the driving wavelength decreases. When energy is inserted mainly on small spatial scales, the network of interacting shocks is very tightly knit, and protostellar cores form independently of each other at random locations throughout the cloud and at random times. Individual shock-generated clumps have lower mass and the time interval between two shock passages through the same point in space is small. Collapsing cores are easily destroyed again and the resulting mass spectrum shows deviations from the observed IMF. All this points toward interstellar gas clouds being driven on large scales.

Altogether, stellar birth is intimately linked to the dynamic behavior of the parental gas cloud, which governs when and where star formation sets in. The chemical and thermodynamic properties of interstellar clouds play a key role in this process. In particular, the value of the polytropic exponent γ, when adopting an EOS of the form $P \propto \rho^\gamma$, strongly influences the compressibility of density condensations as well as the temperature of the gas. The EOS thus determines the amount of clump fragmentation, and so directly relates to the IMF (Vázquez-Semadeni *et al.* 1996) with values of γ larger than unity leading to little fragmentation and high mass cores (Li, Klessen, & Mac Low 2003; Jappsen *et al.* 2005). The stiffness of the EOS in turn depends strongly on the ambient metallicity, density and infrared background radiation field produced by warm dust grains. The EOS thus varies considerably in different galactic environments (see Spaans & Silk 2000, 2005 for a detailed account).

3. Formation of Stellar Clusters in the Solar Neighborhood

Early studies of the balance between heating and cooling processes in collapsing clouds predicted temperatures of the order of 10 K to 20 K, tending to be lower at the higher densities (e.g., Hayashi & Nakano 1965; Hayashi 1966; Larson 1969, 1973b). In their dynamical collapse calculations, these and other authors approximated this somewhat varying temperature by a simple constant value, usually taken to be 10 K. Nearly all subsequent studies of cloud collapse and fragmentation have used a similar isothermal approximation. However, this approximation is actually only a somewhat crude one, valid only to a factor of 2, since the temperature is predicted to vary by this much above and below the usually assumed constant value of 10 K. Given the strong sensitivity of the results of fragmentation simulations like those of Li *et al.* (2003) to the assumed equation of state of the gas, temperature variations of this magnitude may be important for quantitative predictions of stellar masses and the IMF.

As can be seen in Fig. 2 of Larson (1985), observational and theoretical studies of the thermal properties of collapsing clouds both indicate that at densities below about 10^{-18} g cm^{-3}, roughly corresponding to a number density of $n = 2.5 \times 10^5$ cm^{-3}, the temperature generally decreases with increasing density. In this low-density regime, clouds are externally heated by cosmic rays or photoelectric heating, and they are cooled mainly by the collisional excitation of low-lying levels of C$^+$ ions and O atoms; the strong dependence of the cooling rate on density then yields an equilibrium temperature that decreases with increasing density. The work of Koyama & Inutsuka (2000), which assumes that photoelectric heating dominates, rather than cosmic ray heating as had been assumed in earlier work, predicts a very similar trend of decreasing temperature with increasing density at low densities. The three-dimensional magnetohydrodynamic simulations of Glover & Mac Low (2007) also produce a similar result, although in this case the point-to-point scatter is larger. The resulting temperature-density relation can be approximated by a power law with an exponent of about -0.275, which corresponds to a polytropic equation of state with $\gamma = 0.725$. The observational results of Myers (1978) shown in Fig. 2 of Larson (1985) suggest temperatures rising again toward the high end of this low-density regime, but those measurements refer mainly to relatively massive and warm cloud cores and not to the small, dense, cold cores in which low-mass stars form. As reviewed by Evans (1999), the temperatures of these cores are typically only about 8.5 K at a density of 10^{-19} g cm^{-3}, consistent with a continuation of the decreasing trend noted above and with the continuing validity of a polytropic approximation with $\gamma \approx 0.725$ up to a density of at least 10^{-19} g cm^{-3}.

At higher densities, atomic line cooling becomes less effective as the cooling rates start to reach their local thermodynamic equilibrium (LTE) limits and as the line opacities grow larger. Consequently, at densities above 10^{-18} g cm^{-3} the gas becomes thermally coupled to the dust grains, which then control the temperature by their far-infrared thermal emission. In this high-density regime, dominated thermally by the dust, there are few direct temperature measurements because the molecules normally observed freeze out onto the dust grains, but most of the available theoretical predictions are in good agreement concerning the expected thermal behavior of the gas (Larson 1973b; Low & Lynden-Bell 1976; Masunaga & Inutsuka 2000; Larson 2005). The balance between compressional heating and thermal cooling by dust results in a temperature that increases slowly with increasing density, and the resulting temperature-density relation can be approximated by a power law with an exponent of about 0.075, which corresponds to $\gamma = 1.075$. Between the low-density and the high-density regimes, the temperature is predicted to reach a minimum of $5\,\mathrm{K}$ at a density of about $2 \times 10^{-18}\,\mathrm{g\,cm^{-3}}$, at which point the Jeans mass is about $0.3\,M_\odot$. The actual minimum temperature reached is somewhat uncertain because observations have not yet confirmed the predicted very low values, but such cold gas would be very difficult to observe; various efforts to model the observations have suggested central temperatures between $6\,\mathrm{K}$ and $10\,\mathrm{K}$ for the densest observed prestellar cores, whose peak densities may approach $10^{-17}\,\mathrm{g\,cm^{-3}}$ (e.g. Zucconi et al. 2001; Evans et al. 2001; Tafalla et al. 2004). A power-law approximation to the equation of state with $\gamma \approx 1.075$ is expected to remain valid up to a density of about $10^{-13}\,\mathrm{g\,cm^{-3}}$, above which increasing opacity to the thermal emission from the dust causes the temperature to begin rising much more rapidly, resulting in an "opacity limit" on fragmentation that is somewhat below $0.01\,M_\odot$ (Low & Lynden-Bell 1976; Masunaga & Inutsuka 2000).

Adopting the piecewise polytropic equation of state outlined above, Jappsen et al. (2005) showed that the changing γ from a value below unity to one somewhat above unity at a critical density n_c influences the number of protostellar objects. If the critical density increases then more protostellar objects form but the mean mass decreases. Consequently, the peak of the resulting mass spectrum moves to lower masses with increasing critical density. This spectrum not only shows a pronounced peak but also a power-law tail towards higher masses. Its behavior is thus similar to the observed IMF.

A simple scaling argument based on the Jeans mass M_J at the critical density n_c leads to $M_\mathrm{ch} \propto n_c^{-0.95}$. If there is a close relation between the average Jeans mass and the characteristic mass of a fragment, a similar relation should hold for the expected peak of the mass spectrum. The simulations by Jappsen et al. (2005) qualitatively support this hypothesis, but find a weaker density dependency $M_\mathrm{ch} \propto n_c^{-0.5 \pm 0.1}$.

4. Formation of Stellar Clusters in the Early Universe

The formation of the first and second generations of stars in the early universe has far-reaching consequences for cosmic reionization and galaxy formation (Loeb & Barkana 2001; Bromm & Loeb 2004; Glover 2005). The physical processes that govern stellar birth in a metal-free or metal-poor environment, however, are still very poorly understood. Numerical simulations of the thermal and dynamical evolution of gas in primordial protogalactic halos indicate that the metal-free first stars, the so called Population III, are expected to be very massive, with masses anywhere in the range 20–2000 $M_\odot$ (Abel, Bryan, & Norman 2002; Bromm, Coppi, & Larson 2002; Yoshida et al. 2006; O'Shea & Norman 2007), much larger than the characteristic mass scale found in the local IMF.

This means that at some stage of cosmic evolution there must have been a transition from primordial, high-mass star formation to the "normal" mode of star formation that dominates today. The discovery of extremely metal-poor subgiant stars in the Galactic halo with masses below one solar mass (Christlieb et al. 2002; Beers & Christlieb 2005) indicates that this transition occurs at abundances considerably smaller than the solar value. At the extreme end, these stars have iron abundances less than $10^{-5}\,Z_\odot$, and

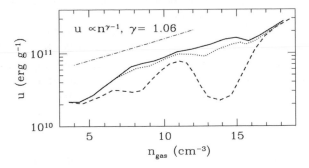

Figure 1. Three equations of state (EOSs) from Omukai *et al.* (2005) for different metallicities used by Clark, Glover, & Klessen (2007). The primordial case (solid line), $Z = 10^{-6}\,Z_\odot$ (dotted line), and $Z = 10^{-5}\,Z_\odot$ (dashed line), are shown alongside an example of a polytropic EOS with an effective $\gamma = 1.06$.

carbon or oxygen abundances that are still $\lesssim 10^{-3}$ the solar value. These stars are thus strongly iron deficient, which could be due to unusual abundance patterns produced by enrichment from pair-instability supernovae (Heger & Woosley 2002) from Population III stars or due to mass transfer from a close binary companion (Ryan *et al.* 2005; Komiya *et al.* 2007). There are hints for an increasing binary fraction with decreasing metallicity for these stars (Lucatello *et al.* 2005).

If metal enrichment is the key to the formation of low-mass stars, then logically there must be some critical metallicity Z_{crit} at which the formation of low mass stars first becomes possible. However, the value of Z_{crit} is a matter of ongoing debate. Some models suggest that low mass star formation becomes possible only once atomic fine-structure line cooling from carbon and oxygen becomes effective (Bromm *et al.* 2001; Bromm & Loeb 2003; Santoro & Shull 2006; Frebel, Johnson, & Bromm 2007), setting a value for Z_{crit} at around $10^{-3.5}\,Z_\odot$. Another possibility is that low mass star formation is a result of dust-induced fragmentation occurring at high densities, and thus at a very late stage in the protostellar collapse (Schneider *et al.* 2002; Omukai *et al.* 2005; Schneider *et al.* 2006; Tsuribe & Omukai 2006). In this model, $10^{-6} \lesssim Z_{\text{crit}} \lesssim 10^{-4}\,Z_\odot$, where much of the uncertainty in the predicted value results from uncertainties in the dust composition and the degree of gas-phase depletion (Schneider *et al.* 2002, 2006).

Clark, Glover, & Klessen (2007) modeled star formation in the central regions of low-mass halos at high redshift adopting an EOS similar to Omukai *et al.* (2005). They focused on a high-density regime with $10^5\,\text{cm}^{-3} \leqslant n \leqslant 10^{17}\,\text{cm}^{-3}$. They find that enrichment of the gas to a metallicity of only $Z = 10^{-5}Z_\odot$ dramatically enhances fragmentation. A typical time evolution is illustrated in Fig. 2. It shows several stages in the collapse process, spanning a time interval from shortly before the formation of the first protostar (as identified by the formation of a sink particle in the simulation) to 420 years afterwards. During the initial contraction, the cloud builds up a central core with a density of about $n = 10^{10}\,\text{cm}^{-3}$. This core is supported by a combination of thermal pressure and rotation. Eventually, the core reaches high enough densities to go into free-fall collapse, and forms a single protostar. As more high angular momentum material falls to the center, the core evolves into a disk-like structure with density inhomogeneities caused by low levels of turbulence. As it grows in mass, its density increases. When dust-induced cooling sets in, it fragments heavily into a tightly packed protostellar cluster within only a few hundred years. One can see this behavior in particle density-position plots in Fig. 3. The simulation is stopped 420 years after the formation of the first stellar object (sink

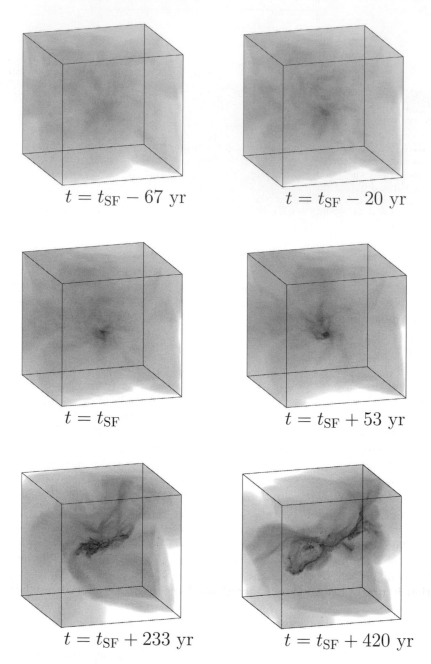

Figure 2. Time evolution of the density distribution in the innermost 400 AU of the proto-galactic halo shortly before and shortly after the formation of the first protostar at t_{SF}. Only gas at densities above 10^{10} cm^{-3} are plotted. The dynamical timescale at a density $n = 10^{13}$ cm^{-3} is of the order of only 10 years. Dark dots indicate the location of protostars as identified by sink particles forming at $n \geqslant 10^{17}$ cm^{-3}. Note that without usage of sink particles to identify collapsed protostellar cores one would not have been able to follow the build-up of the proto-stellar cluster beyond the formation of the first object. There are 177 protostars when we stop the calculation at $t = t_{SF} + 420$ yr. They occupy a region roughly a hundredth of the size of the initial cloud. With 18.7 M$_{\odot}$ accreted at this stage, the stellar density is 2.25×10^9 M$_{\odot}$ pc^{-3}. Data are from Clark, Glover, & Klessen (2007).

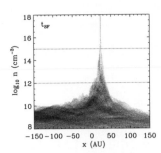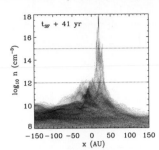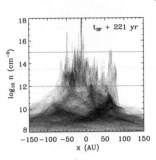

Figure 3. To illustrate the onset of the fragmentation process in the $Z = 10^{-5}\,Z_\odot$ simulation, the graphs show the densities of the particles, plotted as a function of their x-position. Note that for each plot, the particle data has been centered on the region of interest. Results are plotted for three different output times, ranging from the time that the first star forms (t_{sf}) to 221 years afterwards. The densities lying between the two horizontal dashed lines denote the range over which dust cooling lowers the gas temperature. The figure is from Clark, Glover, & Klessen (2007).

particle). At this point, the core has formed 177 stars. The evolution in the low-resolution simulation is very similar. The time between the formation of the first and second protostars is roughly 23 years, which is two orders of magnitude higher than the free-fall time at the density where the sinks are formed. Note that without the inclusion of sink particles, one would only have been able to capture the formation of the first collapsing object which forms the first protostar: the formation of the accompanying cluster would have been missed entirely.

The fragmentation of low-metallicity gas in this model is the result of two key features in its thermal evolution. First, the rise in the EOS curve between densities $10^9\,\mathrm{cm}^{-3}$ and $10^{11}\,\mathrm{cm}^{-3}$ causes material to loiter at this point in the gravitational contraction. A similar behavior at densities around $n = 10^3\,\mathrm{cm}^{-3}$ is discussed by Bromm *et al.* (2001), who call it a loitering phase. The rotationally stabilized disk-like structure, as seen in the plateau at $n \approx 10^{10}\,\mathrm{cm}^{-3}$ in Fig. 3, is able to accumulate a significant amount of mass in this phase and only slowly increases in density. Second, once the density exceeds $n \approx 10^{12}\,\mathrm{cm}^{-3}$, the sudden drop in the EOS curve lowers the critical mass for gravitational collapse by two orders of magnitude. The Jeans mass in the gas at this stage is only $M_{\mathrm{J}} = 0.01\,\mathrm{M}_\odot$. The disk-like structure suddenly becomes highly unstable against gravitational collapse and fragments vigorously on timescales of several hundred years. A very dense cluster of embedded low-mass protostars builds up, and the protostars grow in mass by accretion from the available gas reservoir. The number of protostars formed by the end of the simulation is nearly two orders of magnitude larger than the initial number of Jeans masses in the cloud set-up.

Because the evolutionary timescale of the system is extremely short – the free-fall time at a density of $n = 10^{13}\,\mathrm{cm}^{-3}$ is of the order of 10 years – none of the protostars that have formed by the time that the simulation is stopped have yet commenced hydrogen burning. This justifies neglecting the effects of protostellar feedback in this study. Heating of the dust due to the significant accretion luminosities of the newly-formed protostars will occur (Krumholz 2006), but is unlikely to be important, as the temperature of the dust at the onset of dust-induced cooling is much higher than in a typical Galactic protostellar core ($T_{\mathrm{dust}} \sim 100\,\mathrm{K}$ or more, compared to $\sim 10\,\mathrm{K}$ in the Galactic case). The rapid collapse and fragmentation of the gas also leaves no time for dynamo amplification of magnetic fields (Tan & Blackman 2004), which in any case are expected to be weak and dynamically unimportant in primordial and very low metallicity gas (Widrow 2002).

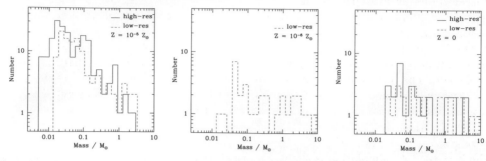

Figure 4. Mass functions resulting from simulations with metallicities $Z = 10^{-5} \, Z_\odot$ (left-hand panel), $Z = 10^{-6} \, Z_\odot$ (center panel), and $Z = 0$ (right-hand panel). The plots refer to the point in each simulation at which 19 $M_\odot$ of material has been accreted (which occurs at a slightly different time in each simulation). The mass resolutions are 0.002 $M_\odot$ and 0.025 $M_\odot$ for the high and low resolution simulations, respectively. Note the similarity between the results of the low-resolution and high-resolution simulations. The onset of dust-cooling in the $Z = 10^{-5} \, Z_\odot$ cloud results in a stellar cluster which has a mass function similar to that for present day stars, in that the majority of the mass resides in the lower-mass objects. This contrasts with the $Z = 10^{-6} \, Z_\odot$ and primordial clouds, in which the bulk of the cluster mass is in high-mass stars. The figure is from Clark, Glover, & Klessen (2007).

The forming cluster represents a very extreme analogue of the clustered star formation that we know dominates in the present-day Universe (Lada & Lada 2003). A mere 420 years after the formation of the first object, the cluster has formed 177 stars (see Fig. 2). These occupy a region of only around 400 AU, or 2×10^{-3} pc, in size, roughly a hundredth of the size of the initial cloud. With $\sim 19 \, M_\odot$ accreted at this stage, the stellar density is $2.25 \times 10^9 \, M_\odot$ pc^{-3}. This is about five orders of magnitude greater than the stellar density in the Trapezium cluster in Orion (Hillenbrand & Hartmann 1998) and about a thousand times greater than that in the core of 30 Doradus in the Large Magellanic Cloud (Massey & Hunter 1998). This means that dynamical encounters will be extremely important during the formation of the first star cluster. The violent environment causes stars to be thrown out of the denser regions of the cluster, slowing down their accretion. The stellar mass spectrum thus depends on both the details of the initial fragmentation process (e.g. as discussed by Jappsen *et al.* 2005; Clark & Bonnell 2005) as well as dynamical effects in the growing cluster (Bonnell *et al.* 2001; Bonnell, Bate & Vine 2004). This is different to present-day star formation, where the situation is less clear-cut and the relative importance of these two processes may vary strongly from region to region (Krumholz, McKee, & Klein 2005; Bonnell & Bate 2006; Bonnell, Larson & Zinnecker 2007).

The mass functions of the protostars at the end of the $Z = 10^{-5} \, Z_\odot$ simulations (both high and low resolution cases) are shown in Fig. 4 (left-hand panel). When the simulation is terminated, collapsed cores hold $\sim 19 \, M_\odot$ of gas in total. The mass function peaks somewhere below 0.1 $M_\odot$ and ranges from below 0.01 $M_\odot$ to about 5 $M_\odot$. This is not the final protostellar mass function. The continuing accretion of gas by the cluster will alter the mass function, as will mergers between the newly-formed protostars (which cannot be followed using our current sink particle implementation). Protostellar feedback in the form of winds, jets and HII regions may also play a role in determining the shape of the final stellar mass function. However, a key point to note is that the chaotic evolution of a bound system such as this cluster ensures that a wide spread of stellar masses will persist. Some stars will enjoy favourable accretion at the expense of others that will be thrown out of the system (as can be seen in Fig. 2), thus having their accretion effectively terminated (see for example, the discussions in Bonnell & Bate 2006 and Bonnell, Larson & Zinnecker 2007). The survival of some of the low mass stars formed in the cluster is therefore inevitable.

In the $Z = 0$ and $Z = 10^{-6}\,Z_\odot$ calculations Clark, Glover, & Klessen (2007) find that fragmentation of the gas occurs as well, albeit at a much lower level than in the $Z = 10^{-5}\,Z_\odot$ run. The mass functions from these simulations are shown in Fig. 4 (middle and right-hand panels), and are again taken when $\sim 19\,M_\odot$ of gas has been accreted onto the sink particles, the same amount as is accreted by the end of the $Z = 10^{-5}\,Z_\odot$ calculations. Both distributions are considerably flatter than the present day IMF, in agreement with the suggestion that Population III stars are typically very massive. The fragmentation in the $Z = 10^{-6}\,Z_\odot$ simulation is slightly more efficient than in the primordial case, with 33 objects forming.

5. Conclusions

In this proceedings paper we have discussed several studies, where the thermodynamic behavior of the interstellar medium plays a crucial role in fragmentation and subsequent stellar birth. These examples range from star formation in the solar neighborhood at the present day to the formation of the very first stars in the early universe, and support the idea that the distribution of stellar masses depends, at least in part, on the thermodynamic state of the star-forming gas. Dips in the effective EOS, such that the relation between temperature T and density ρ changes from decreasing T with increasing ρ to increasing T with increasing ρ, i.e. the transition from a cooling ($\gamma < 1$) to a heating ($\gamma > 1$) regime, define a characteristic mass scale for fragmentation.

The thermodynamic state of interstellar gas is a result of the balance between heating and cooling processes, which in turn are determined by fundamental atomic and molecular physics and by chemical abundances. The derivation of a characteristic stellar mass can thus be based on quantities and constants that depend solely on the chemical abundances of the star forming gas. This is an attractive feature explaining the apparent universality of the IMF in the solar neighborhood as well as the transition from purely primordial high-mass star formation to the more normal low-mass mode observed today. Clearly more work needs to be done to investigate the validity of this hypothesis.

Acknowledgements

We would like to thank Robi Banerjee, Anne-Katharina Jappsen, Richard Larson, Yuexing Li, and Mordecai-Mark Mac Low for stimulating discussions and collaboration.

References

Abel, T., Bryan, G. L., & Norman, M. L. 2002, *Science*, 295, 93
Beers, T. C. & Christlieb, N. 2005, *ARA&A*, 43, 531
Bonnell, I. A., Clarke, C. J., Bate, M. R., & Pringle, J. E. 2001, *MNRAS*, 324, 573
Bonnell, I. A., Vine, S. G. & Bate, M. R. 2004, *MNRAS*, 349, 735
Bonnell, I. A. & Bate, M. R. 2006, *MNRAS*, 370, 488
Bonnell, I. A., Larson, R. B., & Zinnecker, H. 2007, Protostars and Planets V, B. Reipurth, D. Jewitt, and K. Keil (eds.), University of Arizona Press, Tucson, p. 149
Bromm, V., Ferrara, A., Coppi, P. S., & Larson, R. B. 2001, *MNRAS*, 328, 969
Bromm, V., Coppi, P. S., & Larson, R. B. 2002, *ApJ*, 564, 23
Bromm, V. & Loeb, A. 2003, *Nature*, 425, 812
Bromm, V. & Loeb, A. 2004, *New Astron.*, 9, 353
Chabrier, G. 2003, *PASP*, 115, 763

Christlieb, N., Bessell, M. S., Beers, T. C., Gustafsson, B., Korn, A., Barklem, P. S., Karlsson, T., Mizuno-Wiedner, M., & Rossi, S. 2002, *Nature*, 419, 904

Clark, P. C., & Bonnell, I. A. 2005, *MNRAS*, 361, 2

Clark, P. C., Glover, S. C. O., & Klessen, R. S. 2007, *ApJ*, in press; arXiv:0706.0613

Evans, N. J. 1999, *ARA&A*, 37, 311

Evans, N. J., Rawlings, J. M. C., Shirley, Y. L., & Mundy, L. G. 2001, *ApJ*, 557, 193

Frebel, A., Johnson, J. L., & Bromm, V. 2007, astro-ph/0701395

Glover, S. C. O. 2005, *Space Sci. Reviews*, 117, 445

Glover, S. C. O. & Mac Low, M.-M. 2007, *ApJ*, 659, 1317

Hayashi, C. 1966, *ARA&A*, 4, 171

Hayashi, C. & Nakano, T. 1965, *Prog. Theor. Phys.*, 34, 754

Heger, A. & Woosley, S. E. 2002, *ApJ*, 567, 532

Hillenbrand, L. A. & Hartmann, L. W. 1998, *ApJ*, 492, 540

Jappsen, A.-K., Klessen, R. S., Larson, R. B., Li, Y., & Mac Low, M.-M. 2005, *A&A*, 435, 611

Komiya, Y., Suda, T., Minaguchi, H., Shigeyama, T., Aoki, W., & Fujimoto, M. Y. 2007, *ApJ*, 658, 367

Koyama, H. & Inutsuka, S. 2000, *ApJ*, 532, 980

Kroupa, P. 1998, *MNRAS*, 298, 231

Kroupa, P. 2002, *Science*, 295, 82

Krumholz, M. R., McKee, C. F., & Klein, R. I. 2005, *Nature*, 438, 332

Krumholz, M. R. 2006, *ApJ*, 641, L45

Lada, C. J. & Lada, E. A. 2003, *ARA&A*, 41, 57

Larson, R. B. 1969, *MNRAS*, 145, 271

—. 1973b, *Fundamentals of Cosmic Physics*, 1, 1

—. 1985, *MNRAS*, 214, 379

—. 2005, *MNRAS*, 359, 211

Li, Y., Klessen, R. S., & Mac Low, M.-M. 2003, *ApJ*, 592, 975

Loeb, A. & Barkana, R. 2001, *ARA&A*, 39, 19

Low, C. & Lynden-Bell, D. 1976, *MNRAS*, 176, 367

Lucatello, S., Tsangarides, S., Beers, T. C., Carretta, E., Gratton, R. G., & Ryan, S. G. 2005, *ApJ*, 625, 825

Massey, P. & Hunter, D. A. 1998, *ApJ*, 493, 180

Masunaga, H. & Inutsuka, S. 2000, *ApJ*, 531, 350

Myers, P. C. 1978, *ApJ*, 225, 380

Omukai, K., Tsuribe, T., Schneider, R., & Ferrara, A. 2005, *ApJ*, 626, 627

O'Shea, B. W. & Norman, M. L. 2007, *ApJ*, 654, 66

Ryan, S. G., Aoki, W., Norris, J. E., & Beers, T. C. 2005, *ApJ*, 635, 349

Santoro, F. & Shull, J. M. 2006, *ApJ*, 643, 26

Scalo, J. 1998, in ASP Conf. Ser. 142: The Stellar Initial Mass Function (38th Herstmonceux Conference), ed. G. Gilmore & D. Howell (San Francisco: Astron. Soc. Pac.), 201

Schneider, R., Ferrara, A., Natarajan, P. & Omukai, K. 2002, *ApJ*, 571, 30

Schneider, R., Omukai, K., Inoue, A. K., & Ferrara, A. 2006, *MNRAS*, 369, 1437

Tafalla, M., Myers, P. C., Caselli, P., & Walmsley, C. M. 2004, *A&A*, 416, 191

Tan, J. C. & Blackman, E. G. 2004, *ApJ*, 603, 401

Tsuribe, T. & Omukai, K. 2006, *ApJ*, 642, L61

Widrow, L. M. 2002, *Rev. Mod. Phys.*, 74, 775

Yoshida, N., Omukai, K., Hernquist, L., & Abel, T. 2006, *ApJ*, 652, 6

Zucconi, A., Walmsley, C. M., & Galli, D. 2001, *A&A*, 376, 650

Dynamical Evolution of Dense Stellar Systems
Proceedings IAU Symposium No. 246, 2007
E. Vesperini, M. Giersz & A. Sills, eds.

The formation, disruption and properties of pressure-supported stellar systems and implications for the astrophysics of galaxies

Pavel Kroupa

Argelander Institute for Astronomy, Auf dem Hügel 71, D 53121 Bonn, Germany
email: pavel@astro.uni-bonn.de

Abstract. Most stars form in dense star clusters deeply embedded in residual gas. These objects must therefore be seen as the fundamental building blocks of galaxies. With this contribution some physical processes that act in the very early and also later dynamical evolution of dense stellar systems in terms of shaping their later appearance and properties, and the impact they have on their host galaxies, are highlighted. Considering dense systems with increasing mass, it turns out that near $10^6 \, M_\odot$ their properties change fundamentally: stellar populations become complex, a galaxial mass–radius relation emerges and the median two-body relaxation time becomes longer than a Hubble time. Intriguingly, only systems with a two-body relaxation time longer than a Hubble time show weak evidence for dark matter, whereby dSph galaxies form total outliers.

Keywords. stars: formation, stars: luminosity function, mass function, globular clusters: general, galaxies: formation

1. Embedded clusters

Fragmentation: The very early stages of cluster evolution on a scale of a few pc are dominated by gravitational fragmentation of a turbulent magnetized contracting molecular cloud core (Clarke, Bonnell & Hillenbrand 2000; Tilley & Pudritz 2007; Mac Low & Klessen 2004). The existing simulations show the formation of contracting filaments which fragment into denser cloud cores that form sub-clusters of accreting proto-stars. As soon as the proto-stars emit radiation and outflows of sufficient energy and momentum to affect the cloud core these simulations become expensive as radiative transport and deposition of momentum and mechanical energy by non-isotropic outflows are difficult to handle with given present computational means (Stamatellos *et al.* 2007; Dale, Ercolano & Clarke 2007).

Observations of the very early stages at times $\lesssim$ few 10^5 yr suggest proto-clusters to have a hierarchical proto-stellar distribution: a number of sub-clusters with radii $\lesssim 0.2$ pc and separated in velocity space are often seen embedded within a region less than a pc across (Testi *et al.* 2000). Most of these sub-clusters may merge to form a more massive embedded cluster (Scally & Clarke 2002; Fellhauer & Kroupa 2005).

Mass segregation: Whether or not star clusters or sub-clusters form mass-segregated remains an open issue. Mass segregation by birth is a natural expectation because proto-stars near the density maximum of the cluster have more material to accrete. For these, the ambient gas is at a higher pressure allowing proto-stars to accrete longer before feedback termination stops further substantial gas inflow (Zinnecker & Yorke 2007). Initially mass-segregated sub-clusters preserve mass segregation upon merging (McMillan, Versperini & Portegies Zwart 2007). However, in dense proto-clusters (thousands of stars

13

within less than a pc), the energy equipartition time-scale between the stars is very short such that mass segregation may formally occur dynamically within one to a few crossing times (Kroupa 2002a). Currently we cannot say conclusively if mass segregation is a birth phenomenon (e.g. Gouliermis *et al.* 2004), or whether the more massive stars form anywhere throughout the proto-cluster volume. Star clusters that have already blown out their gas are typically mass-segregated (e.g. R136, Orion Nebula Cluster).

Affirming natal mass segregation would impact positively on the notion that massive stars ($\gtrsim 10 \, M_\odot$) only form in rich clusters, and negatively on the suggestion that they can also form in isolation (for recent work on this topic see Li, Klessen & Mac Low (2003); Parker & Goodwin (2007)).

Feedback termination: The star-formation efficiency (SFE), $\epsilon \equiv M_{\rm ecl}/\,(M_{\rm ecl} + M_{\rm gas})$, where $M_{\rm ecl}, M_{\rm gas}$ are the mass in freshly formed stars and residual gas, respectively, is $0.2 \lesssim \epsilon \lesssim 0.4$ (Lada & Lada 2003) implying that the physics dominating the star-formation process on scales less than a few pc is stellar feedback. Within this volume, the pre-cluster cloud core contracts under self gravity thereby forming stars ever more vigorously, until feedback energy suffices to halt the process (*feedback-termination*).

Dynamical state at feedback termination: Each proto-star needs about $t_{\rm ps} \approx 10^5$ yr to accumulate about 95 % of its mass (Wuchterl & Tscharnuter 2003). The proto-stars form throughout the pre-cluster volume as the proto-cluster cloud core contracts. The overall pre-cluster cloud-core contraction until feedback-termination takes $t_{\rm cl,form} \approx$ few $\times \, (2/\sqrt{G})(M_{\rm ecl}/\epsilon)^{-\frac{1}{2}} R^{\frac{3}{2}}$ (a few times the crossing time), which is about the time over which the cluster forms. Once a proto-star condenses out of the hydro-dynamical flow it becomes a ballistic particle moving in the time-evolving cluster potential. Because many generations of proto-stars can form over the cluster-formation time-scale and if the crossing time through the cluster is a few times shorter than $t_{\rm cl,form}$, then the assumption may be made that the very young cluster is mostly virialised when star formation stops and before the removal of the residual gas. It is noteworthy that $t_{\rm ps} \approx t_{\rm cl,form}$ for $M_{\rm ecl}/\epsilon \approx 10^{4.5} \, M_\odot$ (proto-star formation time is comparable to the cluster formation time) which is near the turnover mass in the old-star-cluster mass function.

A critical parameter is thus the ratio $\tau = t_{\rm cl,form}/t_{\rm cross}$. If it is less than unity then proto-stars "freeze out" of the gas and cannot virialise in the potential before the residual gas is removed. Such embedded clusters may be kinematically cold if the pre-cluster cloud core was contracting, or hot if the pre-cluster cloud core was pressure confined, because the young stars do not feel the gas pressure.

In those cases where $\tau > 1$ the embedded cluster is approximately in virial equilibrium and the pre-gas-expulsion stellar velocity dispersion in the embedded cluster, $\sigma \approx \sqrt{G \, M_{\rm ecl}/(\epsilon \, R)}$, may reach $\sigma = 40 \, {\rm pc/Myr}$ if $M_{\rm ecl} = 10^{5.5} \, M_\odot$ which is the case for $\epsilon R < 1$ pc. This is easily achieved the radius of one-Myr old clusters is $R \approx 0.8$ pc with no dependence on mass. Some observationally explored cases are discussed by Kroupa (2005). Notably, using K-band number counts, Gutermuth *et al.* (2005) appear to find evidence for expansion after gas removal. Interestingly, recent Spitzer results suggest a scaling of R with mass, $M_{\rm ecl} \propto R^2$ (Allen *et al.* 2007), so the question how compact embedded clusters form and whether there is a mass–radius relation needs further clarification. I note that such a scaling is obtained for a stellar population that expands freely with a velocity given by the velocity dispersion in the embedded cluster. Is the observed scaling then a result of expansion from a compact birth configuration after gas expulsion?

There are two broad camps suggesting on the one hand side that molecular clouds and star clusters form on a free-fall time-scale (Elmegreen 2000; Hartmann 2003) and on the other that many free-fall times are needed (Krumholz & Tan 2007). The former implies $\tau \approx 1$ while the latter implies $\tau > 1$.

Thus, currently unclear issues concerning the initialisation of N-body models of embedded clusters is the ratio $\tau = t_{\rm cl,form}/t_{\rm cross}$, and whether a mass–radius relation exists for embedded clusters *before* the development of HII regions. To make progress I assume for now that the embedded clusters are in virial equilibrium at feedback termination ($\tau > 1$) and that they form highly concentrated with $R \lesssim 1$ pc independently of mass.

The mass of the most massive star: Young clusters show a well-defined correlation between the mass of the most massive star, $m_{\rm max}$, in dependence of the stellar mass of the embedded cluster, $M_{\rm ecl}$, which appears to saturate at $m_{\rm max} \approx 150\,M_\odot$ (Weidner & Kroupa 2006). This may indicate feedback termination of star formation within the proto-cluster volume coupled to the most massive stars forming latest, or "turning-on" at the final stage of cluster formation (Elmegreen 1983). The physical maximum stellar mass near $150\,M_\odot$ (Weidner & Kroupa 2004; Figer 2005; Oey & Clarke 2005; Koen 2006; Zinnecker & Yorke 2007) must be a result of stellar structure stability, but may be near $80\,M_\odot$ as predicted by theory if the most massive stars reside in near-equal component-mass binary systems (Kroupa & Weidner 2005). It may also be that the calculated stellar masses are significantly overestimated (Martins, Schaerer & Hillier 2005).

The cluster core of massive stars: Irrespectively of whether the massive stars ($\gtrsim 10\,M_\odot$) form at the cluster centre or whether they segregate there due to energy equipartition, they ultimately form a compact sub-population that is dynamically highly unstable. Massive stars are ejected from such cores very efficiently on a core-crossing time-scale, and for example the well-studied Orion Nebula cluster (ONC) has probably already shot out 70 per cent of its stars more massive than $5\,M_\odot$ (Pflamm-Altenburg & Kroupa 2006). The properties of O and B runaway stars have been used by Clarke & Pringle (1992) to deduce the typical birth configuration of massive stars, finding them to form in binaries with similar-mass components in compact small-N groups devoid of low-mass stars. Among others, the core of the ONC is just such a system.

The star-formation history in a cluster: The detailed star-formation history in a cluster contains information about the events that build-up the cluster. Intriguing is the recent evidence for some clusters that while the bulk of the stars have ages different by less than a few 10^5 yr, a small fraction of older stars are often harboured (Palla & Stahler 2000). This may be interpreted to mean that clusters form over about 10 Myr with a final highly accelerated phase, in support of the notion that turbulence of a magnetized gas determine the early cloud-contraction phase (Krumholz & Tan 2007).

A different interpretation would be that as a pre-cluster cloud core contracts on a free-fall time-scale, it traps surrounding field stars which thereby become formal cluster members: Most clusters form in regions of a galaxy that has seen previous star formation. The velocity dispersion of the previous stellar generation, such as an expanding OB association, is usually rather low, around a few km/s to 10 km/s. The deepening potential of a newly-contracting pre-cluster cloud core will be able to capture some of the preceding generation of stars such that these older stars become formal cluster members although they did not form in this cluster. Pflamm-Altenburg & Kroupa (2007) study this problem for the ONC showing that the reported age spread by Palla *et al.* (2007) can be accounted for in this way. This suggests that the star-formation history of the ONC may in fact not have started about 10 Myr ago, supporting the argument by Elmegreen (2000) and Hartmann (2003) that clusters form on a timescale comparable to the crossing time of the pre-cluster cloud core.

For very massive clusters such as ω Cen, Fellhauer *et al.* (2006) show that the potential is sufficiently deep such that the pre-cluster cloud core may capture the field stars of a previously existing dwarf galaxy. Up to 30 % or more of the stars in ω Cen may be captured field stars. This would be able to explain an age spread of a few Gyr in the cluster, and is consistent with the notion that ω Cen formed in a dwarf galaxy that was captured by the Milky Way. The attractive aspect of this scenario is that ω Cen need not have been located at the center of the incoming dwarf galaxy as a nucleus, but within its disk, because it opens a larger range of allowed orbital parameters for the putative dwarf galaxy moving about the Milky Way. The currently preferred scenario in which ω Cen was the nucleus of the dwarf galaxy implies that the galaxy was completely stripped while falling into the Milky Way leaving only its nucleus on its current retrograde orbit. The new scenario allows the dwarf galaxy to be absorbed into the Bulge of the MW with ω Cen being stripped from it on its way in.

Expulsion of residual gas: When the most massive stars are O stars they destroy the proto-cluster nebula and quench further star formation by first ionising most of it. The ionised gas, being now at a temperature near 10^4 K and in serious over-pressure, pushes out and escapes the confines of the cluster volume with the sound speed (near 10 km/s) or faster if the winds being blown off O stars with velocities of thousands of km/s impart sufficient momentum. In reality, this evolution is highly dynamic and can be described as an explosion (the cluster "pops"), and probably occurs non-spherically because the gas seeks low-density channels in the nebula which then allow the hot gas to escape (Dale *et al.* 2005).

If the clusters are more massive than about 10^5 $M_\odot$ such that the velocity dispersion is larger than the sound speed of the ionised gas, then the cluster reacts adiabatically because the stars move in a potential that varies more slowly than the stellar crossing time through the cluster. For clusters without O and massive B stars, nebula disruption probably occurs on the cluster-formation time-scale, $\approx 10^6$ yr, and the evolution is again adiabatic.

Kroupa & Boily (2002) referred to clusters without O stars (stellar mass of the embedded cluster $M_{ecl} \lesssim 10^{2.5}\,M_\odot$) as clusters of type I, those with O stars but with $10^{2.5} \lesssim M_{ecl}/M_\odot \lesssim 10^{5.5}$ as type II clusters, and the very massive clusters ($M_{ecl} \gtrsim 10^{5.5}\,M_\odot$) as type III clusters. A type IV "cluster" may be added for extremely massive "clusters" for which only many supernovae are able to provide sufficient energy to blow out the residual gas ($M_{ecl} \gtrsim 10^7\,M_\odot$). This broad categorisation has easy-to-understand implications for the star-cluster mass function.

If clusters pop and which fraction of stars remain in a post-gas expulsion cluster depend critically on the ratio between the gas-removal time scale and the cluster crossing time. This ratio thus defines which clusters succumb to *infant mortality*, and which clusters suffer *cluster infant weight loss*. The well-studied cases do indicate that the removal of most of the residual gas does occur within a cluster-dynamical time, $\tau_{gas}/t_{cross} \lesssim 1$. Examples noted (Kroupa 2005) are the ONC and R136 in the LMC both having significant super-virial velocity dispersions. Other examples are the Treasure-Chest cluster and the very young star-bursting clusters in the massively-interacting Antennae galaxy which appear to have HII regions expanding at velocities such that the cluster volume may be evacuated within a cluster dynamical time.

A simple calculation of the amount of energy deposited by an O star into its surrounding cluster-nebula also suggests it to be larger than the nebula binding energy (Kroupa 2005). This, however, only gives at best a rough estimate of the rapidity with which gas can be expelled; an inhomogeneous distribution of gas leads to the gas removal occurring preferably along channels and asymmetrically, such that the overall gas-excavation process is highly non uniform and variable (Dale *et al.* 2005). Bastian & Goodwin (2006) note that many young clusters have a radial-density profile signature expected if they are expanding rapidly, supporting the notion of fast gas blow out. Goodwin & Bastian (2006) and de Grijs & Parmentier (2007) also find the dynamical mass-to-light ratio of young clusters to be too large strongly implying they are in the process of expanding after gas expulsion. A more detailed N-body study by Baumgardt, Kroupa & Parmentier (2008) considers the change of τ_{gas}/t_{cross} with the number of OB stars in clusters of varying mass by comparing the feedback energy in radiation and winds with the binding energy of the embedded cluster. These calculations support the sub-division of clusters into IV types suggested above.

Weidner *et al.* (2007) attempted to measure infant weight loss by using a sample of young but exposed Galactic clusters and applying the maximal-star-mass vs cluster mass relation from above to estimate the birth mass of these clusters. The uncertainties are large, but the data firmly suggest that the typical cluster looses at least about 50 per cent of its stars.

Mass loss from evolving stars: An old globular cluster with a turn-off mass near $0.8 \, M_{\odot}$ will have lost 30 per cent of the mass that remained in it after gas expulsion due to stellar evolution (Baumgardt & Makino 2003). As the mass loss is most rapid during the earliest times after re-virialisation after gas expulsion, the cluster expands further during this time. This is nicely seen in the Lagrange radii of realistic cluster-formation models (Kroupa, Aarseth & Hurley 2001).

2. Some implications for the astrophysics of galaxies

In general, the above have a multitude of implications for galactic and stellar astrophysics:

(*a*) The heaviest-star—star-cluster-mass correlation constrains feedback models of star cluster formation (Elmegreen 1983). It also implies that by adding up all IMFs in all young clusters in a galaxy, the *integrated galaxial initial mass function* (IGIMF) is steeper than the invariant stellar IMF observed in star clusters with important implications for the mass–metallicity relation of galaxies (Kroupa & Weidner 2005)

(*b*) The deduction that type II clusters probably "pop" implies that young clusters will appear to an observer to be super-virial, i.e. to have a dynamical mass larger than the luminous mass (Bastian & Goodwin 2006; de Grijs & Parmentier 2007).

(*c*) It also implies that galactic fields can be heated, and may also lead to galactic thick-disks and stellar halos around dwarf galaxies (Kroupa 2002b).

(*d*) The variation of the gas expulsion time-scale among clusters of different type implies that the star-cluster mass function (CMF) is re-shaped rapidly, on a time-scale of a few ten Myr (Kroupa & Boily 2002).

(*e*) Associated with this re-shaping of the cluster CMF is the natural production of population II stellar halos during cosmologically early star-formation bursts (Kroupa & Boily 2002; Parmentier & Gilmore 2007; Baumgardt, Kroupa & Parmentier 2008).

Points *b–e* are considered in more detail in what follows.

2.1. *Stellar associations, open clusters and moving groups*

As one of the important implications of point *b*, a cluster in the age range $1 - 50$ Myr will have an unphysical M/L ratio because it is out of dynamical equilibrium rather than having an abnormal stellar IMF (Bastian & Goodwin 2006; de Grijs & Parmentier 2007).

Another implication is that a Pleiades-like open cluster would have been born in a very dense ONC-type configuration and that, as it evolves, a "moving-group-I" is established during the first few dozen Myr which comprises roughly 2/3rd of the initial stellar population and is expanding outwards with a velocity dispersion which is a function of the pre-gas-expulsion configuration (Kroupa, Aarseth & Hurley 2001). These computations were in fact the first to demonstrate, using high-precision N-body modelling, that the re-distribution of energy within the cluster during the embedded phase and during the expansion phase leads to the formation of a substantial remnant cluster despite the inclusion of all physical effects that are disadvantageous for this to happen (explosive gas expulsion, low SFE $\epsilon = 0.33$, Galactic tidal field and mass loss from stellar evolution and an initial binary-star fraction of 100 per cent). Thus, expanding OB associations may be related to star-cluster birth, and many OB associations ought to have remnant star clusters as nuclei (see also Clarke *et al.* (2005)).

As the cluster expands becoming part of an OB association, the radiation from its massive stars produce expanding HII regions that may trigger star formation Gouliermis *et al.* (e.g. 2007).

A "moving-group-II" establishes later as the "classical" moving group made-up of stars which slowly diffuse/evaporate out of the re-virialised cluster remnant with relative kinetic energy close to zero. The velocity dispersion of moving group I is thus comparable to the pre-gas-expulsion velocity dispersion of the cluster, while moving group II has a velocity dispersion close to zero.

2.2. *The velocity dispersion of galactic-field populations and galactic thick disks*

Thus, the moving-group-I would be populated by stars that carry the initial kinematical state of the birth configuration into the field of a galaxy. Each generation of star clusters would, according to this picture, produce overlapping moving-groups-I (and II), and the overall velocity dispersion of the new field population can be estimated by adding in quadrature all expanding populations. This involves an integral over the embedded-cluster mass function, $\xi_{\mathrm{ecl}}(M_{\mathrm{ecl}})$, which describes the distribution of the stellar mass content of clusters when they are born. Because the embedded cluster mass function is known to be a power-law, this integral can be calculated for a first estimate (Kroupa 2002b, 2005). The result is that for reasonable upper cluster mass limits in the integral, $M_{\mathrm{ecl}} \lesssim 10^5 \, M_\odot$, the observed age–velocity dispersion relation of Galactic field stars can be re-produced.

This theory can thus explain the much debated "energy deficit": namely that the observed kinematical heating of field stars with age could not, until now, be explained by the diffusion of orbits in the Galactic disk as a result of scattering on molecular clouds, spiral arms and the bar (Jenkins 1992). Because the velocity-dispersion for Galactic-field stars increases with stellar age, this notion can also be used to map the star-formation history of the Milky-Way disk by resorting to the observed correlation between the star-formation rate in a galaxy and the maximum star-cluster mass born in the population of young clusters (Weidner, Kroupa & Larsen 2004).

An interesting possibility emerges concerning the origin of thick disks. If the star formation rate was sufficiently high about 11 Gyr ago, then star clusters in the disk with masses up to $10^{5.5} \, M_\odot$ would have been born. If they popped a thick disk with a velocity dispersion near 40 km/s would result naturally (Kroupa 2002b). The notion for the origin of thick disks appears to be qualitatively supported by the observations of Elmegreen, Elmegreen & Sheets (2004) who find galactic disks at a redshift between 0.5 and 2 to show massive star-forming clumps.

2.3. *Structuring the initial cluster mass function*

Another potentially important implication from this theory of the evolution of young clusters is that *if* the gas-expulsion-time-to-crossing-time ratio and/or the SFE varies with initial (embedded) cluster mass, then an initially featureless power-law mass function of embedded clusters will rapidly evolve to one with peaks, dips and turnovers at cluster masses that characterize changes in the broad physics involved.

As an example, Kroupa & Boily (2002) assumed that the function $M_{icl} = f_{st} M_{ecl}$ exists, where M_{ecl} is as above, M_{icl} is the "classical initial cluster mass" and $f_{st} = f_{st}(M_{ecl})$. The "classical initial cluster mass" is that mass which is inferred by classical N-body computations without gas expulsion (i.e. in effect assuming $\epsilon = 1$, which is however, unphysical). Thus, for example, for the Pleiades, $M_{cl} \approx 1000\, M_\odot$ at the present time (age about 100 Myr). A classical initial model would place the initial cluster mass near $M_{icl} \approx 1500\, M_\odot$ by using standard N-body calculations to quantify the secular evaporation of stars from an initially bound and virialised "classical" cluster (Portegies Zwart *et al.* 2001). If, however, the SFE was 33 per cent and the gas-expulsion time-scale was comparable to or shorter than the cluster dynamical time, then the Pleiades would have been born in a compact configuration resembling the ONC and with a mass of embedded stars of $M_{ecl} \approx 4000\, M_\odot$ (Kroupa, Aarseth & Hurley 2001). Thus, $f_{st}(4000\, M_\odot) = 0.38\,(= 1500/4000)$.

By postulating that there exist three basic types of embedded clusters, namely clusters without O stars (type I: $M_{ecl} \lesssim 10^{2.5}\, M_\odot$, e.g. Taurus-Auriga pre-main sequence stellar groups, ρ Oph), clusters with a few O stars (type II: $10^{2.5} \lesssim M_{ecl}/M_\odot \lesssim 10^{5.5}$, e.g. the ONC) and clusters with many O stars and with a velocity dispersion comparable to or higher than the sound velocity of ionized gas (type III: $M_{ecl} \gtrsim 10^{5.5}\, M_\odot$) it can be argued that $f_{st} \approx 0.5$ for type I, $f_{st} < 0.5$ for type II and $f_{st} \approx 0.5$ for type III. The reason for the high f_{st} values for types I and III is that gas expulsion from these clusters may be longer than the cluster dynamical time because there is no sufficient ionizing radiation for type I clusters, or the potential well is too deep for the ionized gas to leave (type III clusters). Type II clusters undergo a disruptive evolution and witness a high "infant mortality rate" (Lada & Lada 2003), therewith being the pre-cursors of OB associations and Galactic clusters.

Under these conditions and an assumed functional form for $f_{st} = f_{st}(M_{ecl})$, the power-law embedded cluster mass function transforms into a cluster mass function with a turnover near $10^5\, M_\odot$ and a sharp peak near $10^3\, M_\odot$ (Kroupa & Boily 2002). This form is strongly reminiscent of the initial globular cluster mass function which is inferred by e.g. Vesperini (1998, 2001); Parmentier & Gilmore (2005); Baumgardt (1998) to be required for a match with the evolved cluster mass function that is seen to have a universal turnover near $10^5\, M_\odot$.

This analytical formulation of the problem has been verified nicely using N-body simulations combined with a realistic treatment of residual gas expulsion by Baumgardt, Kroupa & Parmentier (2008), who show the Milky-Way globular cluster mass function to emerge from a power-law embedded-cluster mass function. Parmentier *et al.* (2008) expand on this by studying the effect that different assumptions on the physics of gas removal have on shaping the star-cluster mass function within about 50 Myr.

The general ansatz that residual gas expulsion plays a dominant role in early cluster evolution may thus bear the solution to the long-standing problem that the deduced initial cluster mass function needs to have this turnover, while the observed mass functions of young clusters are feature-less power-law distributions.

2.4. *The origin of population II stellar halos*

The above theory implies naturally that a major field-star component is generated whenever a population of star clusters forms. About 11 Gyr ago, the MW began its assembly by an initial burst of star formation throughout a volume spanning about 10 kpc in radius. In this volume, the star formation rate must have reached $10\, M_\odot/\mathrm{yr}$ such that star clusters with masses up to $\approx 10^6 \, M_\odot$ formed (Weidner, Kroupa & Larsen 2004), probably in a chaotic, turbulent early interstellar medium. The vast majority of embedded clusters suffered infant weight loss or mortality, the surviving long-lived clusters evolving to globular clusters. The so generated field population is the spheroidal population II halo, which has the same chemical properties as the surviving (globular) star clusters, apart from enrichment effects evident in the most massive clusters. All of these characteristics emerge naturally in the above model, as pointed out by Kroupa & Boily (2002), by Parmentier & Gilmore (2007) and most recently by Baumgardt, Kroupa & Parmentier (2008).

3. Long term, or classical, cluster evolution

The long-term evolution of star clusters that survive infant weight loss and the mass loss from evolving stars is characterised by three physical processes: the drive of the self-gravitating system towards energy equipartition, stellar evolution processes and the heating or forcing of the system through external tides. One emphasis of star-cluster work in this context is on testing stellar-evolution theory and on the interrelation of stellar astrophysics with stellar dynamics given that the stellar-evolution and the dynamical-evolution time-scales are comparable. The reader is directed to other chapters in this book for further details.

Tidal tails: Tidal tails contain the stars evaporating from long-lived star clusters (the moving group II above). Given that energy equipartition leads to a filtering in energy space of the stars that escape at a particular time, one expects a gradient in the stellar mass function progressing along a tidal tail towards the cluster such that the mass function becomes flatter, i.e. richer in more massive stars. This effect is difficult to detect, but for example the long tidal tails found emanating from Pal 5 (Odenkirchen *et al.* 2003) may show evidence for this. As emphasised by Odenkirchen *et al.* (2003), tidal tails have another very interesting use: they probe the gravitational potential of the Milky Way if the differential motions along the tidal tail can be measured. They are thus important future tests of gravitational physics.

Death: Nothing lasts forever, and star clusters that survive initial re-virialisation after residual gas expulsion and mass loss from stellar evolution ultimately "die" after evaporating all member stars leaving a binary or a long-lived highly hierarchical multiple system composed of near-equal mass components (de La Fuente Marcos 1997, 1998). Note that these need not be stars. These cluster remnants are interesting, because they may account for all hierarchical multiple stellar systems in the Galactic field (Goodwin & Kroupa 2005) with the implication that they are not a product of star formation, but rather of star-cluster dynamics.

4. What is a galaxy?

Old star clusters, dwarf-spheroidal (dSph) and dwarf-elliptical (dE) galaxies as well as galactic bulges and giant elliptical (E) galaxies are all stellar-dynamical systems that are supported by random stellar motions, i.e. they are pressure-supported. But why is one class of these pressure supported systems referred to as "star clusters", while the others are "galaxies"? Is there some fundamental physical difference between these two classes of systems?

Considering the radius as a function of mass, it becomes apparent that systems with $M \lesssim 10^6 \, M_\odot$ do not show a mass–radius relation (MRR) and have $R \approx 4$ pc. More massive objects, however, show a well-defined MRR. In fact, Dabringhausen, Hilker & Kroupa

(2008) find that the "massive compact objects" (MCOs), which have $10^6 \lesssim M/M_\odot \lesssim 10^8$, lie on the MRR of giant E galaxies ($\approx 10^{13} M_\odot$) down to normal E galaxies ($10^{11} M_\odot$): $R/\mathrm{pc} = 10^{-3.15} (M/M_\odot)^{0.60\pm0.02}$. Noteworthy is that systems with $M \gtrsim 10^6 M_\odot$ also sport complex stellar populations, while less massive systems have single-age, single-metallicity populations. The median two-body relaxation time is longer than a Hubble time for $M \gtrsim 3 \times 10^6 M_\odot$, and *only* for these systems is there evidence for a slight increase in the dynamical mass-to-light ratio. Intriguingly, $(M/L)_V \approx 2$ for $M < 10^6 M_\odot$, while $(M/L)_V \approx 5$ for $M > 10^6 M_\odot$ with a possible decrease for $M > 10^8 M_\odot$. Finally, the average stellar density maximises at $M = 10^6 M_\odot$ with about $3 \times 10^3 M_\odot/\mathrm{pc}^3$.

Thus,

- the mass $10^6 M_\odot$ appears to be special,
- stellar populations become complex above this mass,
- evidence for dark matter *only* appears in systems that have a median two-body relaxation time longer than a Hubble time,
- dSph galaxies are the *only* stellar-dynamical systems with $10 < (M/L)_V < 1000$ and as such are *total outliers*.

$M \approx 10^6 M_\odot$ therefore appears to be a characteristic mass scale such that less-massive objects show characteristics of star clusters being well-described by Newtonian dynamics, while more massive objects show behaviour more typical of galaxies. Defining a galaxy as a stellar-dynamical object which has a median two-body relaxation time longer than a Hubble time, i.e. essentially a system with a smooth potential, may be an objective and useful way to define a "galaxy" (Kroupa 1998).

Why *only smooth* systems show evidence for dark matter remains at best a striking coincidence, at worst it may be symptomatic of a problem in understanding dynamics in such systems.

References

Allen, L., *et al.* 2007, *Protostars and Planets V*, 361
Baumgardt, H. 1998, *A&A*, 330, 480
Baumgardt, H. & Makino, J. 2003, *MNRAS*, 340, 227
Baumgardt, H., Kroupa, P., & Parmentier, P. 2008, *MNRAS*, submitted
Bastian, N. & Goodwin, S. P. 2006, *MNRAS*, 369, L9
Clark, P. C., Bonnell, I. A., Zinnecker, H., & Bate, M. R. 2005, *MNRAS*, 359, 809
Clarke, C. J., Bonnell, I. A., & Hillenbrand, L. A. 2000, *Protostars and Planets IV*, 151
Clarke, C. J. & Pringle, J. E. 1992, *MNRAS*, 255, 423
Dabringhausen, J., Hilker, M., & Kroupa, P. 2008, *MNRAS*, submitted
Dale, J. E., Bonnell, I. A., Clarke, C. J., & Bate, M. R. 2005, *MNRAS*, 358, 291
Dale, J. E., Ercolano, B., & Clarke, C. J. 2007, *MNRAS*, 1056
de Grijs, R. & Parmentier, G. 2007, *Chinese Journal of Astronomy and Astrophysics*, 7, 155
de La Fuente Marcos, R. 1997, *A&A*, 322, 764
de La Fuente Marcos, R. 1998, *A&A*, 333, L27
Elmegreen, B. G. 1983, *MNRAS*, 203, 1011
Elmegreen, B. G. 2000, *ApJ*, 530, 277
Elmegreen, D. M., Elmegreen, B. G., & Sheets, C. M. 2004, *ApJ*, 603, 74
Fellhauer, M. & Kroupa, P. 2005, *ApJ*, 630, 879
Fellhauer, M., Kroupa, P. & Evans, N. W. 2006, *MNRAS*, 372, 338
Figer, D. F. 2005, *Nature*, 434, 192
Goodwin, S. P. & Bastian, N. 2006, *MNRAS*, 373, 752
Goodwin, S. P. & Kroupa, P. 2005, *A&A*, 439, 565
Gouliermis, D., Keller, S. C., Kontizas, M., Kontizas, E., & Bellas-Velidis, I. 2004, *A&A*, 416, 137

Gouliermis, D. A., Quanz, S. P., & Henning, T. 2007, *ApJ*, 665, 306

Gutermuth, R. A., Megeath, S. T., Pipher, J. L., Williams, J. P., Allen, L. E., Myers, P. C., & Raines, S. N. 2005, *ApJ*, 632, 397

Hartmann, L. 2003, *ApJ*, 585, 398

Jenkins, A. 1992, *MNRAS*, 257, 620

Koen, C. 2006, *MNRAS*, 365, 590

Kroupa, P. 1998, *MNRAS*, 300, 200

Kroupa, P. 2002, *Science*, 295, 82

Kroupa, P. 2002, *MNRAS*, 330, 707

Kroupa, P. 2005, *The fundamental building blocks of galaxies*, in ESA SP-576: The Three-Dimensional Universe with Gaia, 629 (astro-ph/0412069)

Kroupa, P. & Boily, C. M. 2002, *MNRAS*, 336, 1188

Kroupa, P. & Weidner, C. 2003, *ApJ*, 598, 1076

Kroupa, P. & Weidner, C. 2005, in *Massive Star Birth: A Crossroads of Astrophysics*, IAUS 227, 423

Kroupa, P., Aarseth S.J., Hurley, J. 2001, *MNRAS*, 321, 699

Krumholz, M. R. & Tan, J. C. 2007 *ApJ*, 654, 304

Lada, C. J. & Lada, E. A. 2003, ARA&A, 41, 57

Li, Y., Klessen, R. S., & Mac Low, M.-M. 2003, *ApJ*, 592, 975

Mac Low, M.-M. & Klessen, R. S. 2004, *Reviews of Modern Physics*, 76, 125

Martins, F., Schaerer, D., & Hillier, D. J. 2005, *A&A*, 436, 1049

McMillan, S. L. W., Vesperini, E., & Portegies Zwart, S. F. 2007, *ApJ*655, L45

Odenkirchen, M., *et al.* 2003, *AJ*, 126, 2385

Oey, M. S. & Clarke, C. J. 2005, *ApJ*, 620, L43

Palla, F. & Stahler, S. W. 2000, *ApJ*, 540, 255

Palla, F., Randich, S., Pavlenko, Y. V., Flaccomio, E., & Pallavicini, R. 2007, *ApJ*, 659, L41

Parker, R. J. & Goodwin, S. P. 2007, *MNRAS*, 380, 1271

Parmentier, G., Gilmore, G. 2005, *MNRAS*, 363, 326 (2005)

Parmentier, G. & Gilmore, G. 2007, *MNRAS*, 377, 352

Parmentier, G., Goodwinj, S., Kroupa, P., & Baumgardt, H. 2008, *ApJ*, submitted

Pflamm-Altenburg, J. & Kroupa, P. 2006, *MNRAS*, 373, 295

Pflamm-Altenburg, J. & Kroupa, P. 2007, *MNRAS*, 375, 855

Portegies Zwart, S. F., McMillan, S. L. W., Hut, P., & Makino, J. 2001, *MNRAS*, 321, 199

Scally, A., & Clarke, C. 2002, *MNRAS*, 334, 156

Stamatellos, D., Whitworth, A. P., Bisbas, T., & Goodwin, S. 2007, *A&A*, 475, 37

Testi, L., Sargent, A. I., Olmi, L., & Onello, J. S. 2000, *ApJ*, 540, L53

Tilley, D. A. & Pudritz, R. E. 2007, *MNRAS*, 930

Vesperini, E. 1998, *MNRAS*, 299, 1019

Vesperini, E. 2001, *MNRAS*, 322, 247

Weidner, C. & Kroupa, P. 2004, *MNRAS*, 348, 187

Weidner, C. & Kroupa, P. 2005, *ApJ*, 625, 754

Weidner, C. & Kroupa, P. 2006, *MNRAS*, 365, 1333

Weidner, C., Kroupa, P., & Larsen, S. S. 2004, *MNRAS*, 350, 1503

Weidner, C., Kroupa, P., Nürnberger, D. E. A., & Sterzik, M. F. 2007, *MNRAS*, 376, 1879

Wuchterl, G. & Tscharnuter, W. M. 2003, *A&A*, 398, 1081

Zinnecker, H. & Yorke, H. W. 2007, *ARA&A*, 45, 481

Dynamical Evolution of Dense Stellar Systems
Proceedings IAU Symposium No. 246, 2007
E. Vesperini, M. Giersz & A. Sills, eds.

The Early Evolution of Dense Stellar Systems

C. J. Clarke

Institute of Astronomy, Madingley Road, Cambridge, UK
email: cclarke@ast.cam.ac.uk

Abstract. The early evolution of dense stellar systems is dominated by the majority mass component – the gas – and so any credible modeling of the first Myr or so of a cluster's life inevitably involves hydrodynamical simulations. Such simulations have increased considerably in sophistication over the last few years and are now beginning to incorporate the effects of stellar feedback, thus enabling one, for the first time, to model the formation of populous clusters. In this review I focus on two issues that have arisen from the simulations – the relationship between maximum stellar mass and cluster mass, and the issue of the maximum density that is attainable during the cluster formation process. I also report on the first results of new simulations that model feedback from ionising radiation.

Keywords. hydrodynamical simulations, stars: formation, ISM: evolution

1. Introduction

Most of those attending this Symposium will not have considered in any detail the early (hydrodynamic) stages of cluster formation. Indeed, many would probably consider it to be too inherently messy to be appealing, in contrast to the relatively well posed problems concerning a cluster's further evolution as an Nbody system.

There are in fact two potentially valid reasons to feel this way about hydrodynamic simulations of cluster formation. The first is the issue of numerical accuracy; in dissipative systems, one cannot rely on simple constraints such as total energy, and even where codes can readily reproduce simple one dimensional tests, there remain some doubts about their ability to model the sort of complex three dimensional flows encountered in turbulent star forming regions. In this situation, an important check is provided by comparisons between different numerical codes and it is particularly encouraging that, whereas Lagrangian codes (i.e. SPH) have blazed the trail in this area, there is the prospect that in future it will become possible to undertake detailed comparisons with Eulerian (Adaptive Mesh Refinement) codes.

The second reason for considering this evolutionary phase as 'messy' is however intrinsic to the problem, and derives from the fear that – even in the case of perfect numerical accuracy – the evolution may be unduly sensitive to details of the initial gas distribution. This is a particular issue once one considers the effects of feedback from massive stars, given the extreme density sensitivity of the photoionisation process. One then has to worry that even the gross outcome of a simulation (for example, whether a proto-cluster remains bound or not), could depend on a detailed knowledge of the gas filling factor and morphology, which could never be known in practice. It is however still unclear (given the small number of simulations that have explored this issue) whether such simulations can be more than exercises in detailed 'weather forecasting' and how easy it will be to derive general trends and insights from them.

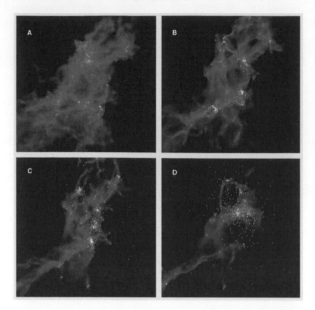

Figure 1. The hierarchical assembly of a star cluster from turbulent fragmentation simulations of Bonnell *et al.* 2003.

Notwithstanding these two caveats, it is evidently of great interest to all those working on the evolution of dense stellar systems to understand what factors control such properties as the stellar mass spectrum and its dependence on environment, as well as considering the conditions of possibly extreme stellar density that may prevail at early times. In this contribution, I paint a thumbnail sketch of the expectations of turbulent fragmentation simulations and show that a key insight of recent years, both observationally and theoretically, is that clusters are assembled in a hierarchical (bottom-up) manner. I then relate this picture to the observed relationship between maximum stellar mass and cluster mass. Next I turn to the question of the maximum density achieved in young star clusters, and show that the maximum density – following adiabatic accretion of gas onto a cluster core – exceeds the mean cluster density by a factor depending on the square of the cluster mass. This result then implies that the runaway collisional growth of stars in an adiabatic core is limited to the case of more massive clusters. Finally, I review the results of pilot simulations that include feedback from massive stars and show that it is currently unclear whether the net effect of feedback on star formation is positive or negative. I also highlight the difficulty in using observational diagnostics to identify examples of triggered star formation.

2. The hierarchical assembly of stellar clusters in turbulent molecular clouds

Fig. 1 contains several frames from the (SPH) simulation of Bonnell, Bate and Vine 2003, which depict the hierarchical assembly of a star cluster from a turbulent cloud. In this simulation, the (piecewise polytropic) gas is subject to an initial turbulent velocity field, whose spectral slope is set so as to replicate the observed size-line width relation in molecular clouds (Larson 1981). These motions are supersonic (Mach ~ 10) on the cloud scale (~ 1 pc) and lead to the development of a filamentary network of shocks on the cloud crossing timescale ($\sim 2 \times 10^5$ years). Since the energy dissipated in shocks

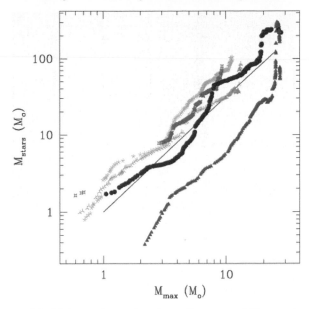

Figure 2. The assembly history of massive stars: the mass of the most massive star in its cluster is plotted as a function of the total mass in stars within 0.1 pc. From Bonnell *et al.* 2004.

is not replenished in these simulations, the subsequent evolution involves not only the fragmentation of over-dense (shocked) regions but also the over-all collapse of the cloud. Thus, whereas, star formation proceeds initially in a small N cluster mode in regions of the highest density (where the Jeans mass is lowest), these clusters subsequently merge and form successively larger structures. Such a picture of hierarchical cluster formation is consistent with the wealth of small scale structure seen in star forming regions (Guillout *et al.* 1998, Gouliermis *et al.* 2000, Elmegreen & Elmegreen 2001).

A striking property of such simulations is that in each cluster, the most massive star is generally located close to the cluster centre (where it intercepts the densest accretion flow and hence grows most rapidly). Following cluster mergers, the most massive star sinks rapidly to the gas-rich core of the new cluster and grows further in mass. Thus, in the simulations, the mass of the most massive star increases with the mass of the parent cluster, although less than linearly: Fig. 2 depicts the growth in mass of several stars in the simulations of Bonnell *et al.* 2004, each of which are the most massive members of their cluster, plotted as a function of the mass of the parent cluster (as represented by the mass contained within 0.1 pc of the most massive star). Although both quantities grow in time, as the most massive star and its neighbours accrete gas, there are phases when new members arrive in the cluster core (near vertical sections in Fig. 2) which flattens the slope of the M_{max} versus M_{clus} relation to $M_{max} \propto M_{clus}^{2/3}$.

It is notable that this slope is close to the relationship $M_{max} \propto M_{clus}^{1/1.35}$ which is expected by random drawing from a Salpeter IMF. In fact, it is found that the IMF within the clusters is indeed close to a Salpeter distribution and that the most massive star is close to that expected from populating this distribution with a finite number of stars appropriate to the cluster mass.

The observational situation is depicted in Fig. 3 – an increase in maximum stellar mass as the cluster scale increases (up to a cluster mass of a few $\times 10^4 M_\odot$ – i.e. ~ 100 OB stars) and then a flattening off with a maximum recorded value of around $150 - 200 M_\odot$.

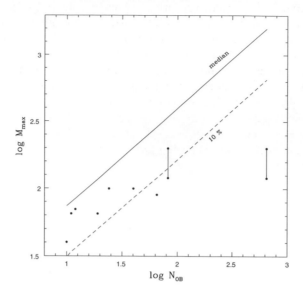

Figure 3. The observed maximum stellar mass of clusters plotted as a function of the cluster richness (number of OB stars). Also shown is the median and tenth percentile of the expected distribution if stars are drawn at random from an untruncated Salpeter IMF. Based on Oey & Clarke 2005.

The *slope* of the relationship at lower masses is consistent with random drawing from a Salpeter IMF (although the values are somewhat low: the fact that the points all lie below the expected median relation is however not statistically significant). *If* one interprets this relationship as being purely a consequence of the statistics of random drawing then, although the median for a large sample of clusters would be expected to lie along the line shown, there would be considerable scatter in both directions and there would be nothing *physical* that would rule out the occasional discovery of a low mass cluster containing a very massive star.

Whereas it is debatable whether the observed M_{max}, M_{clus} relation is statistical or physical at lower masses, it is clear that at higher masses, the apparent saturation of M_{max} *is* physical. Indeed, the probability that in a cluster on the scale of R136 (containing ~ 600 OB stars) the most massive star should be as low as is observed would be 10^{-15} according to a random drawing model (Oey & Clarke 2005). It is however unclear what physics should operate at this mass scale so as to preclude further mass growth (for example, a feedback mechanism that is often invoked in relation to the assembly of massive stars – radiation pressure on dust – becomes effective at the $\sim 10 M_{\odot}$ scale, where the IMF is in fact featureless). Alternatively, the upper mass limit may not relate to star *formation* but to subsequent mass loss (see, e.g. Smith & Owocki 2006, Belkus *et al.* 2007).

Since the simulations (which in this case incorporate no feedback) can so well reproduce the observed M_{max}, M_{clus} relationship at low masses, it is instructive to enquire whether this effect should be regarded as statistical or physical *in the simulations*. This essentially boils down to examining the *scatter* in the histories of star and cluster growth in the M_{max}, M_{clus} plane. In as far as the limited number of accretion histories contained in Fig. 2 can provide any insight into the problem, it would appear that the relationship in the simulations is actually statistical. Although the tracks follow an over-all trend, they do exhibit some scatter (for example, the right hand trajectory exhibits a much larger value of M_{max} at given M_{clus} than the others, presumably reflecting some stochastic variation in the accretion history of its natal cluster). This means that there are not factors which *physically prevent* the acquisition of a higher M_{max} value at low M_{clus}, but just that such an outcome is relatively rare. Pursuing this line of thought, we see that if we were to 'dry merge' a number of clusters (i.e. combine their stellar components without changing their stellar masses), the resulting merged cluster would still have a higher maximum stellar mass than the majority of the constituent clusters from which it formed. This would suggest that we *cannot* appeal to the onset of 'dry mergers' in order to explain the observed saturation of M_{max} at $150 - 200M_\odot$. We however stress that such a conclusion is necessarily preliminary, given the small number of accretion histories analysed to date.

3. The maximum stellar density and the possibility of stellar collisions

Examination of the simulations reveals that clusters form at the intersections of filamentary shocks which channel an accretion flow onto the clusters. The timescale for mass doubling due to this flow is set by the free fall timescale of the large scale structure (parent cloud) which is generally longer than that of the cluster itself. Thus the cluster acquires mass in the adiabatic regime and it is an easy matter to demonstrate that the twin requirements of virial equilibrium and the preservation of the adiabatic invariant $\Sigma p_i.r_i$, (where p_i and r_i are respectively stellar positions and momenta), imply that the cluster radius shrinks as the inverse cube of the cluster mass (i.e. $R \propto M^{-3}$).† This adiabatic mass-radius relationship implies an extremely steep dependence of cluster density on mass ($\rho \propto M^{10}$); thus since the cluster free fall timescale declines steeply (as M^{-5}), any cluster that enters the adiabatic regime will become increasingly adiabatic as time goes on. Thus Bonnell, Bate & Zinnecker (1998) argued that the stellar density would eventually rise to the point where physical collisions between stars would become important. Given the possible problems that have been discussed about the viability of forming the most massive stars by accretion (due to the effect of radiation pressure on dust: see e.g. Wolfire and Cassinelli 1987, Edgar & Clarke 2004) this appeared to be an attractive alternative mechanism for massive star formation.

A simple estimate of collisional cross sections for single O stars leads to the benchmark requirement that the density must exceed 10^8 stars pc^{-3} before the collision time falls below 10^5 years. This exceeds the densities of the densest *observed* star clusters by two orders of magnitude and therefore – if such conditions are ever met – they must belong to a deeply embedded, and observationally inaccessible, phase of cluster formation. We note in passing that Moraux, Lawson and Clarke (2007) have recently *inferred* initial densities of $\sim 10^8$ pc^{-3} for the Eta Cha association, based on the assumption that the observed lack of brown dwarfs and wide binaries in this cluster is a result of dynamical effects. (It should be stressed however that Eta Cha does not contain stars that are sufficiently large in cross section to have collided even at such high densities.)

The above density estimates make the (pessimistic) assumption of (gravitationally focused) encounters between single stars. (Note that the velocity dispersion rises rather gently with density in the adiabatic phase ($v \propto \rho^{1/5}$) and so gravitational focusing can remain important even at high densities. By the same token, velocities can remain less than the stellar escape velocity and thus collisions can lead to mergers rather than

† Note that this argument is equally applicable if, instead of a cluster and parent cloud, we instead consider a dense cluster core gaining mass via inflow from a gas-rich outer cluster.

disruptions). Bonnell & Bate (2002) however found that the main route to stellar collisions was via the gravitational perturbation of close binaries. In practice, collisions are
likely in the case of binaries whose separations are around ~ 10 stellar radii (i.e. around
an A.U. or so in the case of massive stars). Thus, if the cluster contains a significant population of primordial binaries on the scale of 1 A.U., the densities required for collisions
are reduced by around an order of magnitude.

The above description suggests that adiabatic accretion leads to an inexorable rise in
core density and that therefore – given a continuing supply of accretable material – the
eventual realisation of conditions conducive to stellar collisions in the core is inevitable.
Nevertheless, Bonnell & Bate (2002) found instead that, in the case of their simulation
of a cluster containing $\sim 10^3$ stars, the maximum core density saturated at around 5
orders of magnitude times its initial value. At this point, the shrinkage due to adiabatic
accretion was offset by puffing up of the cluster core by two body effects (specifically
the injection into stellar motions of kinetic energy released by the creation or hardening
of binaries). The maximum density attained was not quite sufficient (assuming a cluster
with average density comparable to the Orion Nebula Cluster) to give rise to collisions
unless (like Bonnell & Bate 2002) one applies artificially enhanced collision cross sections.

Recently, Clarke & Bonnell (2007) have developed this insight in order to determine the
properties of host systems whose cores should be able to attain the densities necessary for
stellar collisions. The maximum core density is obtained at the point that the timescale
for core shrinkage (which is the mass doubling timescale in the core, $t_{\dot{M}}$) becomes longer
than the timescale for puffing up by two body processes, a timescale that can (in the
case of a small core, containing $N_c < 100$ stars) be approximated as $\sim N_c$ core crossing
times (Bonnell & Clarke 1999). $t_{\dot{M}}$ is simply the ratio of the core mass to the mass inflow
rate, $\dot{M}$, where $\dot{M}$ is given (in the case of the free fall collapse of a gaseous reservoir of
mass M_g) by M_g/t_{dyn}, with t_{dyn} being the dynamical timescale at the cluster half mass
radius. Putting all this together, one obtains the very simple result:

$$\rho_{max} = \bar{\rho}\left(\frac{M_g}{\bar{m}_c}\right)^2 \tag{3.1}$$

where $\bar{m}_c$ is the mean stellar mass in the core. Substituting $M_g \sim 1000 M_\odot$ and $\bar{m}_c \sim 3 M_\odot$
one recovers the result (obtained numerically by Bonnell & Bate 2002) that the core
density growth should saturate at around five orders of magnitude above the mean value.
One may then compute the mean number of collisions per star, f_{coll}, as being the product
of the collision rate at this maximum density and the core shrinkage timescale. After a
little manipulation, we obtain:

$$f_{coll} = N_c\left(\frac{v}{v_*}\right)^2 \tag{3.2}$$

where v and v_* are respectively the cluster velocity dispersion and escape velocity from
the stellar surface. Given that v is typically a few km s^{-1} and v_* is several hundreds of
km s^{-1}, it follows that stellar collisions are likely to be important only in the case of
cluster cores containing $> 10^4$ stars.

The above estimate is necessarily a rough one and brushes aside a number of issues
such as the feasibility of continued accretion in the face of stellar feedback, as well as the
effects of mass segregation within a dynamically decoupled (adiabatic) core. It also, by
focusing on the fate of a core containing N_c stars, does not make any predictions about
the relationship between N_c and the number of stars in the parent cluster (N_*), apart
from the obvious restriction that $N_c < N_*$. It nevertheless suggests that runaway stellar
collisions should be a possible outcome only in the case of clusters with $N_* >> 10^4$.

A preliminary conclusion, therefore, is that one should *not* expect runaway stellar collisions in the sorts of environments in which they have been discussed hitherto (i.e. the Orion Nebula Cluster, the Arches Cluster or R136 in 30 Doradus). This is consistent with the fact that even in these most populous clusters, there is no evidence for the presence of stars with masses in excess of $150 - 200 M_\odot$ (Figer *et al.* 1998), although it may be argued that runaway collisions may occur in these regions *in the future* through the purely stellar dynamical operation of the mass segregation (Spitzer) instability (Spitzer 1969, Gürkan *et al.* 2004). On the other hand, more populous systems, such as globular clusters or super star clusters (SSCs) are certainly candidate systems for accretion induced runaway stellar collisions, the main requirement being a strong radial inflow of gas into a populous core region. This result has obvious ramifications for the production of intermediate mass black holes (IMBHs) in globular clusters.†

4. The role of feedback

The vast majority of cluster star formation simulations performed to date omit the effects of feedback, whether on the scale of accretion onto individual stars (where radiation pressure on dust may be significant) or on a cluster wide scale, due to either thermal ionisation or the mechanical feedback supplied by stellar winds. Simulations involving feedback are in their infancy and have thus far served more to patent the viability of the algorithms rather than permitting any exploration of parameter space. Nevertheless, current work does allow one to draw a few preliminary conclusions and cautionary notes.

In the case of feedback from the ionising radiation from massive stars, Dale and collaborators have recently succeeded in implementing this effect in SPH simulations (see Dale *et al.* (2007) for a demonstration that, in the dense environments of molecular cloud cores, the simple – on the spot – assumption employed by the algorithm is validated by comparison with Monte Carlo radiative transfer calculations). To date, pilot simulations have applied such feedback to the case of an internal ionising source in the core of a dense, highly inhomogeneous protocluster cloud (Dale *et al.* 2005) and also to the case of an external ionising source irradiating an initially unbound cloud region (Dale *et al.* 2006). In the former case (see Fig. 4), the source ionises and drives a set of thermally driven outflows via low density channels, while at the same time, the lateral expansion of such ionised channels compresses intervening filaments of dense, inflowing gas. The result is a complex combination of positive feedback (induced star formation in the compressed dense filaments) and negative feedback (inhibition of inflow into into the cluster core, due to entrainment of material in outflows), such that current simulations do not have the resolution to determine the sign of the *net* effect. In the case of external irradiation of an unbound cloud, the net effect is certainly positive, inasmuch as gas – which in the absence of feedback, would have just expanded away from the cloud core – is instead impacted by the ionisation front and returned to the cloud core, shocking and producing additional stars in the process. A notable outcome of this simulation is that it is apparently rather hard to distinguish observationally between those stars whose formation is triggered in this way, and those that arise spontaneously in the absence of feedback. Comparison between the feedback simulations and control experiments suggest that there is no kinematic signature of triggering (Dale *et al.* 2006): although the free expansion speed of ionised gas is around 10 km s^{-1}, this is decreased, by mass loading,

† Note that if stellar collisions are ever important – either as a result of adiabatic accretion or the action of the mass segregation instability – it has still to be demonstrated that the collision product could grow into the very massive (hundreds of solar masses or more) regime instead of undergoing avoid violent mass loss: see e.g. Belkus *et al.* (2007).

Figure 4. Simulation of the formation of a star cluster including the feedback of ionising radiation from massive stars. Dale *et al.* 2005.

to a few km s $^{-1}$, which is in any case comparable with the typical speed of turbulent motions in molecular clouds.

Perhaps the most notable – and salutary – result to emerge from these pilot simulations is simply that, in a highly inhomogeneous medium, the gas absorbs a quantity of thermal and kinetic energy that far exceeds the cloud's gravitational binding energy, does *not* imply that the nascent cluster will become gravitationally unbound (Dale *et al.* 2005). This is simply because most of the energy injected is carried off by material traveling at close to 10 km s $^{-1}$, which is considerably greater than the cluster escape velocity, while the bulk of (dense) gas in the simulation (and stars) remains bound. This effect is in addition to the often noted fact that the quantity of energy absorbed by the gas is itself many orders of magnitude less than the energy output of the ionising source, due to radiative losses in the gas. Taken together, these two effects imply that energetic arguments about the viability of cluster disruption should – if not backed up by simulations – be treated with great caution.

In parallel with these first steps towards modeling ionisation feedback, Dale *et al.* have recently embarked on SPH simulations which include the mechanical feedback from stellar winds. Since the mechanical luminosity of the stellar wind rises much more steeply with stellar mass than does its ionising luminosity (Vink *et al.* 2001), it follows that feedback from stellar winds is likely to be of increasing significance in populous clusters which contain the most massive stars. Ultimately, of course, one would wish to combine the effect of photoionisation and mechanical energy injection as in existing galactic scale simulations which assume a smooth density stratification of the interstellar medium (e.g. Tenorio Tagle *et al.* 1999). A medium term challenge, which must be met if one is to provide any credible insights into cluster formation, is to incorporate these effects into the complex and highly dynamical environments in which star clusters form.

References

Belkus, H., Van Bever, J., & Vanbeveren, D. 2007, *ApJ*, 659, 1576

Bonnell, I., Bate, M., & Zinnecker, H. 1998, *M*NRAS, 298, 93

Bonnell, I. & Bate, M. 2002, *M*NRAS, 336, 659

Bonnell, I. A., Bate, M. R., & Vine, S. G., 2003, *M*NRAS, 343, 413

Bonnell, I., Vine, S., & Bate, M. 2004, *M*NRAS, 349, 735

Bonnell, I. A. & Clarke, C. J. 1999, *M*NRAS, 309, 461

Clarke, C. J. & Bonnell, I. A. *M*NRAS, submitted

Dale, J., Bonnell, I., Clarke, C., & Bate, M. 2005, *M*NRAS, 358, 291

Dale, J., Clark, P. C., & Bonnell, I., 2006, *M*NRAS, submitted

Dale, J., Ercolano, B., & Clarke, C. 2007, *M*NRAS, in press (arXiv:0705.3396)

Edgar, R. G. & Clarke, C. J. 2004, *M*NRAS, 349, 678

Elmegreen, B. G. & Elmegreen, D. M. 2001, *A*J, 121, 1507

Figer, D. F, Najarro, F., Morris, M., McLean, I. S., Geballe, T. R., Ghez, A. M., & Langer, N. 1998, *A*pJ, 506, 384

Gouliermis, D., Kontizas, M., Korakitis, R., Morgan, D. H., Kontizas, E., & Dapergolas, A. 2000, *A*J, 119, 1757

Guillout, P., Sterzik, M. F., Schmitt, J. H. M. M., Motch, C., & Neuhaeuser, R. 1998, *A* & *A*, 337, 113

Gürkan, M. A., Freitag, M., & Rasio, F. 2004, *A*pJ, 604, 632

Larson, R. B. 1981, *M*NRAS, 194, 809

Moraux, E., Lawson, W. & Clarke, C. J. 2007, A&A, 473, 163

Oey, S. M. & Clarke, C. J. 2005, *A*pJL, 620, 430

Smith, N. & Owocki, S. P. 2006, *A*pJ, 645, L45

Spitzer, L. J. 1969, *A*pJ, 158, L139

Tenorio-Tagle, G., Silich, S. A., Kunth, D., Terlevich. E., & Terlevich, R. 1999, *M*NRAS, 309, 332

Vink, J., de Koter, A., & Lamers, H. J. G. L. M. 2001, *A* & *A*, 369, 574

Wolfire, M. G. & Casinelli, J. P. 1987, *A*pJ, 319, 850

Dynamical Evolution of Dense Stellar Systems
Proceedings IAU Symposium No. 246, 2007
E. Vesperini, M. Giersz & A. Sills, eds.

Dynamical Masses of Young Star Clusters: Constraints on the Stellar IMF and Star-Formation Efficiency

N. Bastian

Department of Physics and Astronomy, University College London, Gower Street,
London WC1E 6BT, UK
email: bastian@star.ucl.ac.uk

Abstract. Through the use of detailed light profiles and dynamical measurements of young clusters we investigate claims that the stellar initial mass function within clusters varies greatly. We find a strong age dependence in the clusters which have been claimed to have non-standard stellar IMFs, and suggest that the lack of equilibrium of these clusters is responsible for their 'strange' light-to-mass ratios and *not* IMF variations. The most likely culprit is the rapid removal of residual gas left over from the star-formation process which leaves the clusters severely out of dynamical equilibrium. By comparing the observations to N-body simulations we quantify to what degree a cluster is out of equilibrium and consequently its survival chances. We find that $> 60 \%$ of young clusters will be disrupted, due gas removal, within the first 20–50 Myr of their lives.

Keywords. galaxies: star clusters

1. The Stellar Initial Mass Function in Clusters

Many recent works have attempted to constrain the stellar initial mass function (IMF) inside massive clusters by comparing their dynamical mass estimates (found through measuring the velocity dispersion and effective radius) to the measured light. These studies have come to different conclusions, with some claiming standard Kroupa-type (Kroupa 2002) IMFs (e.g. Maraston *et al.* 2004) while others have claimed extreme non-standard IMFs (e.g. the top or bottom of the IMF is over-populated with respect to a Kroupa IMF (Smith & Gallagher 2001). However, the results appear to be correlated with the age of the clusters, as older clusters (>80 Myr) all appear to be well fit by a Kroupa-type IMF whereas younger clusters display significant scatter in their best fitting IMF (Bastian *et al.* 2006). This has led to the suggestion that the younger clusters are out of Virial equilibrium, thus undercutting the fundamental assumption which is necessary to derive dynamical masses. We will return to this point in § 2 and § 3 Focusing on the older clusters, we see that they all have standard IMFs (see Fig. 2), arguing that at least in massive clusters the IMF does not vary significantly.

2. Dynamical Equilibrium of Young Clusters

One explanation of why the youngest clusters are not in dynamical equilibrium is that young clusters are expected to expel their remaining gas (left over from the star-formation process) on extremely rapid timescales, which will leave the cluster severely out of equilibrium (e.g. Goodwin 1997). In order to search for such an effect we compared the luminosity profiles of three young clusters with that of N-body simulations of clusters which are undergoing violent relaxation due to rapid gas loss (Bastian & Goodwin 2006).

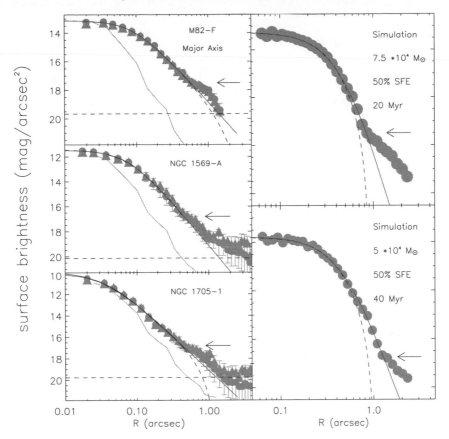

Figure 1. Taken from Bastian & Goodwin (2006): Surface brightness profiles for three young clusters (left – M82-F, NGC 1569-A, and NGC 1705-1) and two N-body simulations which include the rapid removal of gas which was left over from a non-100% star-formation efficiency (right). The solid (red) and dashed (blue) lines are the best fitting EFF (Elson, Fall, & Freeman 1987) and King (1962) profiles respectively. Note the excess of light at large radii with respect to the best fitting EFF profile in both the observations and models. This excess light is due to an unbound expanding halo of stars caused by the rapid ejection of the remaining gas after the cluster forms. *Hence, excess light at large radii strongly implies that these clusters are not in dynamical equilibrium.* For details of the modelling and observations see Bastian & Goodwin (2006) and Goodwin & Bastian (2006).

The simulations (Fig. 1, right panel) make the generic prediction of excess light at large radii (with respect to the best fitting EFF profile (Elson, Fall, & Freeman 1987), due to an unbound expanding halo of stars which stays associated with the cluster for $\sim$ $20 - 50$ Myr. These stars are unbound due to the rapid decrease of potential energy as the gas is removed on timescales shorter than a crossing time (e.g. Goodwin 1997). Observations of the three young clusters also show excess light at large radii (Fig. 1, left panel), strongly suggesting that they are experiencing violent relaxation (Bastian & Goodwin 2006). Hence these clusters are not in dynamical equilibrium.

3. The Star Formation Efficiency and Infant Mortality

Assuming that young clusters are out of equilibrium due to rapid gas loss (the extent of which is determined by the star-formation efficiency – SFE one can fold these effects

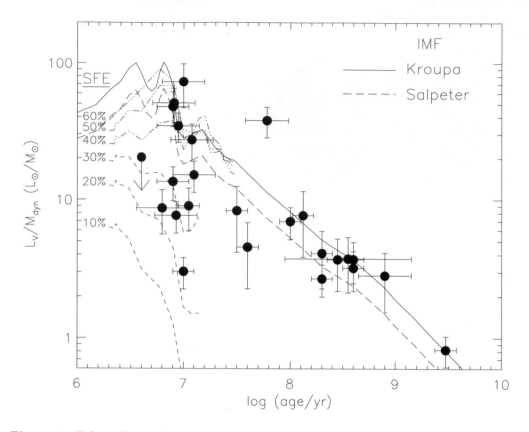

Figure 2. Taken from Goodwin & Bastian (2006): The light-to-mass ratio of young clusters. The data are from Goodwin & Bastian (2006) and references therein. The solid line is the prediction of simple stellar population models (SSPs) with a Kroupa (2002) stellar IMF wile the long-dashed line shows the same but for a Salpeter (1995) IMF. The short-dashed and dash-dotted lines are the SSP model tracks folded with the effects of rapid gas removal following non-100% star-formation efficiencies (SFE). Short-dashed lines represent the SFEs where the clusters will become completely unbound. The SFE in the simulations measures the degree to which the cluster is out-of-virial equilibrium after gas loss, and so is an *effective* SFE (see Goodwin & Bastian 2006).

(see Fig. 3 in Bastian & Goodwin 2006) into SSP models (Goodwin & Bastian 2006). The results are shown as solid and dashed red lines in Fig. 2 for various SFEs, where we have assumed all gas is lost instantaneously at 2 Myr. The dashed lines show the results for SFEs below 30% for which the cluster will become completely unbound. Solid lines represent SFEs above 30% where a bound core may remain. Note that the observed SFEs of the clusters range from 10–60% (Goodwin & Bastian 2006).

We also note that 7 out of the 12 clusters with ages below 20 Myr appear unbound (i.e. SFE < 30%), suggesting that ∼ 60% of clusters will become unbound in the first 20-50 Myr of their lives Goodwin & Bastian (2006), i.e. what has been termed "infant mortality". This is in close agreement with cluster population studies of M51 which found an infant mortality rate of 68% (Bastian *et al.* 2005) and comparable to the open cluster dispersal rate of ∼ 87% (Lada & Lada 1991, see also Whitmore 2003 and Pellerin *et al.* 2007).

4. Conclusions

Through detailed comparisons of the luminosity profiles of young clusters with N-body simulations of clusters including the effects of rapid gas loss, we argue that young clusters are not in Virial equilibrium. This undercuts the fundamental assumption needed to determine dynamical masses. This suggests that the claimed IMF variations are probably due to the internal dynamics of the clusters and not related to the IMF. By limiting the sample to the oldest clusters (which appear to be in equilibrium) we see that they are all well fit by a Kroupa-type IMF arguing that, at least in massive star clusters, the IMF does not vary significantly.

By combining the above N-body simulations with SSP models we can derive the (effective) SFE of clusters. From this we find that $\sim 60\%$ of young clusters appear to be unbound, in good agreement with other estimates of the infant mortality rate. Note however that even if a cluster survives this phase it may not survive indefinitely due to internal and external effects (e.g. Gieles *et al.* 2005).

Acknowledgements

NB gratefully acknowledges Simon Goodwin for his large part in the studies presented here.

References

Bastian, N., Gieles, M., Lamers, H. J. G. L. M., Scheepmaker, R. A., & de Grijs, R. 2005, *A&A* 431, 905
Bastian, N. & Goodwin, S. P. 2006, *MNRAS*, 369, L9
Bastian, N., Saglia, R. P., Goudfrooij, P., Kissler-Patig, M., Maraston, C., Scwheizer, F., & Zoccali, M. 2006 *A&A*, 448, 881
Elson, R. A. W., Fall, M. S., & Freeman, K. C. 1987, *ApJ* 323, 54 (EFF)
Gieles, M., Bastian, N., Lamers, H. J. G. L. M., & Mout, J. N. 2005, *A&A*, 441, 949
Goodwin, S. P. 1997, *MNRAS*, 284, 785
Goodwin, S. P. & Bastian, N. 2006, *MNRAS*, 373, 752
King, I. 1962, *AJ* 67, 471
Kroupa, P. 2002, *Science*, 295, 82
Lada, C. J. & Lada, E. A. 1991, in *The formation and evolution of star clusters*, ed. K. Joes, *ASP Conf. Ser.*, 13, 3
Maraston, C., Bastian N., Saglia R. P., Kissler-Patig, M., Schweizer, F., Goudfrooij, P. 2004, *A&A*, 416, 467
Pellerin, A., Meyer, M., Harris, J., & Calzetti, D. 2007, *ApJL*, 658, L87
Salpeter, E. E. 1955, *ApJ*, 121, 161
Smith, L. J. & Gallagher, J. S. 2001, *MNRAS*, 326, 1027
van Wijngaarden, L. 1968, *J. Engng Maths* 2, 225
Whitmore, B. C. 2003, in *A Decade of HST Science*, ed. M. Livio, K. Noll, & M. Stiavelli (Cambridge: Cambridge University Press), 153

Dynamical Evolution of Dense Stellar Systems
Proceedings IAU Symposium No. 246, 2007
E. Vesperini, M. Giersz & A. Sills, eds.

The Influence of Gas Expulsion on the Evolution of Star Clusters

H. Baumgardt and P. Kroupa

Argelander-Institut für Astronomie, Auf dem Hügel 71, 53121 Bonn, Germany
email: holger@astro.uni-bonn.de, pavel@astro.uni-bonn.de

Abstract. We present new results on the dynamical evolution and dissolution of star clusters due to residual gas expulsion and the effect this has on the mass function and other properties of star cluster systems. To this end, we have carried out a large set of N-body simulations, varying the star formation efficiency, gas expulsion time scale and strength of the external tidal field, obtaining a three-dimensional grid of models which can be used to predict the evolution of individual star clusters or whole star cluster systems by interpolating between our runs. When applied to the Milky Way globular cluster system, we find that gas expulsion is the main dissolution mechanism for star clusters, destroying about 80% of all clusters within a few 10s of Myers. Together with later dynamical evolution, it seems possible to turn an initial power-law mass function into a log-normal one with properties similar to what has been observed for the Milky Way globular clusters.

Keywords. stellar dynamics, methods: n-body simulations, galaxies: star clusters

1. Introduction

It is well known that most, if not all, stars form in star clusters. Star clusters form as so called embedded clusters within the dense cores of giant molecular clouds. The star formation efficiency (SFE), i.e. the fraction of gas that is converted into stars, can be defined as follows:

$$\epsilon = \frac{M_{ecl}}{M_{ecl} + M_{gas}} \tag{1.1}$$

where M_{ecl} is the total mass of stars formed in the embedded cluster and M_{gas} the mass of the gas not converted into stars. Inside molecular cloud cores, the star formation efficiency is usually smaller than $\epsilon < 30\%$ (Lada & Lada 2003), which implies that once the primordial gas is expelled by UV radiation and massive stellar winds from OB stars or supernova explosions, star clusters will become super-virial and their further dynamical evolution will be strongly affected by the gas loss.

If the star formation efficiency or gas expulsion time scale depends on the mass of the cluster, residual gas expulsion can also influence the mass function of star cluster systems. For example, observations of globular cluster systems show that globular clusters follow a bell shaped distribution in luminosity with an average magnitude of $M_V^0 \approx -7.3$ and dispersion $\sigma_V = 1.2$. For a mass-to-light ratio of $M/L_V = 1.5$, this corresponds to a characteristic cluster mass of $M_C = 1.1 \cdot 10^5 M_\odot$.

Their bell-shaped luminosity function sets globular clusters apart from young, massive star clusters in starburst and interacting galaxies and the open clusters of the Milky Way and other nearby spiral galaxies, which generally follow a power-law distribution over luminosities down to the smallest observable clusters, and the question arises whether the luminosity function of globular clusters is of primordial origin or whether globular clusters also started with a power-law mass function and the present-day peak is due to the quicker dynamical evolution and preferential destruction of the low-mass clusters.

2. *N*-body simulations of gas expulsion

We have performed a large parameter study of residual gas expulsion from star clusters, varying the star formation efficiency, the ratio of the gas expulsion time scale to the crossing time of the star cluster and the ratio of the half-mass radius to the tidal radius of the star cluster. This grid of models will be useful in later studies of individual star clusters and also whole star cluster systems since the evolution of the clusters can be determined by interpolation between our grid points, without the need for further simulations. This makes it possible to determine the effect of gas expulsion on whole cluster systems where the large number of clusters prevents simulations for all individual clusters.

In our runs, we assumed that the SFE does not depend on the position inside the cluster, so gas and stars follow the same density distribution initially, which was given by a Plummer model. The gas was not simulated directly, instead its influence on the stars was modelled as a modification to the equation of motion of stars. We used the collisional *N*-body code NBODY4 (Aarseth 1999) on the GRAPE computers of Bonn University to perform the simulations. All simulated clusters contained 20.000 equal-mass stars initially and simulations were run for 1000 initial *N*-body times (equivalent to about 300 initial crossing times).

Gas expulsion was assumed to start at a certain time t_D, which was set equal to one *N*-body time unit. After the delay time t_D, the gas density was decreased exponentially on a characteristic time τ_M, so the total gas left at later times is given by:

$$M_{gas}(t) = M_{gas}(0) \, e^{-(t-t_D)/\tau_M} . \tag{2.1}$$

The influence of the external tidal field was modelled in the so-called 'Near Field Approximation', which assumes that the size of the star cluster is much smaller than its distance from the galactic centre.

Within the framework of our model, the fate of a star cluster can be deduced by specifying only three parameters: The star formation efficiency ϵ, the ratio of the gas expulsion time scale to the crossing time t_{Cross} of the star cluster, and the strength of the external tidal field, quantified by the ratio of the half-mass radius r_h to the tidal radius r_t. This reduction in the number of parameters makes it feasible to run a grid of models covering the complete parameter space. The following values were chosen as grid-points:

ϵ: 5%, 10%, 15%, 20%, 25%, 33%, 40%, 50%, 75%

r_h/r_t: 0.01, 0.033, 0.06, 0.1, 0.15, 0.2

τ_M/t_{Cross}: 0.0, 0.05, 0.10, 0.33, 1.00, 3.0, 10.0

Fig. 1 compares our results for the bound mass fraction as a function of the star formation efficiency with published results from the literature. Shown are cases where the gas is removed instantaneously (right group of points) and cases of slow gas removal. It can be seen that for instantaneous gas removal there is very good agreement between the results of this paper and published results. SFEs of 33% already lead to a final bound cluster, although only a very small mass fraction remains bound in this case. For a SFE of 50%, about 70% of the total cluster mass remains bound. In case of near adiabatic gas removal, we find that the critical SFE needed to produce a bound cluster is between 5% to 10%. This is about 5% smaller than what Geyer & Burkert (2001) found for their

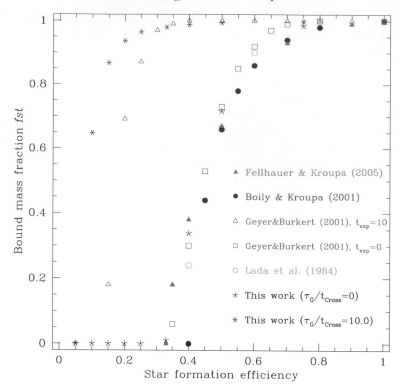

Figure 1. Comparison of the surviving mass fraction derived in this work with results from the literature. For instantaneous gas removal (right group of points) there is very good agreement between both. For slow gas removal (left points, stars and open triangles), the critical SFE needed to produce a bound cluster determined here is about 5% smaller than the one found by Geyer & Burkert (2001). This can be explained by the different initial density profiles and the fact that Geyer & Burkert (2001) assumed linear gas removal while we assume an exponential one.

model N2 with $t_{exp} = 10$. Performing additional N-body runs shows that the difference becomes significantly smaller if we let the gas fraction decrease linearly with time, as was done by Geyer & Burkert (2001). The remaining difference is probably due to the different density profiles. Geyer & Burkert (2001) used King $W_0 = 3$ and $W_0 = 5$ models in their runs, which are significantly less concentrated than the Plummer models we use. More results of our simulations can be found in Baumgardt & Kroupa (2007a).

3. Influence of gas expulsion on star cluster systems

In order to study the influence of gas expulsion on star cluster systems, we assume that pre-cluster molecular cloud cores are distributed with a power-law mass function $dN/dM_{Cl} \sim M_{Cl}^{-\beta_{Cl}}$ between lower and upper mass limits of $M_{Low} = 10^3 M_\odot$ and $M_{Up} = 10^7 M_\odot$. The star formation efficiencies ϵ are assumed to follow a Gaussian distribution with a mean of 25% and dispersion of 10%. The cluster radii are assumed to follow a Gaussian distribution with dispersion $\log \sigma_R/\mathrm{pc} = 0.2$ and various means given by $\log r_h/\mathrm{pc} = \log r_{hm} + k_r \log R_{GC}/\mathrm{kpc}$, i.e. our distributions are allowed to change with galactocentric distance. Most of our simulations had $\log r_{hm} = -0.3$ and $k_r = 0.2$.

We then study the influence of various destruction mechanisms on the mass function of star clusters. In particular, we study the influence of gas expulsion, stellar evolution, two-body relaxation and an external tidal field, disc shocks and dynamical friction. The influence of gas expulsion was modelled by interpolating between the grid of runs made above. Interpolation was done by using the 8 grid points surrounding the position of each cluster and then by linearly interpolating in each coordinate. Stellar evolution reduces the masses of star clusters by about 30%. We applied stellar evolution mass loss after gas expulsion and before the other mechanisms, since most mass lost from star clusters due to stellar evolution is lost within the first 100 Myer. The effects of two-body relaxation and a spherical external tidal field were modelled according to the results of Baumgardt & Makino (2003). According to Baumgardt & Makino (2003), the lifetime of a star cluster moving through an external galaxy with circular velocity V_C on an orbit with pericentre distance R_P and eccentricity e is given by

$$\frac{t_{DisR}}{[\text{Myr}]} = k \left(\frac{N}{ln(0.02\,N)} \right)^x \frac{R_P}{[\text{kpc}]} \left(\frac{V_C}{220\text{km/sec}} \right)^{-1} (1+e). \qquad (3.1)$$

Here N is the number of cluster stars left after gas expulsion, which can be calculated from the cluster mass and the mean mass of the cluster stars as $N = M_C / <m>$. x and k are constants describing the dissolution process and are given by $x = 0.75$ and $k = 1.91$ (Baumgardt & Makino 2003). Disc shocks and dynamical friction were modelled according to the equations 7-72 and 7-26 of Binney & Tremaine (1987).

The gas expulsion timescales are derived by comparing the energy input from massive stars through stellar winds and supernova explosions with the total energy of a gas cloud. From this, we derive the following relation for the gas expulsion timescale τ_M (see Baumgardt & Kroupa (2007b) for a complete discussion):

$$\tau_M = E_{Gas}/\dot{E} = 7.1 \cdot 10^{-8} \, \frac{1-\epsilon}{\epsilon} \frac{M_C}{[M_\odot]} \left(\frac{r_h}{[\text{pc}]} \right)^{-1} \text{Myr}, \qquad (3.2)$$

where M_C is the mass of the cluster and r_h its half-mass radius. The gas expulsion time increases with cluster mass since more massive gas clouds have deeper potential wells, so it takes longer until the gas is expelled from them.

Fig. 2 depicts the resulting evolution of the mass function of clusters if we apply the above calculations to the globular cluster system of the Milky Way and evolve the system for a Hubble time. It can be seen that most low-mass clusters are destroyed as a result of residual gas expulsion. In total, only 21% of all clusters survive residual gas expulsion, mostly those starting with high masses. Due to the efficient destruction of low-mass clusters, the overall mass function develops a turnover and is in agreement with the observed mass function of globular clusters for both inner and outer star clusters.

4. Conclusions

We have performed a large grid of simulations studying the impact of initial gas expulsion on the survival rate and final properties of star clusters, varying the star formation efficiency, ratio of gas expulsion timescale to the crossing time of the cluster and the strength of the external tidal field.

Our simulations show that both the star formation efficiency and the speed with which the gas is removed have a strong influence on the evolution of star clusters. In case of instantaneous gas removal, clusters have to form with SFEs $\geqslant 33\%$ in order to survive gas expulsion. This limit is significantly lowered for gas removal on longer timescales and clusters with SFEs as low as 10% can survive gas expulsion in the adiabatic limit if the external tidal field is weak.

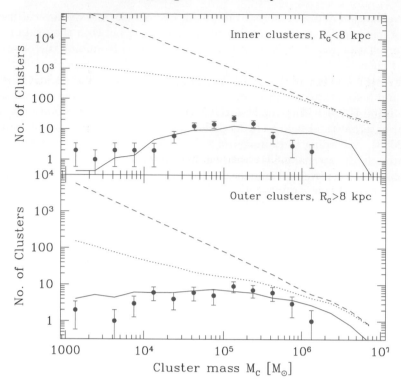

Figure 2. Mass distribution of star clusters before gas expulsion (dashed lines), after gas expulsion (dotted lines) and after a Hubble time (solid lines), compared to the observed distribution of Milky Way clusters (points). Most low-mass clusters are destroyed by residual gas expulsion. The distribution of surviving clusters is in good agreement with the observed one for both inner and outer clusters.

We have then applied these simulations to follow the evolution of the galactic globular cluster system. We assumed that globular clusters start with power-law mass functions, similar to what is observed for the galactic open clusters and young, massive star clusters in interacting galaxies. The dissolution of the clusters was then studied under the combined influence of residual gas expulsion, stellar mass-loss, two-body relaxation and an external tidal field.

We find that residual gas expulsion is the main dissolution mechanism for star clusters, destroying about 80% of them within a few 10s of Myr. It seems possible to turn an initial power-law mass function into a log-normal one, because clusters with masses between 10^4 to $10^5 M_\odot$ lose their residual gas on a timescale shorter than their crossing times as indicated by our feedback analysis. Stars released from these clusters become halo field stars. It seems possible that all halo stars originated from dissolved star clusters.

References

Aarseth, S. 1999, *PASP* 111, 1333
Baumgardt, H. & Kroupa, P. 2007a, *MNRAS* 380, 1589, astro-ph/0707.1944
Baumgardt, H. & Kroupa, P. 2007b, *MNRAS* submitted
Baumgardt, H. & Makino, J. 2003, *MNRAS* 340, 227
Binney, J. & Tremaine, S. 1987, *Galactic Dynamics*, Princeton Univ. Press, Princeton
Geyer, M. P. & Burkert, A. 2001, *MNRAS* 323, 988
Lada, C. J. & Lada E. A. 2003, *ARA&A* 41, 57

Dynamical Evolution of Dense Stellar Systems
Proceedings IAU Symposium No. 246, 2007
E. Vesperini, M. Giersz & A.Sills, eds.

© 2008 International Astronomical Union
doi:10.1017/S174392130801524X

A Dynamical Origin for Early Mass Segregation in Young Star Clusters

Steve McMillan,[1] Enrico Vesperini[1] and Simon Portegies Zwart[2]

[1] Department of Physics, Drexel University, Philadelphia, PA 19104, USA
email: steve@physics.drexel.edu
email: vesperin@physics.drexel.edu

[2] Astronomical Institute 'Anton Pannekoek' and Section Computational Science, University of Amsterdam, Kruislaan 403, 1098SJ Amsterdam, the Netherlands
email: spz@science.uva.nl

Abstract. Some young star clusters show a degree of mass segregation that is inconsistent with the effects of standard two-body relaxation from an initially unsegregated system without substructure, in virial equilibrium, and it is unclear whether current cluster formation models can account for this degree of initial segregation in clusters of significant mass. We show that mergers of small clumps that are either initially mass segregated, or in which mass segregation can be produced by two-body relaxation before they merge, generically lead to larger systems which inherit the progenitor clumps' segregation. We conclude that clusters formed in this way are naturally mass segregated, accounting for the anomalous observations and suggesting that this process of prompt mass segregation due to initial clumping should be taken into account in models of cluster formation and dynamics.

Keywords. stellar dynamics; methods: n-body simulations; stars: formation; globular clusters: general; open clusters and associations: general

1. Introduction

Mass segregation has been observed in many old globular clusters (Sosin & King 1997, Pasquali *et al.* 2004), consistent with the fact that these systems have relaxation times significantly less than a Hubble time. However, a number of studies show significant mass segregation in clusters having actual ages, as measured by the evolutionary state of their stars, substantially less than the time needed to produce the observed segregation by standard two-body relaxation (Hillenbrand 1997, Hillenbrand & Hartmann 1998, Fischer *et al.* 1998, de Grijs *et al.* 2002, Sirianni *et al.* 2002, Gouliermis *et al.* 2004, Stolte *et al.* 2006). Numerical simulations indicate that dynamical evolution from initially unsegregated systems cannot account for the degree of mass segregation observed in these clusters (e.g. Bonnell & Davies 1998).

The obvious explanation is that these clusters were born mass segregated, and recent studies do indeed suggest that massive stars form preferentially in the centers of star-forming regions (Elmegreen & Krakowski 2001, Klessen 2001, Bonnell *et al.* 2001, Stanke *et al.* 2006, Bonnell & Bate 2006). The mechanism invoked to explain this primordial mass segregation relies mainly on the higher accretion rate for stars in the centers of young clusters. However, the efficiency of this mechanism is still a matter of debate (Klein & McKee 2005, Bonnell & Bate 2006) and, more generally, the processes of massive star formation and feedback remain poorly understood (Krumholz *et al.* 2005).

We report results from a numerical study exploring dynamical routes to mass segregation during the early stages of cluster formation (McMillan, Vesperini, & Portegies Zwart 2007). We imagine that stars form in small clumps, which subsequently merge to form larger systems (Bonnell, Bate & Vine 2003; Elmegreen 2006 and references therein). We assume that the clumps are significantly mass segregated at formation, or that they have short enough relaxation times that mass segregation can occur within the merger time scale. In either case, the final clusters inherit the segregation of their progenitor clumps, providing a natural explanation for large systems which are mass segregated yet physically young.

2. Method and initial conditions

We adopt initial conditions in which the cluster consists of N_c clumps, with centers distributed within a sphere of radius $R_{cluster}$. The system of clumps is not in virial equilibrium; rather the clump centers are dynamically "cool," with $q_c = -T/U < \frac{1}{2}$. Our simulations have $N_c = 2$, 4, and 8, and explore the evolution of systems having a range of values of several key bulk parameters of the clump system:

• *Clumping Ratio.* The ratio $\mathcal{R}_c \equiv R_h/R_{cluster}$, where R_h is the half-mass radius of an individual clump, is a convenient measure of clumping. We concentrate on two sets of runs: "strongly clumped" models, with $\mathcal{R}_c = 0.013$, and "moderately clumped" clusters with $\mathcal{R}_c = 0.037$. (The 90% Lagrangian radius for a clump is $\sim 5R_h$.)

• *Virial Ratio.* Our clump systems are initially out of equilibrium, with $0 \leqslant q_c \lesssim 0.25$.

• *Survey of Two-Clump Mergers.* The case $N_c = 2$ gives us greatest control over the parameters of the interaction, and in this case we vary systematically the clump mass ratio $\mu = M_1/M_2$, the impact parameter b, virial ratio q, and the clump mass–radius relation $R_h \sim M^\alpha$. Our survey spans the ranges $\mu = 1, 2, 4$; $b = 0, r_h$; $q_c = 0, 0.1, 0.25$; and $\alpha = 0, \frac{1}{3}, \frac{1}{2}$.

Individual clumps are modeled as systems of $N \sim 10^4$ particles in virial equilibrium, with Plummer density profiles. We neglect initial binaries and concentrate on two stellar mass distributions:

• *Two-component mass functions* consisting simply of a "heavy" and a "light" component, for conceptual ease.

• *Realistic mass functions* as defined by Kroupa, Tout, & Gilmore 1993.

The degree of mass segregation is conveniently quantified by the ratio $f_{seg} = R_h/R_h^{(heavy)}$, where R_h is the half-mass radius of the entire system and $R_h^{(heavy)}$ refers to the heavy component in the two-component case and to stars above $2.5M_\odot$ in the realistic case.

We consider both initially unsegregated clumps and clumps with significant initial mass segregation. Initial mass segregation in the latter case is achieved by letting an unsegregated clump evolve in isolation for long enough for mass segregation to occur by normal two-body relaxation. This mass-segregated system is then used as a template for all clumps in our simulations. This procedure is simply a convenient means of generating a self-consistent system as an initial condition for a mass-segregated clump; our results are insensitive to the precise means by which the segregation comes about.

Our study is based on direct N-body simulations using the `starlab` package (Portegies Zwart *et al.* 2001; http://www.manybody.org), accelerated by GRAPE-6 special-purpose hardware (Makino *et al.* 2003). Throughout, our time unit is the dynamical time scale (Heggie & Mathieu 1986) of one of the initial unsegregated clumps. In these units, the internal clump relaxation time scale is $t_r \sim 0.1N/\ln N \sim 100$ (for $N = 10^4$); the free-fall time for the cluster is $t_{ff} \sim 0.7 \mathcal{R}_c^{-3/2} \sim 90$ (450) for $\mathcal{R}_c = 0.037$ (0.013).

The results of our simulations are presented in more detail by Vesperini, McMillan, & Portegies Zwart (2008).

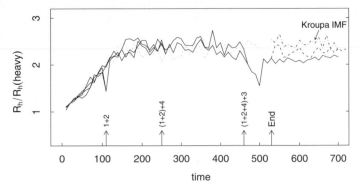

Figure 1. Time evolution of f_{seg} for an initially unsegregated strongly clumped system. Vertical arrows mark various merging events between the clumps (arbitrarily numbered 1–4); the labels above each arrow indicate the clumps involved in the merger. The solid lines at each stage of the merging process show the evolution of f_{seg} for the remaining clumps in the cluster. The dotted lines show f_{seg} for an individual clump evolved in isolation. The dot-dashed line shows the results of a simulation with a Kroupa initial mass function.

3. Results

3.1. *Systems with initial mass segregation*

The goal of the initially segregated runs is to establish a connection between the mass segregation of the original clumps and that of the cluster resulting from the merger. We have performed "hierarchical" simulations, in which clumps merge sequentially in a series of two-body encounters, and "cluster" simulations, in which all clumps are followed simultaneously.

In all cases, once the merger is complete, the degree of mass segregation in the final cluster, as measured by f_{seg}, is approximately equal to that in the original clumps—mass segregation is preserved during the merging process. This is consistent with van Albada (1982) and Funato *et al.* (1992), who found that memory of particles' initial binding energy is not erased during violent relaxation. This result is largely *insensitive* to any of the structural parameters listed in the previous section, with the sole proviso that, for the clumps to merge into an effectively featureless (smooth) cluster within a few free-fall times, the initial clump system must be relatively cool—$q_c \lesssim 0.1$.

Unlike f_{seg}, other bulk properties of the resultant cluster, e.g. central concentration and virial radius, do depend on the properties of the initial clumps—denser clumps tend to produce more concentrated final clusters.

3.2. *Initially unsegregated systems*

We have repeated many of the $N_c = 4, 8$ simulations, without initial mass segregation in the individual clumps. We find that the segregation properties of the end-products are controlled by the ratio $\tau = t_{ff}/t_{seg}$, representing the degree to which significant internal mass segregation can occur in a global free-fall time. For our choice of system parameters, with $t_{seg} \sim 0.1\,t_r$, $\tau \sim 2$ for moderately clumped initial conditions, and $\tau \sim 10$ for the strongly clumped case, so significant mass segregation is expected within a merger time.

Fig. 1 shows the time evolution of f_{seg} for an initially unsegregated, strongly clumped ($N_c = 4$) system. It shows the detailed merger history of the original clumps, illustrating how mass segregation proceeds first within the clumps, then within each new merger product, culminating in the final merged cluster. We clearly see internal mass segregation

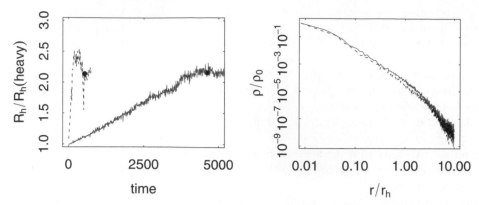

Figure 2. (Left) Time evolution of f_{seg} for one of the simulations discussed in §4. The left (dashed) curve began from strongly clumped unsegregated initial conditions with $N_c = 4$; the right (solid) curve from a single unsegregated Plummer profile. (Right) The density profiles of the two runs at the indicated points are almost indistinguishable.

in the clumps before they merge. The final value of f_{seg} is comparable to those found in the initially segregated simulations.

As an additional point of comparison, the dot-dashed line in Fig. 1 shows the corresponding ratio for an additional simulation with the same overall parameters but using a mass function from Kroupa *et al.* (1993), demonstrating that the effect persists when a realistic cluster mass distribution is used.

For this scenario to work, the clumps must have $\tau > 1$; for $t_{ff} \sim 1$ Myr, this implies $t_r \lesssim 10$ Myr. We note that 4 (out of 5) of the young embedded clusters listed by Baba *et al.* (2004) have relaxation times between 2 and 10 Myr, and the segregated clusters cited in §1 have relaxation times ranging from ~ 6 Myr (Orion) to ~ 40 Myr (NGC 3603). The relaxation time in the final cluster is expected exceed that in a clump by a factor $\approx N_c^{1/2} (0.1/\mathcal{R}_c)^{3/2} = 8.9\,(43)$ for $N_c = 4$, $\mathcal{R}_c = 0.037\,(0.013)$. [The "$N$" in the relaxation time contributes a factor of N_c, the dynamical time scale contributes $(N_c/\mathcal{R}_c^3)^{-1/2}$, and the numerical factor 0.1 comes from an estimate of the relationship between the final half-mass radius and $R_{cluster}$, based on energy conservation and the virial theorem.] We conclude that the above condition on t_r would be met for even a modest number of clumps or moderate clumping.

4. Possible dynamical histories of a young segregated cluster

The end products of the simulations described above are young, yet significantly mass segregated, clusters. Without knowing the actual dynamical history of such a system, one might imagine "observing" one of these simulated clusters to try to reproduce its properties and reconstruct its past dynamical evolution. The traditional way to do this is to perform N-body simulations starting from the initial conditions adopted in the vast majority of numerical studies of star cluster evolution—a spherical system with no primordial mass segregation and a Plummer (or King) density profile. We have carried out this experiment, running a simulation starting from a two-component spherical system in virial equilibrium, with 40,000 particles and a Plummer density profile. We refer to this simulation as our "standard" model.

Fig. 2 (left frame) compares the time evolution of f_{seg} in the standard model with the strongly clumped unsegregated run described in §3. The standard model is scaled so that, at the indicated times, when mass segregation is effectively complete and the degree of mass segregation is similar in each run, the two models have the same half-mass radius. We see that the clumped model achieves "complete" mass segregation much sooner (at least a factor of $\sim 7 - 10$ faster) than does the standard model.

Furthermore, as shown in the right frame of Fig. 2, the density profiles at the indicated times are very similar. Not shown in the figure is the fact that the median radii corresponding to different mass ranges (for cases with a realistic mass spectrum) are also very similar in the two models. Since the standard model takes much longer than the clumped system age to reproduce the same cluster properties, one might incorrectly conclude from this numerical study that the mass segregation found in this cluster must reflect its initial conditions. However, as we have shown, several possible dynamical histories can lead to similar final systems.

These simulations demonstrate that there are a number of viable evolutionary paths, relying on initial mass segregation in clumpy systems or on multiscale dynamical evolution, that can lead to significant mass segregation in a physically young cluster.

Acknowledgements

This work was supported in part by NASA grants NNG04GL50G and NNX07AG95G, NSF grant AST-0708299, and by the Royal Netherlands Academy of Arts and Sciences (KNAW).

References

Bonnell, I. A. & Davies, M. B 1998, *MNRAS*, 295, 691
Bonnell, I. A., Clarke, C. J., Bate, M. R., & Pringle, J. E. 2001, *MNRAS*, 324, 573
Bonnell, I. A., Bate, M. R., & Vine S. 2003, *MNRAS*, 343, 413
Bonnell, I. A. & Bate, M. R. 2006, *MNRAS*, 370, 488
de Grijs, R., Gilmore, G. F., Johnson, R. A., & Mackey, A. D. 2002, *MNRAS*, 331, 245
Elmegreen, B. & Krakowski, A. 2001, *ApJ*, 562, 433
Elmegreen, B. 2006, astro-ph/0605519
Fischer, P., Pryor, C., Murray, S., Mateo, M., & Richtler, T. 1998, *AJ*, 331, 592
Funato, Y., Makino, J. & Ebisuzaki, T. 1992, *PASJ*, 44, 291
Gouliermis, D., Keller, S. C., Kontizas, M., Kontizas, E., & Bellas-Velidis, I. 2004, *A&A*, 416, 137
Heggie, D. C. & Mathieu, R. D. 1986, in *The Use of Supercomputers in Stellar Dynamics* (P. Hut and S. McMillan, eds.; Springer-Verlag, New York)
Hillenbrand, L. A. 1997, *AJ*, 113, 1733
Hillenbrand, L. A. & Hartmann L. E. 1998, *ApJ*, 331, 540.
Klessen R., 2001, *ApJ*, 556, 837
Kroupa, P., Tout, C. A., & Gilmore, G. 1993, *MNRAS*, 262, 545
Krumholz, M. R., Klein, R. I., & McKee, C. F. 2005, *Nature*, 438, 332
Makino, J., Fukushige, T., Koga, M., & Namura, K. 2003, *PASJ*, 55, 1163
McMillan, S. L. W., Vesperini, E., & Portegies Zwart, S. F.2007, *ApJ*, 655, 45
Pasquali, A., De Marchi, G., Pulone, L., & Brigas, M. S. 2004, *A&A*, 428, 469
Portegies, Zwart, S. F., McMillan, S. L. W., Hut, P., & Makino, J. 2001, *MNRAS*, 321, 199
Sirianni, M., Nota, A., De Marchi, G., Leitherer, C., & Clampin, M. 2002, *ApJ*, 579, 275
Sosin, C., & King, I. R. 1997, *AJ*, 113, 1328
Stanke, T., Smith, M. D., Gredel, R., & Khanzadyan, T. 2006, *A&A*, 447, 609
Stolte, A., Brandner, W., Brandl, B., & Zinnecker H. 2006, AJ, 132, 253
van Albada, T. S. 1982, *MNRAS*, 201, 939
Vesperini, E., McMillan, S. L. W., & Portegies Zwart, S. F., 2008, *in preparation*

Dynamical Evolution of Dense Stellar Systems
Proceedings IAU Symposium No. 246, 2007
E. Vesperini, M. Giersz & A. Sills, eds.

A Near-infrared Survey of the Rosette Complex: Clues of Early Cluster Evolution

Carlos G. Román-Zúñiga[1], Elizabeth A. Lada[2] and Bruno Ferreira[2]

[1] Dept. of Astronomy, University of Florida, Gainesville 32611
Present address: Harvard-Smithsonian Center for Astrophysics, Cambridge, MA USA
email: cromanzu@cfa.harvard.edu

[2] Dept. of Astronomy, University of Florida, Gainesville 32611
email: lada@astro.ufl.edu, ferreira@astro.ufl.edu

Abstract. The majority of stars in our galaxy are born in embedded clusters, which can be considered the fundamental units of star formation. We have recently surveyed the star forming content of the Rosette Complex using FLAMINGOS in order to investigate the properties of its embedded clusters. We discuss the results of our near-infrared imaging survey. In particular, we on the first evidence for the early evolution and expansion of the embedded clusters. In addition we present data suggesting a temporal sequence of cluster formation across the cloud and discuss the influence of the HII region on the star forming history of the Rosette.

Keywords. stars: early-type, stars: formation, infrared: stars, ISM: evolution, surveys, Galaxy: open clusters and associations

---

1. Introduction

Observational studies of embedded clusters are crucial to understand the early evolution of stellar systems. Embedded clusters are abundant in the many active Giant Molecular Clouds of the Milky Way, and tell us childhood stories that are common to the majority (60 to 90%) of the stars in the galaxy (Lada *et al.* 1991; Carpenter 2000). Their observational properties place essential constraints for studies of larger and more evolved systems.

The Rosette Complex, located at d = 1.6 ± 0.2 kpc in the constellation of Monoceros, is a valuable laboratory for the study of embedded clusters: it presents a convenient (and certainly impressive) layout of interaction between a HII region generated by a large OB association (NGC 2244) and a family of small clusters deeply embedded in an adjacent molecular cloud remnant. It has been suggested that the shock front generated by the HII region could have triggered the formation of the young clusters in the Rosette Molecular Cloud (Phelps & Lada 1997, Williams, Blitz & Stark 1995), following a process known as Sequential Star Formation (Elmegreen & Lada (1977)). To test this hypothesis, it is crucial to investigate the influence of the local environment on the properties of the young clusters, as it would help us to understand the different initial conditions under which cluster formation occurs. Furthermore, studying the Rosette Complex allows to extend the embedded population surveys of nearby (d < 500 pc) clouds beyond the local parsec neighborhood, something that until nowadays, has rarely been done.

As part of the NOAO survey program *Toward a Complete Near-Infrared Spectroscopic and Imaging Survey of Giant Molecular Clouds* (P.I.: Elizabeth A. Lada), we chose the Rosette Complex for a new systematic near-infrared study, focused on the investigation of its embedded cluster population (Román-Zúñiga *et al.* 2008.) Our main goals were to investigate the distribution of star formation in the Rosette, determine the fraction of stars that form in embedded clusters and compare these results to those found in other local clouds. We present here a brief summary of our study.

2. Exploring the Rosette Complex

One essential difficulty in our survey was the heavily populated background towards the Rosette Complex: at the anticenter of the Galaxy the field is composed of an abundant disk population and a tapestry of faint background galaxies; in the Near-infrared where much of the nebulosities of the cloud disappear, it is quite difficult to notice the cluster population against the sea of stars and galaxies. Our observations are sensitive to objects almost three times fainter than those observed with previous surveys like 2MASS, a clear advantage for the detection of young members, but unfortunately, at 1.6 kpc it was a difficult task to separate galaxies, young low mass stars and objects with high color dispersion.

For those reasons we decided to apply (for this particular study) two major constraints to our survey: first, we limited our sample to a population of objects brighter than 15.75 (approximately $0.1 \, M_\odot$ for unreddened Rosette stars) and presenting a near infrared excess (NIRX) in a JHK color-color diagram. For embedded stars, the color excess is a signature of circumstellar material and therefore of youth because it is known that 50% of the circumstellar disks in an embedded clusters will decrease in brightness by the second million year of age. Thus, the detection of infrared excess stars allows to highlight the most recent episode of formation in the star forming region. Second, we applied a technique of nearest neighbors specifically aimed to the detection of star clusters as regions with surface densities higher than the background calculated in a near control field (Ferreira & Lada 2008, in preparation).

The combination of high surface density and circumstellar emission allowed us to confirm the existence of the seven embedded clusters identified visually by Phelps & Lada (labeled PL01 to Pl07), and also indicated the existence of four more clusters, two heavily embedded in the molecular cloud (labeled REFL08 and REFL09), one hiding to the west of the OB association NGC 2244 (NGC 2237, as suggested by 2MASS data Li 2005) and a small cluster near a northwest faint clump (REFL10) at the Rosette Nebula region (see Fig. 1). The two smallest clusters, PL06 and RLEF10 have about 10 NIRX objects, indicating a total of around 30 members if the average JHK NIRX fraction in young clusters is close to 30%. This is consistent with the minimum required for dynamical equilibrium (Adams & Myers 2001; Lada & Lada 2003), and let us think that we detected almost every active spot of formation in the cloud.

We analyzed the 11 Rosette clusters individually, and calculated their sizes, average extinction and NIRX fraction within the boundaries defined by contours of surface density. The properties of the clusters yield several interesting results: 1) the embedded clusters in the Rosette have an average JHK NIRX fraction around 30-40%, which is consistent with those of clusters with ages of less than 2 Myr (e.g. slightly older than the Trapezium in Orion and younger then IC 348 in Perseus). This agrees with the spectroscopic age of 2 ± 0.4 Myr estimated for NGC 2244 (Park & Sung 2002). 2) The average equivalent radii of the Rosette clusters are slightly larger than the average calculated by Lada & Lada (2003) in a large sample of galactic embedded clusters, and the equivalent radii of individual clusters are negatively correlated with their mean extinction and NIRX fraction; this strongly suggests that the more evolved clusters are already less concentrated, and noticeably less embedded, all consistent with a picture of rapid infant evolution. In the Rosette, the embedded phase timescale would be similar to the T Tauri phase, as

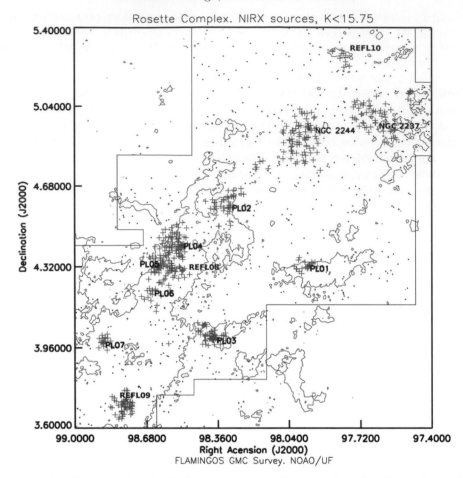

Figure 1. Identification of embedded clusters in the Rosette Complex. Both dot and plus symbols are sources identified as having near-infrared excess, but plus symbols indicate those objects with nearest neighbor surface densities above the background average, calculated in near control fields. The solid line contour indicate the base level of ^{13}CO emission of 0.8 K·km·s^{-1} that we used to define the extension of the main molecular cloud regions. The solid thin line indicates the limits of the NOAO survey coverage.

evidenced by the fact that a few stars in the more evolved embedded clusters are already visible in optical plates.

We found that 85% of the young stars in the Rosette Molecular Cloud belong to embedded clusters. This is in excellent agreement with previous studies in other clouds like Orion and Perseus (Lada *et al.* 1991, Carpenter 2000), confirming that the cluster mode of formation is highly dominant in galactic star forming complexes.

Finally, to investigate the influence of the HII region in the evolution of the star forming properties of the complex, we analyzed the distribution of young stars as a function of distance from the center of NGC 2244. We found that the radial distribution appear to show four major groups of clusters: the first group contains NGC 2244, NGC 2237 and RLEF10 in the Rosette Nebula, likely the older episode of formation in the cloud. The second group is located at the edge of the cloud that is in direct contact with the photodissociation region (the cloud 'Ridge', Blitz & Thadeus 1980) and contains clusters PL01 and PL02; at the cloud ridge the molecular cloud is being compressed and destroyed systematically, possibly stopping the star formation process more quickly. The third group is located at the central core of the cloud, which is the region of main interaction between the nebula and the molecular cloud (the shock front), and contains clusters PL04, PL05, PL06, PL03 and REFL08, which account for almost 50% of the total embedded population; at the central part of the cloud – where clusters formed from the most massive molecular gas clumps (Williams, Blitz & Stark 1995) – the efficiency of formation could have been enhanced by the influence of the cloud. Finally, a fourth group of two clusters, PL07 and REFL09 is located at the back of the cloud, far away from the influence of the HII region. Were these lonely clusters formed spontaneously, without any influence from the Rosette Nebula?

We also found that if we calculate the average NIRX fraction for each of these groups, then this fraction appears to increase with distance from the nebula, indicating a possible sequence of cluster ages. The sequence of ages for embedded clusters in the Rosette Complex is really in agreement with the model of Elmegreen & Lada (1977) because in that model, clusters form one after another, with one generation evolving and sweeping the material for the formation of the next one. In the Rosette all the clusters in the molecular cloud are still highly embedded and cannot be old enough to allow for a sequence of triggered formation. Instead, it appears that the clusters in the Rosette possibly have a sequence of ages that may be primordial and possibly reflects the early evolution of the molecular cloud. Then, the HII region affects the embedded population depending on how close is from each cluster. It will be necessary to constraint the relative ages of clusters using spectroscopy and mid-infrared photometry, to refine these interesting results. However, once we have a better idea of the real time scales involved in the formation of the Rosette Population we will be able to determine clearly its star forming history and to recollect what seems to be a very interesting story, one that could unlock many of the secrets of cluster formation.

References

Blitz, L. & Thadeus, P. 1980, *ApJ*, 241, 676
Carpenter, J. M. 2000, *AJ*, 120, 3139
Elmegreen B. G. & Lada, C. J. 1977, *ApJ*, 214, 725
Lada, E. A., DePoy, D. L., Evans, N. J., & Gatley, I. 1991, *ApJ*, 371, 171
Lada, C. J. & Lada, E.A. 2003, *ARA&A*, 41, 57
Li, J. Z. 2005, *ApJ*, 625, 242
Park, B. & Sung, H. 2002, *AJ*, 123, 892
Phelps, R. & Lada, E. A. 1997, *ApJ*, 477, 176
Román-Zúñiga, C. G., Elston, R., Ferreira, B. & Lada, E. A. 2008, *ApJ*, 695 (in press)
Williams, J. P., Blitz, L., & Stark, A. A. 1995, *ApJ*, 451, 252

Dynamical Evolution of Dense Stellar Systems
Proceedings IAU Symposium No. 246, 2007
E. Vesperini, M. Giersz & A. Sills, eds.

Changing Structures in Galactic Star Clusters

S. Schmeja[1], M. S. N. Kumar[1], D. Froebrich[2] and R. S. Klessen[3]

[1]Centro de Astrofísica da Universidade do Porto, Rua das Estrelas, 4150-762 Porto, Portugal
email: sschmeja@astro.up.pt, nanda@astro.up.pt

[2]Centre for Astrophysics and Planetary Science, University of Kent,
Canterbury, CT2 7NH, UK
email: df@star.kent.ac.uk

[3]Zentrum für Astronomie der Universität Heidelberg, Institut für Theoretische Astrophysik,
Albert-Ueberle-Str. 2, 69120 Heidelberg, Germany
email: rklessen@ita.uni-heidelberg.de

Abstract. We investigate the structures of embedded and open clusters using statistical methods, in particular the combined parameter $\mathcal{Q}$, which permits to quantify the cluster structure. Star clusters build up from several subclusters evolving from a structured to a more centrally concentrated stage. The evolution is not only a function of time, but also of the mass of the objects. Massive stars are usually centrally concentrated, while lower-mass stars are more widespread, reflecting the effect of mass segregation. Using this method we find that in IC 348 and the Orion Nebula Cluster the spatial distribution of brown dwarfs does not follow the central clustering of stars, giving important clues to their formation mechanism by supporting the ejected embryo scenario.

Keywords. open clusters and associations: general, stars: formation, methods: statistical

1. Introduction

Stars are usually born in clusters embedded in the dense regions of giant molecular clouds (see Lada & Lada 2003 for a review). The embedded phase of a cluster lasts only a few Myrs; most clusters will likely dissolve before reaching an age of 10 Myrs, less than 10 per cent of embedded clusters may survive the emergence from the parental molecular cloud to become gravitationally bound open clusters (Lada & Lada 2003).

We investigate the structures of a number of embedded and open clusters, as well as the results of numerical simulations of clustered star formation. While the structure of an embedded cluster may hold clues to its formation mechanism, the structure of an open cluster will reflect its dynamical evolution.

There are two basic types of clusters with regard to their structure (Lada & Lada 2003): On one hand, hierarchical clusters showing a stellar surface density distribution with multiple peaks and possible fractal substructure; on the other hand, centrally condensed clusters exhibiting highly centrally concentrated stellar surface density distributions with relatively smooth radial profiles that can be approximated by simple power-law functions or King profiles.

2. Method

We analyse the cluster structures using a minimum spanning tree (MST), the unique set of straight lines ("edges") connecting a given set of points without closed loops, such that the sum of the edge lengths is a minimum (e.g. Gower & Ross 1969). Besides the

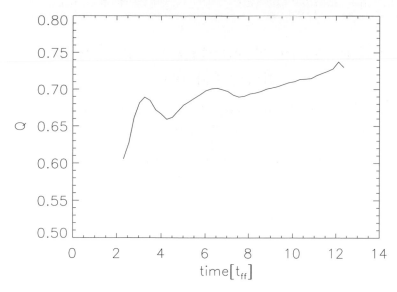

Figure 1. The temporal evolution of the clustering measure $\mathcal{Q}$ in a nascent star cluster for a numerical model with a Mach number of $\mathcal{M} = 6$ and a wave number of $3 \leqslant k \leqslant 4$. The time is given in units of the free-fall time ($t_{ff} \approx 10^5$ yr).

mean edge length ℓ_{MST} we use in particular the parameter $\mathcal{Q}$, introduced by Cartwright & Whitworth (2004) as the ratio of the normalized mean correlation length $\bar{s}$ and the normalized mean edge length $\bar{\ell}_{MST}$. The dimensionless measure $\mathcal{Q}$ allows us to quantify the cluster structure and to distinguish between a centrally condensed cluster ($\mathcal{Q} > 0.8$) and a hierarchical cluster showing possible fractal substructure ($\mathcal{Q} < 0.8$). The method has been applied successfully to observed embedded clusters as well as to the results of numerical simulations (Cartwright & Whitworth 2004; Schmeja & Klessen 2006).

3. Results and discussion

3.1. *Structures changing with time*

We performed numerical SPH simulations of the fragmentation and collapse of turbulent, self-gravitating gas clouds and the resulting formation and evolution of a star cluster as described in Schmeja & Klessen (2004). We investigated models with different turbulent velocities and driving scales and analysed the temporal evolution of the cluster structure. Although the absolute values of $\mathcal{Q}$ differ in the different models, in all models $\mathcal{Q}$ increases with time (Fig. 1), indicating that the cluster builds up from a hierarchical structure consisting of several subclusters and quickly evolves towards a more centrally concentrated state, although in no model does $\mathcal{Q}$ rise significantly above the "dividing value" of 0.8 (Schmeja & Klessen 2006).

The analysis of near and mid-infrared data of embedded clusters in the Perseus, Ophiuchus and Serpens star-forming regions shows that the youngest Class 0 and 1 sources are usually distributed over a smaller area and show a lower $\mathcal{Q}$ value indicating a hierarchical structure possibly reflecting the fractal structure of the molecular cloud. The more evolved Class 2 and 3 objects are spread out over a larger area but show a more centrally condensed configuration in agreement with the simulations.

We also investigated a large sample of open clusters selected from the catalogue of Dias *et al.* (2006). Although open and embedded clusters seem to cover roughly the

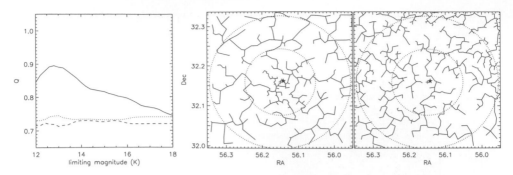

Figure 2. Embedded cluster IC 348: Changing of Q with the limiting magnitude (K band) for the cluster stars and two off-cluster control fields (left panel); MST of IC 348 for stars ($m_K < 15$; middle panel) and brown dwarfs ($m_K > 15$; right panel).

same range in Q, the mean Q value of the sample of embedded clusters is larger than the mean Q value of the open clusters, and a Kolmogorov-Smirnov test suggests (with a false alarm probability $< 1\%$) that the Q values of embedded and open clusters are drawn from different populations. So we may suspect that a cluster evolving from the embedded phase to an open cluster regresses to a more hierarchical configuration due to dynamical interactions. The central condensation may be most significant in the late embedded phase, when gravity is the dominant force.

3.2. *Structures changing with stellar mass*

Almost all open and embedded clusters in our analysis show signs of mass segregation, except the youngest ones like NGC 1333 or Serpens. Considering only the brightest sources usually results in a significantly higher Q value. The left panel of Fig. 2 shows the dependence of Q on the limiting magnitude for the cluster IC 348. While the brightest (i.e. most massive) stars show a high degree of central concentration, including the faintest (substellar) objects yields a value of $Q \approx 0.72$, the value expected for a random distribution.

In particular we investigate this behaviour in view of the formation of brown dwarfs. One of several theories, the so-called "ejected embryos" scenario (Reipurth & Clarke 2001; Kroupa & Bouvier 2003), suggests the formation of brown dwarfs as a result of premature ejection of protostellar embryos from multiple systems. Such a mechanism may be expected to produce different spatial distributions of stars and substellar objects. We analysed deep near-infrared K-band data obtained from the literature of the two nearby and well-sampled star-forming regions IC 348 and the Orion Trapzium Cluster. The parameters Q and ℓ_{MST} are evaluated for stars and substellar objects as a function of cluster core radius. The ℓ_{MST} values for stars within and outside the cluster core radius vary by a factor of ~ 2, whereas they show roughly similar values for brown dwarfs in both IC 348 and Orion/Trapezium. So the stellar population displays a centrally concentrated distribution whereas the substellar population is distributed more homogeneously in space within twice the cluster core radius. Although the substellar objects seem to experience less influence from the cluster potential well, they are still within the limits of the cluster and not significantly displaced from their birth sites. The spatially homogeneous distribution of substellar objects with respect to stars is best explained by assuming higher initial velocities, distributed in a random manner, and going through multiple interactions in the early phase. The overall spatial coincidence of the objects with the cluster locations can be understood if these objects are nevertheless travelling slowly enough to

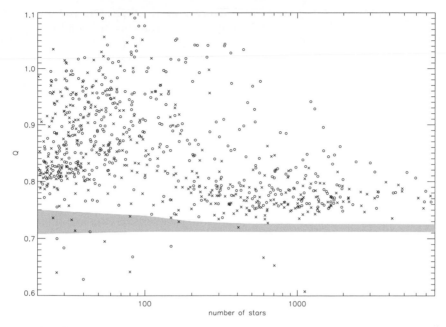

Figure 3. The Q parameter of known and suspected clusters plotted versus the number of objects from 2MASS data (circles: open clusters, crosses: cluster candidates). The grey-shaded area shows the parameter range expected for a random distribution of objects. Since observed globular clusters usually suffer from non-resolved central regions, thereby shifting Q towards lower values, they are not shown here.

feel the gravitational influence of the cluster. The observed effect is more pronounced in the older IC 348 cluster, which is in agreement with the N-body simulations of Goodwin *et al.* (2005) showing that a significant spatial spread of brown dwarfs with respect to stars may only occur after 5 Myr. The observational data therefore support the scenario of formation of brown dwarfs as "ejected stellar embryos". Higher ejection velocities are necessary but spatial displacements are not needed to explain the observational data. (See Kumar & Schmeja 2007 for details.)

3.3. *Structures changing with cluster type*

Quite obviously, different types of star clusters show different structures. While globular clusters are expected to be highly centrally condensed ($Q \gtrsim 1$), open clusters are much less so. (The Q value of globular clusters can be very high, but in observational data it is usually limited by the membership sampling in the central region.) A random distribution, on the other hand, always produces a value around $Q \approx 0.72$. We use this fact to identify and classify clusters and cluster candidates selected from 2MASS data (Froebrich *et al.* 2007). Fig. 3 shows the Q parameters of the clusters from the sample of Froebrich *et al.* (2007) plotted versus the number of objects: While there are many overlaps, the different cluster types clearly occupy different areas in the parameter space, which allows us to determine a probability whether a cluster candidate is a globular cluster, an open cluster or no cluster at all.

4. Summary

We show that the apparent structure of a star cluster strongly depends on its evolutionary stage, the limiting magnitude and the cluster type. Nascent star clusters evolve from a hierarchical, fractal structure with usually several subclusters to a more centrally condensed stage, which may be most pronounced while the cluster is still very young and gravity the dominant force. More evolved open clusters, on the other hand, show, in a statistical sense, less central concentration, which indicates that due to the dynamical evolution the cluster may again develop a more hierarchical structure (and eventually disperse). In almost all but the youngest clusters only the more massive stars show a significant central concentration, whereas low-mass stars and substellar objects do not follow the cluster potential and are distributed in a more homogeneous way, supporting the scenario of brown dwarfs formed as "ejected stellar embryos".

Acknowledgements

S.S. wishes to thank Annabel Cartwright and Ant Whitworth for valuable discussions. S.S. and R.S.K. acknowledge financial support from the Emmy Noether Programme of the German Science Foundation (grant KL 1358/1) for parts of this work. S.S. and M.S.N.K. are supported by a research grant POCTI/CFE-AST/55691/2004 approved by FCT and POCTI, with funds from the European community programme FEDER.

References

Cartwright, A. & Whitworth, A. P. 2004, *MNRAS* 348, 589
Dias, W. S., Assafin, M., Flório, V., Alessi, B. S., & Líbero, V. 2006, *A&A* 446, 949
Froebrich, D., Scholz, A. & Raftery, C. L. 2007, *MNRAS* 374, 399
Goodwin, S. P., Hubber, D. A., Moraux, E. & Whitworth, A.P. 2005, *Astron. Nachr.* 326, 1040
Gower, J. C. & Ross, G. J. S. 1969, *Applied Statistics* 18, 54
Kroupa, P. & Bouvier, J. 2003, *MNRAS* 346, 369
Kumar, M. S. N. & Schmeja, S. 2007, *A&A* 471, L33
Lada, C. J. & Lada, E. A. 2003, *ARAA* 41, 57
Reipurth, B. & Clarke, C. J. 2001, *AJ* 122, 432
Schmeja, S. & Klessen, R. S. 2004, *A&A* 419, 405
Schmeja, S. & Klessen, R. S. 2006, *A&A* 449, 151

Dynamical Evolution of Dense Stellar Systems
Proceedings IAU Symposium No. 246, 2007
E. Vesperini, M. Giersz & A. Sills, eds.

© 2008 International Astronomical Union
doi:10.1017/S1743921308015275

The Formation and Dynamics of the SMC Cluster NGC 346

Linda J. Smith[1,2]

[1] Space Telescope Science Institute and European Space Agency, 3700 San Martin Drive,
Baltimore, MD 21218, USA

[2] Department of Physics and Astronomy, University College London, Gower St., London
WC1E 6BT, UK

Abstract. The young resolved cluster NGC 346 in the SMC provides us with the opportunity to study the details of cluster formation and the efficiency of feedback mechanisms at low metallicity. I describe the latest results from a large-scale study of this cluster and its H II region N66. HST/ACS images reveal that NGC 346 is composed of a number of sub-clusters which appear to be coeval with ages of 3 ± 1 Myr, strongly suggesting formation by the hierarchical fragmentation of a giant molecular cloud (Nota *et al.* 2006; Sabbi *et al.* 2007a). HST Hα images show that the central cluster and the sub-clusters still contain some of their residual gas. We present high resolution spectroscopy of the ionized gas, and find that it shows little evidence for gas motions. This suggests that, at the low SMC metallicity, the cluster O star winds are not powerful enough to sweep away the residual gas. Instead, we find that stellar radiation is the dominant process shaping the interstellar environment of NGC 346.

Keywords. Galaxies: individual (SMC), galaxies: star clusters, ISM: individual (N66), ISM: H II regions, stars: formation, stars: winds, outflows

1. Introduction

The Small Magellanic Cloud (SMC) is an ideal laboratory to study the formation and evolution of stellar clusters at low metallicity. Its close proximity (60 kpc; Hilditch, Howarth & Harries 2005) means that the clusters are spatially resolved and photometry of the individual stars is well determined. During the first ~ 1 Myr in the evolution of a cluster, stellar winds remove the gas left over from star formation, and depending on the efficiency of star formation, this can unbind the cluster (e.g. Bastian & Goodwin 2006). The early evolution of clusters may, however, be different at low metallicity because of the reduced stellar wind power. The investigation of this effect in recently formed clusters in the SMC is important because of the insight it provides on the formation and duration of clusters in young, low metallicity galaxies.

2. NGC 346

The young star cluster NGC 346 represents the most active star-forming region in the SMC. It is located towards the northern end of the SMC bar and contains at least 30 O stars (Massey, Parker & Garmany 1989) which ionize N66, the largest H II region in the SMC. In Fig. 1, an image of NGC 346 in the light of Hα is shown. This image is taken from Nota *et al.* (2006) and was obtained with the Advanced Camera for Surveys (ACS) on board the *Hubble Space Telescope* (HST). Fig. 1 shows that even though NGC 346 is ~ 3 Myr old (Bouret *et al.* 2003), it still contains some of its residual gas. This suggests that a supernova explosion has yet to occur in the central region, and this is in agreement with the gas motions (see below) and the low diffuse X-ray flux (Nazé *et al.*

Figure 1. HST/ACS Hα image of the SMC cluster NGC 346, covering an area of 200″ by
200″ (or 60 ×60 pc) from Nota *et al.* (2006). North is up and east is to the left.

2002). There are, however, at least two supernova remnants close to NGC 346 (Ye, Turtle
& Kennicutt 1991; Reid *et al.* 2006).

Nota *et al.* (2006) and Sabbi *et al.* (2007a,b) present photometric analyses of NGC 346
based on HST F555W and F814W images. We find that the young stellar population of
NGC 346 has an age of 3 ± 1 Myr and that a rich population of low mass (0.6–3 M$_\odot$)
pre-main sequence stars also exists, that were probably formed at the same time as the
central cluster.

We find that the young stellar population is not uniformly distributed within the
ionized nebula. We identify at least 15 sub-clusters that differ in size and stellar content.
The locations of these sub-clusters are shown in Fig. 2 which is taken from Sabbi *et al.*
(2007a). The central cluster contains at least three sub-clusters (Sc1, 2 and 3), and has a
half-mass radius of 9 pc and a total stellar mass of $\sim$ few $\times 10^5$ M$_\odot$ (Sabbi *et al.* 2007b).

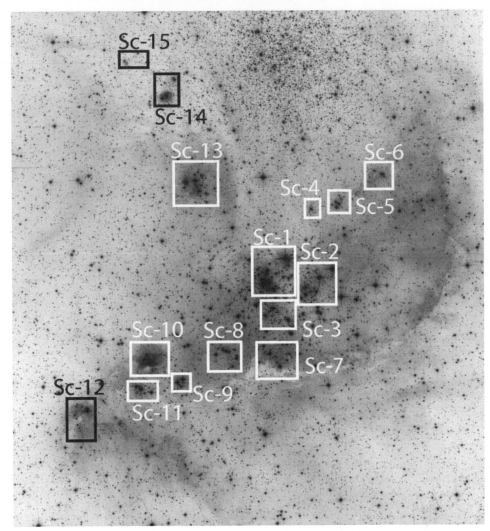

Figure 2. HST/ACS F555W image of the SMC cluster NGC 346 showing the 15 identified sub-clusters.

Within the uncertainties of the isochrone fitting process, we find that the sub-clusters are coeval with an age of 3 ± 1 Myr. They are still embedded in nebulosity and coincide with clumps of neutral gas and CO emission (Rubio *et al.* 2000; Contursi *et al.* 2000).

The pre-main sequence stars detected in the HST images are concentrated in the centre of NGC 346 and are also present, in smaller numbers, in most of the small compact sub-clusters (e.g. Sc-10, 12, 13 and 14). Simon *et al.* (2007) have recently presented *Spitzer* observations of NGC 346. They find 111 embedded young stellar objects (YSOs), showing that star-formation is still ongoing in this region today. The most massive YSOs in their sample (if they are single objects) are located in the central sub-clusters, presenting strong evidence for primordial mass segregation. They also find that all but one of the 15 sub-clusters identified by Sabbi *et al.* (2007a) contain YSOs.

Recently, Sabbi *et al.* (2007b) have determined the present day mass function (PDMF) for NGC 346 from their deep HST photometry. We find that the PDMF is Salpeter ($\Gamma = -1.43 \pm 0.18$) over the mass range of 0.8–60 $M_\odot$. The PDMF slope varies as

L. J. Smith

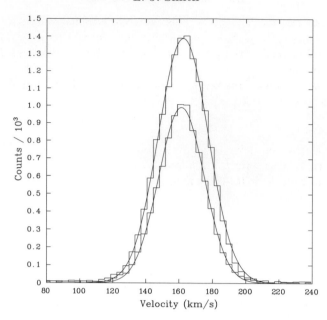

Figure 3. AAT+UCLES Hα profiles for the slit crossing the sub-cluster Sc-6 in NGC 346. The two profiles, with Gaussian fits superimposed, represent data from the extremes of the 1 arcmin long slit, and show that the nebular gas is quiescent.

a function of the radial distance from the centre of the cluster, indicating that mass segregation, probably of primordial origin, is present.

The crossing-time from the central sub-cluster to the outer sub-clusters (Sc 6, 12 and 15) is ~ 2 Myr (for a sound speed of 10 $\mathrm{km\,s}^{-1}$) and thus NGC 346 is about one crossing-time old. The observed sub-cluster structure and the coevality of the sub-clusters strongly suggest that NGC 346 was formed by the collapse, and subsequent hierarchical fragmentation of a giant molecular cloud (GMC) into multiple seeds of star formation (Elmegreen 2000; Klessen & Burkert 2000; Bonnell & Bate 2002; Bate, Bonnell & Bromm 2003; Bonnell, Bate & Vine 2003). In this model, the fragmentation of the GMC is due to supersonic turbulent motions in the gas. The turbulence induces the formation of shocks, and produces filamentary structures. The chaotic nature of the turbulence results in local density enhancements in the filamentary structures. High density regions that become self-gravitating then collapse to form stars, and this happens simultaneously at different locations within the cloud (Bonnell *et al.* 2003).

To test these theories and to study the effect of stellar winds at the low metallicity of the SMC, we have obtained high spectral resolution observations of the ionized gas in NGC 346/N66. These data were obtained with the University College London Echelle Spectrograph (UCLES) on the Anglo-Australian Telescope (AAT), and cover the Hα and [O III] λ5007 emission lines at a spectral resolution of 6 $\mathrm{km\,s}^{-1}$. We observed a total of six slit positions across N66 covering a number of sub-clusters. Surprisingly, we find that the ionized gas is quiescent with no evidence for large-scale gas motions. In general, we find that the Hα profiles are single with a velocity dispersion of ~ 14 $\mathrm{km\,s}^{-1}$ and a constant velocity along the length of each slit position (1 arc min). An example is shown in Fig. 3, where the Hα profiles for the two extremes of the slit crossing Sc-6 are plotted. Even at the centre of NGC 346 where the most massive O stars are located, we detect no significant ionized gas motions. These results strongly suggest that at the low metallicity of the SMC, stellar winds are much reduced, and the dominant form of interaction is via stellar radiation rather than by winds. This is in accord with Bouret *et al.* (2003) who measured the mass loss rates of six O stars within the central cluster and found that their winds are considerably weaker than their Galactic counterparts. We also find that there are very small differences ($\leqslant 3$ km s^{-1}) in the velocities of the sub-clusters, as measured by their ionized gas motions. This suggests that if the sub-clusters merge with the main cluster in the centre, it will take at least ~ 7 Myr.

Acknowledgements

I would like to thank and acknowledge my collaborators on the NGC 346 project: A. Nota, L. Angeretti, L. Carlson, J. Gallagher, M. Meixner, M. S. Oey, A. Pasquali, E. Sabbi, M. Sirianni, M. Tosi & R. Walterbos.

References

Bate, M. R., Bonnell, I. A., & Bromm, V. 2003, *MNRAS* 339, 577

Bastian, N. & Goodwin, S. P. 2006, *MNRAS* 369, L6

Bonnell, I. A. & Bate, M. R. 2002, *MNRAS* 336, 659

Bonnell, I. A., Bate, M. R., & Vine, S. 2003, *MNRAS* 343, 413

Bouret, J.-C., Lanz, T., Hillier, D. J., Heap, S. R., Hubeny, I., Lennon, D. J., Smith, L. J., & Evans, C. J. 2003, *ApJ* 595, 1182

Contursi, A. *et al.* 2000, *A&A*, 362, 310

Elmegreen, B. G. 2000, *AJ* 530, 227

Hilditch, R. W., Howarth, I. D., & Harries, T. J. 2005, *MNRAS* 357, 304

Klessen, R. S. & Burkert, A. 2000, *ApJS* 128, 287

Massey, P., Parker, J. W., & Garmany, C. D. 1989, *AJ* 98, 1305

Nazé, Y., Hartwell, J. M., Stevens, I. R., Corcoran, M. F., Chu, Y.-H., Koenigsberger, G., Moffat, A. F. J., & Niemela, V.S. 2002, *ApJ* 580, 225

Nota, A. *et al.* 2006, *ApJ* 640, L29

Reid, W. A., Payne, J. L., Filipović, M. D., Danforth, C. W., Jones, P. A., White, G. L., & Staveley-Smith, L. 2006, *MNRAS* 367, 1379

Rubio, M., Contursi, A., Lequeux, J., Probst, R., Barbá, R., Boulanger, F., Cesarsky, D., & Maoli, R. 2000, *A&A* 359, 1139

Sabbi, E. *et al.* 2007a, *AJ* 133, 44

Sabbi, E. *et al.* 2007b, *AJ in press*

Simon, J. D. *et al.* 2007, *ApJ* 669, 327

Ye, T., Turtle, A. J., & Kennicutt, R. C. 1991, *MNRAS* 249, 722

Dynamical Evolution of Dense Stellar Systems
Proceedings IAU Symposium No. 246, 2007
E. Vesperini, M. Giersz & A. Sills, eds.

© 2008 International Astronomical Union
doi:10.1017/S1743921308015287

Clustered Star Formation in the Magellanic Clouds

Dimitrios A. Gouliermis

Max-Planck-Institute for Astronomy, Königstuhl 17, 69117 Heidelberg, Germany

email: dgoulier@mpia.de

Abstract. The Large and Small Magellanic Cloud (LMC, SMC) offer an outstanding variety of young stellar associations, in which large samples of low-mass stars (with $M \leqslant 1 \, M_\odot$) currently in the act of formation can be resolved and explored sufficiently with the *Hubble Space Telescope*. Previous observations with the *Wide-Field Planetary Camera 2* (WFPC2) provided the first evidence of the existence of low-mass pre-main sequence (PMS) stars in the vicinity of star forming associations in the Magellanic Clouds (MCs) (Gouliermis *et al.* 2006a), and recent results from deeper observations with the *Advanced Camera for Surveys* (ACS) enhanced dramatically the picture of these systems with the discovery of large numbers of PMS stars. The associations LH 95 (Gouliermis *et al.* 2002, 2007a) in the LMC, and NGC 346 (Gouliermis *et al.* 2006b) and NGC 602 (Gouliermis *et al.* 2007b) in the SMC, are currently under investigation with the use of observations from both *Hubble* and *Spitzer Space Telescope*. I present the impact of our recent results in terms of the star formation history and Initial Mass Function (IMF) of these interesting systems, using as example the case of NGC 602.

Keywords. Magellanic Clouds — Hertzsprung-Russell diagram — stars: low-mass — stars: pre–main-sequence — stars: formation — stars: evolution — HII regions — galaxies: star clusters — open clusters and associations: individual (LH 95, NGC 346, NGC 602)

---

1. Pre–Main-Sequence Stars in Associations of the Magellanic Clouds

The loci of the low-mass PMS stars in the color-magnitude diagrams (CMD) of these associations show a broadening, which implies continuous star formation for up to ~ 10 Myr. However, there are important factors, such as differential reddening, binarity and variability, that affect significantly the positions of these stars in the CMD, giving false evidence of an age-spread (Hennekemper *et al.* 2007). Still, our simulations showed that a second or third generation of star formation cannot be excluded for these PMS stars (Hennekemper *et al.* 2007; Schmalzl *et al.* 2007). Specifically, in the case of NGC 602 (Fig. 1), if a mean extinction, as it is measured by the displacement of the upper MS stars in comparison to the ZAMS is assumed, then the observed PMS widening can be successfully reproduced in a synthetic CMD if at least two epochs of star formation are considered. On the other hand, if the extinction is higher (which is the case for highly embedded clusters), then a single star formation event would be sufficient to explain the star formation history of this system. However, in this case the system should have an age of the order of ~ 10 Myr, much older than what has been previously claimed (~ 4 Myr).

Naturally, the uncertainty in the age of the PMS stars has significant implications to the construction of the IMF of the low-mass stars in the system. Consequently, although for the massive part of the IMF a mass-luminosity relation was considered based on young MS isochrones, the low-mass part was constructed by counting the PMS stars between evolutionary tracks, since there is no specific indication about their age. The constructed IMF is shown in Fig. 1 (left). Further details are available from Schmalzl *et al.* (2007).

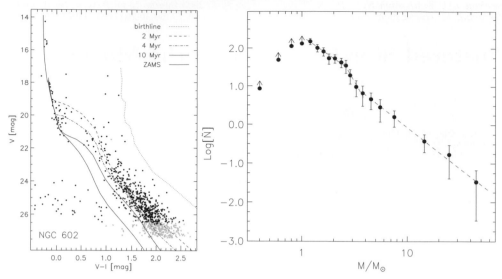

Figure 1. *Left*: The $V - I$, V CMD of NGC 602 constructed from photometry with ACS, after the contribution of the general SMC field has been statistically subtracted. It is shown that the stellar association consists of bright MS stars and a large number of low-mass PMS stars. The observed wide spread of the positions of the latter in the CMD, cannot be explained by differential reddening or binarity alone. An alternative explanation is that, assuming that the extinction is relatively low, multi-epoch star formation has taken place, so that the observed widening of the PMS stars is evidence of an age-spread in NGC 602. Three PMS isochrones are overplotted as the most representative of such sequential star formation. *Right*: The Initial Mass Spectrum of NGC 602 constructed with the use of different methods for the massive MS and the low-mass PMS stars. We find that a slope $\Gamma \simeq -1.4 \pm 0.2$ represents well the IMF of NGC 602. Both plots from Schmalzl *et al.* (2007).

Acknowledgements

The author kindly acknowledges the support of the German Research Foundation (DFG) through the individual grant GO 1659/1-1. Based on observations made with the NASA/ESA Hubble Space Telescope, obtained from the data archive at the Space Telescope Science Institute, which is operated by the Association of Universities for Research in Astronomy, Inc. under NASA contract NAS 5-26555. Research supported by the German Research Foundation (DFG), and the German Aerospace Center (DLR).

References

Gouliermis, D., Keller, S. C., de Boer, K. S., Kontizas, M., & Kontizas, E. 2002, *A&A* 381, 862
Gouliermis, D., Brandner, W., & Henning, T. 2006a, *ApJ* (Letters) 636, L133
Gouliermis, D. A., Dolphin, A. E., Brandner, W., & Henning, T. 2006b, *ApJS* 166, 549
Gouliermis, D. A., *et al.* 2007a, *ApJ* (Letters) 665, L27
Gouliermis, D. A., Quanz, S. P., & Henning, T. 2007b, *ApJ* 665, 306
Hennekemper, E., *et al.* 2007, Accepted for publication in *ApJ* (arXiv:0710.0774)
Schmalzl, M., *et al.* 2007, Submitted to *ApJ*

Dynamical Evolution of Dense Stellar Systems
Proceedings IAU Symposium No. 246, 2007
E. Vesperini, M. Giersz & A. Sills, eds.

© 2008 International Astronomical Union
doi:10.1017/S1743921308015299

The Fraction of Runaway OB Stars in the SMC Field

J. B. Lamb[1] and M. S. Oey[1]

[1]Department of Astronomy, University of Michigan, Ann Arbor, MI 48109-1042, USA
email: joellamb@umich.edu

Abstract. The fraction of field OB stars that originate from clusters can help probe the dynamical evolution of clusters. Field stars represent a significant fraction (20-30%) of the OB population in galaxies, and estimates for the fraction of field OB stars that are runaways range from the classical value of <10% (Blaauw 1961) to contemporary results suggesting >90% (de Wit *et al.* 2005). We obtained Magellan IMACS observations on the kinematics of field OB stars in the SMC to examine the line-of-sight velocities of this population. Using these observations, we will estimate the fraction of runaways to serve as a probe of cluster evolution.

Keywords. binaries: general, stars: early-type, stars: kinematics, stars: statistics, open clusters and associations: general, galaxies individual (Magellanic Clouds), galaxies: stellar content

Cluster evolution is heavily influenced by the massive stars formed in the cluster. Ionizing radiation and hot star winds can unbind clusters on the order of dynamical timescales ($\leqslant 1$ Myr) by dispersing gas from the cluster. In some cases, a mass loss as little as 10% can dissociate a cluster (Hills 1980). These feedback effects also affect the rate and efficiency of star formation on timescales of ~ 1 Myr (Dale *et al.* 2005). In addition, on time scales > 3 Myr, supernova feedback can further affect the binding and star formation of the cluster (Dray *et al.* 2005). The fraction of OB stars that leave clusters as runaways is important, not only because of the stellar dynamical effects, but also because these runaways deprive clusters of their feedback effects.

Runaways are created primarily through two possible mechanisms. The Dynamical Ejection Scenario (DES; Poveda, Ruiz & Allen 1967) involves the interactions of primarily binary-binary pairs in the cores of star clusters (Leonard & Duncan 1990). These interactions typically result in the less massive star of one of the binaries being ejected from the system. The other runaway-generating mechanism is the Binary Supernova Scenario (BSS; Blaauw 1961), in which the supernova of one of the stars in an early-type binary leaves the other star as a runaway, sometimes with the compact remnant still bound. As in DES, the less massive star of the binary pair ends up as the dominant runaway. Due to binary evolution, BSS-generated runaways should have large rotational velocities (Hoogerwerf, de Bruijne & de Zeeuw 2001). Portegies Zwart (2000) finds that the rotational velocity is expected to be higher with increasing mass.

For DES, the OB stars are ejected at a very young age, < 1 Myr. This deprives the cluster of nearly all massive star feedback from the runaway star. However, for BSS, the supernova requirement necessitates that the more massive star in the binary pair must complete its life cycle for ejection to occur. This requirement leads to "older" stars ($3 - 40$ Myr) being ejected for BSS. Thus, for BSS, much of the massive star feedback from the runaways is still imparted to the cluster. Therefore, not only the fraction of runaways, but also the method of runaway ejection, affects the evolution of clusters. However, neither of these mechanisms is well understood. Few papers provide comparisons of the likelihood of one ejection scenario versus the other, but one estimate is that BSS dominates by a 2:1 ratio (Hoogerwerf, de Bruijne & de Zeeuw 2001).

Field stars represent 20-30% of the OB population in galaxies (Oey, King, & Parker 2004). The percentage of these field stars that are thought to be runaways varies considerably, with estimates from 10% (Blaauw 1961) to 90% (de Wit *et al.* 2005). Combining these estimates yields the result that anywhere from 2% to 27% of OB stars created in clusters become runaways. However, studies of specific clusters show evidence of ejection rates as high as 75% (Pflamm-Altenburg & Kroupa 2006).

All previous empirical studies of runaway stars focus on samples in the local Galactic neighborhood where gas and dust cause line-of-sight confusion and uncertainties in distances (e.g., Gies 1987). Observing the SMC circumvents these issues since it lies outside the Galactic plane, allowing us to obtain a virtually spatially complete sample of the SMC bar. We choose target stars from a sample of candidate field OB stars with estimated spectral types B0 V, B0.5 I and earlier (Oey, King, & Parker 2004; and in these Proceedings). Multi-slit spectra of ~ 150 stars within the SMC bar were taken with the Magellan IMACS multi-object spectrograph.

Preliminary data for two fields ($15' \times 15'$ each) containing a total of sixteen targets were found to have a radial velocity range of 102 km/s with a velocity resolution in the 15–20 km/s range. The SMC systemic velocity of 158 km/s gives the sample a velocity range of -49 km/s to $+54$ km/s ± 20 km/s. Comparing with known velocity dispersions such as typical OB associations (10 km/s), SMC HI (10-40 km/s; Stanimirovic, Staveley-Smith & Jones 2004), and runaway populations (30 km/s; Stone 1991) gives an initial impression that runaway stars may be present in this sample. Five of the sixteen stars have velocities exceeding the 30 km/s threshold velocity for runaway status.

With our complete dataset, we will estimate the fraction of runaways among field OB stars. We will test the applicability of the two ejection methods: DES ejection predicts the lowest mass stars to have the highest runaway velocities (Leonard & Duncan 1990),while BSS predicts the opposite (Stone 1991). Finally, we will measure the rotational velocities to test the predicted correlation of rotational speed with mass, as expected for BSS (Portegies Zwart 2000). We will obtain an estimate for the fraction of stars produced by each ejection method, and examine the consequences for cluster evolution.

References

Blaauw, A. 1961, *Bull. Astron. Inst. Netherlands* 15, 265
Dale, J. E., Bonnell, I. A., Clarke, C. J., & Bate, M. R. 2005, *MNRAS* 358, 291
de Wit, W. J., Testi, L., Palla, F., & Zinnecker, H. 2005, *A&A* 437, 247
Dray, L. M., Dale, J. E., Beer, M. E., Napiwotzki, R., & King, A. R. 2005, *MNRAS* 364, 59
Gies, D. R. 1987, *ApJS* 64, 545
Hills, J. G. 1980, *ApJ* 225, 986
Hoogerwerf, R., de Bruijne, J. H. J., & de Zeeuw, P. T. 2001, *A&A* 365, 49
Leonard, P. J. T. & Duncan, M. J. 1990, *ApJ* 99, 608
Oey, M. S., King, N. L., & Parker, J. 2004, *AJ* 127, 1632
Pflamm-Altenburg, J. & Kroupa, P. 2006, *MNRAS* 373, 296
Portegies Zwart, S. F. 2000, *ApJ* 544, 437
Poveda, A., Ruiz, J., & Allen, C. 1967 *Bol. Obs. Tonantzinla Tacubaya* 28, 68
Stanimirovic, S., Staveley-Smith, L., & Jones, P. A. 2004, *ApJ* 604, 176
Stone, R. C. 1991, *AJ* 102, 333

Dynamical Evolution of Dense Stellar Systems
Proceedings IAU Symposium No. 246, 2007
E. Vesperini, M. Giersz, & A. Sills, eds.

The Relation Between Field Massive Stars and Clusters

M. S. Oey[1], N. L. King[2], J. Wm. Parker[3] and J. B. Lamb[1]

[1] University of Michigan, Department of Astronomy, Ann Arbor, MI   48109-1042, USA

[2] Lowell Observatory, 1400 W. Mars Hill Rd., Flagstaff, AZ 86001, USA

[1] Southwest Research Institute, Suite 426, 1050 Walnut St., Boulder, CO 80302, USA

Abstract. Massive "field" stars are those that appear in apparent isolation, in contrast to those in clusters. Whereas cluster stars are formed together in large aggregates, simultaneously, field stars have multiple origins. Some massive field stars may be the "tip of the iceberg" on small groups of physically associated stars, while others appear to be "runaway" stars that are dynamically ejected from clusters. What is the intrinsic relation between clusters and field stars, and what is the faction of runaway stars? Since massive stars are the most luminous stellar population, their demographics are accessible in the nearest external galaxies. We present our current efforts to understand these issues for the Small Magellanic Cloud.

Keywords. open clusters and associations: general, stars: early-type, stars: kinematics

1. Introduction

Runaway OB stars are an important probe of the binary fractions and dynamical properties of their parent clusters. Dynamical ejection of massive stars, dominated by binary-binary interactions (e.g., Mikkola 1983; Leonard & Duncan 1988), are diagnostic of cluster core densities and binary parameters like binary fraction, hardness, and mass ratios. Supernova "slingshot" ejection also contributes to the runaway population, and is also strongly dependent on properties of the parent binary population (e.g., Stone 1982; Portegies Zwart 2000). A number of studies have compared predicted massive star ejection rates and properties with those of observed runaways (e.g., Gies & Bolton 1986; Hoogerwerf *et al.* 2000). However, there are relatively few observations of runaway OB stars as populations, and therefore their statistical properties are uncertain.

To study the properties of any runaway population, it is essential to understand the entire OB field and to accurately distinguish contributions from different origins. In addition to runaways, an empirically-defined sample of field OB stars also includes stars in small groups that have no other observed massive star. There may also be high-mass stars that formed in true isolation, although if they exist, they may be extremely rare.

To date, all studies of runaway OB stars consider samples in the broader solar neighborhood (e.g., Gies & Bolton 1986; Hoogerwerf *et al.* 2000; de Wit *et al.* 2005). As a complementary effort, we are carrying out a comprehensive study of the field massive stars in the nearby Small Magellanic Cloud (SMC). Our sample of photometrically identified field OB candidates is defined by applying a friends-of-friends algorithm (Battinelli 1991) to the *UBVR* survey of the SMC by Massey (2002). A second sample of O-star candidates is identified by combining these data with the 1625 Å imaging of the SMC Bar by the *Ultraviolet Imaging Telescope* (Parker *et al.* 1998). Candidate stars having no other candidate within clustering lengths of 28 pc and 34 pc for the OB and O-star samples, respectively, are defined to be field massive stars. *These represent essentially complete samples of field OB star candidates across the spatial extent of these surveys.*

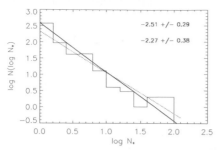 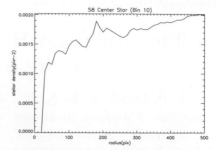

Figure 1. Left panel: Clustering law for OB sample (from Oey *et al.*2004). Right panel: Stellar density vs radius near the O star AZV 58 from *HST*/ACS F814W imaging.

This analysis also yields the SMC clusters and the distribution in the number of candidates N_* per cluster, for both samples. Both are broadly consistent with the universal $N(N_*) \propto N_*^{-2}$ clustering law (e.g., Hunter *et al.* 2003). The individual field stars, corresponding to $N_* = 1$, fall smoothly onto this distribution, strongly suggesting that the majority of these stars are the "tip of the iceberg" on small stellar groups of lower-mass stars, having no other candidate massive stars (Fig. 1 left panel; see Oey, King & Parker 2004 for details).

Nevertheless, a significant fraction ($\lesssim 50\%$) of our field massive stars could be runaways and still be statistically consistent with the smooth appearance of the clustering law. Our *HST*/ACS SNAP program yielded imaging of 7 field O stars in F555W and F814W, to confirm whether the majority indeed have lower-mass companions. Our preliminary results of the stellar density profiles show that about half of the fields show the existence of small groups, while half appear to be truly isolated stars (Fig. 1 right panel).

We are also carrying out a complete spectroscopic survey of the SMC Bar with the Magellan IMACS multi-object spectrograph. These will yield radial velocity measurements to constrain the fraction of runaways (see Lamb & Oey, this volume), as well as confirm exact spectroscopic types. Although full space velocities are not available, the statistics of the radial velocities will yield strong constraints on the properties of both the runaway and *in situ* field stars.

Acknowledgements

We acknowledge NASA ADP grant NAG5-9248 and NASA HST-GO-10629.01.

References

Battinelli, P. 1991, *A&A* 244, 69
de Wit, W. J., Testi, L., Palla, F., & Zinnecker, H. 2005, *A&A* 437, 247
Gies, D. R. & Bolton, C. T. 1986, *ApJS* 61, 419
Hoogerwerf, R., de Bruijne, J. H. J., & de Zeeuw, P. T. 2001, *A&A* 365, 49
Hunter, D. A., Elmegreen, B. G., Dupuy, T. J., & Mortonson, M. 2003, *AJ* 126, 1836
Leonard, P. J. T. & Duncan, M. J. 1988, *AJ* 96, 222
Massey, P. 2002, *ApJS*, 141, 81
Mikkola, S. 1983, *MNRAS* 203, 1107
Oey, M. S., King, N. L., & Parker, J. W. 2004, *AJ* 127, 1632
Parker, J. W., *et al.* 1998, *AJ* 116, 180
Portegies Zwart, S. F. 2000, *ApJ* 544, 437
Stone, R. C. 1982, *AJ* 87, 90

Dynamical Evolution of Dense Stellar Systems
Proceedings IAU Symposium No. 246, 2007
E. Vesperini, M. Giersz & A. Sills, eds.

Imprints of Stellar Encounters in the ONC

C. Olczak and S. Pfalzner

I. Physikalisches Institut, University of Cologne, 50937 Cologne, Germany
email: olczak@ph1.uni-koeln.de

Abstract. External destruction of protoplanetary discs acts mainly due to two mechanisms: gravitational drag by stellar encounters and evaporation by stellar winds and radiation. It is an important question whether any of these mechanisms is important in the stellar evolution process. We focus on the effect of stellar encounters and investigate if there are any observables that could trace this mechanism in young stellar clusters. An analysis of observational data of the Orion Nebula Cluster (ONC) and accompanying n-body simulations both provide evidence for encounters of star-disc systems in the ONC, eventually leading to substantial disk disruption.

Keywords. stellar dynamics, methods: data analysis, n-body simulations, stars: kinematics, circumstellar matter

1. Introduction

In this investigation we are dealing with the question if there might be *observational* evidence for star-disc systems which have been subject to an encounter. We set up numerical simulations with an dynamical model of the Orion Nebula Cluster (ONC) to validate observational data.

2. Observational Data

We have compiled data from Jones & Walker (1988), Hillenbrand (1997), and Hillenbrand *et al.* (1998) and excluded stars lacking proper motion or infrared excess data. The resulting database contains ~ 450 stars, about two thirds of which show disc signatures.

Fig. 1 shows the observed velocity distribution for stars with ages $\leqslant 1\,$Myr. The age boundary excludes possible non-members of the ONC and is roughly equal to the mean age of the cluster. The 3D velocity dispersion of the ONC is $\sigma \approx 4.3\,$km/s. Potential candidates of recent encounters are stars with velocities $v \geqslant 3\sigma$. Several such stars (most of low mass, $\sim 0.2\,$M$_\odot$) are present in the compiled data ($\sim 5\,\%$), among also apparent disc-less objects. The adopted age boundary is clarified by plotting stellar ages against velocities for the high-velocity stars only (Fig. 2, left). The diagram shows a bimodal

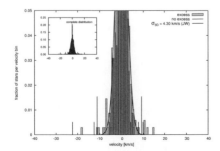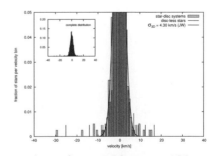

Figure 1. *Left:* Distribution of observed proper motions of stars with ages $\leqslant 1\,$Myr. *Right:* Velocity distribution from simulations of the ONC at $1\,$Myr.

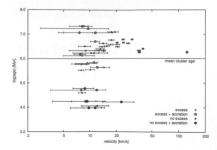 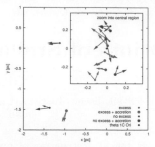

Figure 2. *Left:* Age-velocity distribution of high-velocity $(v \geqslant 3\sigma)$ stars in the ONC. *Right:* Positions and velocity vectors of young $(\leqslant 1\,\mathrm{Myr})$ high-velocity stars.

distribution: older stars have much larger proper motions and mostly lack signs of disc emission. These objects are considered to be highly probable foreground stars.

For the younger probable high-velocity members of the ONC the projected positions are plotted on the right-hand side of Fig. 2. It is apparent that most stars are located close to the cluster center $(r \lesssim 0.4\,\mathrm{pc})$ where densities are highest and encounters most probable. The massive stars in the center act as additional gravitational foci (Pfalzner, Olczak & Eckart 2006). The high velocities of the selected stars strongly favor very recent interactions since after $\tau_{esc} \leqslant 0.4\,\mathrm{pc}/3\sigma \approx 0.05\,\mathrm{Myr}$ they should have left the inner cluster region.

3. Cluster Simulations

The basic dynamical model of the ONC is described in Olczak, Pfalzner & Spurzem (2006). Here we use the same concept but have included a background potential and primordial binaries in particular for a more realistic model of the stellar dynamics. The code used is NBODY6++.

The resulting velocity distribution is shown on the right-hand side of Fig. 1. It reproduces well the features seen in the observational data: the bulk of the stars forms an approximate Gaussian velocity distribution. Moreover, about the same fraction of stars $(\sim 4\,\%)$ shows much higher velocities $v \geqslant 3\sigma$. In the simulations stellar discs are affected solely by encounters. Thus the resemblance of the two velocity distributions probably points to a dynamical origin of the observed high-velocity non-excess ONC stars.

4. Conclusions

Numerical simulations of the ONC support the scenario that the observationally established small sample of young, low-mass, high-velocity stars close to the cluster center consists of potential candidates for recent encounters. The lack of signatures of circumstellar discs in half of the stars eventually tracks destructive encounters, probably with high-mass stars.

References

Jones, B. F. & Walker, M. F. 1988, *AJ* 95, 1755.

Hillenbrand, L. A. 1997, *AJ*, 113, 1733.

Hillenbrand, L. A., Strom, S. E., Calvet, N., Merrill, K. M., Gatley, I., Makidon, R. B., Meyer, M. R., & Skrutskie, M. F. 1998, *AJ* 116, 1816.

Olczak, C., Pfalzner, S., & Spurzem, R. 2006, *ApJ* 642, 1140.

Pfalzner, S., Olczak, C., & Eckart, A. 2006, *A&A* 454, 811.

Dynamical Evolution of Dense Stellar Systems
Proceedings IAU Symposium No. 246, 2007
E. Vesperini, M. Giersz & A. Sills, eds.

Capture-Induced Binarity of Massive Stars in Young Dense Clusters

S. Pfalzner and Ch. Olczak

I. Physikalisches Institut, University of Cologne, 50937 Cologne, Germany
email: pfalzner@ph1.uni-koeln.de

Abstract. Observations show that for massive stars the binary frequency seems to be higher than for lower mass stars in young dense clusters. This suggests that in clusters like the ONC different mechanisms are at work in the formation of high-mass binary or multiple systems than for low-mass stars. We investigate the stellar dynamics in young dense clusters to determine the role of capture in binary formation in high-mass stars. It turns out that in contrast to lower mass stars capture is a frequent process for massive stars. However, this does not necessarily lead to long lasting binary systems but is often of transient nature. Nevertheless, capture processes could account for 15-25% of the observed 'binaries' of the OB-stars (75%) in Orion.

Keywords. open clusters and associations, binaries: general, stellar dynamics, methods: n-body simulations

1. Introduction

Observations indicate that in young stellar clusters the binary fraction for massive stars is higher than for solar mass stars. For the Orion Nebula Cluster (ONC) there is a binary frequency of $\sim 50\%$ for solar-mass stars compared to 70-100% for the massive O- and B-stars. In principle there are only two explanations for the higher binary frequency of massive stars in comparison to intermediate mass stars: either massive stars are more likely to be binaries primordially, or their binarity increases within the first Myr of their existence in a significant way. Here we restrict ourselves to the second possibility, i.e. we start with the assumption that the primordial number of binaries is the same for solar-mass and massive stars and ask whether dynamical processes can lead to a sufficient amount of additional binaries to explain the difference in observed binary frequency between solar-mass and massive stars. As a model cluster we chose the Orion Nebula Cluster, because it is one of the densest clusters in the Galaxy, so if capture processes play any role one should find indications for it here.

2. Method

For simplicity we start with a system initially consisting only of single stars - the influence of both primordial binaries and discs around the stars are excluded from this first study. In the cluster simulations we followed the dynamical development of ~ 4000 stars in an virial equilibrium situation, i.e. $Q_{vir} = 0.5$, with a spherical density distribution $\rho(r) \sim r^{-2}$ using NBODY6++. Gas components and the potential of the background molecular cloud OMC 1 were neglected in these simulations. The most massive star was assigned a mass $M^* = 50$ M$_\odot$ and all other stars' masses according to the mass distribution given by Kroupa (2001). The quality of the dynamical models was judged by comparing them to observational data at 1–2 Myr, marking the range of the mean ONC age.

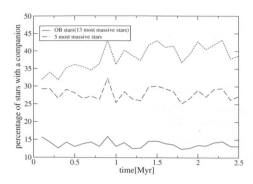

Figure 1. Average binary rate as a function of time for the 5 most massive stars (solid line), all 13 OB stars counting two massive stars in a binary only once (dashed line) and counting them both (dotted line).

3. Result

All capturing encounter events with an eccentricity $\epsilon < 1$ are recorded. These so formed bound systems are transient binary systems (TBS). As to be expected these capturing encounters mostly happen early on in the cluster development, close to the cluster centre and mainly involve one of the most massive stars. Fig. 1 shows the percentage of the 5 and 13 (number of OB stars in the ONC) most massive stars of the cluster to have at least one companion as a function of time. If a TBS is taken into account only once if both stars were massive, 10–15% of the OB stars and 25–30% of the five most massive stars would form at 2 Myrs and would appear as binaries. The thin dashed line shows the case where these massive TBSs are considered twice. This is equivalent to the likelihood of a specific star to be in a TBS and it is $\sim$ 40 % at 2 Myr. So for the most massive stars of the cluster there is a high likelihood of being a TBS and appear as a binary. It is actually sufficient to explain the difference in binary rates between massive and solar-mass stars.

But what are the properties of these capture-formed TBSs? At 1–2 Myrs, the most likely age of the ONC, the average periastron of the most massive TBS is between 50-200 AU. The mass ratio in these TBSs develops from an initial preference of low-mass companions to companions with high mass. At an cluster age of 2–5Myr the maximum of q lies in the range of 0.6-0.8. The most massive star usually captures just one of the 10 most massive stars as companion and not necessarily the second most massive star. The average eccentricity in these TBSs is $\epsilon \sim$ 0.5–0.6. The bound state lasts on average several times 10^6 yrs in contrast to bound state durations of on average $< 10^5$ yrs for lower mass stars. So in very young clusters these massive stars would *appear* as binaries but are actually just running through a succession of TBS.

4. Conclusion

In cluster environments similar to the ONC, massive stars have a much higher probability of involvement in a capturing encounter than solar-mass stars. Assuming a cluster age of 1–2 Myr, at least 10–15% of the OB stars in the ONC are in a bound state caused by capturing processes.

Reference

Kroupa, P. 2001, MNRAS 322, 231

Dynamical Evolution of Dense Stellar Systems
Proceedings IAU Symposium No. 246, 2007
E. Vesperini, M. Giersz & A. Sills, eds.

© 2008 International Astronomical Union
doi:10.1017/S1743921308015330

On the Origin of Complex Stellar Populations in Star Clusters

J. Pflamm-Altenburg[1,2] and P. Kroupa[1,2]

[1] Argelander Institut für Astronomie (AIfA), Auf dem Hügel 71, D-53121 Bonn, Germany
email: jpflamm@astro.uni-bonn.de, pavel@astro.uni-bonn.de
[2] Rhine Stellar Dynamics Network (RSDN)

Abstract. The existence of complex stellar populations in some star clusters challenges the understanding of star formation. E.g. the ONC or the sigma Orionis cluster host much older stars than the main bulk of the young stars. Massive star clusters (ω Cen, G1, M54) show metallicity spreads corresponding to different stellar populations with large age gaps. We show that (i) during star cluster formation field stars can be captured and (ii) very massive globular clusters can accrete gas from a long-term embedding inter stellar medium and restart star formation.

Keywords. stars: formation, globular clusters: general

1. Field star capture

The collapsing cloud is described by a Plummer potential with a fixed total gas mass and a time-dependent Plummer parameter, infinite at the beginning and decreasing within the collapse time-scale. The field stars are simulated by test particles which are initially uniformly distributed in a sphere and have a Gaussian velocity distribution.

1.1. *ONC*

Palla *et al.* (2005) found 4 stars out of 84 low mass stars in the ONC being 10 Myr older than the main bulk of stars and followed that star formation is prolonged. Extrapolating, ≈ 53 such older stars are expected in the whole ONC (Pflamm-Altenburg & Kroupa 2007b). The number of simulated captured stars within a radius of 2.5 pc of the centre of the ONC is plotted in Fig. 1 (left) for different collapse time-scales and initial background velocity dispersions. Age spreads in young star clusters may therefore not be due to prolonged star formation but can be explained by stellar capture.

1.2. *R136*

Brandl *et al.* (1996) found 110 faint red sources in a field of 3×3 pc^2 of the centre of R136 in the LMC of unknown physical nature and excluded that they can be red giants as their required age would be larger than 350 Myr. As membership probabilities can not be determined for stars in R136 all stars within the central sphere with a radius of 1.7 pc are calculated after the collapse with a time-scale of 10 Myr has stopped (Pflamm-Altenburg & Kroupa 2007a, *in preparation*). The faint red sources can indeed be captured old red giants.

2. Gas accretion

Globular clusters more massive than $\approx 10^6$ M$_\odot$ such as ω Cen, G1 and M54 show a spread in metallicity corresponding to different stellar population with age gaps of

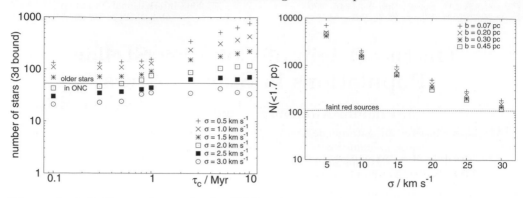

Figure 1. *Left*: Captured stars in the ONC within a radius of 2.5 pc for different velocity dispersions, σ, and collapse time-scales, τ_c, (Pflamm-Altenburg & Kroupa 2007b). *Right*: Field stars in the central sphere of 1.7 pc radius in R136 for different velocity dispersions and final Plummer parameters, b, (Pflamm-Altenburg & Kroupa 2007a, *in preparation*).

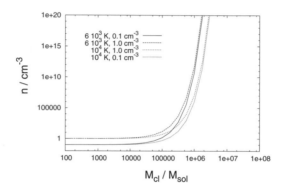

Figure 2. Particle density of the ISM at the origin of the cluster potential as a function of the total cluster mass (Pflamm-Altenburg & Kroupa 2007c, *in preparation*). Note the instability near 10^6 $M_\odot$.

several hundred Myr or a few Gyr. M54 and ω Cen are confirmed and supposed, respectively, to have been embedded in the ISM of a dwarf galaxy. To explore the effect of the cluster potential on a long-term embedding co-moving ISM we calculate the hydrostatic solution of an isothermal non-self-gravitating gas with an additional Plummer potential (Pflamm-Altenburg & Kroupa 2007c, *in preparation*). The central particle density in the cluster potential is plotted in Fig. 2 for different temperatures and densities of the warm component of the ambient ISM. Star clusters more massive than $\approx 10^6$ $M_\odot$ may cause an instability of the ISM, start gas accretion and form stars. Gas accretion can explain multiple stellar populations in massive star clusters and the change of the mass-radius relation at cluster masses of $\approx 10^6$ $M_\odot$.

References

Brandl B., Sams B. J., Bertoldi F., Eckart A., Genzel R., Drapatz S., Hofmann R., Loewe M., & Quirrenbach A. 1996, *ApJ* 466, 254

Palla F., Randich S., Flaccomio E., & Pallavicini R. 2005, *ApJL* 626, L49

Pflamm-Altenburg J., & Kroupa P. 2007b, *MNRAS* 375, 855

Dynamical Evolution of Dense Stellar Systems
Proceedings IAU Symposium No. 246, 2007
E. Vesperini, M. Giersz & A. Sills, eds.

Star Formation in Young Cluster NGC 1893

Saurabh Sharma[1], A. K. Pandey[2,1], D. K. Ojha[3], W. P. Chen[2], S. K. Ghosh[3], B. C. Bhatt[4], G. Maheswar[1] and Ram Sagar[1]

[1] Aryabhatta Research Institute of Observational Sciences (ARIES), Nainital, 263 129, India;
saurabh@aries.ernet.in

[2] Institute of Astronomy, National Central University, Chung-Li 32054, Taiwan

[3] Tata Institute of Fundamental Research, Mumbai (Bombay) – 400 005, India

[4] CREST, Indian Institute of Astrophysics, Hosakote 562 114, India

Abstract. We have carried out a multi-wavelength study of the star forming region NGC 1893 to make a comprehensive exploration of the effects of massive stars on low mass star formation. Using deep optical $UBVRI$ broad band, $H\alpha$ narrow band photometry and slit-less spectroscopy along with archival data from the surveys such as 2MASS, MSX, IRAS and NVSS, we have studied the region to understand the star formation scenario in the region.

Keywords. techniques: photometric, stars: formation, stars: luminosity function, mass function, stars: pre–main-sequence

1. Introduction

Young open clusters provide important information relating to star formation process and stellar evolution, because such clusters contain massive stars as well as low mass pre-main sequence (PMS) stars. Advancement in detectors along with various surveys such as the 2MASS, DENIS, ISO and Spitzer have permitted detailed studies of low-mass stellar population in regions of high mass star formation.

The very young open cluster NGC 1893 (Massey *et al.* 1995) is considered to be the center of the Aur OB2 association. NGC 1893 can be recognized as an extended region of loosely grouped early-type stars, associated with the H II region IC 410 with two pennant nebulae, Sim 129 and Sim 130 (Gaze & Shajn 1952) and obscured by several conspicuous dust clouds. NGC 1893 contains at least five O-type stars, two of which, HD 242908 and LS V +33°16 are main-sequence O5 stars (Marco & Negueruela 2002). The detection of PMS objects in NGC 1893 is interesting because it is one of the youngest known open cluster and has a moderately large population of O-type stars, representing thus a good laboratory for the study of massive star formation and the impact of massive stars on the formation of lower-mass stars (Marco & Negueruela 2002).

2. Results

Deep optical $UBVRI$ and narrow band $H\alpha$ photometric data were taken using 1 meter Sampurnanand Telescope of ARIES, Nainital, India. The cluster region was also observed in the slitless mode with a grism as the dispersing element using the Himalayan Faint Object Spectrograph Camera (HFOSC) instrument mounted on 2-meter Himalayan Chandra Telescope at Hanle. Archival data from the surveys such as 2MASS, MSX, IRAS and NVSS are also used to understand the star formation scenario in and around the cluster region.

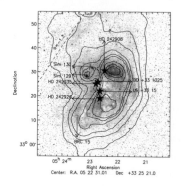

Figure 1. The $1° \times 1°$ R-band image of the field containing NGC 1893, Sim 129 & Sim 130 is reproduced from Digitized Sky Survey (DSS). North is up and east is to the left. Contours of 100-μm flux from IRAS 100-μm image is over-plotted. The IRAS 100-μm image is smoothened. The contours are drawn at 60 (outer most), 80, 100, 150, 180, 200, 220, 250, 280 and 300 $MJy\ sr^{-1}$. Locations of five O-type stars towards NGC 1893 are identified and shown using filled star symbols.

Reddening $(E(B-V))$ in the direction of cluster is found to be varying between 0.40 to 0.60 mag. The post-main-sequence age and distance of the cluster are found to be ∼4 Myr and 3.25 ±0.20 kpc respectively. Using the NIR two colour diagram and excess $H\alpha$ emission we identified candidate YSOs which are aligned from the cluster to the direction of nebula Sim 129 (see Fig. 1). The stars closer to Sim 129 and 130 show an average age of ∼1.5 Myr which is smaller than the estimated age of the cluster (∼4 Myr). The O-type stars at the center of the cluster may be responsible for the trigger of star formation in the region. The morphology of the cluster seems to be influenced by the star formation in the region. The PMS YSOs in the cluster region have masses ∼$1-3.5M_\odot$. The position of the YSOs on the CMDs indicates that the majority of these stars have ages between ∼1 Myr to 5 Myr indicating a possibility of non-coeval star formation in the cluster.

The power law slope of the K-band luminosity function for the cluster is found to be 0.34 ± 0.08 which is consistent with the average value (∼0.4) obtained for young star clusters (Lada *et al.* 1991; Lada & Lada 1995; Lada & Lada 2003). The slope of the initial mass function 'Γ' for PMS stars (mass range $0.6 < M/M_\odot \leqslant 2.0$) is found to be -0.88 ± 0.09, which is shallower than the value (-1.71 ± 0.20) obtained for MS stars having mass range $2.5 < M/M_\odot \leqslant 17.7$. However for the entire mass range $(0.6 < M/M_\odot \leqslant 17.7)$ the 'Γ' comes out to be -1.27 ± 0.08. The effect of mass segregation can be seen on the MS stars. The estimated dynamical evolution time is found to be greater than the age of the cluster, therefore the observed mass segregation in the cluster may be the imprint of the star formation process.

References

Gaze, V. F. & Shajn, G. A., 1952, *Izv. Krym. Astrofiz. Obs.*, 9, 52

Lada, E. A., Evans, N. J. II, Depoy, D. L., & Gatley, I., 1991, *ApJ*, 371, 171

Lada, E. A. & Lada, C. J., 1995, *AJ*, 109, 1682

Lada, C. J. & Lada, E. A., 2003, *ARA&A*, 41, 57

Marco, A. & Negueruela, I., 2002, *A&A*, 393, 195

Massey, P., Johnson, K. E., & DeGioia-Eastwood, K., 1995, *ApJ*, 454, 151

Dynamical Evolution of Dense Stellar Systems
Proceedings IAU Symposium No. 246, 2007
E. Vesperini, M. Giersz & A. Sills, eds.

© 2008 International Astronomical Union
doi:10.1017/S1743921308015354

On the Origin of the Orion Trapezium System

Hans Zinnecker

Astrophysikalisches Institut Potsdam
An der Sternwarte 16, D-14482 Potsdam, Germany
email: hzinnecker@aip.de

Abstract. Numerical SPH simulations of supersonic gravo-turbulent fragmentation of a proto-cluster cloud ($1000\,M_\odot$) suggest that the cloud develops a few subclusters (star+gas systems) which subsequently merge into a single cluster entity. Each subcluster carries one most massive star (likely multiple), thus the merging of subclusters results in a central Trapezium-type system, as observed in the core of the Orion Nebula cluster.

Keywords. star formation, young star cluster, multiple systems, Orion Nebula

1. Introduction

The origin of massive stars and their multiplicity has become one of the hottest topics in stellar astrophysics (see, for example, the recent review of Zinnecker & Yorke 2007). Many if not most of the high-mass stars appear to originate in dense star clusters (e.g. Zinnecker *et al.* 1993, Lada & Lada 2003), although isolated massive star formation may also occur (de Wit *et al.* 2005). Here, in this short contribution, we are concerned to explain the occasional occurrence of Trapezium-type systems in the centers of dense clusters (Abt 1986, Garcia & Mermilliod 2001). Such Trapezium type configurations of massive stars, unstable as they are, may be evidence for primordial mass segregation in young OB clusters (Bonnell & Davies 1998, McMillan *et al.* 2007).

2. SPH simulations of hierarchical young star cluster formation

The stellar cluster forms through the hierarchical fragmentation of a turbulent molecular cloud. The relevant numerical evolutionary simulations of a $1000\,M_\odot$ cloud by Bonnell *et al.* (2003) are shown in Fig. 1 (top 4 panels): each panel shows a region of 1 parsec per side. The logarithm of the gas column density is plotted. The stars are indicated by the (yellow) dots. The four panels capture the evolution of the $1000\,M_\odot$ system at times of 1.0, 1.4, 1.8 and 2.4 initial free-fall times ($t_{ff} = 2 \times 10^5$ yr). The turbulence causes shocks to form in the molecular cloud, dissipating kinetic energy and producing filamentary structure which fragment to form dense cores and individual stars (panel A). The stars fall towards local potential minima and hence form subclusters (panel B). These subclusters evolve by accreting more stars and gas, ejecting stars, and by merging with other subclusters (panel C). There is one massive star per subcluster. The final state of the simulation is a single, centrally condensed cluster with little substructure but with 4 massive stars, one from each subcluster (Trapezium-system + 400 stars) (panel D).

Acknowledgments

I would like to thank Douglas Heggie for first introducing me to the basics of N-body dynamics in the context of massive stars in young dense clusters (Zinnecker 1986). I also

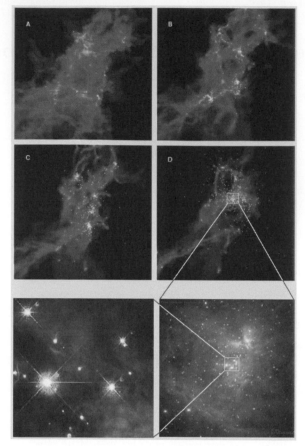

Figure 1. Collapse of a $1000 \, M_\odot$ molecular cloud and the hierarchical merging of subclusters (top 4 panels, see text). Comparison with the Orion Trapezium Cluster (bottom 2 panels; left: Orion Trapezium system HST/WFPC2 nb-optical image (Bally *et al.* 1998), right: Orion Trapezium cluster IRTF infrared JKL image (McCaughrean *et al.* 1994)).

thank Enrico Vesperini for inviting me to this symposium. It is a pleasure to acknowledge the collaboration with Ian Bonnell and Mathew Bate on star cluster formation. Their simulation provided the inspiration for the idea presented here.

References

Abt, H. A. 1986, *ApJ*, 304, 688
Bally, J. Sutherland, R. S., Devine, D., & Johnstone, D. 1998, *AJ*, 116, 293
Bonnell, I. A. & Davies, M. B. 1998, *MNRAS*, 295, 691
Bonnell, I. A., Bate, M. R., & Vine St., G. 2003, *MNRAS*, 343, 413
de Wit, W. J., Testi, L., Palla, F., & Zinnecker, H. 2005, *A&A*, 437, 247
García, B. & Mermilliod, J. C. 2001, *A&A*, 368, 122
Lada, C. J. & Lada, E. A. 2003, *Ann. Rev. A&A*, 41, 57
McCaughrean, M. J., Zinnecker, H., & Rayner, J. T. 1994, *ASSL*, 190, 13
McMillan, S. L. W., Vesperini, E., & Portegies Zwart, S. F. 2007, *ApJ*, 655, L45
Zinnecker, H. 1986, IAU Symp. 116, 271
Zinnecker, H., McCaughrean, M. J., & Wilking, B. A. 1993, In: *Protostars and Planets III*, pp. 429
Zinnecker, H. & Yorke, H. W. 2007, *Ann. Rev. A&A*, 45, 481, arXiv:0707.1279 (astro-ph)

Part 2

Open Clusters

Dynamical Evolution of Dense Stellar Systems
Proceedings IAU Symposium No. 246, 2007
E. Vesperini, M. Giersz & A. Sills, eds.

© 2008 International Astronomical Union
doi:10.1017/S1743921308015366

Open Clusters: Open Windows on Stellar Dynamics

Robert D. Mathieu[1]

[1]Department of Astronomy, University of Wisconsin – Madison, WI, 53706, USA
email: mathieu@astro.wisc.edu

Abstract. The extensive stellar radial-velocity surveys of the WIYN Open Cluster Study now allow comprehensive studies of the solar-type hard-binary populations in open clusters as a function of age. We first describe an *empirical* "initial" hard-binary population as derived from the young open cluster NGC 2168 (M35). Given the limited analyses so far, the cluster binary population is indistinguishable from that of the field. We then compare the hard-binary population in the old open cluster NGC 188 to the binary population in the sophisticated N-body simulations of the old cluster M67 by Hurley *et al.* The binary populations in the cluster and the simulation show significant differences in binary frequency and fraction of circularized binaries, while otherwise showing similar orbital eccentricity distributions. Since the simulations were designed to match the encounter products in M67, such as blue stragglers, the large reduction in binary fraction indicated by the empirical results likely will also require changes in the simulation physics producing blue stragglers and other anomalous stars arising from stellar dynamics. We present three case studies of stars in open clusters which very likely are products of dynamical encounters between binaries and either single stars or other binaries: the M67 blue straggler S1082, the M67 sub-subgiant S1113, and the horizontal branch star 6819-3002 in the intermediate-age open cluster NGC 6819. Finally, we remind the reader of recent empirical results on the rates of tidal interactions, using tidal circularization periods in open clusters. Every indication is that current theories underestimate the effectiveness of tidal circularization, a result that need to be incorporated into dynamical simulations of dense stellar systems.

Keywords. Stars: spectroscopic binaries, blue stragglers, evolution; Galaxy: open clusters; Physical data and processes: stellar dynamics

1. Introduction

It might be argued that open clusters are not an appropriate subject for a meeting on the dynamical evolution of dense stellar systems, since by most dynamical measures open clusters are not particularly dense. Nonetheless open clusters are highly accessible laboratories for many of the processes that take place in dense stellar systems. Furthermore, current numerical techniques permit full-scale dynamical simulations of open clusters, even including reasonable approximations of dissipative stellar interactions. As such, they are superb testbeds for the codes that will be used to model dense stellar systems in the near future.

The WIYN Open Cluster Study (WOCS; Mathieu 2000) has been intensively observing a set of select open clusters to obtain complete and accurate descriptions of their populations, spatial distributions, and internal kinematics. The contribution of the University of Wisconsin – Madison team has been primarily highly precise ($\sigma \sim 0.4$ km s^{-1}) radial-velocity surveys providing multiple measurements for stars over time spans of a decade and longer. As cases in point, as of August 2007 the WOCS database includes 8572 radial-velocity measurements of 1092 stars in the field of the open cluster NGC 188, 7698 measurements of 1517 stars in the field of NGC 6819, and 7104 measurements of

1597 stars in the field of NGC 2168 (M35). More than 100 velocity variables have been identified in each of these clusters, with most having spectroscopic orbit solutions.

Such data are now ready to speak to broad questions in stellar dynamics, stellar evolution, and most interestingly the interfaces between these two classical fields. Here I give a first glimpse at the imminent prospects.

2. Hard Binary Populations in Open Clusters

Two of the more advanced WOCS studies are those of the young cluster NGC 2168 (0.15 Gyr) and the very evolved cluster NGC 188 (7 Gyr). Here we demonstrate the current possibilities for detailed comparisons of clusters and simulations, in this case in the context of hard-binary populations.

2.1. *"Initial" Hard-Binary Population*

The initial conditions of most N-body simulations are not truly primordial, of course, in that the star- and cluster-formation events are rarely included. (Although it is exciting to see that simulations from cluster birth to death will be in our very near future (e.g., C. Clarke, this volume).) Rather, many simulations start from a configuration that includes a mass spectrum of main-sequence stars bound within a near-equilibrium cluster. A critical initial condition is the binary population, including descriptions of binary frequency, period distribution, orbital eccentricity distribution, secondary mass function, and the like. To date the initial binary population often has been characterized by equilibrium expectations or integrations of equilibrium expectations with field binary observations (e.g., Kroupa 1995). The WOCS study of the young open cluster NGC 2168 provides an *empirical* initial binary population. We summarize here several key findings – more details can be found in Braden *et al.* (this volume, and *in preparation*).

We take the field solar-type binary population of Duquennoy & Mayor (1991; DM) as the null hypothesis against which we compare the NGC 2168 binary population. In Fig. 1 we compare the NGC 2168 period and eccentricity distributions with those of DM, scaled appropriately by total sample size. For periods of less than 100 days the field and cluster frequency distributions are indistinguishable. At greater than 100 days the cluster distribution does not continue to rise as does the field. This is likely an effect of incompleteness in binary orbit solutions; we should achieve completeness through periods of 1000 days within a year or two.

The M35 orbital eccentricity distribution shows a significant population of circular orbits, almost certainly the result of tidal circularization processes (see Section 4). For $e > 0.1$, we compare the cluster eccentricity distribution with a Gaussian fit to the DM eccentricity distribution (in the DM case, for binaries with periods between the tidal circularization cutoff period and 1000 days). Both show distributions that decrease at low and high eccentricities. The DM distribution peaks at an eccentricity of 0.3, somewhat shorter than the peak in the cluster distribution. Whether this difference is meaningful remains to be seen; certainly to zeroth order it is the similarity of the cluster and field eccentricity distributions that is most significant.

Finally in Fig. 2 we show the radial spatial distribution of single and binary solar-type stars in NGC 2168, which are not evidently different. Mathieu (1983) and McNamara & Sekiguchi (1986) found that the most massive stars in the cluster are mass segregated, but the degree of segregation lessens for masses of 1-1.5 $M_\odot$. More detailed dynamical modeling is merited to see whether the similar radial distributions of the single and binary solar-type stars are in equilibrium with the more centrally concentrated distributions of the massive stars.

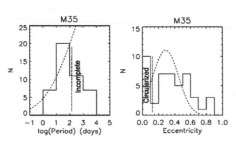

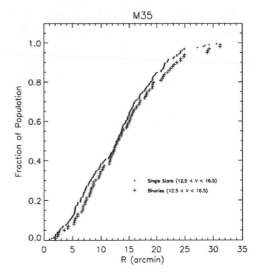

Figure 1. Left: The period distribution of spectroscopic binaries in the young open cluster NGC 2168. The dashed line represents the field period distribution of Duquennoy & Mayor (1991), scaled by sample size. Right: The orbital eccentricity distribution of short-period binaries in NGC 2168. Again, the dashed line represents the field eccentricity distribution of DM, scaled by sample size.

Figure 2. The cumulative radial distributions of solar-type single stars and binary stars in the young open cluster NGC 2168. The two distributions are indistinguishable.

2.2. *Evolved Hard-Binary Populations*

The most detailed N-body simulation of an open cluster is that of Hurley *et al.* (2005), who sought to create a cluster of age 4.5 Gyr that could be considered as a match (statistically speaking) to the open cluster M67. Importantly, Hurley *et al.* had little *a priori* knowledge of the current M67 binary population; their primary interfaces between their models and M67 were the current color-magnitude diagram, mass function, and population of blue stragglers. As such, the current binary population of M67 represents a powerful *post facto* test of the dynamical models.

We are in the process of defining the current M67 binary population. Here in its place we use the binary population of NGC 188, an older open cluster with an age of 7 Gyr. We summarize here several key findings – more details can be found in Geller *et al.* (this volume, and 2008).

The period distributions of both NGC 188 and the M67 simulation are shown in Fig. 3. Here we have adapted the M67 simulation to match our observational constraints – specifically, a mass range of greater than $\sim$0.9 M$_\odot$, a 15 pc diameter region, and the inclusion only of binaries with periods less than 10^4 days. Finally in Fig. 3 is shown the scaled DM distribution as a reference for comparison.

Clearly, the binary frequency of the M67 simulation is much higher than the actual binary population of NGC 188 (by a factor 6 at a period of 10 days). We suspect that this difference is primarily linked to the high initial binary frequency of the simulation rather than dynamical evolution. Importantly, this high binary frequency was needed to reproduce the M67 blue straggler population, so significantly lowering it to match the observed binary populations of either NGC 2168 or NGC 188 also will require changes to the physics of blue straggler creation.

The decrease in frequency for periods greater than 100 days of the binary population of the M67 simulation is the result of dynamical evolution. Again, the NGC 188 population may not be complete at periods longer than 100 days, so noting the similar drop in the NGC 188 distribution is likely premature. In addition to adding a decade of completeness

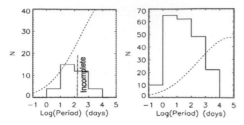

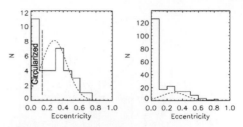

Figure 3. The period distribution of hard binaries in NGC 188 (left) and from simulated observations of an N-body model of M67 (Hurley *et al.* 2005). The dashed line represents the field period distribution of DM, scaled by sample size.

Figure 4. The orbital eccentricity distribution of hard binaries in NGC 188 (left) and simulated observations of an N-body model of M67 (Hurley *et al.* 2005). The dashed line represents the field eccentricity distribution of Duquennoy & Mayor (1991), scaled by sample size.

to the NGC 188 period distribution, we hope to soon use photometric techniques to get a handle on the *total* binary frequency of NGC 188. This also will allow us to constrain the dynamical evolution of binaries in the cluster (Frinchaboy & Nielsen, this volume.)

However, we can already examine the evolution of hard-binary frequencies as a function of cluster age. Specifically, we consider the frequency of solar-type binaries within 4 core radii whose radial-velocity measurements produce a standard deviation greater than 2 km s^{-1}. The frequency is $33\% \pm 3\%$ in NGC 2168 (0.15 Gyr), $37\% \pm 3\%$ in NGC 6819 (2.4 Gyr), and $35\% \pm 3\%$ in NGC 188 (7 Gyr). Thus we find remarkably little evolution of the hard binary frequency over 7 Gyr. (The frequency is $58\% \pm 4\%$ in the M67 simulation, much higher as already noted.)

In Fig. 4 we show the orbital eccentricity distributions for both NGC 188 and for simulated observations of the M67 model. Clearly there is a striking excess of circularized binaries in the simulation. The frequency of circularized binaries in NGC 188 compares well with those in other old open clusters, including M67 (Mathieu *et al.* 2004). Curiously, current tidal circularization theories *under*estimate the tidal circularization cutoff periods in these older clusters. Thus the presence of so many circularized binaries in the M67 simulation is somewhat of a surprise that still needs to be understood.

For non-zero eccentricities, the agreement of the two eccentricity distributions is reasonably good. Indeed, both NGC 188 and the M67 simulation show eccentricity distributions much like that found by DM for field solar-type binaries. This is notable in that the cluster binaries have been immersed in a denser stellar environment than the field binaries, with consequent dynamical encounters and evolution.

Finally, we note that the binaries and the blue stragglers of both NGC 188 and the M67 simulation are both centrally concentrated with respect to the single stars (or, more strictly, the stars not showing measurable velocity variation; Geller *et al.*). Curiously, though, the giant stars in the M67 simulation are centrally concentrated while the giants in both NGC 188 and M67 are not (Geller *et al.*, this volume; Mathieu 1983, Tinsley & King 1976).

3. An Array of Anomalous Stars – The Interface of Stellar Dynamics and Stellar Astrophysics

I suspect that in every astronomy survey course around the world a figure is shown to the students that overlays main sequences and giant branches from a set of clusters of varying ages, thereby demonstrating stellar evolution. Often, the cluster sequences are

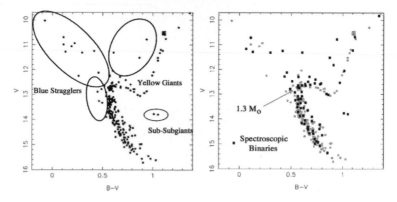

Figure 5. Left: A color-magnitude diagram of M67. All stars shown are three-dimensional kinematic cluster members. Right: The same figure with all known spectroscopic binaries shown as black square symbols.

simplified to show only the traces of the main sequence, the subgiants if appropriate, and the giants. While perhaps pedagogically useful, this approach can mislead our students into thinking that stellar evolution theory is essentially complete and, while impressive in its success, no longer at the forefront as a field of research.

In reality, the color-magnitude diagram of M67 looks like Fig. 5a. A substantial fraction of the cluster members do not lie on the classical single-star isochrone. Some of the members are not on the single-star locus simply because they are the combined light of two or more stars in multiple systems. But many of the stars are not so easily explained, as labeled in the figure. The blue stragglers of course have been known (if not understood) for a half-century. Also indicated in the figure is the possibility (indeed, the likelihood) that "blue stragglers" also exist below a cluster turnoff, either to the blue of or embedded within the main sequence. In addition to the blue stragglers there are "yellow giants" (e.g., Mathieu & Latham 1986) and the recently discovered sub-subgiants (Mathieu *et al.* 2003).

Another important way to look at a cluster color-magnitude diagram is shown in Fig. 5b . Here we have shown all of the currently known spectroscopic binaries in M67. The figure shows graphically the fact that at least one third, and presumably a higher fraction, of the stars in a cluster color-magnitude diagram are binaries. As such, they have a much higher cross-section for dynamical encounters with other single stars and binaries, and a consequently higher chance of taking an evolutionary path different from isolated single stars. This interface of stellar dynamics and stellar evolution represent the exciting forefronts of both classical fields. Some examples...

3.1. *The Blue Straggler S1082*

In 1992, Goranskij *et al.* suggested that the star S1082 in M67 was an eclipsing binary. This discovery went largely unnoticed for a decade. Then both van den Berg *et al.* (2001) and Sandquist *et al.* (2003) studied the system extensively. Fig. 6 shows the outcome of the Sandquist *et al.* analysis of the system. The system is likely a hierarchical triple. The eclipsing binary has a period of 1.07 days, while the tertiary has a period of roughly 1000 days. Interestingly, both the tertiary and the primary of the eclipsing binary are blue stragglers. The secondary of the eclipsing binary lies on the main sequence, within substantial error bars.

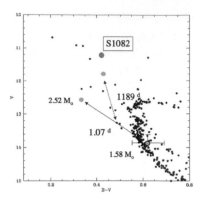

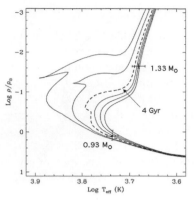

Figure 6. Location of the three spectro-scopic components of the eclipsing blue straggler S1082 in the M67 color-magnitude diagram. (Adapted from Sandquist *et al.* 2003.)

Figure 7. A set of isochrones in the log $\rho - T_{eff}$ diagram. Also shown are the locations of the primary and secondary stars of the sub-subgiant binary S1113. The lower density datum represents the primary star, the other datum is the secondary star. The curves in the figure are solar-metallicity Yale isochrones (Kim *et al.* 2002) for ages 1, 2, 4, 6, 8, and 10 Gyr. The 4 Gyr isochrone, comparable with the age of M67, is shown as a dashed line. (Taken from Mathieu *et al.* 2003.)

With the addition of the stellar masses, the system becomes truly fascinating. The mass of the blue straggler primary of the eclipsing binary is 2.52 $M_\odot$, encouragingly greater than the cluster turnoff mass of 1.3 $M_\odot$. But most remarkably, the mass of the secondary is 1.58 $M_\odot$– even though it lies on the main-sequence below the turnoff! Apparently we are not dealing with a simple single star in equilibrium.

At this point discussing an evolutionary scenario is more for the pleasure of speculation than for secure understanding. Remembering that the turnoff mass of M67 is 1.3 $M_\odot$, the presence of 4 $M_\odot$ in the eclipsing binary suggests an origin involving at least three stars. Including the blue straggler tertiary, van den Berg *et al.* (2001) suggested that the system formed from dynamical interactions of five stars or more, likely in multiple binary-binary encounters. In this scenario, during an encounter a resonant interaction of three stars leads two stars to collide and merge directly to the blue straggler primary of the eclipsing binary, with the third star becoming the current secondary. A fourth star becomes the widely separated current tertiary. That this star is also a blue straggler is challenging. Was it already such in one of the encountering binaries? What is the likelihood of this relative to a blue straggler lifetime? Or was it too formed from a collision during the formation of the current system?

From the point of view of stellar evolution, the most intriguing result may be the secondary star, lying on the main sequence with a mass substantially larger than the turnoff mass. Equally notable, the radius of the secondary is slightly larger than the radius of the more massive primary star. Both findings indicate that the secondary may not be in a thermal equilibrium state. This would require a rather recent formation, also suggested by the asynchronous rotation of at least the secondary despite the very short-period orbit.

Ultimately for this meeting the significance of S1082 is that it is one of the most convincing cases in support of dynamical encounters, and indeed stellar collisions, occurring in stellar systems.

3.2. *The Sub-Subgiant S1113*

The M67 stars S1063 and S1113 have attracted attention for their location in the cluster color-magnitude diagram roughly 1 mag below the subgiant branch. Both stars are three-dimensional kinematic members of the cluster. Mathieu *et al.* (2003) gave them the name of "sub-subgiants", while Albrow *et al.* (2001) call similarly located stars in the globular cluster 47 Tuc "red stragglers".

Both M67 sub-subgiants are spectroscopic binary stars; here we present S1113 in some detail. S1113 is a double-lined spectroscopic binary with a period of 2.8 days, a circular orbit, and a mass ratio of 0.7. The system is photometrically variable, also with a period of 2.8 days. Presuming a spot origin for the variability, the identity of the orbital and photometric periods indicates synchronous rotation, as expected for such a short-period orbit.

Presuming synchronous rotation, the mean density of each star can be obtained directly from observables (P, v sin i, and M sin^3 i). In Fig. 7 we place both stars in a log ρ − log T$_{eff}$ diagram. Interestingly, both stars can be fit to a 4 Gyr isochrone, with the secondary a 0.9 M$_\odot$ main-sequence star and the primary a 1.3 M$_\odot$ giant. However, such an interpretation of the system requires a luminosity ratio of 9, while the measured luminosity ratio is only 3. If the secondary is in fact a main-sequence star, then the primary star is underluminous − hence its location below the subgiant branch of the cluster. This low luminosity indicates a primary radius that is currently well within the associated Roche lobe.

As yet the case of S1113 awaits a satisfactory evolutionary scenario. Mathieu *et al.* (2003) conjecture that its evolutionary history includes mass transfer episodes, mergers, and/or dynamical stellar exchanges. To this might be added close encounters that may have stripped the primary envelope.

3.3. *The Horizontal Branch Star 6819-3002*

At first glance, star 3002 (WOCS identification number) in the red clump (i.e., the horizontal branch) of the intermediate-age open cluster NGC 6819 is not evidently note-worthy. It has a period of 17.7 days and a circular orbit; the latter is not a surprise given a giant primary at such an orbital period. Presuming that the primary has the turnoff mass of 1.6 M$_\odot$, the current separation is about 0.2 AU, so that at present the primary star fits comfortably within its Roche lobe.

However, given its present position in the red clump, the primary star would have already evolved up to the tip of the giant branch. Thus in the past the star would have had a radius approaching 50 R$_\odot$ (using the Yale isochrones), and far exceeded its Roche lobe if it were in the current binary. Presumably mass transfer would have precluded the star becoming a normal cluster horizontal branch star.

Thus we suggest that this system may be the result of a dynamical encounter, and specifically we hypothesize that the current horizontal branch primary star was exchanged into the system during a single star − binary encounter.

Gosnell *et al.* (2008) tested this hypothesis using the FEWBODY code of Fregeau *et al.* (2004) to run nearly 100,000 single star − binary encounters. They explored much of the available parameter space, and found that the current binary in fact could be formed dynamically by an exchange encounter (after subsequent tidal circularization). However, producing a product binary that remains bound to the cluster is a significant additional constraint that rules out most encounter paths to the current binary. Only 0.11% of the simulated encounters produced a binary similar to 3002 that would still be bound to the cluster. The next step, currently underway, is to determine the probability of a horizontal-branch single star – binary encounter given an assumed binary population and cluster mass function.

In summary, an essential point of all three cases is that we are identifying candidate products of dynamical encounters in real clusters and approaching the capability to explore numerically specific formation paths, as compared to mere qualitative conjectures. Of course, accurate injection of hydrodynamics awaits, but this too on the horizon. Perhaps the most significant challenge will soon no longer be technical, but rather proper understanding of *a posteriori* simulations.

4. Tidal Circularization Rates in Solar-type Binary Stars

Tidal effects play major roles in the close stellar encounters possible in resonant binary-single star and binary-binary encounters, and indeed tidal rates can be the difference between a close fly-by, the formation of a close binary, or a merger. As such, tidal physics may impact significantly on the accuracy of dynamical simulations.

For two decades, open clusters have been prime laboratories for the study of tidal circularization rates. Given a coeval sample of binaries in a cluster, theory predicts that the shortest period binaries will be tidally circularized while longer period binaries will retain their primordial orbital eccentricities. The orbital period at the boundary between these two eccentricity domains, or the *tidal circularization period*, is set by tidal circularization rates and, in most theories, the cluster age.

The most recent and extensive study of tidal circularization periods in open clusters is that of Meibom & Mathieu (2005). Fig. 8, taken from their paper, summarizes the

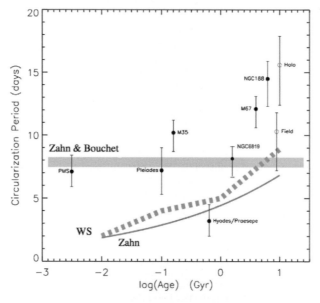

Figure 8. Distribution of tidal circularization periods with age for eight solar-type binary populations. The solid curve shows the predicted cutoff period as a function of time based on main-sequence tidal circularization using the revised equilibrium tide theory by Zahn (1989). The broad dashed band represents the predicted cutoff period for initially supersynchronous 1 $M_\odot$ stars calculated in the framework of the dynamical tide model including resonance locking (Witte & Savonije 2002). The horizontal gray band represents the prediction by Zahn & Bouchet (1989), in which tidal circularization is significant only during the pre-main-sequence phase. (Taken from Meibom & Mathieu 2005.)

essential results. With the exception of the Hyades/Praesepe population, the distribution of tidal circularization periods with age shows an increase from the pre-main-sequence to the late main sequence. The theoretical models of main-sequence tidal circularization using either the equilibrium tide theory (Zahn 1989) or the dynamical tide theory with resonance locking (Witte & Savonije 2002) predict longer cutoff periods with increasing population age, in agreement with the trend in the distribution of circularization periods. However, the predicted circularization periods fall significantly below the observed circularization periods. This suggests that the efficiency of the dissipation in these models is too low. Alternative, Zahn & Bouchet (1989) suggest that all tidal circularization takes place during the pre-main-sequence phase when the stars are large with deep convective zones, and that no further circularization takes place during the main-sequence stage. This theoretical model does not predict the observed increase in tidal circularization periods with increasing cluster age.

The essential result of these studies is that actual tidal circularization rates in main-sequence solar-type stars are substantially higher than theoretical prediction. As such, the use of current theory in dynamical simulations must be done with some caution.

5. Closing Thoughts

For the last half-century, N-body simulations lagged behind observations of actual star clusters, the latter having the advantage of "the great analog computer in the sky". Thus attempts to compare N-body simulations with star clusters were limited by the models.

I would suggest that we are entering a period during which the simulations will catch up with, and indeed in certain areas will pull ahead of, the observations. At least for open clusters, the true numbers of stars in even the more massive clusters (at birth) are accessible to current computational capabilities. Equally importantly these numbers now can include high frequencies of primordial binaries. Normal stellar evolution is being computed in "real" time. And the next frontier of hydrodynamical interactions between stars is being crossed on numerous fronts, including mass transfer, tidal effects, and collisions.

One domain of stellar dynamics where the simulations have always provided more insight than have observations has been dynamical encounters between binaries and other single or binary stars. While we have known for a long time that open clusters have high binary frequencies, we have had very little information on the orbital properties of the binaries. As shown in this paper, that gap is being closed, at least for the hard binaries. It is worth mentioning, though, that the majority of post-encounter binaries and single stars are not identifiable as such from their specific properties. Thus observational studies of dynamical encounters in clusters will likely require statistical comparisons of binary populations as a function of age.

I would like to close with a more philosophical thought on the scientific method in our field, and specifically on the issue of *a priori* versus *a posteriori* modeling. One of the truly beautiful aspects of the Hurley *et al.* experiment for M67 was that their simulations made *a priori* testable predictions about the cluster, as demonstrated in Section 2. True, the Hurley *et al.* models were *a posteriori* in the sense that they sought to match their outcomes to an existing cluster, M67. However, the interfaces available to them for that comparison were primarily the color-magnitude diagram (especially the products of dynamical encounters such as blue stragglers) and the current mass functions. The interface with the binary populations was not accessible, simply because our work was not yet done, so they made *a priori* predictions about the current binary population.

That aspects of their predicted binary population may not agree with reality, based on NGC 188, is not nearly as significant as the fact that we are now in a place where N-body simulations can make meaningful, detailed predictions of observables. Associated with this capability is the fundamental question of how to proceed. At the time of the Hurley *et al.* calculations, creating an M67 was still a substantial task in terms of real time; thus very soon (even now?) such calculations will be easily doable. Given this capability, I suggest that we will want to transition from trying to match specific real clusters such as M67 *a posteriori* to building *a priori* arrays of clusters that we compare as an ensemble to observed clusters.

For N-body simulators, learning about observations of clusters is a bit like losing the innocence of youth – it can be difficult to maintain an *a priori* independence when one knows the answer. Arguably to ignore observations would be (computationally) inefficient in any case. Nonetheless, perhaps in these early days we will be able to continue to enjoy the pleasure of simulations like those of Hurley *et al.* with predictions that are both *a priori* and testable, mimicking the way that we teach our students that science "really works"!

6.   Acknowledgements

I am deeply indebted to the many fine students who have done much of the work reported here, including Ella Braden, Michael DiPompeo, Christopher Dolan, Aaron Geller, Natalie Gosnell, Soeren Meibom, Meagan Morscher, Keivan Stassun, and Sylvana Yelda. I also want to acknowledge the superb staff of the WIYN Observatory who have made the acquisition of many thousand of superb stellar spectra both efficient and fun. This work was funded by National Science Foundation grant AST-0406615.

References

Albrow, M. D., Gilliland, R. L., Brown, T. M., Edmonds, P. D., Guhathakurta, P., & Sarajedini, A. 2001, *ApJ*, 559, 1060

Duquennoy, A. & Mayor, M. 1991, *A&A* 248, 485

Fregeau, J. M., Cheung, P., Portegies Zwart, S. F., & Rasio, F. A. 2004, *MNRAS*, 352, 1

Geller, A. M., Mathieu, R. D., Harris, H. C., & McClure, R. D. 2008, *AJ*, submitted

Goranskij, V. P., Kusakin, A. V., Mironov, A. V., Moshkaljov, V. G., & Pastukhova, E. N. 1992, *A&AT*, 2, 201

Gosnell, N. M., DiPompeo, M. A., Braden, E. K., Geller, A. M., & Mathieu, R. D., 2008, *BAAS*, in press

Hurley, J. R., Pols, O. R., Aarseth, S. J., & Tout, C. A. 2005, *MNRAS*, 363, 293

Kim, Y. C., Demarque, P., Yi, S. K., & Alexander, D. R. 2002, *ApJS*, 143, 499

Kroupa, P. 1995, *MNRAS*, 277, 1507

Mathieu, R. D. 2000, ASP Conf. Ser., 198, 517

Mathieu, R. D., Meibom, S., & Dolan, C. J. 2004, *ApJ*, 602, L121

Mathieu, R. D. & Latham, D. W. 1986, *AJ*, 92, 1364

Mathieu, R. D., van den Berg, M., Torres, G., Latham, D., Verbunt, F., & Stassun, K. 2003, *AJ*, 125, 246

McNamara, B. J. & Sekiguchi, K. 1986, *ApJ*, 310, 613

Meibom, S. & Mathieu, R. D. 2005, *ApJ*, 620, 970

Sandquist, E. L., Latham, D. W., Shetrone, M. D., & Milone, A. A. E. 2003, *AJ*, 125, 810

Tinsley, B. M. & King, I. R. 1976, *AJ*, 81, 835

van den Berg, M., Orosz, J., Verbunt, F., & Stassun, K. 2001, *A&A*, 375, 375

Witte, M. G. & Savonije, G. J. 2002, *A&A*, 386, 222

Zahn, J.-P. 1989, *A&A*, 220, 112

Zahn, J.-P. & Bouchet, L. 1989, *A&A*, 223, 112

Dynamical Evolution of Dense Stellar Systems
Proceedings IAU Symposium No. 246, 2007
E. Vesperini, M. Giersz & A. Sills, eds.

N-body Models of Open Clusters

Jarrod R. Hurley

Centre for Astrophysics and Supercomputing,
Swinburne University of Technology,
P.O. Box 218, VIC 3122, Australia
email: jhurley@swin.edu.au

Abstract. N-body simulations of open cluster evolution with primordial binaries are reviewed. In particular, recent results arising from models with initial N in the range of 20 000–100 000 bodies are compared to earlier idealized models with $N \sim 2\,000$. Efforts to model real clusters are discussed, including how limitations of the models such as simplified initial conditions will be addressed in the near future.

Keywords. stellar dynamics, methods: n-body simulations, binaries: close, open clusters and associations: general

---

1. Introduction

From the time of the first N-body simulation of star cluster evolution recorded by von Hoerner (1960) the number of particles, N, that can be followed in a reasonable timeframe has risen from ten to of order 100 000 (Baumgardt & Makino 2003; Hurley, Aarseth & Shara 2007). This is due to increased hardware performance, such as the introduction of the GRAPE-4 special-purpose computers (Makino *et al.* 1997), as well as the development of improved computational algorithms. Further developments are required before direct models of globular clusters with the N-body method become feasible. In the meantime, much can be learnt from understanding the evolution of the open cluster type models performed to date.

In the pre-GRAPE era of the early 1990's the N limit was of the order of 2 000 stars. However, a major development at this time was the introduction of primordial binaries in the N-body models. This was important for understanding the evolution of real clusters as observations clearly indicate that open clusters contain a significant primordial binary population (e.g. Mermilliod & Mayor 1990). Two studies at this time by Heggie & Aarseth (1992) and McMillan & Hut (1994) stand out as landmarks because of their in-depth analysis of the effect of the primordial binaries on the cluster evolution and, in reverse, the effect of the evolution on the make-up of the binary population. These were idealized simulations in that only equal-mass stars were considered but they did include the effect of the tidal field of the Galaxy as well as primordial binary frequencies of 3–6%, in the case of Heggie & Aarseth (1992) and up to 20% for McMillan & Hut (1994). This work was extended by de la Fuente Marcos (1996) who looked at models with 33% primordial binaries and stellar masses distributed according to an initial mass function (but no stellar evolution). The models of Kroupa (1995) with $N = 400$ stars and 100% binaries also deserve mention†. The first model to move away from the idealized regime to what has become known as the 'kitchen-sink' regime was that of Aarseth (1996). This was a model starting with 10 000 stars and a 5% binary frequency evolved with the NBODY4

---

† This is by no means an extensive history of N-body simulations – for that the interested reader is referred to Aarseth (2003).

Table 1. Overview of N-body models used in this work.

Model	N_s	N_b	a_{max} [au][1]	M_i [$M_\odot$]	$r_{h,i}$ [pc]	t_f [Myr][2]
B1	9000	9000	200	14405	3.9	5770
B6	9000	9000	10	14010	4.0	5150
S7	30000	0	–	14570	4.2	8460

Notes:
[1] Maximum binary separation.
[2] Time when only 1 000 stars remain bound.

code that is still at the forefront of cluster modelling today. The tidal field of the Galaxy, an initial mass function (IMF) and stellar/binary evolution were all considered.

Recently, Heggie, Trenti & Hut (2006) have begun revisiting the idealized models of Heggie & Aarseth (1992) by extending the models to $N = 16\,000$ and including binary frequencies from 0–100%. This is supplemented in this paper by comparing the results from the pioneering work of Heggie & Aarseth (1992) and McMillan & Hut (1994) to recent 'kitchen-sink' models of binary-rich open clusters performed with NBODY4 (Aarseth 1999). Also included is a brief summary of ongoing efforts to understand observations of actual open clusters and a presentation of preliminary simulations aimed at improving the initial conditions of the cluster models.

2. Binary-rich models

To study the general evolution properties of binary-rich open clusters Hurley, Aarseth, Tout & Pols (in preparation) have evolved a series of N-body models with $N = 20$–$30\,000$ stars and 50% primordial binaries (some models with 40% and 10% binaries were also considered). Parameters varied between the models include the initial density profile of the stars and the distribution of binary binding energies. The aim is to complement the overview of the evolution of single-star open clusters provided in Hurley *et al.* (2004) using models with $N = 30\,000$ stars and 0% binaries.

Here two models from the binary-rich series will be used to make some early comparisons to the findings of Heggie & Aarseth (1992) and McMillan & Hut (1994). These are model B1 – the reference model – and model B6 which differs in setup from B1 only by the maximum orbital separation allowed for the primordial binaries. A single star model from Hurley *et al.* (2004), namely their Model 7 (labeled S7 here), is also used for comparison. An overview of the starting parameters of these three models is given in Table 1 including the model label, the number of single (N_s) and binary (N_b) stars, the cluster mass (M_i) and the half-mass radius ($r_{h,i}$). The maximum orbital separation (a_{max}) is also given, where applicable.

For each model the stellar masses were chosen from the IMF of Kroupa, Tout & Gilmore (1993) with a lower mass limit of $0.1 M_\odot$ and an upper limit of $50 M_\odot$. The component masses of binaries were set by choosing a mass-ratio, q, from a uniform distribution, $n(q) = 1$. A metallicity of $Z = 0.02$ was assumed in each case.

Orbital separations for the binaries were distributed according to the suggestion of Eggleton, Fitchett & Tout (1989, EFT) with a peak at 30 au. In model B1 an upper limit of $a_{max} = 200$ au was applied – safely in excess of the hard/soft boundary of approximately 70 au for the starting model. By comparison, model B6 took $a_{max} = 10$ au so that all primordial binaries were hard, i.e. tightly bound.

Each model was evolved using the NBODY4 code on a 32-chip GRAPE-6 board (Makino 2002). Stellar and binary evolution are included in NBODY4 as described in Hurley *et al.* (2001). The tidal field of the Galaxy was modeled by placing the model cluster on a

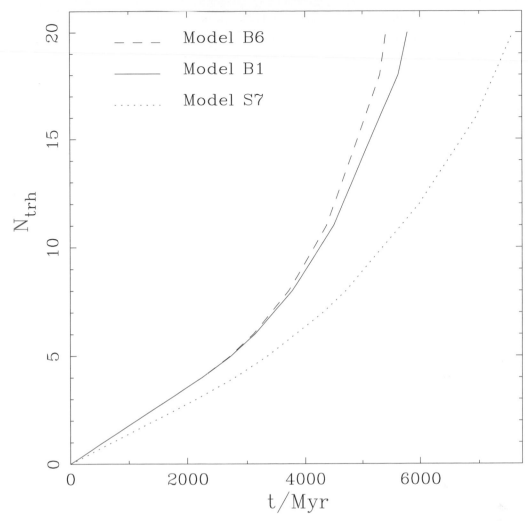

Figure 1. Number of half-mass relaxation times elapsed as a function of cluster age. This is calculated using the 'co-moving' instantaneous half-mass relaxation time which, for an evolved cluster, is typically a factor of 2-3 shorter than the initial half-mass relaxation time.

circular orbit at 8.5 kpc from the center of a point-mass galaxy. Full details of the setup and evolution of these simulations will be provided in Hurley, Aarseth, Tout & Pols (in preparation).

The lifetimes of models B1, B6 and S7 are given in Table 1 as t_f, the age at which the cluster membership has been reduced to 1 000 stars. Lifetimes can also be compared by looking at Fig. 1 which demonstrates the relative dynamical ages of the models. The dissolution timescale clearly decreases when primordial binaries are included (comparing B1 to S7). This is not surprising as the presence of binaries in the center of the cluster leads to an increase in the escape of stars through velocity kicks imparted in three- and four-body interactions. The escape rate of single stars from B1 is typically 30% greater than for S7 at comparable ages. This result varies little if B6, or a model with 10% primordial binaries, is instead compared to S7, in agreement with the findings of Heggie & Aarseth (1992: see their Fig. 12). As the relaxation timescale of a tidally-limited cluster decreases with decreasing cluster mass, the presence of primordial binaries shortens the

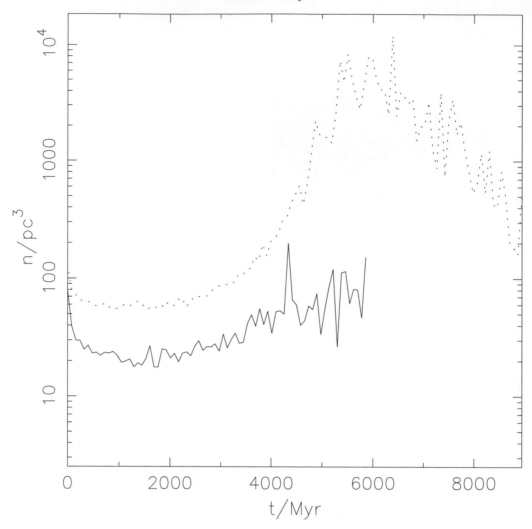

Figure 2. Core density evolution of models B1 (solid line) and S7 (dotted line).

relaxation timescale as the cluster evolves. Therefore, a cluster with primordial binaries is dynamically more evolved than a single star model at the same age (as exhibited in Fig. 1). Comparing the dissolution times of B1 and B6 the lifetime is shorter for the latter. However, the difference is small compared to that between B1/B6 and S7 and may be partly statistical. Analysis of the larger family of models will be able to confirm this. Previously, McMillan & Hut (1994) found that the dissolution timescale was insensitive to details of the primordial binary distribution (provided that the primordial fraction was non-zero). This was the result of mass-segregation saturating the core with 20–30% hard binaries regardless of the (non-zero) initial fraction or the relative distribution of initial binding energies.

Fig. 2 looks at the evolution of the number density of stars in the core of models with (B1) and without (S7) primordial binaries. There is an obvious difference with the single-star model able to achieve a much higher core density. This result is as expected from earlier models which showed that primordial binaries are efficient at reversing core-collapse and inflate the size of the core relative to single-star models (Heggie & Aarseth

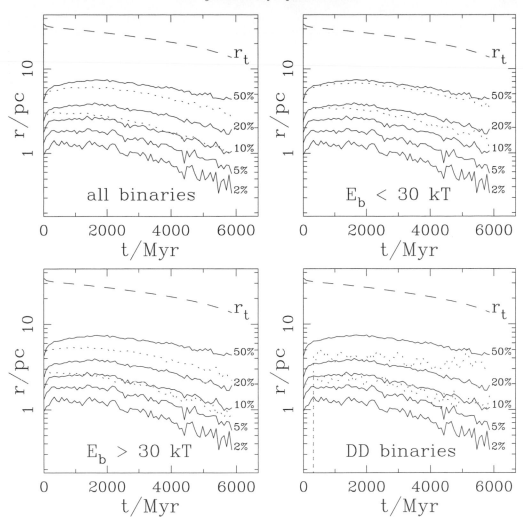

Figure 3. Time evolution of the Lagrangian radii in model B1. In each panel the five solid curves plot the Lagrangian radii containing the innermost 2, 5, 10, 20 and 50% of single stars, by mass. The dashed line denotes the tidal radius of the model cluster. Dotted lines are the 20 and 50% Lagrangian radii for: all binaries (upper-left panel); binaries with $E_b < 30\,kT$ (upper-right panel); binaries with $E_b > 30\,kT$ (lower-left panel); and, binaries containing two degenerate stars (lower-right panel).

1992; McMillan & Hut 1994). The age at which core-collapse is halted is about $4\,000\,\mathrm{Myr}$ for B1 and $6\,000\,\mathrm{Myr}$ for S7. Reference to Fig. 1 shows that this corresponds to a dynamical age of roughly ten half-mass relaxation times in both cases. This is similar to the dynamical age at core-collapse shown by the models of McMillan & Hut (1994) with $N = 1\,000$–$2\,000$ and 0–20% binaries (see their Fig. 2). However, the depth of core-collapse reached by the models presented here with 50% binaries appears to be greater than for the models of McMillan & Hut (1994) with 20% binaries or less, which is somewhat counter-intuitive. In models B1 and B6, as well as a wider range of recent NBODY4 models with primordial binary fractions of 5% or more, the ratio of core-radius to half-mass radius at the point identified with the end of the core-collapse phase is in the 0.04–0.07 range. This is compared to a ratio of ~ 0.01 for single-star models such as S7. In the

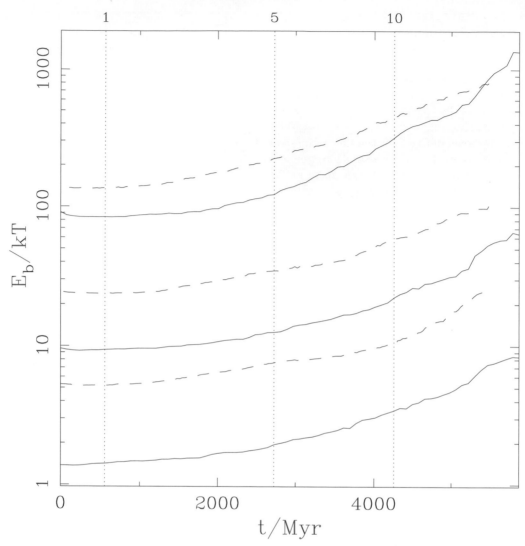

Figure 4. Energy quartiles for binary binding energies in models B1 (solid lines) and B6 (dashed lines). For each model, at any particular time, 25% of the binaries have binding energies below that of the lower curve, 50% have energies below the middle curve and 75% lie below the upper curve. So the hardest binaries lie above the upper curve. For reference, the times at which 1, 5 and 10 half-mass relaxation times have elapsed (for B1) are shown by the dotted lines.

primordial binary models of McMillan & Hut (1994) the core/half-mass radii ratios are typically 0.1 at core-collapse. Heggie & Aarseth (1992) find 0.03 for their model starting with $N = 2\,500$, 3% binaries and a tidal field.

The evolution of the spatial distribution of the binary population in model B1 is investigated in Fig. 3. This can be compared directly to Fig. 4 of McMillan & Hut (1994) and Fig. 8 of Heggie & Aarseth (1992). Note, however, that the latter is for a model without a tidal field. It should also be noted that both of these earlier studies were for models with equal-mass stars in which the mass of a binary was twice that of a single star. In the more recent models with an IMF the average binary mass is only slightly greater than the average single star mass so the effects of mass-segregation, as regards the binary population, will be exaggerated in the earlier models. Looking first at the spatial distribution of all binaries in Fig. 3 we see that the binary population is more concentrated towards the center of the cluster than are the single stars. This is clear evidence of mass-segregation and the effect increases with age. The binary population is then split into binaries with binding energies less than or greater than $30\,kT$, representing approximately the populations of loosely and tightly bound binaries, respectively†. This demonstrates that the spatial distribution of binaries depends upon their binding energy – the hardest binaries are more centrally concentrated (as shown by McMillan & Hut 1994). The distribution of double-degenerate binaries (primarily composed of two white dwarfs) is also shown in Fig. 3 for the sake of interest. These show strong signs of mass-segregation from early in the cluster evolution, being born from the most massive binaries, with the effect becoming weaker at late times as the cluster dissolves.

Related to the segregation of binaries towards the cluster center is the increase in core binary fraction with time highlighted by Hurley, Aarseth & Shara (2007). They looked at a range of models including a model similar to B1 and a model starting with $N = 100\,000$ stars and 5% binaries. Interestingly, the evolution of the core binary fraction in the latter model compares very well to that seen by Heggie & Aarseth (1992) in their model with $N = 2\,500$ stars and 3% binaries. In both the central binary fraction peaks at between 20–30% near the end of core-collapse and then steadily decreases back towards the primordial value thereafter. The results of the $N = 100\,000$ model also show that the critical binary fraction observed by McMillan & Hut (1994) – where clusters starting with less than 10% binaries exhausted their binary population before dissolution – does not scale to larger models.

Fig. 4 is a reproduction of Fig. 6 of McMillan & Hut (1994) showing the energy quartiles for the binary binding energies as the cluster evolves. Clearly visible is the hardening of binaries as the cluster evolves. This is true for both models B1 and B6 although the rate of hardening for the hardest binaries is less in the latter. Heggie & Aarseth (1992) reported a factor of 8 increase in the median binary binding energy as their model clusters evolved from zero-age towards dissolution. They also found that this factor decreased to 3–4 if they considered an extended initial energy range and a higher primordial binary fraction. Models B1 and B6 show a factor of 5–6 increase. In comparison some of the models presented by McMillan & Hut (1994) show an increase by a factor of 10 or more. Further interrogation of the series of binary-rich open clusters models will proceed and spatial and energy evolution of the binary populations to be looked at in detail.

3. Real Clusters

Much effort has been expended in the past decade to improve the realism of N-body codes such as `NBODY4`. Examples include the incorporation of stellar evolution, binary evolution, three- and four-body effects, and external tidal fields (see Aarseth 2003). This has paid off by allowing the generation of direct models that can be compared to observations of the stellar content and global properties of open clusters.

Young clusters such as the Pleiades and Hyades have been modelled: Kroupa, Aarseth & Hurley (2001) looked at the formation of the Pleiades and the consequences for its binary population while Portegies Zwart *et al.* (2001) looked more generally at the evolution of stellar content in young open clusters. Hurley *et al.* (2005) have presented a direct model of the old open cluster M67. They investigated in detail the formation of

† The unit of kT is a thermodynamic quantity commonly used to scale the binding energies of binaries. The mean stellar kinetic energy corresponds to $(3/2)kT$ which is used to determine the boundary between hard and soft binaries. Note that the binding energy E_b given throughout this work is the absolute value of this quantity.

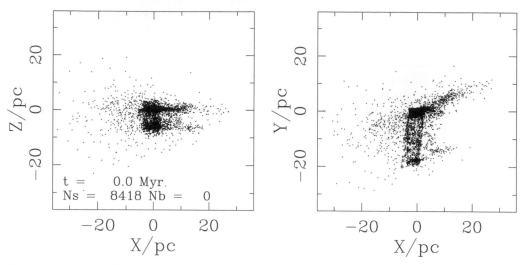

Figure 5. Spatial distribution in the X-Z (left panel) and X-Y (right panel) planes of stars in the proto-cluster input to `NBODY4` at a simulation age of 0 Myr.

blue straggler stars in M67 and also provided a census of the X-ray binaries expected and the white dwarf content. Comparison of the results of detailed models such as these with observations of particular clusters can teach us about the initial binary properties of open clusters as well as the intervening dynamical evolution. As such, this relatively new practice of targeting specific clusters should continue. Future candidates include NGC 6791, which is even older than M67, and NGC 6819 which has a well observed white dwarf sequence (Kalirai *et al.* 2001).

4. Initial Conditions

While it is true that the range of physical processes included in the cluster models has improved tremendously, as outlined in the previous section, there is still much that can be added. This is certainly the case for the initial conditions of the N-body models which remain somewhat naive. Typically the presence of gas is neglected, which will be important for young clusters, and pre-main-sequence stellar evolution, as well as the associated possibility of staggered star formation, is not included. In terms of the distributions of stars (positions and velocities) normal practice is to use a King or Plummer density profile and assume virial equilibrium (see Aarseth 2003). Such assumptions are based on observations of evolved clusters and are not necessarily correct for clusters at, or soon after, the formation stage. However, the error induced may be minimal if, for example, young clusters attain virial equilibrium on a timescale much shorter than their lifetime.

In an upcoming publication Hurley & Bekki (in preparation) aim to begin addressing some of these shortcomings by interfacing the results of galaxy-scale simulations of star cluster formation with the N-body codes that follow the long-term cluster evolution. A preliminary calculation along these lines is presented here. Fig. 5 shows the spatial characteristics of a proto-cluster formed from the collapse of a turbulent molecular cloud in a low-mass dwarf galaxy which in turn is embedded in a massive dark matter halo. This is output from the chemodynamical code of Bekki & Chiba (2007). The example proto-cluster contains $\sim 8\,400$ stars each with a mass close to $0.5 M_\odot$ and is used as input to the `NBODY4` code. It is found that the cluster, which was far from being in virial equilibrium

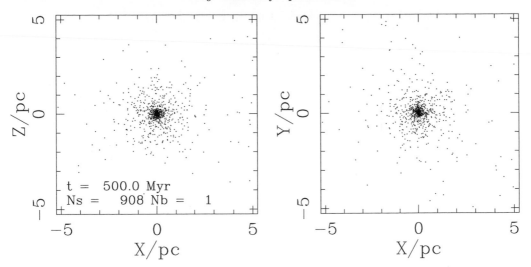

Figure 6. As for Fig. 5 but after 500 Myr of NBODY4 evolution.

to begin with, reaches a state of virial equilibrium after ~ 50 Myr of evolution (with zero-age taken as the start of the NBODY4 simulation). After 500 Myr approximately 900 stars remain bound in a relaxed and regular (in terms of appearance) cluster (as shown in Fig. 6). The results here are certainly promising and will be presented in more detail, and for a wider range of scenarios, in the upcoming publication.

5. Summary

The properties of early idealized models of open clusters with $N \sim 2\,000$ and primordial binaries generally scale well when compared to the new generation of more realistic models. A notable exception is the depth of core-collapse which warrants further investigation. Also, the critical primordial binary fraction below which the binary population of an open cluster is exhausted before cluster dissolution, found by McMillan & Hut (1994), is not observed in models with larger N. On a final point it is noted that Heggie & Aarseth (1992) demonstrated the effectiveness of comparing the results of cluster evolution models produced by complementary but differing simulation methods. This fine example needs to be continued using current statistical and N-body models (see Fregeau, these proceedings).

References

Aarseth, S. J. 1996, in: P. Hut & J. Makino (eds.), *Proc. IAU Symp. 174, Dynamical evolution of star clusters: confrontation of theory and observations* (Dordrecht: Kluwer), p. 161

Aarseth, S. J. 1999, *PASP*, 111, 1333

Aarseth, S. J. 2003, *Gravitational N-body Simulations: Tools and Algorithms* (Cambridge: Cambridge University Press)

Aarseth, S., Hénon, M., & Wielen, R. 1974, *A&A*, 37, 183

Baumgardt, H. & Makino, J. 2003, *MNRAS*, 340, 227

Bekki, K. & Chiba, M. 2007, *ApJ*, 665, 1164

deLaFuenteMarcos, R. 1996, *A&A*, 314, 453

Eggleton, P. P., Fitchett, M., & Tout, C. A. 1989, *ApJ*, 347, 998

Heggie D. C. & Aarseth, S. J. 1992, *MNRAS*, 257, 513

Heggie, D. C., Trenti, M., & Hut, P. 2006, *MNRAS*, 368, 677

Hurley, J. R., Aarseth, S. J., & Shara M. M. 2007, *ApJ*, 665, 707

Hurley, J. R., Pols, O. R., Aarseth, S. J., & Tout, C. A. 2005, *MNRAS*, 363, 293

Hurley, J. R., Tout, C. A., Aarseth, S. J., & Pols, O.R. 2004, *MNRAS*, 355, 1207

Hurley, J. R., Tout, C. A., Aarseth, S. J., & Pols, O. R. 2001, *MNRAS*, 323, 630

Kalirai, J. S., Richer, H. B., Fahlman, G. G., Cuillandre, J.-C., Ventura, P., D'Antona, F., Bertin, E., Marconi, G., & Durrell, P. R. 2001, *AJ*, 122, 266

Kroupa, P. 1995, *MNRAS*, 277, 1491

Kroupa, P., Aarseth, S., & Hurley, J. 2001, *MNRAS*, 321, 699

Kroupa, P., Tout, C. A., & Gilmore, G. 1993, *MNRAS*, 262, 545

Makino, J. 2002, in: M.M. Shara (ed.), *ASP Conference Series 263, Stellar Collisions, Mergers and their Consequences* (San Francisco: ASP), p. 161

Makino, J., Taiji, M., Ebisuzaki, T., & Sugimoto, D. 1997, *ApJ*, 480, 432

McMillan, S. & Hut, P. 1994, *ApJ*, 427, 793

Mermilliod, J.-C. & Mayor, M. 1990, *A&A*, 237, 61

Portegies Zwart, S. F., McMillan, S. L. W., Hut, P., & Makino, J. 2001, *MNRAS*, 321, 199

von Hoerner, S. 1960, *Z. Astrophys.*, 50, 184

Dynamical Evolution of Dense Stellar Systems
Proceedings IAU Symposium No. 246, 2007
E. Vesperini, M. Giersz & A. Sills, eds.

© 2008 International Astronomical Union
doi:10.1017/S174392130801538X

Monte Carlo Simulations of Star Clusters with Primordial Binaries. Comparison with N-body Simulations and Observations

Mirek Giersz[1] and Douglas C. Heggie[2]

[1] Nicolaus Copernicus Astronomical Center, Polish Academy of Sciences, Warsaw, Poland
email: mig@camk.edu.pl

[2] University of Edinburgh, School of Mathematics and Maxwell Institute of Mathematical Sciences, King's Buildings, Edinburgh, UK
email: d.c.heggie@ed.ac.uk

Abstract. We outline the steps needed in to calibrate the Monte Carlo code in order to perform large scale simulations of real globular clusters. We calibrate the results against N-body simulations for $N = 2500$, 10000 and for the old open cluster M67. The calibration is done by choosing appropriate free code parameters.

Keywords. stellar dynamics, methods: numerical, stars: evolution, binaries: general, open clusters and associations:individual:M67

1. Introduction

Most of the modeling of individual star cluster has been focused on static models based on the King model (see Meylan & Heggie 1997). There have been also a small number of studies able to follow the dynamical evolution of a system. They were mainly based on variants of a Fokker-Planck technique and small N-body simulations (e.g. Grabhorn *et al.* 1992, Drukier 1993, Murphy *et al.* 1998, Giersz & Heggie 2003 and Hurley *et al.* 2005).

In this presentation we show the further developments of the Monte Carlo code needed to properly follow the evolution of the real star clusters. The dynamical ingredients of the code are basically the some as those described in Giersz (2006 and references therein). This code is based on an original code by Stodółkiewicz (1986), which in turn was based on the code devised by Hénon (1971). The extensions to the code were mainly connected with: (i) upgrading the prescription of stellar evolution to the algorithm of Hurley *et al.* (2000), (ii) adding the procedures for the internal evolution of binary stars (Hurley *et al.* 2002), both using the McScatter interface (Heggie, Portegies Zwart & Hurley 2006), (iii) a better treatment of the escape process in the presence of a static tidal field. Then the new version of the code was calibrated against the results of N-body simulations so as to allow us to construct dynamical evolutionary models of real star clusters.

2. The calibration of the Monte Carlo technique

For Monte Carlo simulations the standard N-body units are adopted. The unit of time, however, is proportional to $N/ln(\gamma N))$, where N is a number of stars and γ is a parameter. Additionally, to properly follow the relaxation process two other parameters have to be chosen: the range of deflection angles, β_{min} and $\beta_{max} = 2\beta_{min}$ and the overall time step, τ, which has to be a small fraction of the half-mass relaxation time. Properly

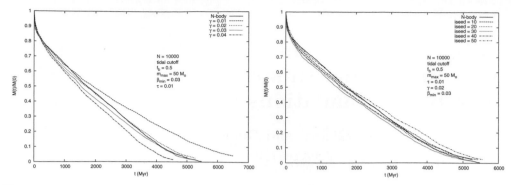

Figure 1. Evolution of the total mass for N-body and Monte Carlo models. Left panel for different γ, right panel for different iseed – initial random number sequence. Parameters of the models are described in the figures.

chosen above parameters together with additional parameters, which characterize mass functions, dynamical interactions of binaries and escape process from the system should make possible to reproduce N-body simulations and follow evolution of real star clusters.

For calibrate the Monte Carlo code the N-body simulations with $N = 2500$, 10000 and 24000 were used. The initial parameters of all N-body and Monte Carlo runs were the same as used by Hurley *et al.* (2005) for simulations of the old open cluster M67 ($N = 24000$).

2.1. Models with tidal cutoff

First, we concentrated on calibration of Monte Carlo models for which the influence of the tidal field of a parent galaxy is characterized by the tidal energy cutoff – all stars which have energy larger than $E_{tid_c} = -GM/r_{tid}$ are immediately removed from the system – G is the constant of gravity, M is the total mass and r_{tid} is the tidal radius.

As was pointed out by Hénon (1975) the value of γ strongly depends on the mass function and distribution of stars in the system. The γ for equal mass stars is rather well known (Giersz & Heggie 1994). For unequal mass case and primordial binaries is much less known. The dependence of results of the Monte Carlo simulations on γ and initial random number sequence is presented in Fig. 1.

The best value of γ inferred from simulations with $N = 2500$ and 10000 is equal to 0.02 (see Fig. 1 left panel). The τ and β_{min} are equal to 0.01 and 0.03, respectively. As can be seen from Fig. 1 (right panel) the spread between models (statistical fluctuations) with exactly the same parameters, but with different initial random number sequence (iseed) is very substantial. The spread between results with different β_{min} and τ is well inside the spread connected with different iseed. The statistical fluctuation of the models are larger for smaller N as one can expect.

2.2. Models with full tidal field

The process of escape from a cluster for steady tidal field is extremely complicated. Some stars which fulfill the energy criterion (binding energy of a star is greater than critical energy $E_{tid_f} = -1.5(GM/r_{tid})$) can be still trapped inside a potential wall. Those stars can be scattered back to lower energy before they escape from the system. According to the theory presented by Baumgardt (2001) the energy excess of those stars is decreasing with the increasing number of stars. So the cluster lifetimes do not scale linearly with relaxation time as expected from the standard theory. To account for this process in the Monte Carlo code an additional free parameter, a, was introduced. The

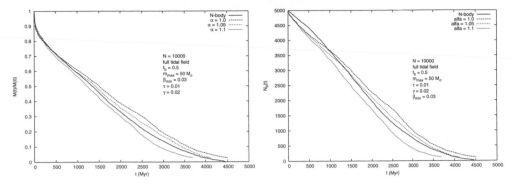

Figure 2. Left panel: evolution of the total mass for N-body and Monte Carlo models for different α. Right panel: evolution of the number of binaries for N-body and Monte Carlo models for different α. Parameters of the models are described in the figures.

critical energy for escaping stars was approximated by: $E_{tid_f} = -\alpha(GM/r_{tid})$, where $\alpha = 1.5 - a(ln(\gamma N)/N)^{1/4}$. So the effective tidal radius for Monte Carlo simulations is $r_{t_{eff}} = r_{tid}/\alpha$ and it is smaller than r_{tid}. This leads that for Monte Carlo simulations a system is slightly too concentrated comparable to N-body simulations, but the evolution of the total mass is reasonably well reproduced.

In Fig. 2 we show the evolution of the total mass and the number of binaries for different α for $N = 10000$.

The value of α inferred from the comparison between N-body and Monte Carlo simulations for $N = 10000$ is equal to about 1.05. The other free parameters for the case of full tidal field are the same as for the tidal cutoff case: $\gamma = 0.02$, $\tau = 0.01$ and $\beta_{min} = 0.03$. Again the spread between models with different β_{min} and τ is well inside the spread connected with different iseed. The statistical spread also does not substantially influence the determination of α. As can be seen in Fig. 2. (right panel) the Monte Carlo code can well reproduce N-body simulations not only in terms of the global parameters of the system, but also in terms of the properties connected with binary activities. Despite the fact that the total number of binaries in the system reasonably well agree with N-body simulations the total binding energy of the binaries increases too quickly in the Monte Carlo simulations. This is connected with the fact that the present Monte Carlo code does not follow directly the 3- and 4-body interactions as N-body code does, but uses cross sections. The coalescence of binaries induced by dynamical interactions and the exchange interactions are missing in the present Monte Carlo simulations.

2.3. *Model of M67*

The data from N-body simulations of M67 (Hurley *et al.* 2005) was used in addition to $N = 2500$ and 10000 to finally calibrate the Monte Carlo code, namely a. The inferred formula is $\alpha = 1.5 - 3.0(ln(\gamma N)/N)^{1/4}$. The comparison of results from N-body and Monte Carlo simulations for M67 confirmed the values of γ, τ and β_{min} found for smaller N systems. The results of comparison are summarized in Table 1. Taking into account the intrinsic statistical fluctuations of both methods the results presented in the Table 1 show reasonably good agreement. At the time of 4 Gyr when the comparison was done, the both models consists of only a small fraction of the initial number of stars (about 10%) making the fluctuations even stronger. The Monte Carlo model is slightly too concentrated compared to the N-body one. This can be attributed to the parameter α, which leads to smaller effective tidal radius than the tidal radius inferred from N-body simulations. As can be seen from Fig. 3 the Monte Carlo code also well reproduces

Table 1. Monte Carlo and N-body results for M67 at 4 Gyr

	N-body	MC
$M/M_\odot$	2037	1984
f_b	0.60	0.59
$r_t\ pc^{-1}$	15.2	15.1
$r_h\ pc^{-1}$	3.8	3.03
$M_L/M_\odot$	1488	1219
$M_{L10}/M_\odot$	1342	1205
$r_{h,L10}\ pc^{-1}$	2.7	2.67

L – stars with mass above $0.5M_\odot$ and burning nuclear fuel
L10 – the same as L but for stars contained within 10 pc

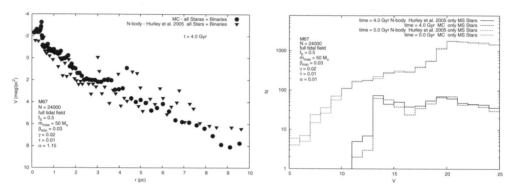

Figure 3. Left panel: surface brightness profile for N-body and Monte Carlo models. Right panel: luminosity function for time equal to 0 and 4 Gyr for N-body and Monte Carlo models. Parameters of the models are described in the figures.

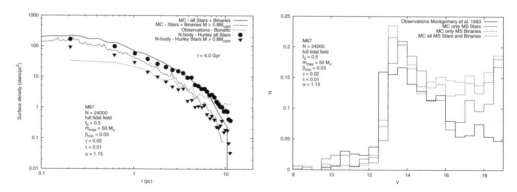

Figure 4. Left panel: surface density profile for Monte Carlo and N-body models and for observations (Bonatto & Bica 2005). Right panel: luminosity function for the Monte Carlo model for time = 4 Gyr and for observations (Montgomery *et al.* 1993) for main sequence stars only, binaries only and both main sequence stars and binaries. Parameters of the models are described in the figures.

the results of N-body simulations regarding the surface brightness profile and luminosity function.

To finally validate the Monte Carlo model of the old open cluster M67 the brief and very preliminary comparison with the observational data (Montgomery *et al.* 1993 and Bonatto & Bica 2005) was performed (Fig. 4). Both models (N-body and Monte Carlo) do not reproduce well the observations. They are too centrally concentrated. Also for the

Monte Carlo model the luminosity function is too shallow for a dim stars and too high for a stars around $V = 13$ *mag*. In order to achieve a better agreement with observation the initial parameters adopted by Hurley *et al.* (2005) have to be slightly changed. Definitely, more work, simulations and observations are needed.

3. Conclusions

It was shown that the Monte Carlo code can be successfully calibrated against small N-body simulations. Calibration was done by choosing the free parameters describing the relaxation process, such as: coefficient in the Coulomb logarithm $\gamma = 0.02$, minimum deflection angle β_{min}, time step τ, and coefficient in the formula for the critical energy of escaping stars, $\alpha = 1.5 - 3.0(ln(\gamma N)/N)^{1/4}$. The calibrated code successfully reproduced the N-body simulations of the old open cluster M67 (Hurley *et al.* 2005), which was the main objective of the calibration procedure. The code is able to provide as detailed data as the observations do. However, it showed also some weaknesses, e.g. some important channels of blue stragglers formation are not present (coalescence of binaries due to their dynamical interactions) and too crude treatment of escape process. The work is in progress to cure these problems (e.g. a few body direct integrations). It was shown also that the Monte Carlo code can be used to model evolution of real star clusters and successfully compare results with observations (see Heggie 2008, in this volume). The very high speed of the code makes it as an ideal tool for getting information about the initial parameters of star clusters. It is worth mentioning that to complete the model of the M67 cluster only about seven minutes is needed!

Acknowledgements

We would like to acknowledge Jarrod Hurley's assistance in implementation of stellar and binary evolution packages into Monte Carlo code. This work was partly supported by the Polish National Committee for Scientific Research under grant 1 P03D 002 27.

References

Baumgardt, H. 2001, *MNRAS* 325, 1323
Bonatto, C. & Bica, E. 2005, *A&A* 937, 483
Drukier, G. A. 1993, *MNRAS* 265, 773
Giersz, M. 2006, *MNRAS* 371, 484
Giersz, M. & Heggie, D. C. 1994, *MNRAS* 268, 257
Giersz, M. & Heggie, D. C. 2003, *MNRAS* 339, 486
Grabhorn, R. P., Cohn, H. N., Lugger, P. M., & Murphy, B. W. 1992, *ApJ* 392, 86
Heggie, D. C., Portegies Zwart, S. & Hurley, J. R. 2006, *New Astron.* 12, 20
Hénon, M. H. 1971, *Ap&SS* 14, 151
Hénon, M. H. 1975, in: Hayli, A., (ed.), *Dynamics of Stellar Systems*, Proc. IAU Symposium No. 69, (Reidel: Dordrecht), p. 133
Hurley, J. R., Pols, O. R., & Taut, C. A. 2000, *MNRAS* 315, 543
Hurley, J. R., Taut, C. A., & Pols, O. R. 2002, *MNRAS* 329, 897
Hurley, J. R., Pols, O. R., Aarseth, S. J., & Taut, C. A. 2005, *MNRAS* 363, 293
Meylan, G. & Heggie, D. C. 1997, *A&AR* 8, 1
Montgomery, K. A., Marschall, L. A., & Janes, K. A. 1993, *AJ* 106, 181
Murphy, B. W., Moore, C. A., Trotter, T. E., Cohn, H. N., & Lugger, P. M. 1998, *BAAS* 30, 1335
Stodółkiewicz, J. S. 1986, *AcA* 36, 19

Dynamical Evolution of Dense Stellar Systems
Proceedings IAU Symposium No. 246, 2007
E. Vesperini, M. Giersz, & A. Sills, eds.

© 2008 International Astronomical Union
doi:10.1017/S1743921308015391

Defining the Binary Star Population in the Young Open Cluster M35 (NGC 2168)

Ella K. Braden[1], Robert D. Mathieu[1], Sören Meibom[2]

[1]Department of Astronomy, University of Wisconsin-Madison, Madison, WI 53706
email: braden@astro.wisc.edu, mathieu@astro.wisc.edu
[2]Harvard-Smithsonian Center for Astrophysics, Cambridge, MA 02138
email: smeibom@cfa.harvard.edu

Abstract. We present current results from the ongoing WIYN Open Cluster Study radial-velocity survey for 1410 stars in the young (150 Myr) open cluster M35 (NGC 2168) and establish a benchmark for initial conditions in young open clusters. We find for periods $\lesssim 1000$ days a minimum binary frequency of $0.36 - 0.51$. We also analyze the spatial, period and eccentricity distributions of the binary systems and find that the period and eccentricity distributions are well approximated by scaled field distributions from Duquennoy & Mayor (1991). With our large sample size and long baseline, we have a unique understanding of the binary population in this young cluster, making it ideal for defining initial conditions for dynamical simulations.

Keywords. Stars: binaries: spectroscopic, Galaxy: open clusters: individual (M35 (NGC 2168))

Initial conditions are crucial to the outcome of dynamical simulations (e. g. Hurley *et al.* 2005). We directly measure the binary frequency and period and eccentricity distributions in the dynamically young cluster M35, a fundamental target of the WIYN Open Cluster Study (Mathieu 2000). Our spectra of 1410 stars in the magnitude range $12.5 < V < 16.5$ ($1.5 M_\odot < M < 0.7 M_\odot$; the turnoff is $V \simeq 9.5$) yield radial velocities with a precision of 0.4 km s^{-1} (Geller *et al.* 2008, in preparation).

We determine cluster membership probabilities for all stars with 3 or more observations using a double Gaussian fit to both the field and cluster velocity distributions, using the center of mass velocities for our 71 binaries with orbital solutions and the average velocity for all other stars. We find 387 members; 243 single and 144 velocity variable systems (of which we have orbital solutions for 45); see Braden, Mathieu & Miebom (2008, in preparation). We find minimum and maximum spectroscopic ($\log(P) \lesssim 3$) binary frequencies of 0.36 ± 0.03 and 0.51 ± 0.05, depending on the membership criteria applied to velocity variable systems.

In Fig. 1 we present the observed period distribution and a scaled distribution of Duquennoy & Mayor (1991); hereafter DM91. Where our sample is complete (to $P \simeq 100$ days), the period distribution is well fit by the DM91 relation. Our observed eccentricity distribution is presented in Fig. 1. The increased number of low-eccentricity orbits is due to tidal circularization of short-period binaries (Meibom & Mathieu 2005). The rest of the distribution is bell-shaped like DM91's sample. This distribution is different than the initial distributions used in cluster simulations to date (e. g. Hurley *et al.* 2005).

Fig. 2 shows the cumulative radial spatial distribution of single and velocity-variable stellar populations in M35. We find that within 4 core radii, single and velocity variable stellar populations cannot be distinguished at the 99% confidence level; the effects of dynamical evolution have not separated the binary and single-star populations in approximately 1 relaxation time (Mathieu 1983). Fig. 2 shows the cumulative radial spatial distributions for proper-motion selected member stars in the $8.0 < V < 14.5$ range (from

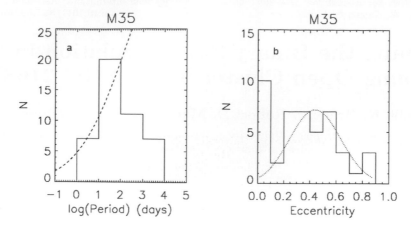

Figure 1. a) Period distribution of binary members systems and the scaled empirical relation of DM91. In the decades where our sample is complete (up to $\log(P) = 2$), the observed number of binary systems matches the relation found by DM91. b) Eccentricity distribution of binary member systems with orbital solutions and a scaled fit to the data in DM91.

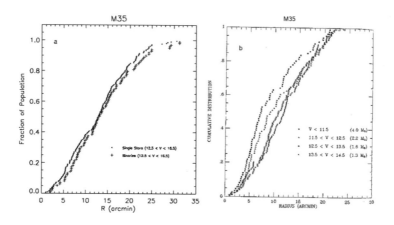

Figure 2. Radial spatial distribution of a) single and binary member systems selected spectroscopically and b) proper-motion selected members divided by magnitude (from Mathieu 1983).

Mathieu 1983) for comparison. While mass segregation is pronounced for the most massive systems ($>2M_\odot$), intermediate-mass stars and the solar-type binary population both show very similar spatial distributions.

References

Duquennoy, A. & Mayor, M. 1991, *A & A*, 248, 485
Hurley, J. R., Pols, O. R., Aarseth, S. J., & Tout, C. A. 2005 *MNRAS*, 363, 293
Mathieu, R. D. 1983 PhD thesis, Univ. Calif. Berkeley
Mathieu, R. D. 2000 *PASP*, 198, 517
Meibom, S. & Mathieu, R. D. 2005 *ApJ*, 620, 970

Dynamical Evolution of Dense Stellar Systems
Proceedings IAU Symposium No. 246, 2007
E. Vesperini, M. Giersz & A. Sills, eds.

Tidal Tails of the Nearest Open Clusters

Yaroslav Chumak[1] and Alexey Rastorguev[2]

[1]Sternberg Astronomical Institute, Universitetskiy prospect 13, Moscow,119992, Russia
email:chyo@mail.ru

[2]Department of Physics, Moscow State University, GSP-2, Leninskiye Gory, Moscow, 119992, Russia
email:rastor@sai.msu.ru

Abstract. We show that an extended population of stars escaping an evolved cluster and moving along its galactic orbit forms at the final phases of its dynamical evolution. Here we present some results of the numerical simulations for nearest open clusters: Hyades, Pleiades, Praesepe, Alpha Persei, Coma, IC 2391, and IC 2602. We calculated the models of the stellar tails for nearest open clusters and estimated some parameters: sizes, densities, locations relative to the solar neighborhood. Stars of the nearest tails can be observed as moving clusters.

Keywords. open clusters and associations: general, methods: n-body simulations

In our simulations (Chumak, Rastorguev, & Aarseth 2005), (Chumak & Rastorguev 2006a) it was shown that stars escaped from a cluster at different times, move very close to cluster's orbit. Their relative velocities are small but relative distances along the cluster orbit change from parsecs, to more than 1.5 kpc. As a consequence, escaped stars form a stretched stellar tail, which can exist for a fairly long time even after full disintegration of the open cluster as its relic. We performed simulations of the dynamical evolution of seven nearest clusters in the tidal field of the Galaxy (Chumak & Rastorguev 2006b). Our computations of the dynamical evolution have been based on the Aarseth's code NBODY6 (Aarseth 2003), with known cluster age estimates and real Galactic orbits. Initial conditions have been chosen in such a way that current parameters of simulated clusters corresponded to their observed parameters (Nordstrom *et al.* 2004).

Fig. 1 (left panel) shows the position of stellar tail relative to the cluster and its orbit for typical cluster with an age 1233 Myr. The right panel shows the close neighborhood of the cluster, ±200 pc.

The results can be briefly summarized as follows:

1. The length of the cluster tails along the Galactic orbit, its thickness (along the Z axis), and its width (along the X axis) for a typical cluster (with $N = 1500$, initial virial radius $RV = 3.5$ pc, an age of 800 Myr) can reach $LY = 900$ pc, $LZ = 20$ pc, and $LX = 100$ pc, respectively.

2. The stars in the tail may be considered as moving clusters, since their mean velocity vector is close to the velocity vector of the parent cluster and their internal velocity dispersion is relatively low($\sim$1 km s^{-1}).

3. Hyades. The Hyades cluster lies inside the circumsolar sphere with the radius of 100 pc, and the cluster tail crosses this sphere and extends far beyond in both directions (see Fig. 2). The approximate length of the tail along the Y axis that falls within the circumsolar sphere is $\sim$160 pc. This sphere contains approximately 150 tail stars.

4. Coma. The tail of Coma cluster also crosses the circumsolar sphere. The approximate length of the tail along the Y axis that falls within the circumsolar sphere is $\sim$80 pc. The sphere contains about 80 tail stars.

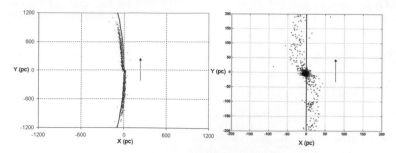

Figure 1. Typical stellar tail for open clusters.

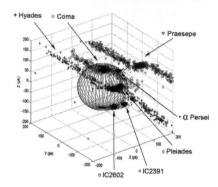

Figure 2. Nearest open star clusters with their tails.

5. Pleiades. The Pleiades cluster lies near the circumsolar sphere (see Fig. 2). However, since the cluster is young, its tail has not yet been completely formed. Our simulations show that the circumsolar sphere may contain several stars.

6. Praesepe, IC 2391, IC 2602, and α Persei clusters. Although Praesepe tail has the length of $\sim$700 pc, it is located entirely outside the circumsolar sphere(see Fig. 2). The clusters IC 2391, IC 2602, and α Persei have virtually no tails and lie fairly far from the circumsolar sphere. Mean relative velocities of the tail stars are low ($\sim$1 km s^{-1}) in the coordinate system associated with the Galactic orbit of the cluster, hence the tail stars may be considered as moving clusters. Thus, dense moving clusters with full space velocity vector close to that of the Hyades and Coma clusters and a sparsely populated cluster with the total velocity vector close to that of the Pleiades can be observed in the circumsolar sphere of radius 100 pc.

Acknowledgements

We thank Dr. O.V. Chumak (Sternberg Astronomical Institute, Moscow State University) and Prof. S. Aarseth (Cambridge University, Great Britain) for the invaluable help and permanent consultations. This work was partly supported by the Russian Foundation for Basic Research (grants 05- 02-16526 and 07-02-00961-) and the Russian Federation President's grant for "Leading Scientific Schools" NSh-389.2003.2.

References

Aarseth, S. 2003, *Gravitational N-Body Simulations* (Cambridge Univ. Press, Cambridge)
Chumak, Ya. O., Rastorguev A. S., & Aarseth, S. 2005, *Astron. Lett.* 31, 6, 308
Chumak, Ya. O. & Rastorguev A. S. 2006, *Astron. Lett.* 32, 3, 157
Chumak, Ya. O. & Rastorguev, A. S. 2006, *Astron. Lett.* 32, 7, 446
Nordstrom B., *et al.* 2004, *A & A* 419, 989

Dynamical Evolution of Dense Stellar Systems
Proceedings IAU Symposium No. 246, 2007
E. Vesperini, M. Giersz & A. Sills, eds.

© 2008 International Astronomical Union
doi:10.1017/S174392130801541X

The WIYN Open Cluster Study Photometric Binary Survey: Initial Findings for NGC 188

P. M. Frinchaboy[1] and D. Nielsen[2]

[1]National Science Foundation Astronomy & Astrophysics Postdoctoral Fellow,
University of Wisconsin–Madison, Department of Astronomy,
4506 Sterling Hall, 475 N. Charter Street, Madison, WI 53706, USA

[2]Department of Physics & Astronomy, Colby College,
860 Mayflower Hill Drive, Waterville, ME 04901
email: frinchaboy@wisc.edu

Abstract. The WIYN open cluster study (WOCS) has been working to yield precise optical ($UBRVI$) photometry for all stars in the field of a selection of "prototypical" open clusters. Additionally, WOCS has been using radial velocities to obtain orbit solutions for cluster member hard-binary stars (with period less than 1000 days). Recently, WOCS has been expanded to include the near-infrared (JHK_s; 2MASS plus new deep ground-based) and mid-infrared ([3.6], [4.5], [5.8], [8.0] micron) photometry from $Spitzer$/IRAC observations. This multi-wavelength data (0.3–8.0 microns) allows us to identify binaries photometrically, with mass ratios from 1.0–0.3, across a wide range of primary masses. The spectral energy distribution (SED) fitter by Robitaille *et al.* (2007) is used to fit the fluxes of 10–12 bands to Kurucz stellar models. *This technique allows us to explore the soft binary population for the first time.* Using this photometric technique, we find that NGC 188 has a binary fraction of 36-49% and provide a star-by-star comparison to the WOCS radial velocity-based hard binary study.

Keywords. open clusters and associations: individual (NGC 188), binaries: general

---

We have combined optical ($UBVRI$, Stetson, McClure & VandenBerg 2004), with NIR (JHK_s 2MASS data, Skrutskie *et al.* 2006) and new deep mid-IR photometry from the Spitzer IRAC for NGC 188. We have restricted our sample to overlap the kinematically-studied WOCS sample containing main sequence (MS) stars ($15.2 < V < 16.5$), with good photometry ($\sigma_{mag} < 0.1$) in all bands, and those with membership probability $\geqslant 80\%$ from the proper motion (PM) analysis of Platais *et al.* (2003) (see Fig. 1a). The spectral energy distribution (SED) fitter by Robitaille *et al.* (2007) is used to fit the fluxes of 10–12 bands to Kurucz (1979) stellar models. The fitted Kurucz models consist of the fluxes of single and two combined MS stars (binaries) with varying mass ratios using $T_{\rm eff}$, $\log(g)$, and masses from Padova isochrones (Girardi *et al.* 2002). This multi-wavelength data (0.3–8.0 μm) allows us to identify binaries *photometrically*, with mass ratios (MR) from 1.0–0.3, across a wide range of primary masses, especially on the faint, lower MS where RV surveys are prohibitive.

We find that NGC 188 has a binary fraction of 36–49%. For the ($15.2 < V < 16.5$) sample, we found 63 of 145 "binary" fits yielding a binary fraction of 43%. However since binaries with MR lower than 0.3 are difficult to distinguish from MS stars, we also determined the binary fraction excluding "binary" fits with MR $\leqslant 0.3$ and found 52 of 145 "binary" fits, as shown in Table 1. We have also compared our results to the spectroscopic binaries (SB) for NGC 188 from Geller *et al.* (*in preparation*), which results in a SB fraction of 31–33%. Due to incompleteness in the Geller *et al.* (*in preparation*) sample, we also analysed the sub-sample ($15.2 < V < 16.0$; Fig. 1a) and found similar binary fractions, shown in Table 1. Direct star-by-star comparison of the method (see Fig. 1c) shows that

Table 1. Statistics of Binaries in NGC 188 using Photometric and Spectroscopic Techniques

Proper Motion Member Sample (Prob > 80% and V > 13.5)	# of Stars	Photom. Binaries	Spectr. Binaries	Spec & Phot Binaries
All Binaries (V < 16.5)	145	63 (43%)	45 (31%)	31 (21%)
Binaries MR > 0.3 (V < 16.5)	145	52 (36%)	45 (31%)	29 (20%)
All Binaries (V < 16.0)	102	50 (49%)	33 (33%)	25 (25%)
Binaries MR > 0.3 (V < 16.0)	102	47 (46%)	33 (33%)	24 (24%)

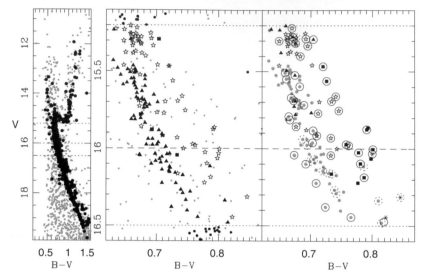

Figure 1. Optical color-magnitude diagram (CMD) for NGC 188 using Stetson, McClure & VandenBerg (2004) data. **a)** Black points have membership probabilities ⩾80%, and grey "non-members" (<80) from Platais *et al.* (2003). The dotted lines delineate the sub-samples of MS stars analysed. **b)** Colors as a) with ⋆ denotes photometric binaries, △ non-binaries, □ possible binaries ($MR < 0.3$). **c)** Grey symbols are non-binaries, black are photometric binaries, having: △: $> 0.95 M_\odot$, ⋆: $(0.9 < M_\odot < 0.95)$, □: $(0.85 < M_\odot < 0.9)$, and *: $(< 0.85 M_\odot)$. Black solid circles denote spectroscopic binaries from Geller *et al.* (*in preparation*), while dotted circles denote stars with insufficient spectroscopic observations to determine if the star is a binary.

we find roughly 2/3 of the SBs using our photometric method, verifying the reliability of our technique.

Acknowledgements

Any opinions, findings, and conclusions or recommendations expressed in this material are those of the author(s) and do not necessarily reflect the views of the National Science Foundation. This project was supported by an NSF Astronomy and Astrophysics Postdoctoral Fellowship under award AST-0602221 and the NSF REU program under NSF Award # 0453442. This work is based on observations made with the Spitzer Space Telescope (GO-3 0800), which is operated by the Jet Propulsion Laboratory, California Institute of Technology under a contract with NASA. Support for this work was provided by NASA through an award issued by JPL/Caltech.

References

Girardi, L., *et al.* 2002, *A&A* 391, 195

Kurucz, R.L. 1979, *ApJS* 40, 1

Platais, I., Kozhurina-Platais, V., Mathieu, R.D., Girard, T M. & van Altena, W.F. 2003, *AJ* 126, 2922

Robitaille, T.P., Whitney, B.A., Indebetouw, R., Wood, K. 2007, *ApJS* 169, 328

Skrutskie, M.F., *et al.* 2006, *AJ* 131, 1163

Stetson, P.B., McClure, R.D. & VandenBerg, D.A. 2004, *PASP* 116, 1012

Dynamical Evolution of Dense Stellar Systems
Proceedings IAU Symposium No. 246, 2007
E. Vesperini, M. Giersz & A. Sills, eds.

© 2008 International Astronomical Union
doi:10.1017/S1743921308015421

Dynamics of the Open Cluster NGC 188: A Comparison to an *N*-body Simulation of M67

Aaron M. Geller[1], Robert D. Mathieu[1], Hugh C. Harris[2] and Robert D. McClure[3]

[1]Department of Astronomy, University of Wisconsin – Madison, WI, 53706, USA;
geller@astro.wisc.edu

[2]United States Naval Observatory, Flagstaff, AZ, 86001, USA

[3]Dominion Astrophysical Observatory, Victoria, B.C., V8X 4M6, Canada

Abstract. We present a detailed dynamical study of the old (7 Gyr) open cluster NGC 188. Our combined radial-velocity data set spans a baseline of 35 years, a magnitude range of $12 \leqslant V \leqslant 16.5$, and a $1°$ diameter region on the sky. Our magnitude limits include solar-mass main-sequence stars, subgiants, giants, and blue stragglers, and our spatial coverage extends radially to 11.5 core radii. We have measured radial velocities for 1014 stars in the direction of NGC 188 with a precision of 0.4 km s^{-1}, and have calculated radial-velocity membership probabilities for stars with $\geqslant 3$ measurements. We find 420 stars to be high-probability cluster members, including 137 spectroscopic binaries. These detectable binaries all have orbital periods of less than 10^4 days, and thus are hard. We have derived orbit solutions for 67 member binary stars, and use our 35 main-sequence binaries with orbit solutions to compare the eccentricity and period distributions with simulated observations of the Hurley *et al.* (2005) model of M67 (4.5 Gyr). We also compare the spatial distributions of cluster member populations.

Keywords. Stars: spectroscopic binaries, blue stragglers, Galaxy: open clusters

1. Introduction

We use precise radial-velocity (RV) observations of the old (7 Gyr) open cluster NGC 188 to perform a detailed study of the hard binary population and the spatial distribution of cluster member populations. We then use these results to conduct the first in-depth comparison of observations to the Hurley *et al.* (2005) M67 simulation (4.5 Gyr). Our combined WIYN 3.5 m and DAO NGC 188 RV data set covers a magnitude range of $12 \leqslant V \leqslant 16.5$ (~0.9 M$_\odot$ to the brightest stars in the cluster), and a $1°$ diameter region on the sky. We have thus limited the Hurley *et al.* (2005) M67 simulation to match this range of observations. Detailed information regarding the analysis and results of our NGC 188 observations can be found in Geller *et al.* (2007, in preparation). Here we simply summarize key results from this project.

2. Results

We have measured RVs for 1014 stars in the direction of NGC 188 with a precision of 0.4 km s^{-1}, and have calculated RV membership probabilities for stars with $\geqslant 3$ measurements. Our careful membership determination reveals 283 single member stars and 137 member binaries, yielding an observed binary fraction of 33%. We also identify 17 blue straggler (BS), 13 of which are RV variables, and 54 giant stars, 20 of which are RV variables. In the magnitude and spatially limited M67 simulation, we find 137 single stars and 249 RV variables, yielding a binary fraction of 65%. We also find 17 BS stars, 6 of which are RV variables, and 30 giants, 15 of which are RV variables. We analyze these stellar populations in the following figures.

2.1. *The Hard Binary Population*

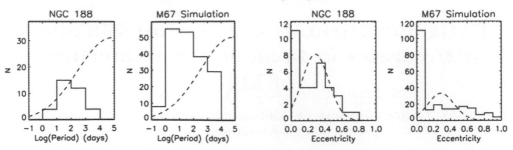

Figure 1. We show the main-sequence binary period (left) and eccentricity (right) distributions for NGC 188 and the M67 simulation in histogram form. We have also over-plotted the Duquennoy & Mayor (1991) field binary distributions normalized to each cluster as a fiducial line. The hard-soft boundary in these clusters is on the order of 10^4 days. The binary population of the M67 simulation is markedly different from the NGC 188 binary population as well as the field. Note the decrease in binaries with periods >100 days in the model, possibly due to dynamical interactions within the open cluster. Currently we observe a similar departure from the field distribution in this period range in NGC 188. However, we do not yet know if this is due to dynamics or incompleteness in our sample. The M67 model eccentricity distribution shows a much larger fraction of circular binaries than NGC 188, and the distribution follows a much different shape than both NGC 188 and the field binary population.

2.2. *Spatial Distribution*

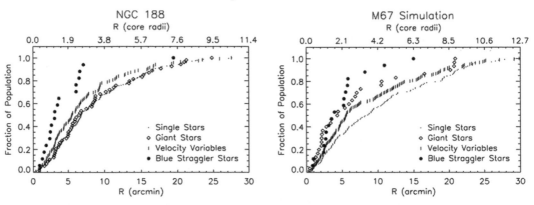

Figure 2. We plot the spatial distributions of member star populations in NGC 188 and the M67 model. We have limited to M67 simulation to include the same mass range of stars as observed in NGC 188. The single, BS, giant, and RV variable star populations are compared. For NGC 188 we conclude that both the BS stars and binaries are centrally concentrated, providing solid evidence for mass segregation. In the M67 simulation, we find the giants ($\overline{M} \sim 1.37$ M$_\odot$), BS stars ($\overline{M} \sim 1.56$ M$_\odot$), and RV variables ($\overline{M} \sim 1.68$ M$_\odot$) to be centrally concentrated. The M67 model average single star mass is $\sim$1.17 M$_\odot$. This central concentration is evidence for mass segregation in the M67 model. However, it is not clear why the giant stars do not appear centrally concentrated in NGC 188 while they do in the M67 model.

Acknowledgements

This work was funded by the Wisconsin Space Grant Consortium and the National Science Foundation grant number AST-0406615.

References

Duquennoy, A. & Mayor, M. 1991, *A&A*, 248, 485
Hurley, J. R., Pols, O. R., Aarseth, S. J., & Tout, C. A. 2005, *MNRAS*, 363, 293

Dynamical Evolution of Dense Stellar Systems
Proceedings IAU Symposium No. 246, 2007
E. Vesperini, M. Giersz & A. Sills, eds.

© 2008 International Astronomical Union
doi:10.1017/S1743921308015433

NIR Spectroscopy of the Most Massive Open Cluster in the Galaxy: Westerlund 1

S. Mengel[1] and L. E. Tacconi-Garman[1]

[1]ESO, Karl-Schwarzschild-Str. 2, 85748 Garching, Germany
email: smengel@eso.org, ltacconi@eso.org

Abstract. Using ISAAC/VLT, we have obtained individual spectra of all NIR-bright stars in the central $2' \times 2'$ of the cluster Westerlund 1 (Wd 1) with a resolution of $R \approx 9000$ at a central wavelength of $2.30\,\mu m$. This allowed us to determine radial velocities of ten post-main-sequence stars, and from these values a velocity dispersion. Assuming virial equilibrium, the dispersion of $\sigma = 8.4$ km/s leads to a total dynamical cluster mass of $1.25 \times 10^5 M_\odot$, comparable to the photometric mass of the cluster. There is no extra-virial motion which would have to be interpreted as a signature of cluster expansion or dissolution.

Keywords. open clusters and associations: individual: (Westerlund 1), galaxies: star clusters, supergiants

1. Observations and Results

We used ISAAC/VLT in its highest spectral resolution mode ($R \approx 9000$) to step the $2'$ long slit across the centre of the cluster for $2'$, in steps of $0.3''$. At each position, the integration time was 3×1.77s. The central wavelength of $2.31\,\mu m$ was optimized for the analysis or red supergiant (RSG) spectra, which show deep CO absorption features in this regime. The data were sky subtracted, flat fielded, distortion corrected and wavelength calibrated using standard *iraf* routines and, where necessary, calibration observations. Individual spectra were extracted from those slit positions where a star with photospheric features in our wavelength range was included. After the extraction of the 1d spectrum, we corrected for telluric absorption. Fig. 1 shows the ten spectra which were extracted from our scan, four red supergiants (RSGs), five yellow hypergiants (YHGs), and one sgB[e] star.

We used the lines in these spectra to determine the radial velocities and their dispersion, which is $\sigma = 8.4$ km/s. This, together with the half-light radius (estimated from archival NTT/SOFI images), assuming that the Virial Theorem is valid, yields a dynamical cluster mass of $1.3 \times 10^5 M_\odot$ (see also Mengel & Tacconi-Garman 2007).

Comparing the light-to-mass-ratio (V-band luminosity and dynamical mass) to that expected for a cluster of this age and luminosity, and a Kroupa IMF, solar metallicity, allows a conclusion regarding the dynamical state of the cluster: Even though it could in principle also be an indication of an excess of low-mass stars (compared to the Kroupa IMF), the more likely explanation for low L/M is that the high "dynamical mass" arises from extra-virial motion, and hence an expanding cluster. Fig. 2 shows a plot taken from Goodwin & Bastian (2006), where they determine the velocity dispersion of the stars in their N-body simulated clusters and plot the resulting L/M ratios as dashed lines. Clusters with a low star-formation efficiency show large amounts of extra-virial motion, and do not survive as bound clusters.

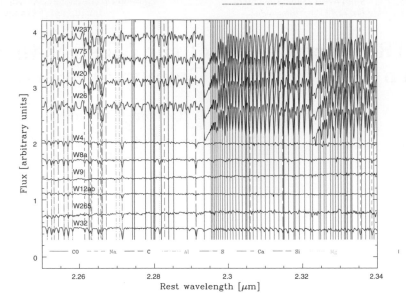

Figure 1. Normalized, vertically offset spectra of all detected cluster members with absorption features in our observed wavelength regime. Top four are the RSGs, five of the six below are YHGs (W9 is a sgB[e] star).

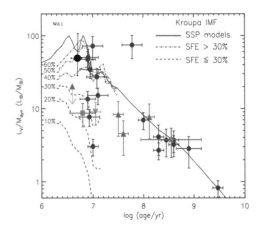

Figure 2. L_V/M_{dyn} as a function of age, for "static" clusters (solid) and clusters adjusting dynamically after gas expulsion (from Goodwin & Bastian (2006)), dashed. Wd 1 (black dot) does not seem to be expanding dramatically.

The L/M ratio for Westerlund 1 is consistent with the expectation from the evolutionary synthesis model, which means that the cluster may survive for a long time, possibly several Gyr.

The most important caveat in our measurement is the low number of stars, which is why we want to improve statistics by obtaining radial velocities also for the hot stars in the cluster, from absorption lines around Brγ.

References

Goodwin, S. P & Bastian, N. 2006, *MNRAS* 373, 752
Mengel, S. & Tacconi-Garman, L. E. 2007, *A&A* 466, 151

Dynamical Evolution of Dense Stellar Systems
Proceedings IAU Symposium No. 246, 2007
E. Vesperini, M. Giersz & A. Sills, eds.

© 2008 International Astronomical Union
doi:10.1017/S1743921308015445

The Population of Open Clusters in the Nearest kpc from the Sun

S. Röser[1], N. V. Kharchenko[1,2,4], A. E. Piskunov[1,3,4], E. Schilbach[1] and R.-D. Scholz[4]

[1] Astronomisches Rechen-Institut, Moenchhofstr. 12-14, 69120 Heidelberg, Germany
email: roeser@ari.uni-heidelberg.de

[2] Main Astronomical Observatory, Kiev, Ukraine

[3] Institute of Astronomy of the Russian Acad. Sci., Moscow, Russia

[4] Astrophysikalisches Institut Potsdam, Potsdam, Germany

Abstract. We present a *volume-limited sample (d < 850 pc) of open clusters* in the Galaxy identified from our studies on galactic open clusters based on data from the all-sky catalogue ASCC-2.5 with absolute proper motions and B, V magnitudes of 2.5 million stars. The astrophysical properties of this sample are discussed.

Keywords. open clusters and associations: general, Galaxy: kinematics and dynamics, techniques: photometric, astrometry

1. The basic material

The basic idea underlying our studies is to use a complete homogeneous sky-survey comprising astrometric, photometric, and, partially, spectroscopic data to investigate all open clusters with the same objective methods. This sky-survey is the All-Sky Compiled Catalogue of 2.5 million stars (Kharchenko 2001) with absolute proper motions in the Hipparcos system, with B, V magnitudes in the Johnson photometric system, and supplemented with spectral types and radial velocities if available. The ASCC-2.5 was used to identify known open clusters and compact associations from the Lund Catalogue (Lyngå 1987), the Dias *et al.* (2004) on-line data collection, and the Ruprecht, Balazs & White (1981) list of associations. In the ASCC-2.5 we found 520 of about 1700 known clusters (Kharchenko *et al.* 2005a), and discovered 130 new open clusters (Kharchenko *et al.* 2005b). A pipeline was developed to determine cluster membership based on kinematic and photometric criteria as well as to obtain a uniform set of cluster structural, kinematic and evolutionary parameters (see Kharchenko *et al.* 2004 and Kharchenko *et al.* 2005a).

2. Spatial distribution and kinematics

This sample of 650 open clusters from ASCC-2.5 contains a *volume-limited sub-sample of 256 clusters* within a distance of 850 pc from the Sun (see Piskunov *et al.* 2006). The symmetry plane of the clusters' distribution is determined to be at $Z_0 = -22 \pm 4$ pc, and the scale height of open clusters is only 56 ± 3 pc. Within the completeness limit, the total surface density and volume density in the symmetry plane are $\Sigma = 114$ kpc^{-2} and $D(Z_0) = 1015$ kpc^{-3}, respectively.

From the parameters of the spatial distribution, we estimate a total number of 10^5 open clusters currently in the Galactic disk. The lifetime and formation rate of clusters obtained from the age distribution of field clusters within the completeness limit are found to be 322 ± 31 Myr and 0.23 ± 0.03 kpc^{-2}Myr^{-1}, respectively. This implies a total number of cluster generations in the history of the Galaxy between 30 to 40. Assuming a typical open cluster of the Pleiades type, we derive the total surface density of disk stars passed through the phase of open cluster members to be about 4×10^6 kpc^{-2}. Compared to the local density of disk stars of about 7×10^7 kpc^{-2}, the input of open clusters into the total population of the Galactic disk is found to be less than about 10%.

We determined kinematic parameters describing the basic motions of the system of open clusters in the Galactic disk: the motion with respect to the Sun, and the differential rotation around the Galactic centre. The Solar motion components $U_\odot(+9.44 \pm 1.14$ km/s$)$, $V_\odot(+11.90 \pm 0.72$ km/s$)$, $W_\odot(+7.20 \pm 0.42$ km/s$)$ were derived from the complete kinematic data of 259 clusters located within 850 pc. Oort's constants A ($+14.5 \pm 0.8$ km/s/kpc) and B (-13.0 ± 1.1 km/s/kpc) of the Galactic rotation were computed from the proper motions μ_l, μ_b of 581 clusters with distances $d \leqslant 2500$ pc. These results are comparable to those obtained from young field stars in the Solar neighbourhood though, the clusters cover distances which typically twice as large as Hipparcos-based samples of field stars. A combination of accurate cluster ages and kinematical parameters provides a possibility to study the temporal variation of the cosmic velocity dispersion. On average, the dispersion of each velocity component increases by a factor of two over a time of 3 Gyr (maximum age of the clusters of our sample).

Fluctuations in the spatial and velocity distributions are attributed to the existence of four open cluster complexes (OCCs) of different ages containing up to a few tens of clusters. Members in an OCC show the same kinematic behaviour, and a narrow age spread. We find, that the youngest cluster complex, OCC 1 ($\log t < 7.9$, 23 clusters), with 19° inclination to the Galactic plane, is apparently a signature of Gould's Belt. The most abundant OCC 2 complex (27 members) has moderate age ($\log t \approx 8.45$). The clusters of the Perseus-Auriga group (8 members), having the same age as OCC 2, but different kinematics, are seen in breaks between Perseus-Auriga clouds. The oldest ($\log t \approx 8.85$) group (9 members) was identified due to its large motion in the Galactic anticentre direction (for a detailed discussion see Piskunov *et al.* 2006).

Acknowledgements

This study was supported by DFG grant 436 RUS 113/757/0-2, and RFBR grant 06-02-16379.

References

Dias, W. S., Lépine, J. R. D., Alessi, B. S., & Moitinho, A. 2004, *Open clusters and Galactic structure*, Version 2.0, http://www.astro.iag.usp.br/~wilton

Kharchenko, N. V. 2001, *Kinematics and Physics of Celestial Bodies* 17, 409 (ASCC-2.5, Cat. I/280A)

Kharchenko, N. V., Piskunov, A. E., Röser, S., Schilbach, E., & Scholz, R.-D. 2004, *Astronomische Nachrichten* 325, 740

Kharchenko, N. V., Piskunov, A. E., Röser, S., Schilbach, E., & Scholz, R.-D. 2005a, *A&A* 438, 1163

Kharchenko, N. V., Piskunov, A. E., Röser, S., Schilbach, E., & Scholz, R.-D. 2005b, *A&A* 440, 403

Lyngå, G. 1987, *Catalogue of open clusters data*, Fifth edition, CDS, Strasbourg (Cat. VII/92)

Piskunov, A. E., Kharchenko, N. V., Röser, S., Schilbach, E., & Scholz, R.-D. 2006, *A&A* 445, 545

Ruprecht, J., Balazs, B., & White, R. E. 1981, *Catalogue of Star Clusters and Associations*, Supplement 1, Associations, Akademiai Kiado, Publ. House Hungarian Acad. Sciences, Budapest

Dynamical Evolution of Dense Stellar Systems
Proceedings IAU Symposium No. 246, 2007
E. Vesperini, M. Giersz & A. Sills, eds.

Tidal Radii and Masses of Galactic Open Clusters

Elena Schilbach[1], Nina V. Kharchenko[2], Anatoly E. Piskunov[3], Siegfried Röser[1] and Ralf-Dieter Scholz[4]

[1] Astronomisches Rechen-Institut, Moenchhofstr. 12-14, 69120 Heidelberg, Germany
email: elena@ari.uni-heidelberg.de

[2] Main Astronomical Observatory, Kiev, Ukraine

[3] Institute of Astronomy of the Russian Acad. Sci., Moscow, Russia

[4] Astrophysikalisches Institut Potsdam, Potsdam, Germany

Abstract. For 236 of 650 Galactic open clusters identified in the ASCC-2.5 catalogue, we determine tidal radii from a three-parameter fit of King's profiles to the observed integrated density distribution of cluster members. The results are used to calibrate the observed sizes of the remaining clusters to a uniform scale of tidal radii of open clusters in the Solar neighbourhood. The tidal masses are computed from tidal radii. Within a distance of 850 pc where our sample is complete, the observed distributions of cluster masses can be explained by a general mass loss in open clusters with increasing age.

Keywords. open clusters and associations: general, Galaxy: kinematics and dynamics

1. Data and methods

For the determination of tidal radii of open clusters, we used the homogeneous set of cluster parameters derived for 650 open clusters with reliable membership based on data of the ASCC-2.5 catalogue (for more details see Röser *et al.*, this volume). For each cluster, we constructed integrated density profiles of cluster members corrected for the background. The profiles were fitted with three-parameter King's profiles in the integrated form

$$n(r) = \pi \, r_c^2 k \left\{ \ln[1 + (r/r_c)^2] - 4 \frac{[1 + (r/r_c)^2]^{1/2} - 1}{[1 + (r_t/r_c)^2]^{1/2}} + \frac{(r/r_c)^2}{1 + (r_t/r_c)^2} \right\} \qquad (1.1)$$

where $n(r)$ is the number of stars within a circle of radius r, and r_c, r_t, k are the unknowns i.e., core radius, tidal radius, and normalization factor, respectively. As a result, we obtained King's parameters together with their *rms* errors for 236 open clusters of our sample (Piskunov *et al.* 2007).

Since King's method assumes spherical systems, we checked whether ellipticity of open clusters can impact the determination of tidal radii. We found that the orientation of clusters in the Galaxy is random and the average ellipticity is too small to produce a prominent systematic bias. We also checked if a distance dependent bias is present. To estimate the effect, we constructed a semi-empirical model of clusters based on apparent luminosity functions of cluster members and field stars. "Moving" clusters away from the Sun, we found that, on average, the standard approach produces smaller tidal radii with increasing distance. The original tidal radii were corrected for this bias.

The subsample of 236 clusters with tidal radii obtained with the King model is not complete to a given distance from the Sun. A relation between the measured tidal radii

E. Schilbach *et al.*

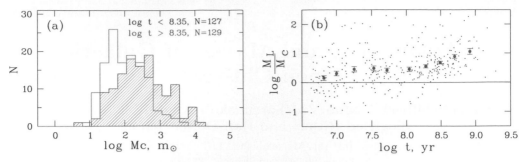

Figure 1. Masses of open clusters. **(a)**: Distribution of tidal masses of 256 open clusters within a distance of 850 pc. The hatched histogram is for clusters younger than 225 Myr, open – for clusters older than 225 Myr. **(b)**: Comparison of the tidal masses M_c with masses M_L by Lamers *et al.* 2005. Large circles with error bars indicate a running average of $\log M_L/M_c$ with a $(\log t)$-bin of 0.5 and a step of 0.25.

r_t^m and the semi-major axis of the projected distribution of cluster members of 236 clusters was used to compute calibrated tidal radii r_t^c for all 650 clusters. No systematic differences were found between r_t^m and r_t^c depending on cluster distances or ages.

2. Tidal masses of open clusters

For each cluster, the tidal mass M_c was computed from the tidal radius as

$$M_c = \frac{4\,A\,(A-B)\,r_t^3}{G} \tag{2.1}$$

where A, B are Oort's constants valid at the galactocentric distance of the cluster, and G is the gravitational constant.

Within a distance of 850 pc where our sample is complete, about 70% of the clusters have masses $log(M_c/M_\odot)$ between 1.5 and 2.8. However, the mass distributions of clusters younger and older than 225 Myr show significant differences (Fig. 1a). The asymmetric shape and shift of the histogram of the older group indicate a general mass loss in open clusters with increasing age. On average, the dependence of tidal masses from the cluster age can be expressed as $\log M_c = (-0.34 \pm 0.07) \times (\log t - 6) + (2.99 \pm 0.15)$.

A hint at MF evolution in open clusters is also obtained from a comparison of our tidal masses with mass estimates M_L based on star counts (Lamers *et al.* 2005). M_L were derived by extrapolation of the normalized Salpeter IMF to low mass stars down to $0.15\,m_\odot$. Both mass estimates are in agreement for the youngest clusters where dynamical evolution had no time to change the IMF (Fig. 1b). However, at $\log t > 7.25$ the difference becomes significant. This indicates that, already in relatively young clusters, the actual MF differs from the Salpeter IMF, and the difference is increasing with cluster age.

Acknowledgements

This work is supported by DFG grant 436 RUS 113/757/0-2 and RFBR grant 06-02-16379.

References

Lamers, H. J. G. L. M., *et al.* 2005, *A&A* 441, 117
Piskunov, A., Schilbach, E., Kharchenko, N., Röser, S., & Scholz, R. 2007, *A&A* 468, 151

Part 3

Globular Clusters

Dynamical Evolution of Dense Stellar Systems
Proceedings IAU Symposium No. 246, 2007
E. Vesperini, M. Giersz & A. Sills, eds.

© 2008 International Astronomical Union
doi:10.1017/S1743921308015469

Modelling Individual Globular Clusters

Douglas C. Heggie[1] and Mirek Giersz[2]

[1] School of Mathematics and Maxwell Institute for Mathematical Sciences, University of
Edinburgh, Edinburgh, EH9 3JZ, UK
email: d.c.heggie@ed.ac.uk

[2] Nicolaus Copernicus Astronomical Center, Polish Academy of Sciences, Warsaw, Poland
email: mig@camk.edu.pl

Abstract. Astronomers have constructed models of globular clusters for over 100 years. These models mainly fall into two categories: (i) static models, such as King's model and its variants, and (ii) evolutionary models. Most attention has been given to static models, which are used to estimate mass-to-light ratios and mass segregation, and to combine data from proper motions and radial velocities. Evolutionary models have been developed for a few objects using the gaseous model, the Fokker-Planck model, Monte Carlo models and N-body models. These models have had a significant role in the search for massive black holes in globular clusters, for example.

In this presentation the problems associated with these various techniques will be summarised, and then we shall describe new work with Giersz's Monte Carlo code, which has been enhanced recently to include the stellar evolution of single and binary stars. We describe in particular recent attempts to model the nearby globular cluster M4, including predictions on the spatial distribution of binary stars and their semi-major axis distribution, to illustrate the effects of about 12 Gyr of dynamical evolution. We also discuss work on an approximate way of predicting the "initial" conditions for such modelling.

Keywords. methods: numerical, globular clusters: general, globular clusters: individual (M4)

---

1. Introduction

1.1. *Some astrophysical questions*

There are many reasons for constructing a dynamical model of a globular cluster, but they fall into two broad categories. First there are problems that can be tackled by constructing static (equilibrium) models, such as

(*a*) Inferring the mass from the surface brightness profile, radial velocities, proper motions and mass functions. In this way one can estimate the total mass of stars below the observational limit, such as faint white dwarfs (e.g. Drukier *et al.* 1988)

(*b*) Inferring the global mass function from local mass functions (e.g. Richer *et al.* 2004): then one can address the question of whether this is the same for all clusters.

(*c*) Measuring cluster distances by comparison of radial velocities and proper motions: the model is used to correct for rotation, or to link the different locations and stellar components which are observed by the different techniques (e.g. van de Ven *et al.* 2006).

Then there is a second range of questions which require dynamic *evolutionary* models, i.e. questions such as

(*a*) Inferring the primordial mass function from local, present-day mass functions: the model is used to correct for preferential escape of low-mass stars (e.g. Baumgardt & Makino 2003).

(*b*) Inferring primordial parameters of the binaries from their present-day statistical properties: primordial abundance, period distribution, etc (e.g. Kroupa *et al.* 2001) (*c*) Determining the effect of dynamics on the estimation of mass through the virial theorem, which is affected by mass segregation (Fleck *et al.* 2006)

Note that these last few references do not deal with specific objects (which is the focus of the rest of this review) but with general trends.

This paper begins with a review of the methods and observational constraints which have been used to construct models of individual globular clusters, often with a view to answering the above types of question. Then we focus on one particular method, the Monte Carlo method, and its application to the nearby globular cluster M4.

1.2. *Methods and constraints*

1.2.1. *The methods*

A number of techniques have been used to construct static models, to answer questions of the first type:

(*a*) Plummer's model (Plummer 1911)

(*b*) King's model (King 1966; Peterson & King 1975)

(*c*) Anisotropic models (Michie & Bodenheimer 1963)

(*d*) Multi-mass models (Gunn & Griffin 1979; Pryor *et al.* 1986; Dubath *et al.* 1990)

(*e*) Non-parametric models (Gebhardt & Fischer 1995)

(*f*) Schwarzschild's method (van de Ven *et al.* 2006)

(*g*) Jeans' equations (Leonard *et al.* 1992)

Even this list may not be exhaustive.

Much less work has been done on dynamical evolutionary models of *specific* globular clusters, but the methods include

(*a*) Gas/fluid models (Angeletti *et al.* 1980 [M3])

(*b*) Fokker-Planck models (Cohn and co-workers: Grabhorn *et al.* 1992 [N6624], Dull *et al.* 1997 [M15]; Drukier 1993, 1995 [N6397]; Phinney 1993 [M15])

(*c*) Monte Carlo models (Giersz & Heggie 2003 [ω Cen])

(*d*) *N*-body models

The last of these should really not be on this list. Though it should be the method of choice, it has not been used for the purposes which are the focus of this paper. An example is the modelling of M15 by Baumgardt *et al.* (2003), who constructed a small version which, when scaled up, corresponded approximately to the conditions expected for M15. The difficulty is that unscaled *N*-body models are practically limited to *N* of order 10^5 at present (e.g. Baumgardt & Makino 2003; Hurley 2007), whereas the median for the globular clusters at the present time is about 5×10^5. Though it might be hoped that the gap would be bridged by the next generation of computers, it must be recognised that all globular clusters at the present day have lost substantial numbers of stars, and the primordial median must have been higher. Note that it has become possible only relatively recently to carry out full simulations of open star clusters. The initial mass of M67, for instance, is estimated at about 19 000 $M_\odot$, i.e. about 10 times its present mass, and this simulation, with a realistic complement of primordial binaries, took of order 1 month (Hurley *et al.* 2005).

1.2.2. *The observational constraints*

Whichever method is chosen to model a globular cluster, there are a number of observational constraints to be satisfied. In historical order of first use we have

(*a*) Surface brightness and/or star counts, starting with Von Zeipel (1908);

(*b*) Radial velocities, whether central averaged values (Illingworth 1976) or radial velocities of individual stars (Gunn & Griffin 1979)

(*c*) Pulsar accelerations (Phinney 1993; Grabhorn 1993)

(*d*) *Deep* luminosity/mass functions, starting essentially with the advent of studies by several authors using HST (1995)

(*e*) Accurate proper motions (van den Bosch *et al.* 2006). (We refer here to models built to satisfy constraints imposed by observations of internal proper motions, and not to the use of proper motions to establish membership.)

1.2.3. *The Monte Carlo Model*

This paper focuses on an application of the Monte Carlo code developed essentially by Giersz (1998, 2001, 2006). In this approach we assume spherical symmetry and dynamic equilibrium, and characterise each star by its energy E and angular momentum J. The code repeatedly alters E and J to mimic the effects of gravitational encounters, using the theory of relaxation. The same theory underpins Fokker-Planck codes, and the basic Monte Carlo code provides essentially a Monte Carlo solution of the Fokker-Planck equation.

The Monte Carlo code is rather suitable for the addition of a number of other process, chiefly:

(*a*) The galactic tidal field, which is treated as a cutoff

(*b*) Binaries, whose interactions are treated using cross sections (from Spitzer (1987) for interactions with single stars, and expressions based on Mikkola (1983, 1984a,b) for interactions between binaries)

(*c*) Stellar evolution of single stars (Hurley *et al.* 2000) and binary stars (Hurley *et al.* 2002).

Each of these requires some comment, though further details are given by Giersz & Heggie in this volume, and in Sec. 4 below.

(*a*) There are significant differences between a tidal field and a tidal cutoff, as these lead to somewhat different scalings of the dissolution time with N (Baumgardt 2001). We have attempted to mimic this with a mass-dependent lowering of the escape energy.

(*b*) The Monte Carlo code described here has no triples, and so hierarchical triples (which are a common product of binary-binary encounters) have to be bypassed. Furthermore, the use of cross sections hinders the inclusion of star-star collisions during 3- and 4-body encounters. Finally, the cross sections are not well known for unequal masses.

(*c*) Stellar evolution is implemented via the McScatter interface (Heggie *et al.* 2006). Besides the stellar evolution packages of Hurley *et al.* (referenced above) it also interfaces to the stellar evolution package SeBa in starlab (Portegies Zwart & Verbunt 1996). At present, however, the latter is limited to solar metallicity, which is unsuitable for the study of old globular clusters.

In view of the above approximations and uncertainties, the testing and calibration of the Monte Carlo code against the results of N-body models (in the regime of small enough N) is an essential safeguard. Such studies are described by Giersz & Heggie in this volume.

2. The globular cluster M4

This, one of the very nearest known globular clusters, was selected at the meeting MODEST-5 (Hamilton, Canada, 2004) as a target for concerted observational and theoretical effort, but so far little theoretical work has been carried out. Table 1 summarises some data for this fascinating object.

The proximity of M4 makes deep observational study possible. (See, for example, the poster by Sommariva *et al.* in this volume). For theoretical purposes too it is well placed for study because its binary population appears to be modest, and its initial mass may

Table 1. Properties of M4

Distance from sun	[a] 1.72 kpc
Distance from GC	5.9 kpc
Mass	[a] 63 000$M_\odot$
Core radius	0.53 pc
Half-light radius	2.3 pc
Tidal radius	21 pc
Half-mass relaxation time (R_h)	660 Myr
Binary fraction	[a] 1–15%
[Fe/H]	−1.2
Age	[b] 12 Gyr
A_V	[a] 1.33

References: All data are from the current version of the catalogue of Harris (1996), except [a] Richer *et al.* (2004) (though this is not always the original reference for the quoted number) and [b] Hansen *et al.* (2004).

not have been very high, as we shall see. One complication for the Monte Carlo code, however, is that the orbit appears to be very elliptical (Dinescu *et al.* 1999), whereas we must assume a steady tidal field.

Table 2 describes the initial conditions which we adopted for this exercise. The primary observational data which we attempted to fit were

(*a*) The surface brightness profile (Trager *et al.* 1995)

(*b*) The radial velocity dispersion profile (Peterson *et al.* 1995)

(*c*) The V-luminosity function (Richer *et al.* 2004, from which we considered the results for the innermost and outermost of their four annuli)

though several other observational comparisons will be described below. We do not have a systematic way of arriving at a best choice of initial parameters, though a possible approach is described towards the end of this paper. We began with a scaled-up version of the models we developed for the old open cluster M67 (see Giersz & Heggie in this volume), but found that a binary population of $f_b = 50\%$ tended to produce a model with too low a concentration. Reducing the binary concentration to 5 or 10% produced a satisfactory surface brightness profile, but was somewhat too massive, because of an excess of low-mass stars, corresponding to a poor fit with the luminosity function. According to Baumgardt & Makino (2003) it might be possible to correct this by devising a model which lost mass at a higher rate, but instead we elected to change the slope of the low-mass IMF from the canonical value of $\alpha = 1.3$ (Kroupa 2007) to $\alpha = 0.9$. (There is some justification for a lower value for low-metallicity populations.) By some experimentation we arrived at a model which gave a fair fit to all three kinds of observational data; see Table 3, and Figs. 1–4. Much of the disagreement in the total luminosity is due to our assumed distance to the cluster, which is significantly smaller than the value of 2.2kpc given by Harris (1996), though our surface brightness profile is also a little faint on average. The disagreement in the inner luminosity function at faint magnitudes may be attributable to the fact that the theoretical result assumes 100% completeness, while the observational data are uncorrected for completeness. A typical plot of a completeness correction is given by Hansen *et al.* (2002, Fig. 3).

It is worth noting that no arbitrary normalisation has been applied in these comparisons between our model and the observations. The surface brightness profile, for example, is computed directly from the V-magnitudes of the stars in the Monte Carlo simulation.

Fig. 5 shows the colour-magnitude diagram of the model. This is of interest, not so much for comparison with observations, but for the presence of a number of interesting

Table 2. Initial parameters for M4

Fixed parameters	
Structure	Plummer model
Stellar IMF	Kroupa double power law
Binary mass distribution	Kroupa *et al.* (1991)
Binary mass ratio	Uniform
Binary semi-major axis	Uniform in log, $2(R_1 + R_2)$ to $50\,\mathrm{AU}$
Binary eccentricity	Thermal, with eigenevolution (Kroupa 1995)
Metallicity Z	0.002
Age	$12\,\mathrm{Gyr}$

Free parameters	
Mass	M
Tidal radius	r_t
Half-mass radius	r_h
Binary fraction	f_b
Slope of the lower mass function	α (Kroupa = 1.3)

Table 3. Monte Carlo and King models for M4

Quantity	MC model $(t=0)$	MC model $(t=12\mathrm{Gyr})$	King model (Richer *et al.* 2004)
Mass $(M_\odot)$	3.40×10^5	4.61×10^4	
Luminosity $(L_\odot)$	6.1×10^6	2.55×10^4	6.25×10^4
Binary fraction f_b	0.07	0.057	0
Low-mass MF slope α	0.9	0.03	0.1
Mass of white dwarfs $(M_\odot)$	0	1.81×10^4	$3.25 \times 10^{4*}$
Mass of neutron stars $(M_\odot)$	0	3.24×10^3	
Tidal radius r_t (pc)	35.0	18.0	
Half-mass radius r_h (pc)	0.58	2.89	

∗: this is the quoted mass of "degenerates"

features. The division of the lower main sequence is simply an artifact of the way binary masses were selected (a total mass above $0.2M_\odot$ and a component mass above $0.1M_\odot$.) Of particular interest are the high numbers of merger remnants on the lower white dwarf sequence. There are very few blue stragglers. Partly this is a result of the low binary frequency, but it is also important to note that some formation channels are unrepresented in our models (in particular, collisions during triple or four-body interactions, though if a binary emerges from an interaction with appropriate parameters, it will be treated as merged.) These numbers also depend on the assumed initial distribution of semi-major axis, which is not yet well constrained by observations in globular clusters.

Photometric binaries are visible in Fig. 5, and these are compared with observations in the inner field of Richer *et al.* (2004) in Fig. 6. In this figure, the model histogram has been normalised to the same total number of stars as the observational one. We made no attempt to simulate photometric errors, but the bins around abscissa $= -0.75$ suggest that the binary fractions in the model and the observations are comparable. Fig. 7 shows that binaries have evolved dynamically as well as through their internal evolution. In particular the softest pairs been almost destroyed.

By 12 Gyr the binaries exhibit segregation towards the centre of the cluster, but perhaps in more subtle ways than might be expected (Figs. 8,9). When *all* binaries are considered, there is little segregation relative to the other objects in the system. (Most binaries in our model are of low mass.) But if one restricts attention to bright binaries,

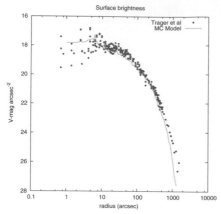

Figure 1. Surface brightness profile of our Monte Carlo model, compared with the data of Trager *et al.* (1995).

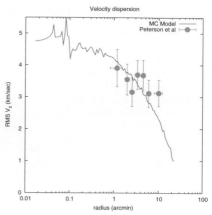

Figure 2. Velocity dispersion profile of our Monte Carlo model, compared with the data of Peterson *et al.* (1995).

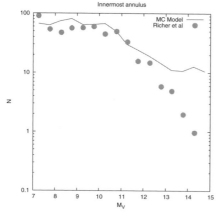

Figure 3. Luminosity function of our Monte Carlo model at the median radius of the innermost annulus in Richer *et al.* (2004), compared with their data.

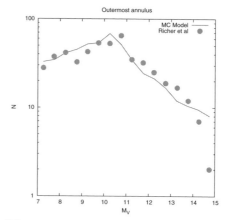

Figure 4. Luminosity function of our Monte Carlo model at the median radius of the outermost annulus in Richer *et al.* (2004), compared with their data.

which we here take to mean those with $M_V < 7$ (i.e. brighter than about two magnitudes below turnoff), the segregation is very noticeable (Fig. 9), with a half-mass radius smaller by almost a factor of 2 than for bright single stars. Still, bright binaries are not nearly as mass-segregated as neutron stars (Fig. 8), which, incidentally, receive no natal kicks in our model.

These data do not reveal one very interesting feature of our model, which is that it exhibited core collapse at about 8 Gyr. Subsequently its core radius is presumably sustained by binary burning. Even non-primordial binaries may be playing a role here. To the best of our knowledge it has not previously been suggested that M4, which is classified as a "King" cluster by Trager *et al.* (1993), is a post-collapse cluster. This raises the long-dormant question of how it is possible for some clusters to exhibit collapsed cores if they also come with significant populations of primordial binaries.

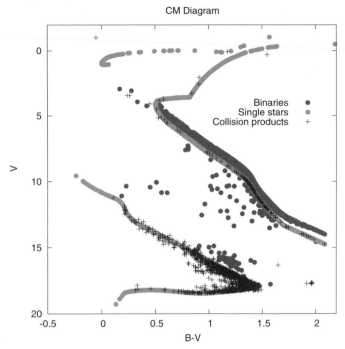

Figure 5. The colour-magnitude diagram at 12 Gyr. Green: single stars; red: binaries; blue pluses: collision or merger remnants.

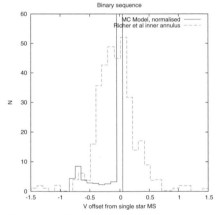

Figure 6. Histogram of V-offset from the main sequence, compared with the corresponding data from the innermost annulus studied by Richer *et al.* See text for details.

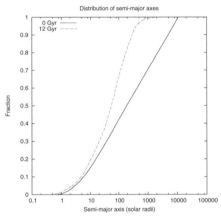

Figure 7. Distribution function of the semi-major axes of the binaries at 0 Gyr and 12 Gyr. Units: solar radii.

3. The search for initial conditions

Each Monte Carlo model for this cluster takes a few days. Therefore the problem of finding appropriate initial conditions is a significant one. One needs a good starting guess, and then a rapid method for iterative improvement. Our techniques for dealing with these issues are still primitive, but have evolved in the course of this research. At the iterative stage we have employed scaled-up small models (with as few as 10^4 objects sometimes), which are designed to relax at the same rate as a full-scale model. This requires a scaling

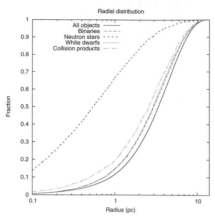

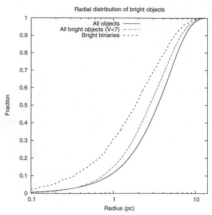

Figure 8. Radial distribution functions at 12 Gyr. Units: parsecs. The key identifies the class of object included. The distributions of white dwarfs and binaries are almost identical.

Figure 9. Radial distribution functions at 12 Gyr, showing the extent of segregation between bright single and bright binary stars ($M_V < 7$). Also shown for comparison is the distribution for all objects, as in Fig. 8. Units: parsecs.

of the length-scale, which does violence to binary interactions, but nevertheless it has been very useful. In this section we focus on the issue of finding starting values.

Consider first the problem of forwards evolution. Useful formulae for $M(t)$ are given by Lamers *et al.* (2005), which we have generalised to other IMF's and metallicities. To this we have added formulae for $r_h(t)$, based on very simple notions of adiabatic expansion (in response to mass-loss from stellar evolution) and tidal truncation. For the evolution of the core we have fitted simple expressions to the results on the time of core collapse given by Baumgardt & Makino (2003), extended by new N-body simulations to a wider range of initial concentrations. Very simple expressions for $r_c(t)$, consistent with the initial and final values, can then be employed as a first approximation. Finally we drew from Baumgardt & Makino a relation between M and α (the slope of the lower mass function.) Putting these relations together, we have constructed a tool which we refer to as *Quick Cluster Evolution*, following S. Portegies Zwart. To apply this to generate initial conditions for our simulations, we can run QCE iteratively in reverse.

Further development of this tool should include the addition of binary heating, which certainly influences the evolution of the half-mass radius unless the primordial binary fraction is low enough, and more concentrated initial conditions than the King models to which we have restricted QCE so far.

4. Discussion

It is shown by Giersz & Heggie (this volume) that Monte Carlo models can provide similar results to N-body models, in the range where comparison is possible, with similar physics (binaries, stellar evolution, etc.), except for a number of restrictions:

(*a*) Use of a tidal cutoff, instead of the tidal field. Though Sec. 1.2.3 summarises our current approach to this problem, other treatments are possible, and worth trying.

(*b*) Use of a *static* tide. The effects of tidal shocks have been studied by a number of authors (e.g. Kundic & Ostriker 1995), and it would be possible to add the effects as another process altering the energies and angular momenta of the stars in the simulation.

(*c*) Rotation: it has been shown (Kim *et al.* 2004) that, to the extent that rotating and non-rotating models can be compared, rotation somewhat accelerates the rate of core collapse. Rotation is hard to implement in this Monte Carlo model, however.

(*d*) Use of cross sections for triple/quad interactions: this limitation could be overcome by direct integration of the interactions, as is done by Fregeau & Rasio (2007) in their version of the Monte Carlo scheme. It will be important to do so, as it would remove the dependence on cross sections which are not well known for unequal masses, and would also permit us to include the collisions between stars which commonly occur in long-lived few-body interactions (Hut & Inagaki 1985).

(*e*) Neglect of triples: these are also commonly produced in binary-binary encounters (Mikkola 1984a), and it is desirable to include these as a third species (beyond single and binary stars). Their observable effects may be small, but of course there is one intriguing example in the very cluster we have focused on here (Thorsett *et al.* 1993).

Despite these limitations, not all of which are easily curable, Monte Carlo models are feasible in reasonable time for globular clusters, which are too large for direct *N*-body models. They yield predictions for mass segregation, luminosity functions, distributions of binary parameters, anisotropy, and many other kinds of data, which can hardly be obtained in any other way. (The only comparable method of which we are aware is the hybrid code of Giersz & Spurzem 2003.) Even when *N*-body simulations eventually become possible, Monte Carlo models will remain as a quicker way of exploring the main issues, just as King models have continued to dominate the field of star cluster modelling even when more advanced methods (e.g. Fokker-Planck models) have become available.

In addition to some of the possible improvements mentioned above, it is our intention to extend the approach to a number of other objects, including a "collapsed-core" cluster such as NGC6624. We welcome all suggestions for observational or theoretical comparisons, either on M4 or on other objects.

Acknowledgements

J. Hurley's help has been invaluable, especially in regard to stellar and binary evolution, and he went to much trouble to assist us. K. Meyer has been largely responsible for developing QCE, following initial work with S. Portegies Zwart. Her work was funded by the Carnegie Trust for the Universities of Scotland. We thank H.B. Richer for comments on a previous draft.

References

Angeletti, L., Dolcetta, R., & Giannone, P. 1980, Astrophys. Sp. Sci., 69, 45
Baumgardt, H. 2001, *MNRAS*, 325, 1323
Baumgardt, H., Hut, P., Makino, J., McMillan, S., & Portegies Zwart, S. 2003, *ApJL*, 582, L21
Baumgardt, H. & Makino, J. 2003, *MNRAS*, 340, 227
Dinescu, D. I., Girard, T. M., & van Altena, W. F. 1999, *AJ*, 117, 1792
Drukier, G. A. 1993, *MNRAS*, 265, 773
Drukier, G. A. 1995, *ApJS*, 100, 347
Drukier, G. A., Fahlman, G. G., Richer, H. B., & Vandenberg, D. A. 1988, *AJ*, 95, 1415
Dubath, P., Meylan, G., Mayor, M., & Magain, P. 1990, *A&A*, 239, 142
Dull, J. D., Cohn, H. N., Lugger, P. M., Murphy, B. W., Seitzer, P. O., Callanan, P. J., Rutten, R. G. M., & Charles, P. A. 1997, *ApJ*, 481, 267
Fleck, J.-J., Boily, C. M., Lançon, A., & Deiters, S. 2006, *MNRAS*, 369, 1392
Fregeau, J. M. & Rasio, F. A. 2007, *ApJ*, 658, 1047
Gebhardt, K., & Fischer, P. 1995, *AJ*, 109, 209
Giersz, M. 1998, *MNRAS*, 298, 1239

Giersz, M. 2001, *MNRAS*, 324, 218

Giersz, M. 2006, *MNRAS*, 371, 484

Giersz, M. & Heggie, D. C. 2003, *MNRAS*, 339, 486

Giersz, M. & Spurzem, R. 2003, *MNRAS*, 343, 781

Grabhorn, R. P. 1993, Ph.D. Thesis, Indiana University

Grabhorn, R. P., Cohn, H. N., Lugger, P. M., & Murphy, B. W. 1992, *ApJ*, 392, 86

Gunn, J. E. & Griffin, R. F. 1979, *AJ*, 84, 752

Hansen, B. M. S., *et al.* 2002, *ApJL*, 574, L155

Hansen, B. M. S., *et al.* 2004, *ApJS*, 155, 551

Harris, W. E. 1996, *AJ*, 112, 1487

Heggie, D. C., Portegies Zwart, S., & Hurley, J. R. 2006, *New Astronomy*, 12, 20

Hurley, J. R., Pols, O. R., & Tout, C. A. 2000, *MNRAS*, 315, 543

Hurley, J. R., Tout, C. A., & Pols, O. R. 2002, *MNRAS*, 329, 897

Hurley, J. R., Pols, O. R., Aarseth, S. J., & Tout, C. A. 2005, *MNRAS*, 363, 293

Hurley, J. R. 2007, *MNRAS*, 379, 93

Hut, P. & Inagaki, S. 1985, *ApJ*, 298, 502

Illingworth, G. 1976, *ApJ*, 204, 73

Kim, E., Lee, H. M., & Spurzem, R. 2004, *MNRAS*, 351, 220

King, I. R. 1966, *AJ*, 71, 64

Kroupa, P. 1995, *MNRAS*, 277, 1507

Kroupa, P. 2007, ArXiv Astrophysics e-prints, arXiv:astro-ph/0703124

Kroupa, P., Gilmore, G., & Tout, C. A. 1991, *MNRAS*, 251, 293

Kroupa, P., Aarseth, S., & Hurley, J. 2001, *MNRAS*, 321, 699

Kundic, T. & Ostriker, J. P. 1995, *ApJ*, 438, 702

Lamers, H. J. G. L. M., Gieles, M., Bastian, N., Baumgardt, H., Kharchenko, N. V., & Portegies Zwart, S. 2005, *A&A*, 441, 117

Leonard, P. J. T., Richer, H. B., & Fahlman, G. G. 1992, *AJ*, 104, 2104

Michie, R. W. & Bodenheimer, P. H. 1963, *MNRAS*, 126, 269

Mikkola, S. 1983, *MNRAS*, 205, 733

Mikkola, S. 1984a, *MNRAS*, 207, 115

Mikkola, S. 1984b, *MNRAS*, 208, 75

Peterson, C. J. & King, I. R. 1975, *AJ*, 80, 427

Peterson, R. C., Rees, R. F., & Cudworth, K. M. 1995, *ApJ*, 443, 124

Phinney, E. S. 1993, in Djorgovski, S. G., Meylan G., eds, Structure and Dynamics of Globular Clusters, ASPCS 50, 141

Plummer, H. C. 1911, *MNRAS*, 71, 460

Portegies, Zwart, S. F., & Verbunt, F. 1996, *A&A*, 309, 179

Pryor, C., Hartwick, F. D. A., McClure, R. D., Fletcher, J. M., & Kormendy, J. 1986, *AJ*, 91, 546

Richer, H. B., *et al.* 2004, *AJ*, 127, 2771

Spitzer, L. 1987, Dynamical Evolution of Globular Clusters, Princeton, NJ, Princeton University Press, 1987

Thorsett, S. E., Arzoumanian, Z., & Taylor, J. H. 1993, *ApJL*, 412, L33

Trager, S. C., Djorgovski, S., & King, I. R. 1993, in Djorgovski, S.G., Meylan G., eds, Structure and Dynamics of Globular Clusters, ASPCS 50, 347

Trager, S. C., King, I. R., & Djorgovski, S. 1995, *AJ*, 109, 218

van de Ven, G., van den Bosch, R. C. E., Verolme, E. K., & de Zeeuw, P. T. 2006, *A&A*, 445, 513

van den Bosch, R., de Zeeuw, T., Gebhardt, K., Noyola, E., & van de Ven, G. 2006, *ApJ*, 641, 852

Von Zeipel, H. 1908, Annales de l'Observatoire de Paris, 25, 1

Dynamical Evolution of Dense Stellar Systems
Proceedings IAU Symposium No. 246, 2007
E. Vesperini, M. Giersz & A. Sills, eds.

© 2008 International Astronomical Union
doi:10.1017/S1743921308015470

The Simple Underlying Dynamics of Globular Clusters

Ivan R. King

Department of Astronomy, University of Washington,
Box 351580, Seattle, WA 98195-1580, USA
email: king@astro.washington.edu

Abstract. Although the overall dynamics of globular clusters involves many complexities, much of their dynamics can be understood on the basis of some simple and straightforward physical arguments. Those arguments are presented here, in an effort to lead the reader through the basic stages of cluster dynamics in a quick and easy way.

Keywords. globular clusters: general, open clusters and associations: general

1. Introduction

The dynamics of globular clusters has many complexities: binaries, collapsed cores, tidal tails, etc., but underlying it all is a basic dynamical structure that is very simple. My aim in this paper is to lead the reader through this underlying structure in a straightforward and logical way.

This is therefore a simple paper, with a simple thesis, but one that I think is very important to an astronomer who wishes to understand the dynamics of globular clusters. The picture that I will draw here includes *the essence of cluster dynamics* – the inevitable submission to the physical laws and processes that make a cluster what it is. The physical arguments presented here are simple ones, such as any first-year graduate student should be able to follow easily. Yet I fear that in this era of facile computation some graduate students are never exposed to the simple initiation in intuitive physics that I am going to present.

What I will do is to use all of the clarity of hindsight to show how the fundamental characteristics of globular clusters, that is, their velocity distributions and density distributions, are the inevitable outcome of some simple physical processes that govern their behavior, and that it is these processes that make the clusters what they are.

More specifically, I will show how the characteristic family of density profiles of globular clusters follows inevitably and naturally from the relaxing effect of stellar encounters. First, the velocity distribution and the spatial density distribution must match in a way that keeps the cluster in dynamical equilibrium. Then we will see how relaxation through stellar encounters brings the velocity distribution as close to Gaussian as it can get, and how the result is the type of density distributions that clusters have, differing only in central concentration.

The approach that I take here will draw heavily on material from two papers of many decades ago (King 1965 and King 1966). The second of these presented some simple dynamical models of clusters; they turned out to fit the density profiles of globular clusters surprisingly well. (I say surprisingly, because they were the first thing that I tried.) But here, instead of just saying "This works", I will present the models as the inevitable unfolding of the underlying physics. In spite of all the complexity that cluster dynamics appears to have, I want to show that whatever may happen with the details,

131

clusters *had* to follow the models that are derived in the King (1966) paper, and that this becomes obvious if we look at the physics in the right way.

Finally, to end the introduction, some simplifications: I will confine my discussion to the dynamics of a spherical cluster in which all the motions are isotropic. Dealing with a flattened, rotating cluster is not necessary to a basic understanding; and I will dispense with anisotropy too, on the grounds that where it has been measured it has been found to be quite small (King & Anderson 2001, 2002), so that it is unlikely to be of fundamental importance.

2. Dynamical Equilibrium

A self-gravitating star cluster must be in equilibrium. This requires not just satisfying the virial theorem; the velocity distribution has to correspond to the spatial density distribution in just such a way that as the individual stars move around on their orbits, stars take the place of other stars in a way that maintains the density distribution in a constant state.

There is a straightforward algorithm for going from the velocity distribution at the center of a cluster to a complete picture of the velocities and densities throughout the cluster – that is, to a cluster model. Jeans' theorem says that the phase-space density of a system must be expressible as a function of the isolating integrals of the equations of motion of a star. A consequence of this is that from an assumed velocity distribution at the cluster center, we can get the isotropic velocity distribution that holds at every other point. Knowing the velocities, we can then integrate over them to get the density as a function of potential, which comes in through our use of Jeans' theorem. If we put that information into Poisson's equation, we get the relationship between the potential V and the radial coordinate r, and closing the circle between ρ, V, and r gives us $\rho(r)$, which is the other half of the model. So this algorithm for building a cluster model from a velocity distribution is quite straightforward.

To be more explicit, here are the details. In general, the motion of a star in a spherical potential has two isolating integrals, the energy and the magnitude of the angular momentum. But if we restrict ourselves to isotropic velocity distributions, the only integral that remains is the energy,

$$E = \frac{1}{2}\,v^2 + V(r).$$

According to Jeans' theorem, then, the joint distribution of density and velocity, $f(r,v)$, is given everywhere by $f(r,v) = F(E)$, where F is some fixed function. This allows us to integrate the density at any point from the formal relationship

$$\rho = 4\pi \int_0^{v_e} f(r,v)\,v^2\,dv$$

$$= \frac{4\pi}{\sqrt{2}} \int_V^0 F(E)\frac{dE}{\sqrt{E-V}}. \tag{2.1}$$

(Note: The limits of integration have been adjusted to recognize the existence of an escape velocity.)

Equation (2.1) allows us to evaluate the right side of Poisson's equation,

$$\frac{d^2 V}{dr^2} + \frac{2}{r}\frac{dV}{dr} = 4\pi G\rho,$$

for any value of V. The equation can then be solved numerically to give the relationship between V and r. Along with the relationship between ρ and V given by Eq. (2.1), this gives us $\rho(r)$, which is the explicit spatial density distribution that corresponds to the velocity distribution that we started with. Thus a velocity distribution implies a unique and explicit density distribution. The two distributions together constitute what we call a cluster model.

3. The Choice of Velocity Distribution

We have just shown that any reasonable velocity distribution will lead to a valid cluster model. But the question is, which velocity distribution to choose. The obvious answer would seem to be the one toward which the physics of stellar encounters makes the stars of the cluster tend: the Gaussian distribution. But that turns out to go too far. If we put a Gaussian velocity distribution into the algorithm that I have just described, the resulting spatial density distribution is the isothermal sphere, whose mass is strongly infinite – strongly, in the sense that far from the center, the total mass within a radius r increases linearly with r.

It is easy to see that something like this had to be so, because a Gaussian distribution has no velocity limit and thus implies an infinite escape velocity. Or, looking at it the other way round, a cluster can never achieve a Gaussian distribution, because its stars of highest velocity would then be above the escape velocity, and would therefore escape. (An interesting sidelight is that nearly 70 years ago Ambartsumian [1938, 1985] estimated the escape rate by making a very simple approximation. He said to imagine that a cluster achieves a full Gaussian distribution each relaxation time, and that the stars that are above the escape velocity then escape. This crude calculation gave the first estimate of the rate at which stars escape from a cluster, and it was a good one.)

Getting back to our cluster-model problem, what we need is a velocity distribution that is lower than Gaussian at the higher velocities. One could imagine fiddling a Gaussian arbitrarily to reduce its tail, but there is a more natural approach to the problem.

3.1. *The steady-state distribution*

What our cluster is most likely to do is to settle down into a steady state – that is, its velocity distribution will try to take on a steady state, or rather it will get as close as it can get to a steady state, given the fact that stars are slowly escaping. The spatial density distribution will then have whatever form corresponds to that.

The steady state follows from the equation that governs the changes that stellar encounters make in the velocity distribution, the Fokker-Planck equation. Let the stars encountered have a Gaussian velocity distribution,

$$f_{\rm b}(v) = \frac{j_{\rm b}^3}{\pi^{3/2}} e^{-j_{\rm b}^2 v^2},$$

and express the time in units of the "reference time", given by

$$t_{\rm R} = \frac{1}{2\pi G^2 m_{\rm b}^2 n_{\rm b} j_{\rm b}^2 \ln(N/2)}.$$

(It is interesting to note that whereas earlier treatments of the theory of stellar encounters had used a quantity called the relaxation time, with an arbitrary definition such as, for example, "the time in which the expected cumulative energy change for an average star is equal to the average kinetic energy of a star", in the case of the Fokker-Planck equation the quantity $t_{\rm R}$ arises naturally, simply as a collection of all the constants in such a way that defining $\theta = t/t_{\rm R}$ as a dimensionless time will remove all those constants from the equation. The value of t_R is almost the same as that of the relaxation time, within a constant of the order of unity, but to emphasize the fact again, it falls naturally out of the physics.)

In these simplified variables the Fokker-Planck equation takes the form

$$\frac{\partial f}{\partial \theta} = \frac{1}{x^2} \frac{\partial}{\partial x} \left[\alpha(x) \left(\frac{\partial f}{\partial x} + 2 \frac{m}{m_b} x f \right) \right],$$

where $x = j_b v$ and $\alpha(x)$ is defined by

$$\alpha(x) = \frac{4}{\sqrt{\pi} x} \int_0^x e^{-y^2} y^2 \, dy.$$

To find the steady-state solution we write

$$f(x, \theta) = g(x) \cdot h(\theta).$$

The separation equation is

$$\frac{1}{h} \frac{dh}{d\theta} = -\lambda = \frac{1}{g} \cdot \frac{1}{x^2} \frac{d}{dx} \left[\alpha \left(\frac{dg}{dx} + 2 \frac{m}{m_b} x g \right) \right].$$

The eigenvalue λ (whose minus sign was chosen with foreknowledge), must be a constant, since one equal sign shows it to be independent of x and the other shows it to be independent of θ.

The time-dependence solution is clearly $h = h_0 e^{-\lambda \theta}$. Thus the solution that corresponds to the smallest eigenvalue will decay most slowly and will at late times become dominant.†

The eigenvalue is found by solving the equation

$$\frac{d}{dx} \left[\alpha \left(\frac{dg}{dx} + 2 \frac{m}{m_b} x g \right) \right] + \lambda x^2 g = 0.$$

Spitzer and Härm (1958), who were the first to write the Fokker-Planck equation in this form, solved it numerically. It is more instructive, though, to derive an analytic approximation to the steady-state solution of the Fokker-Planck equation, because that leads to a useful approximate formula for the velocity distribution itself (King 1965). It goes as follows:

The equation for g can be integrated formally to give

$$\frac{dg}{dx} + 2 \frac{m}{m_b} x g = -\frac{\lambda}{\alpha(x)} \int_0^x g(y) \, y^2 \, dy.$$

If we expand $g(x)$ as a power series in the small quantity λ,

$$g(x) = g_0(x) + \lambda g_1(x) + \lambda^2 g_2(x) + \ldots,$$

and if we substitute in the differential equation and then equate the coefficients of like

---

† A note for the students (and perhaps for professors too): In the courses that I took in graduate school, it was standard practice just to do a separation of variables – no explanation; that's just "what one does". So I will interject the explanation here: It can be shown that the equation below is a member of the class called Sturm-Liouville equations. These have the property that their eigenfunctions are a complete basis set, so that any initial velocity distribution can be expressed as a linear combination of them. In that expansion the coefficients are the corresponding h functions. Thus as time goes on, the term with the slowest time decay becomes dominant. This is the justification for saying that the steady-state solution is really the solution that the cluster will choose.

powers of λ, the results are

$$\frac{dg_0}{dx} + 2\frac{m}{m_{\mathrm{b}}}xg_0 = 0$$

$$\frac{dg_{i+1}}{dx} + 2\frac{m}{m_{\mathrm{b}}}xg_{i+1} = -\frac{1}{\alpha(x)}\int_0^x g_i(y)\,y^2\,dy, \qquad i > 0.$$

If we do the same with the boundary conditions, then $g(0) = 1$ gives $g_0(0) = 1$, and $g_i(0) = 0$ if $i > 0$, while the requirement that $(dg/dx)_{x=0} = 0$ gives $(dg_i/dx)_{x=0} = 0$ for all i.

An integrating factor for our equations is $\exp(mx^2/m_{\mathrm{b}})$. The results are rather complicated, except for the case $m = m_{\mathrm{b}}$, which gives an elegant and simple result. In that case the first equation gives

$$g_0 = e^{-x^2}.$$

In the equation for g_1, the indefinite integral is the same as the one in the definition of α, so that the equation simplifies as follows. (To avoid all that annoying filling in of steps in one's head, I have laid it all out, step by step.)

$$\frac{dg_1}{dx} + 2xg_1 = -\frac{\sqrt{\pi}}{4}x$$

$$e^{x^2}\left(\frac{dg_1}{dx} + 2xg_1\right) = -\frac{\sqrt{\pi}}{4}xe^{x^2}$$

$$\frac{d}{dx}\left(e^{x^2}g_1\right) = -\frac{\sqrt{\pi}}{4}xe^{x^2}$$

$$g_1 = -\frac{\sqrt{\pi}}{4}e^{-x^2}\int_0^x xe^{x^2}\,dx$$

$$= -\frac{\sqrt{\pi}}{8}e^{-x^2}\left(e^{x^2} - 1\right)$$

$$= -\frac{\sqrt{\pi}}{8}\left(1 - e^{-x^2}\right).$$

We will thus take as our approximate solution

$$g(x) = g_0(x) + \lambda g_1(x)$$

$$= e^{-x^2} - \frac{\sqrt{\pi}}{8}\lambda\left(1 - e^{-x^2}\right).$$

The value of λ is set by the condition $g(x_{\mathrm{e}}) = 0$. This gives

$$\lambda = \frac{8}{\sqrt{\pi}}\frac{e^{-x_{\mathrm{e}}^2}}{1 - e^{-x_{\mathrm{e}}^2}}$$

and

$$g(x) = \frac{e^{-x^2} - e^{-x_{\mathrm{e}}^2}}{1 - e^{-x_{\mathrm{e}}^2}}.$$

For $x_{\mathrm{e}}^2 = 6$, the value for which Spitzer and Härm did their numerical integration, this approximation for λ comes out $1/89.4$; their numerical solution gave $1/88.0$. And our approximate $g(x)$ is everywhere within $\frac{1}{2}\%$ of their numerical solution.

The distribution function has a simple and convenient form: a Gaussian lowered by a constant that brings it to zero at $x = x_{\mathrm{e}}$, and renormalized so that $g(0) = 1$.

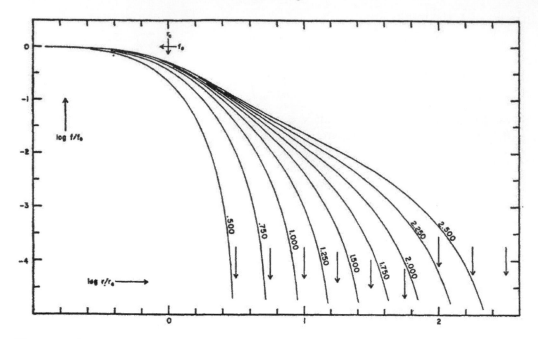

Figure 1. Surface densities in the family of models discussed in the text. All the curves are normalized to the same central surface brightness and core radius. The vertical arrows indicate the limiting radii of successive models.

Notice that x_e has not been specified. In fact, we can evaluate escape-rate eigenvalues and velocity-distribution eigenfunctions for any value of x_e, and we get a one-parameter family of solutions, as shown in Fig. 1.

4. Escape Velocity and the Tidal Limit

It is now time to attach a physical meaning to another quantity. All of our velocity distributions have been limited so as to go to zero at an "escape velocity". Similarly, the spatial distributions that follow from solving Poisson's equation have a limiting radius. But we can of course attach a physical meaning to this radius in a real cluster; it is just the tidal limit of the cluster in the gravitational field of the Milky Way. (Strictly speaking, the tidal limit is a rigorously defined quantity only for a cluster in a circular orbit, but the idea can be stretched a little to cover non-circular orbits.)

Now consider the problem of making a globular cluster. First you take your celestial ice-cream scoop and decide how much mass to use. Then you decide how closely to cram it together; this gives the core radius of the cluster. Then let the Milky Way impose its tidal force; that sets the tidal limit. These conditions define the actual cluster. The cluster model, on its side, is described by three parameters: r_c, r_t, and the mass. Since two of these are just scale factors, the only free parameter is the ratio of r_t to r_c, whose logarithm is often called the central concentration. In the picture that we have just drawn, clusters differ from each other only in scale factors and in this one profile parameter, so that they are as similar to each other as they can possibly be.

These models do fit the profiles of actual globular clusters reasonably well, as shown in Fig. 2.

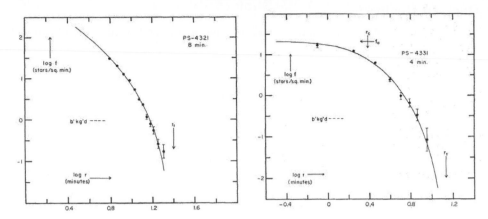

Figure 2. Fitting of model curves to star counts in actual globular clusters. Error bars shown are statistical uncertainties in the counts, from the Poisson distribution. *(left)* M13; near the center the stars were too crowded to count. *(right)* NGC 5053.

5. Additional Favorable Characteristics

In addition to their inevitability, these models have an unusual stability, in two ways, neither of which had to be so *a priori*. You may have noticed that it was only at the cluster center that we imposed the requirement that the velocity distribution be the steady-state solution of the Fokker-Planck equation. In a configuration that is going to have long-term stability, however, the condition should apply everywhere. But lo and behold! If we use Jeans' theorem to find out what the velocity distribution is at some other radial distance, it turns out to be the steady-state solution of the Fokker-Planck equation that is appropriate for *that* point. So the steady state holds everywhere, and the rate of change of the cluster is less than it would be if this were not the case.

The second unexpected equilibrium is in the escape rate of stars. If a different fraction escaped at each distance from the center, the cluster would need to need to readjust, so as to reestablish its dynamical equilibrium. In fact, however, over a large part of the cluster the outward decrease of the escape rate (due to the lower density) is almost exactly balanced by the increased fraction of stars escaping per relaxation time, which is due to the lower escape velocity at radii closer to the tidal limit.

These two circumstances make the dynamical equilibrium much more stable than it would otherwise be, but I have never seen any obvious reason why either of these favorable characteristics should have come about.

6. A Summing Up

This is the basic picture that I wanted to present. Globular clusters look so much like each other because they are all molded by the same simple physics. Dynamical systems like to settle down into as steady a state as they can. First, the velocity distribution and the spatial density distribution must be "matched", in the sense that each sustains the other and they remain in an equilibrium. Second, stellar encounters drive the velocity distribution as close to Gaussian as it can get. The form that the velocity distribution takes on is the steady-state solution of the Fokker-Planck equation, and it is that distribution that gives the cluster the characteristic density profile that it has. Third, the remaining physical constraint is imposed by the tidal force of the Milky Way, and within these restrictions the clusters are as similar as they can be.

7. Back to Reality

But I owe you more realism than this. The dynamical models that I have been talking about are oversimplified in one terribly important way: they consist of stars that all have the same mass, whereas a cluster is a mixture of objects of different mass. How do we cope with this? Well, it turns out not to be a serious complication at all. Gravitational forces are additive, so all that we need to do to make a model for a mixture of masses is to write Poisson' equation as

$$\frac{d^2V}{dr^2} + \frac{2}{r}\frac{dV}{dr} = 4\pi \sum_i G\rho_i,$$

where the ρ_i refer to the different mass groups. Given the velocity distribution for each mass group, we can evaluate its density at each value of V, add up the densities, and do the same integration of Poisson's equation as before.

The only new problem that has come in is getting the velocity distributions for the stars of each mass. I have shown that for stars that all have the same mass the velocity distribution is well approximated by a Gaussian minus a constant. For other masses the solution is not quite so easy, but the velocity distribution is still pretty well approximated by a Gaussian minus a constant. And for the various Gaussians we just take dispersions that correspond to equipartition between the various masses. Because the gravitational potential is pretty well dominated by stars that don't have a very large range in mass, the spatial distribution comes out looking not very different from that of stars of a single mass, and the picture doesn't change very much. And in going through this line of reasoning we have built ourselves a model of how the different stellar masses are distributed relative to each other, so that someone who observes a mass function in one part of a cluster now knows how to go from that local mass function to the global mass function for the whole cluster. A lot more can be learned from this simple picture, like how fast stars of different mass escape, or how tidal shocks affect stars of different mass.

It was stated above that a model made from a mixture of masses usually has a brightness profile that is similar to that of some member of the single-mass family. There is one striking exception, however: a model with a very small core. Such models arise when the central concentration is high and there is a significant number of stars that have a significantly higher mass than any of the other groups – for example, if about 1% the stars in a globular-cluster model have masses suitable for neutron stars ($\sim 1.4M_\odot$). In that case the greater masses of the neutron stars cause them to bunch up in the center and create a lump-like concentration of mass there. In such a model the number of neutron stars can be adjusted to make the brightness profile of the model closely resemble that of a collapsed-core cluster. This ability of the models is valuable to note when, for instance, transforming a local mass function to a global mass function in a collapsed-core cluster.

Now I will say a little about one other aspect of cluster dynamics, and then I'll leave it to others in this volume to treat the difficult problems in the dynamics of globular clusters.

8. Inequalities of the Characteristic Times

I have not called attention to it explicitly, but the ability to treat the dynamics of globular clusters in this simple way is very much dependent on two inequalities:

$$t_{\text{cr}} \ll t_{\text{rlx}} \ll t_{\text{evol}}.$$

The quantities involved are the three characteristic times that pertain to the dynamical behavior of the cluster. The first, t_{cr}, is the crossing time, which is the time that a star takes to go from one side of the cluster to the other. (This is clearly an order of magnitude rather than a rigorously defined quantity.)

Next comes the relaxation time t_{rlx}. As I've said earlier, the reference time, in the Fokker-Planck equation, is physically more natural; but the name "relaxation time" is so firmly entrenched in astronomical terminology that I will use that name here. In these inequalities, of course, the small factor by which the two differ doesn't matter.

The third characteristic time, the "evolution time", is the time in which the cluster will change greatly as a result of dynamical evolution. It is clear that this too is a loose definition. Since the principal effect of dynamical evolution is a loss of stars, it might be better to refer to something like a decimation time, but I have never seen that name used.

In a typical globular cluster the inequalities – or maybe because of the double inequality sign I should refer to them as "incomparabilities" – are of the order of a factor of a hundred in each case – although obviously this has to be taken as a loose statement, because the relaxation time, for example, differs a lot from one place in the cluster to another.

The basic steps that I have led you through in reducing cluster dynamics to such simplicity have tacitly assumed these incomparabilities. In talking about how the velocity distribution and the density distribution correspond to each other, I have relied on the crossing time being short compared with the relaxation time. Otherwise we would not be able to talk about an equilibrium at all; by the time a star had traveled across the cluster, the velocity distribution would have changed. And similarly, if the evolution time were comparable with the relaxation time, it wouldn't make sense to talk about a steady-state solution of the Fokker-Planck equation, because as stars were lost, the actors would be changing in the middle of the play.

There is still another circumstance that makes the dynamics of globular clusters simpler than it would otherwise be: the range of masses of stars is not as large as it would be in a younger cluster. Instead of having to deal with a range of stellar masses all the way from the red dwarfs of the lower main sequence up to 50 or 100 solar masses, we go only up to about 0.8 $M_\odot$, because all the stars more massive than that limit have gone through their main-sequence lives, and nearly all of them are now white dwarfs, whose mass is right in the middle of the range of main-sequence masses.

9. Unsolved Problems

In addition to the successes of understanding clusters, I want to conclude by referring to two unsolved problems of cluster dynamics that have bothered me for years.

9.1. *Open clusters*

The first of these is the problem of the dynamics of open clusters, where the inequality $t_{cr} \ll t_{rlx}$ is not satisfied. It is easy to show that the ratio of the two times depends mainly on the number of stars in the cluster. At 100 stars, which corresponds to a medium-poor open cluster, the cluster relaxes in just about the time it takes a star to cross it. There is no way then to talk about a dynamical equilibrium.

I have felt for a long time that the way to approach modeling open clusters is to apply Prigogine's principle of a stationary state of higher order (Prigogine 1967), in which the most probable state is the one in which entropy increases at a minimum rate, compared with all neighboring states – but I am not aware of anyone ever taking up this problem. (For a discussion of the dynamics of open clusters, see King 1981.)

9.2. *A better treatment of relaxation in a cluster*

At the time I was investigating basic cluster dynamics, I published a paper (King 1960) that took the relaxation process a step closer to reality, by recognizing that the stars were encountering *each other*, rather than a separate reservoir of stars with Gaussian-distributed velocities. The result made only trivial changes in the resulting velocity distribution, but it made an important change in the physical circumstances. Because the relaxation was produced by encounters between the stars under consideration, the energy of escaping stars had to be removed from other stars in the same distribution. Although the escaping stars carried away zero energy, they left the negative binding energy of the cluster to be shared among fewer stars, with a resulting contraction of the cluster. The steady-state solution described in §3.1 was then replaced by a self-similar solution, in which the changes in velocity corresponding to the contraction resulted in an escape rate about 50% higher than in the steady-state solution.

The paper has had absolutely no effect on astronomy. NASA ADS lists 6 citations, all of which merely note the existence of the paper, without any mention of its content. I have always wondered if this reasoning was worth pursuing – or even worthy of being verified by N-body simulations.

10. Conclusion

When I was invited to present a paper at this meeting, I thought about what kind of contribution might be most useful, and I chose this review of the basic dynamical nature of globular clusters. There is nothing here that has not been seen before, and for that I apologize to the reader. But as abstruse details accumulate, the simple truths are often lost sight of, and I thought that presenting a reminder of them might be useful.

Acknowledgements

I would like to acknowledge support from many grants from STScI, especially GO-10146, and also a travel grant from the IAU.

References

Ambartsumian, V. A. 1938, *Uchenniye Zapiskiy, Leningrad State Univ.*, 22, 19 (in Russian)
Ambartsumian, V. A. 1985, in: J. Goodman & P. Hut (eds.), *Dynamics of Star Clusters*, IAU Symp. No. 113 (Reidel: Dordrecht), p. 521 (English translation of Ambartsumian 1938)
King, I. R. 1960, *AJ*, 65, 122
King, I. R. 1965, *AJ*, 70, 376
King, I. R. 1966, *AJ*, 71, 64
King, I. R. 1981, in: J. Hesser (ed.), *Star Clusters*, IAU Symp. No. 85 (Reidel: Dordrecht), p. 139
King, I. R. & Anderson, J. 2001, in: S. Deiters, B. Fuchs, A. Just, R. Spurzem, & R. Wielen (eds.) *Dynamics of Star Clusters and the Milky Way*, ASPCS Vol. 228, p. 19
King, I. R. & Anderson, J. 2002, in: F. van Leeuwen, J. D. Hughes, & G. Piotto (eds.) *ω Centauri, A Unique Window into Astrophysics*, ASPCS Vol. 265, p. 21
Prigogine, I. 1967, *Introduction to the Thermodynamics of Irreversible Processes, 3rd edition* (New York: Wiley)
Spitzer, L. & Härm, R. 1958, *ApJ*, 127, 544

Dynamical Evolution of Dense Stellar Systems
Proceedings IAU Symposium No. 246, 2007
E. Vesperini, M. Giersz & A. Sills, eds.

© 2008 International Astronomical Union
doi:10.1017/S1743921308015482

Observational Evidence of Multiple Stellar Populations in Globular Clusters

Giampaolo Piotto[1]

[1]Dipartimento di Astronomia, Università di Padova,
vicolo dell'Osservatorio, 2, I-35122, Padova, Italy
email: giampaolo.piotto@unipd.it

Abstract. An increasing number of photometric observations of multiple stellar populations in Galactic globular clusters is seriously challenging the paradigm of GCs hosting single, simple stellar populations. These multiple populations manifest themselves in a split of different evolutionary sequences as observed in the cluster color-magnitude diagrams. In this paper we will summarize the observational scenario.

Keywords. Globular Clusters, Stellar Populations, Photometry, Astrometry

1. Introduction

Globular star clusters (GC) occupy a prominent role in modern astrophysics. They are among the oldest Population II objects for which accurate ages can be derived, hence they set an independent constraint on Cosmology. As such, they must have formed before the bulk of the parent galaxy, and then can provide insight into galaxy formation. Given the large number of stars in them, they represent the ideal laboratory for testing and calibrating stellar evolutionary models, as well as for (N-body) dynamical studies. Finally, being though consisting of basically coeval and chemically homogeneous stars, they have been used as ideal templates for "simple stellar populations" (SSP) with which to test and calibrate synthetic models of stellar populations, a critical tool for studying galaxies at low as well as at high redshift.

Color-magnitude diagrams (CMD) like that of NGC 6397 shown in Fig. 1 fully support the paradigma of GCs hosting simple stellar populations. However, there is a growing body of observational facts which challenge this traditional view. Since the eighties we know that GCs show a peculiar pattern in their chemical abundances (see Gratton *el al.* 2004 for a recent review). While they are generally homogenous insofar Fe-peak elements are considered, they often exhibit large anticorrelations between the abundances of C and N, Na and O, Mg and Al. These anticorrelations are attributed to the presence at the stellar surfaces of a fraction of the GC stars of material which have undergone H burning at temperatures of a few ten millions K (Prantzos *et al.* 2007; less for the C and N anticorrelation). This pattern is peculiar to GC stars; field stars only show changes in C and N abundances expected from typical evolution of low mass stars (Gratton *et al.* 2000; Sweigart & Mengel 1979; Charbonnel 1994); it is primordial, since it is observed in stars at all evolutionary phases (Gratton *et al.* 2001); and the whole stars are interested (Cohen *et al.* 2002).

In addition, since the sixties (Sandage and Wildey 1967, van den Bergh 1967), we know that the horizontal branches (HB) of some GCs can be rather peculiar. In some GCs the HB can be extended to very hot temperatures implying the loss of most of the stellar envelope (see compilation by Recio-Blanco *et al.* 2006); the distribution of the stars along the HB can be clumpy, with the presence of one or more gaps (Ferraro *et al.* 1998, Piotto

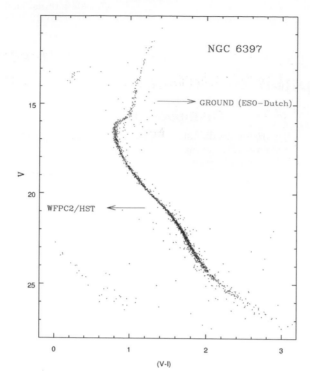

Figure 1. The CMD of NGC 6397 shows very narrow sequences and a well defined TO, supporting the idea that, in general, GCs are populated by coeval and chemically homogeneous stars.

et al. 1999). This problem, usually known as the *the second parameter* problem, still lacks of a comprehensive understanding: many mechanisms, and many parameters have been proposed to explain the HB peculiarities, but none apparently is able to explain the entire observational scenario. It is well possible that a combination of parameters is responsible for the HB morphology (Fusi Pecci *et al.* 1993). Surely, the total cluster mass seems to have a relevant role (Recio-Blanco *et al.* 2006).

It is tempting to relate the second parameter problem to the complex abundance pattern of GCs. Since high Na and low O abundances are signatures of material processed through hot H-burning, they should be accompanied by high He-contents (D'Antona & Caloi 2004). In most cases, small He excesses up to dY 0.04 (that is Y 0.28, assuming the original He content was the Big Bang one) are expected. While this should have small impact on colors and magnitudes of stars up to the tip of the RGB, a large impact is expected on the colors of the HB stars, since He-rich stars should be less massive. E.g., in the case of GCs of intermediate metallicity ([Fe/H] -1.5), the progeny of He-rich, Na-rich, O-poor RGB stars should reside on the blue part of the HB, while that of the "normal" He-poor, Na-poor, O-rich stars should be within the instability strip or redder than it. Actually mean HB colors are also influenced by small age differences of 2-3 Gyr. However, within a single GC a correlation is expected between the distribution of masses (i.e. colors) of the HB-stars and of Na and O abundances.

In summary, a number of apparently independent observational facts seems to suggest that, at least in some GCs, there are stars which have formed from material which must have been processed by a previous generation of stars.

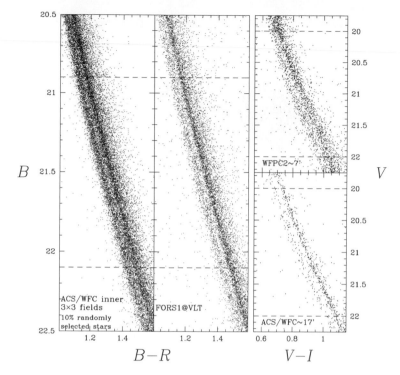

B $B-R$ $V-I$ V

Figure 2. The double MS of Omega Centauri. The MS split is visible from the cluster center to its outskirts, though the ratio of the number of stars populating the two sequences decreases from the cluster core to its outer envelope (from Bellini *et al.* 2008, *in preparation*).

The questions is: do we have some direct, observational evidence of the presence of multiple populations in GCs? Very recent discoveries, made possible by high accuracy photometry on deep HST images, allowed us to positively answer to this question. In this paper, we will summarize these new observational facts, and briefly discuss their link to the complex abundance pattern and to the anomalous HBs.

2. Direct Observational Evidence of Multiple Populations in GCs

The first, direct observational evidence of the presence of more than one stellar population in a GCs was published by Bedin *et al.* (2004). Bedin *et al.* found that, for a few magnitudes below the turn-off (TO), the main sequence (MS) of ω Centauri splits in two (Fig. 2). Indeed, the suspect of a MS split in ω Cen was already raised by Jay Anderson in his PhD thesis, but the result was based on only one external WFPC2 field, and this finding was so unexpected that he decided to wait for more data and more accurate photometry to be sure of its reality. Indeed, Bedin *et al.* (2004) confirmed the MS split in Jay Anderson field and in an additional ACS field located 17 arcmin from the cluster center. Now, we know that the multiple MS is present all over the cluster, though the ratio of blue to red MS stars diminishes going from the cluster core to its envelope (Sollima *et al.* 2007, Bellini *et al.* 2008, *in preparation*).

The more shocking discovery on the multiple populations in ω Cen, however, came from a follow-up spectroscopic analysis that showed that the blue MS has twice the metal abundance of the dominant red branch of the MS (Piotto *et al.* 2005). The only isochrones

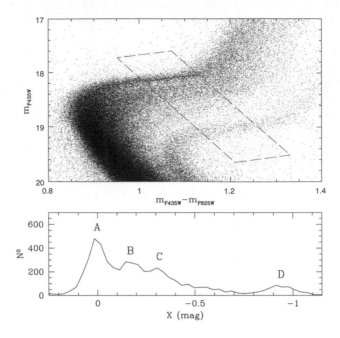

Figure 3. The multiple SGB in Omega Centauri. There are at least 4 distinct SGBs, plus a small fraction of stars spreaded between SGB-C and SGB-D (from Villanova *et al.* 2007).

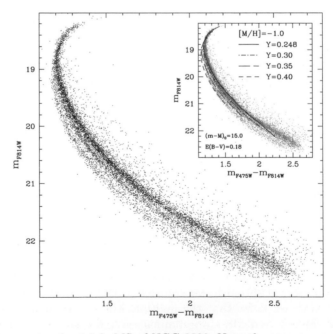

Figure 4. The spectacular triple MS of NGC 2808. Note the narrowness of the turnoff. The isochrone fitting in the inset shows that the bluest sequence can be reproduced only with a He content as high as $Y = 0.40$ (from Piotto *et al.* 2007).

that would fit this combination of color and metallicity were extremely enriched in helium ($Y \sim 0.38$) relative to the dominant old-population component, which presumably has primordial helium.

Indeed, the scenario in ω Cen is even more complex. As is already evident in the CMD of Bedin *et al.* (2004), this object has at least three MSs, which spread into a highly multiple sub-giant branch (SGB) with at least four distinct components (Fig. 3) characterized by different metallicities and ages (Sollima *et al.* 2005, Villanova *et al.* 2007; the latter has a detailed discussion.) A fifth, more dispersed component is spread between the SGB-C and SGB-D of Fig. 3.

These results reinforced the suspicion that the multiple MS of ω Cen could just be an additional peculiarity of an already anomalous object, which might not even be a GC, but a remnant of a dwarf galaxy instead. In order to shed more light on the possible presence of multiple MSs in Galactic GCs, we undertook an observational campaign with *HST*, properly devised to search multiple sequences at the level of the upper-MS, turn-off (TO), and SGB. The new data allowed us to show that the multiple evolutionary sequence phenomenon is not a peculiarity of ω Centauri only.

As shown in Fig. 4, also the CMD of NGC 2808 is splitted into three MSs (Piotto *et al.* 2007). Because of the negligible dispersion in Fe peak elements (Carretta *et al.* 2006), Piotto *et al.* (2007) proposed the presence of three groups of stars in NGC 2808, with three different He contents, in order to explain the triple MS of Fig. 4. These groups may be associated to the three groups with different Oxygen content discovered by Carretta *et al.* (2006). These results are also consistent with the presence of a multiple HB, as discussed in D'Antona and Coloi (2004) and D'Antona *et al.* (2006). Finally, we note that the narrowness of the TO region displayed by Fig. 4 suggests that the three stellar populations of NGC 2808 must have a small age dispersion, much less than 1 Gyr.

Also NGC 1851 must have at least two, distinct stellar populations. In this case the observational evidence comes from the split of the SGB in the CMD (Fig. 5) of this cluster (Milone *et al.* 2007). Would the magnitude difference between the two SGBs be due only to an age difference, the two star formation episodes should have been separated by at least 1 Gyr. However, as shown by Cassisi *et al.* (2007), the presence in NGC 1851 of two stellar populations, one with a normal α-enhanced chemical composition, and one characterized by a strong CNONa anticorrelation pattern could reproduce the observed CMD split. In this case, the age spread between the two populations could be much smaller, possibly consistent with the small age spread implied by the narrow TO of NGC 2808. In other terms, the SGB split would be mainly a consequence of the metallicity difference, and only negligibly affected by (a small) age dispersion. Cassisi *et al.* (2007) hypothesis is supported by the presence of a group of CN-strong and a group of CN-weak stars discovered by Hesser *et al.* 1982, and by a recent work by Yong and Grundahl (2007) who find a NaO anticorrelation among NGC 1851 giants.

NGC 1851 is considered a sort of prototype of bimodal HB clusters. Milone *et al.* (2007) note that the fraction of fainter/brighter SGB stars is remarkably similar to the fraction of bluer/redder HB stars. Therefore, it is tempting to associate the brighter SGB stars to the CN-normal, s-process element normal stars and to the red HB, while the fainter SGB should be populated by CN-strong, s-process element-enhanced stars which should evolve into the blue HB. In this scenario, the faint SGB stars should be slightly younger (by a few 10^7 to a few 10^8 years) and should come from processed material which might also be moderately He enriched, a fact that would help explaining why they evolve into the blue HB. By studying the cluster MS, Milone *et al.* (2007) exclude an He enrichment larger than $\Delta Y = 0.03$), as expected also by the models of Cassisi *et al.* (2007). Nevertheless, this small He enrichment, coupled with an enhanced mass loss, would be sufficient to move

NGC1851

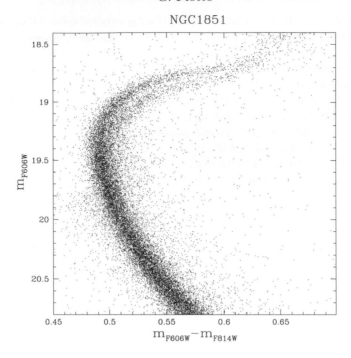

Figure 5. The double SGB in NGC 1851. The two SGBs are separated by about 0.12 magnitudes in F606W (from Milone *et al.* 2007)

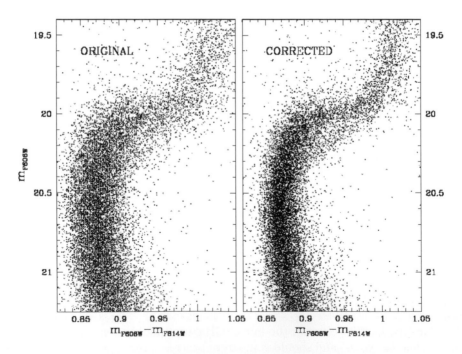

Figure 6. After the correction for differential reddening, also NGC 6388 shows a double SGB (Piotto *et al.* 2008, *in preparation*).

stars from the red to the blue side of the RR Lyrae instability strip. Direct spectroscopic measurements of the SGB and HB stars in NGC 1851 are badly needed.

There is at least another cluster which undoubtedly shows a split in the SGB: NGC 6388 (Piotto *et al.* 2008, *in preparation*). Figure 6 shows that, even after correction for differential reddening, the SGB of NGC 6388 closely resembles the SGB of NGC 1851. NGC 6388, as well as its twin cluster NGC 6441, are two extremely peculiar clusters. Since Rich *et al.* (1997), we know that, despite their high metal content, higher than in 47 Tucanae, they have a bimodal HB, which extends to extremely hot temperatures (Busso *et al.* 2007), totally un-expected for this metal rich cluster. NGC 6388 stars also display a NaO anticorrelation (Carretta *et al.* 2007). Unfortunately, available data do not allow us to study the MS of this cluster, searching for a MS split. Hopefully, new data coming from the HST program GO11233 should help to constrain the MS width, and therefore the He dispersion. In this context, it is worth noting that Caloi and D'Antona (2007), in order to reproduce the HB of NGC 6441, propose the presence of three populations, with three different He contents, one with an extreme He enhancement of Y=0.40. Such a strong enhancement should be visible in a MS split, as in the case of ω Cen and NGC 2808. A strong He enhancement and a consequent MS split may also apply to NGC 6388, because of the many similitudes with NGC 6441..

One more cluster, M54, shows a complex CMD (see, e.g., Siegel *et al.* 2007). This cluster has been shown, however, in too many papers to cite here, to be a part of the Sagittarius dwarf galaxy that is in process of merging into the Milky Way, and very possibly the actual nucleus of that galaxy. Actually, it is still matter of debate which parts of the CMD of M54 represent the cluster population and which ones are due to the Sagittarius stars. M54 may have a complex stellar populations as ω Cen, though this fact will be much harder to demonstrate.

Finally, we note that the multiple population phenomenon in star clusters may not be confined only to Galactic GCs. Mackey & Broby Nielsen (2007) suggest the presence of two populations with an age difference of $\sim$300 Myr in the 2 Gyr old cluster NGC 1846 of the Large Magellanic Cloud (LMC). In this case, the presence of the two populations is inferred by the presence of two TOs in the CMD (Fig. 7). These two populations may either be the consequence of a tidal capture of two clusters or NGC 1846 may be showing something analogous to the multiple populations identified in the Galactic GCs. NGC 1846 might not be an exception among LMC clusters. Vallenari *et al.* (1994) already suggested the possibility of the presence of two stellar populations in the LMC cluster NGC 1850. A quick analysis of the CMDs of about 50 clusters from ACS/HST images shows that about 10% of them might show evidence of multiple generations (Milone *et al.* 2008, *in preparation*).

3. Discussion

So far, we have identified four Galactic globular clusters for which we have a direct evidence of multiple stellar populations, and they are all quite different:

(*a*) In ω Centauri ($\sim 4 \times 10^6 \ M_\odot$), the different populations manifest themselves both in a MS split (interpreted as a split in He and metallicity abundances) and in a SGB split (interpreted in terms of He, metallicity, and age variations > 1 Gyr) which implies at least four different stellar groups within the same cluster, which formed in a time interval greater than 1 Gyr. Omega Centauri has also a very extended HB (EHB), which extends far beyond 30.000K.

(*b*) In NGC 2808 ($\sim 1.6 \times 10^6 \ M_\odot$), the multiple generation of stars is inferred from the presence of three MSs (also in this case interpreted in terms of three groups of stars,

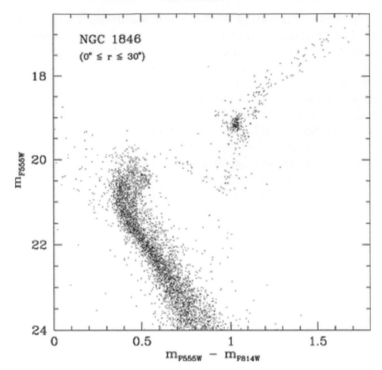

Figure 7. The double main sequence turn off of the Large Magellanic Cloud cluster NGC 1846 (from Mackey & Broby Nielsen 2007).

with different He content), possibly linked to three stellar groups with different oxygen abundances, and possibly to the multiple HB. The age difference between the 3 groups is significantly smaller than 1 Gyr. It also has an EHB, extended as much as the HB of ω Cen. It shows an extended NaO anticorrelation.

(c) In the case of NGC 1851 ($\sim 1.0 \times 10^6\ M_\odot$), we have evidence of two stellar groups from the SGB split. It is difficult to establish the age difference between the two stellar populations without a detailed chemical abundance analysis. However, available observational evidence seems to imply that the SGB split may be due to a difference in CN, Na, O, and s-process elements, while the age difference could be small (e.g. as small as in the case of NGC 2808). If the SGB split would be due only to age, the two star formation episodhes should have been happened with a time separation of ~ 1 Gyr. From the analysis of the cluster CMD, there seems to be no MS split, which would imply a small He spread, if any ($\Delta Y < 0.03$). The cluster has no EHB, but it shows a bimodal HB. It shows a NaO anticorrelation.

(d) In NGC 6388 ($\sim 1.6 \times 10^6\ M_\odot$) we have evidence of two stellar groups from a SGB split. With the available observational data it is not possible to establish whether there is a split in the MS of this GC. NGC 6388 has an EHB, possibly as extended as in the cases of NGC 2808 and ω Cen. It shows an extended NaO anticorrelation.

Another massive ($\sim 2.0 \times 10^6\ M_\odot$) GC, M54, is suspected to host multiple populations, though the analysis is strongly hampered by the contamination of the Sagittarius galaxy. Also M54 has an EHB, similar to the HB of NGC 2808 and ω Cen (Rosenberg *et al.* 2004).

At least one LMC intermediate age cluster shows a population split at the level of the TO: NGC 1846. This is a massive clusters, among the most massive LMC clusters according to Chrysovergis *et al.* (1999), though probably not as massive as the above Galactic GCs (a more accurate mass estimate for this cluster is needed). Other LMC clusters are suspected to shows a similar TO splitting.

Many GCs are clearly not simple, single-stellar-population objects. The emerging evidence is that the star-formation history can vary strongly from GC to GC, and that, GCs are able to produce very unusual objects, as no such He-rich MS stars have ever been found elsewhere. At the moment, we can note that the three GCs in which multiple generations of stars have been clearly identified (Omega Cen, NGC 2808, and NGC 1851), and the two other GCs suspected to contain more than one stellar generation (NGC 6388 and NGC 6441: Fig 6, Caloi & D'Antona 2007, Busso *et al.* 2007) are among the ten most massive GCs in our Galaxy. This evidence suggests that cluster mass might have a role in the star-formation history of GCs.

Reconstruction of this star-formation history requires a a better understanding of the chemical enrichment mechanisms, but the site of hot H-burning remains unclear. There are two requisites: (i) temperature should be high enough; and (ii) the stars where the burning occur should be able to give back the processed material to the intracluster matter at a velocity low enough that it can be kept within the GC itself (a few tens of km/s). Candidates include: (i) Massive ($M > 10M_\odot$) rotating stars (Decressin *et al.* 2007); (ii) the most massive among the intermediate mass stars undergoing hot bottom burning during their AGB phase (Ventura *et al.* 2001). The two mechanisms act on different timescales (10^7 and 10^8 yr, respectively), and both solutions have their pros and cons. The massive star scenario should avoid mixture of O-poor, Na-rich material with that rich in heavy elements from SNe, while it is not clear how the chemically processed material could be retained by the proto-cluster in spite of the fast winds and SN explosions always associated to massive stars. Producing the right pattern of abundances from massive AGB stars seems to require considerable fine tuning. In addition, both scenarios require that either the IMF of GCs was very heavily weighted toward massive stars, or that some GCs should have lost a major fraction of their original population (Bekki and Norris 2006), and then may even be the remnants of tidally disrupted dwarf galaxies, as suggested by the complexity in the CMD of ω Cen and M54.

The observational scenario is becoming more complex, but, the new results might have indicated the right track for a comprehensive understanding of the formation and early evolution of GCs. We are perhaps for the first time close to compose what has been for decades and still is a broken puzzle.

Acknowledgements. I wish to warmly thank J. Anderson, Andrea Bellini, Luigi R. Bedin, Ivan R. King, Antonino P. Milone, without whom most of the results presented in this review would not have been possible. I wish also to thank Sandro Villanova for his help in the spectroscopic investigation of Omega Centauri main sequence and turn off stars. A special thanks to Alvio Renzini and Raffaele Gratton for the many enthusiastic discussions on the subject of multipopulations in globular clusters.

References

Bedin, L. R., Piotto, G., Anderson, J., Cassisi, S., King, I. R., Momany, Y., & Carraro, G. 2004, *ApJ* (Letters), 605, L125

Bekki, K. & Norris, J. E. 2006, *ApJ* (Letters), 637, L109

Busso *et al.* (2007), *A&A*, 474, 105

Caloi, V. & D'Antona, F. 2007, *A&A*, 463, 949

Carretta, E., Bragaglia, A., Gratton, R. G., Leone, F., Recio-Blanco, A., & Lucatello, S. 2006, *A&A*, 450, 523

Carretta, E. *et al.* 2007, *A&A*, 464, 957

Cassisi, S., Salaris, M., Pietrinferni, A., Piotto, G., Milone, A. P., Bedin, L. R., & Anderson, J. 2007, arXiv0711.3823

Charbonnel, C. 1994, *A&A*, 282, 811

Chrysovergis, M., Kontizas, M., & Kontizas, E. 1989, *A&A*S, 77, 357

Cohen, J. G., Briley, M. M., & Stetson, P. B. 2002, *AJ*, 123, 2525

D'Antona, F. & Caloi, V. 2004, *ApJ*, 611, 871

D'Antona, F., Bellazzini, M., Caloi, V., Pecci, F. Fusi, Galleti, S., & Rood, R. T. 2006, *ApJ*, 631, 868

Decressin, T., Meynet, G., Charbonnel, C., Prantzos, N., & Ekström, S. 2007, *A&A*, 464, 1029

Ferraro, F. R., Paltrinieri, B., Fusi Pecci, F., Rood, R. T., & Dorman, B. *ApJ*, 500, 311

Fusi Pecci, F., Ferraro, F. R., Bellazzini, M., Djorgovski, S., Piotto, G., & Buonanno, R. 1993, *AJ*, 105, 1145

Gratton, R., Sneden, C., Carretta, E., & Bragaglia, A., *A&A*, 354, 169

Gratton, R. *et al.* 2001, *A&A*, 369, 87

Gratton, R., Sneden, C., & Carretta, E. 2004, *ARAA*, 42, 385

Hesser, J. E., Bell, R. A., Harris, G. L. H., & Cannon, R. D. 1982, *AJ*, 87, 1470

Mackey, A. D. & Broby Nielsen, P. 2007, *MNRAS*, 379, 151

Milone, A. P. *et al.* 2007, in press, arXiv0709.3762

Piotto, G., Zoccali, M., King, I. R., Djorgovski, S. G., Sosin, C., Rich, R. M., & Meylan, G. 1999, *AJ*, 118, 1727

Piotto, G., *et al.* 2005, *ApJ*, 621, 777 (P05)

Piotto, G., *et al.* 2007, *ApJ* (Letters), 661, L53 (P07)

Prantzos, N., Charbonnel, C., & Iliadis, C. 2007, *A&A*, 470, 179

Recio-Blanco, A., Aparicio, A., Piotto, G., de Angeli, F., & Djorgovski, S. G. 2006, *A&A*, 452, 875

Rich, R. M., Sosin, C., Djorgovski, S. G., Piotto, G., King, I. R., Renzini, A., Phinney, E. S., Dorman, B., Liebert, J., & Meylan, G., 1997, *ApJ* (Letters), 484, L25

Rosenberg, A., Recio-Blanco, A.,& García-Marn, M. 2004, *ApJ*, 603, 135

Sandage, A. & Wildey, R. 1967, *ApJ*, 150, 469

Siegel *et al.* 2007, *ApJ* (Letters), 667, L57

Sollima, A., Pancino, E., Ferraro, F. R., Bellazzini, M., Straniero, O., & Pasquini, L. 2005, *ApJ*, 634, 332

Sollima, A., Ferraro, F. R., Bellazzini, M., Origlia, L., Straniero, O., & Pancino, E. 2007, *ApJ*, 654, 915

Sweigart, A. V., & Mengel, J. G. 1979, *ApJ*, 229, 624

Ventura, P., D'Antona, F., Mazzitelli, I., & Gratton, R. 2001, *ApJ* (Letters), 550, L65

Vallenari, A., Aparicio, A., Fagotto, F., Chiosi, C., Ortolani, S., & Meylan, G. 1994, *A&A*, 284, 447

Villanova, S., *et al.* 2007, *ApJ*, 663, 296

van den Bergh, S. *AJ*, 72, 70

Yong, D. & Grundahl, F. 2007, arXiv0711.1394

Dynamical Evolution of Dense Stellar Systems
Proceedings IAU Symposium No. 246, 2007
E. Vesperini, M. Giersz & A. Sills, eds.

Effects of Stellar Collisions on Star Cluster Evolution and Core Collapse

Sourav Chatterjee, John M. Fregeau and Frederic A. Rasio

Department of Physics and Astronomy, Northwestern University, Evanston, IL 60208, USA
email: s-chatterjee@northwestern.edu, fregeau@northwestern.edu, rasio@northwestern.edu

Abstract. We systematically study the effects of collisions on the overall dynamical evolution of dense star clusters using Monte Carlo simulations over many relaxation times. We derive many observable properties of these clusters, including their core radii and the radial distribution of collision products. We also study different aspects of collisions in a cluster taking into account the shorter lifetimes of more massive stars, which has not been studied in detail before. Depending on the lifetimes of the significantly more massive collision products, observable properties of the cluster can be modified qualitatively; for example, even without binaries, core collapse can sometimes be avoided simply because of stellar collisions.

Keywords. scattering, stellar dynamics, methods: numerical, (stars:) blue stragglers, (Galaxy:) globular clusters: general

1. Introduction

In dense stellar systems like massive young clusters, galactic centers, and old globular clusters (GC), the high densities of stars can give rise to many direct, single–single physical collisions, in addition to collisions mediated by dynamical interactions of binaries (Fregeau *et al.* 2004). In these systems stellar evolution can also be significantly modified through physical collisions. For example, in young star clusters, collisions can give rise to runaway growth of merger products, producing many exotic stellar populations, such as intermediate-mass black holes (Gürkan *et al.* 2006; Trenti 2006). The cores of typical old Galactic GCs can also attain high enough densities so that most core stars undergo collisions during their lifetime (Hills & Day 1976; Fregeau *et al.* 2004). These collisions not only change the evolution of individual stars, but the increased stellar masses also change the overall GC properties like the core radius (r_c) and the half-mass radius (r_h). Moreover, at least some of the observed exotic stellar populations in dense clusters, like blue straggler stars (BSS) (Fekadu *et al.* 2007; Sills *et al.* 2001; Sills *et al.* 1997), and compact binaries like ultracompact X-ray binaries (UCXB), are likely created through collisions (Lombardi *et al.* 2006).

A dense cluster of stars naturally evolves towards eventual core collapse through relaxation. At their present ages, most Galactic GCs are expected to have collapsed cores. However, observations show that the measured values of r_c/r_h for most Galactic GCs are higher than predicted by theoretical models (Vesperini & Chernoff 1994). Many scenarios have been proposed to explain this apparent discrepancy. For example, the core can be supported against deep collapse by dynamically extracting the binding energy of hard primordial binaries (Trenti *et al.* 2007; Fregeau & Rasio 2007). However, it is hard to explain most of the high observed values of r_c/r_h in Galactic GCs purely through this "binary burning" process. Other mechanisms to halt core collapse like ejection of stellar mass BHs from the core (Mackey *et al.* 2007) or stellar captures by a central

Table 1. Initial Conditions of Simulated GCs

	Model	IMF	N	n_{binary}	r_c (pc)	Virial radius (pc)	ρ_c $(M_\odot/pc^3)$
Cases 1,2	Plummer	Single-Mass	10^6	0	0.3	0.85	10^6
Case 3	King, $w_0 = 6$	Kroupa $(0.1 - 2.0 M_\odot)$	10^6	0	0.87	2.89	1.7×10^4

intermediate-mass black hole (Trenti *et al.* 2007; Trenti 2006) have also been discussed at this meeting (see contributions by Mackey and Trenti in this volume). It has also been suggested that r_c/r_h could simply keep increasing over long timescales in clusters having fairly long relaxation times via ongoing mass segregation (Merritt *et al.* 2004).

Here we study stellar collisions as a possible mechanism for supporting clusters against core collapse. In a regime where collisions are important, they can produce many stars significantly more massive than those in the background population. Thus the subsequent evolution of collision products will be much faster than for normal stars in the cluster. Although the stellar evolution and observable properties of collision products have been extensively studied before (Sills *et al.* 2001; Sills *et al.* 1997), the feedback effects of collisions on the overall dynamical evolution of GCs has received less attention. The shorter evolution timescales of massive collision products may support the core against collapse even without any primordial binaries or other mechanisms for energy production, simply via indirect heating through stellar evolution mass loss (Goodman & Hernquist 1991; Goodman & Hut 1989; Lee 1987).

Using N-body simulations, we have begun studying numerically how collisions and the subsequent evolution of collision products can alter the overall properties of GCs. We use the Northwestern group's Hénon-type Monte-carlo code, which provides a detailed, star-by-star representation of clusters with up to $N \sim 10^6 - 10^7$ stars (Fregeau & Rasio 2007; Joshi *et al.* 2001; Joshi *et al.* 2000). In §2 we will first present two simple limiting cases, bracketing reality and illustrating the dramatic changes in global cluster properties depending on how the evolution of collision products is treated in the models. We also present the evolution of a more realistic GC model with a conventional "rejuvenation" prescription for determining the lifetimes of collision products. We discuss the implications of our study and planned future work in §3.

2. Results

All simulations shown here use $N = 10^6$ stars initially. We assume totally conservative collisions with the interaction cross section given by the usual "sticky sphere" approximation. There is no hydrodynamic mass loss during the collisions. However, mass loss occurs through normal evolution of the stars via (instantaneous) compact object formation at the end of the main sequence (MS) lifetime. The collision products can be significantly more massive and hence evolve faster, enhancing the overall stellar mass loss rate. Determining theoretically the true MS lifetime or the effective age of a collision product is a hard task. These quantities depend sensitively on the amount of hydrogen mixed into the core of the collision product as a result of the collision. This in turn can be highly variable because the mixing depends on the details of the collision kinematics. To avoid these uncertainties, in our study we use three very simple prescriptions for determining the effective ages and the remaining MS lifetimes of collision products. **Case 1:** Infinite lifetime for all stars, including the collision products. **Case 2:** Zero lifetime for the collision products. Other stars have infinite lifetime, unless they collide. **Case 3:** A more realistic "rejuvenation" prescription for determining the age and MS lifetime of the

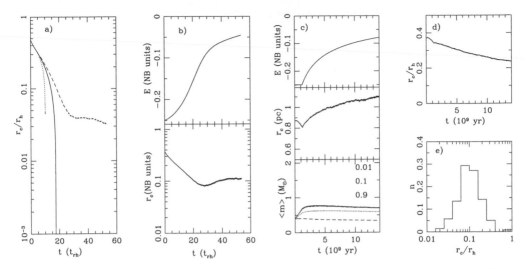

Figure 1. *a)* Evolution of r_c/r_h. Solid line: no collisions. Dotted line: *case* 1, where all stars have infinite lifetime (including collision products). Dashed line: *case* 2, where collision products have zero lifetime. In *case* 2 collapse is avoided through mass loss due to evolution of the collision products. *b)* Evolution of E (top panel) and the core radius (bottom panel). The steepest slope in the E evolution curve corresponds to core collapse. Onset of collapse can also be seen directly in the r_c evolution plot. *c)* Evolution of E (top panel), evolution of r_c in physical units (middle panel), and Evolution of the average mass contained in three Lagrange radii $(0.01, 0.1, 0.9)$ of the GC (bottom panel) for *case* 3. *d)* Evolution of r_c/r_h for *case* 3. *e)* Histogram of r_c/r_h of all Galactic GCs with known r_c and r_h values.

collision products: we assume that stars burn H linearly on the MS so the effective age of a collision product is uniquely determined by the total amount of H coming from the progenitors. The age and the MS lifetime of each collision product is determined by the following simple equation, based on Hurley *et al.* (2002),

$$t = \frac{t_{MS}}{M}\left(\frac{t_1 M_1}{t_{MS1}} + \frac{t_2 M_2}{t_{MS2}}\right). \tag{2.1}$$

Here, t_i, M_i, and t_{MSi} are the age, mass, and the MS lifetime of parent star i, while t, M, and t_{MS} are for the collision product. Note that the coefficient on the right hand side of Eq. 2.1 is different from the one used in Hurley *et al.* (2002). We use Eq. 2.1 so that two zero age main sequence (ZAMS) stars collide to produce a ZAMS star, whereas two stars close to their MS turn-off collide to produce another star close to its turn-off.

2.1. *Limiting cases: Single mass Plummer model (cases 1 & 2)*

The initial cluster properties are listed in Table 1. Fig. 1a shows the evolution of r_c/r_h for the two limiting cases as well as the case without collisions for comparison purposes. Clearly, conservative collisions make the cluster more bound manifested by the faster collapse of the core than the no collision case in **case** 1. In the other limiting case, **case 2**, on the onset of collapse the densities reach a very high value increasing the collision rate. The mass loss after the collisions can stop the collapse and reach a steady core radius with a low rate of further collisions. This mechanism can be better illustrated by Fig. 1b. We define energy E in such a way that it remains constant unless mass is lost from the cluster. The onset of collapse can be seen in the simultaneous decrease of r_c and the increased slope of the E evolution curve.

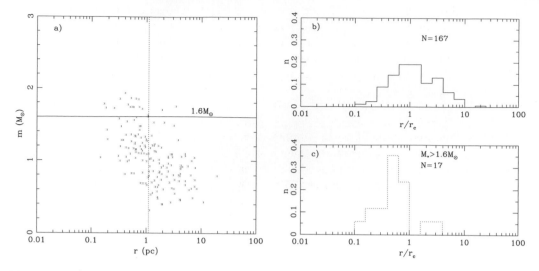

Figure 2. *a)* Position vs mass scatter plot for all collision products still at their MS life at 14 Gyr. Two times the present age turn-off mass ($1.6M_\odot$) is shown as a horizontal solid line to guide the eye. The position of the core is shown with the vertical dotted line. *b)* Histogram showing the positions of the same population as in *(a)*. *c)* Histogram showing the positions of the collision products still in their MS life at present age having masses $\geqslant 1.6M_\odot$.

2.2. *More realistic case: rejuvenation-based prescription; King sphere, $w_0 = 6$ (Case 3)*

The initial cluster properties are listed in Table 1. Here we first evolve the cluster with only stellar evolution for 10^7 years, so that the very massive stars evolve and disappear from the system. We then switch on dynamics.

Till a little more than 1 Gyr E remains constant, indicating no collisions (Fig. 1c). Sometime between 1 & 2 Gyr, the core starts to collapse (Fig. 1c) reducing the core radius and increasing the average mass in the core region. Thus collision rate increases significantly increasing rate of mass loss. This is manifested in the sudden steep increase in E. Thus through mass loss core collapse is avoided and r_c/r_h approaches a steady value (Fig 1d). Fig 1d,e shows that, even with the simple rejuvenation prescription, the final value of r_c/r_h compares well with the observed values for most Galactic GCs. The slow continued increase of E at later times comes as normal stars as well as some low mass collision products go off the MS and disappear.

Fig. 2a shows the positions of the collision products that are still in their MS life at 14 Gyr. Most of these collision products are contained within the core. BSS candidates are most easily identified above ~ 2 times the turn-off mass of the cluster. The more massive subset of collision products in our simulated cluster ($M \geqslant 1.6M_\odot$) are mostly found inside the core (Fig. 2c) (apart from only 2 just outside the core).

3. Discussion

Although determination of stellar evolution after a collision is a subject of continuing research, the effects of collisions on cluster dynamics have not been studied in detail previously. This is a first report of an ongoing systematic study on this effect. Using typical initial conditions for old GCs and simple assumptions for rejuvenation we show that collisions between single stars not only alter the stellar properties and produce exotic stellar populations like some of the BSS, they also affect the overall GC properties. Most importantly, collisions can support the core of a cluster against collapse in typical old Galactic GCs. Even with our simple but reasonable assumptions, the values of r_c/r_h obtained for our simulated GCs compare well with the observed r_c/r_h values for Galactic GCs. Furthermore, we obtain a population of BSS candidates contained within the core, also consistent with observations (Leigh *et al.* 2007). Note that we do not find any BSSs well outside the core, consistent with the current understanding that those BSSs are most likely formed via primordial binary mergers (Mapelli *et al.* 2006).

We have adopted many extreme simplifications for this first look at the problem. For example, at this stage of our study, once a star evolves off the MS, it is removed from the simulation, leaving no remnant. Since remnants are normally only a few percent of the total progenitor mass, we expect that this approximation will not affect the overall GC properties significantly, so far as the increase in energy is concerned from mass loss. However, some remnants from very massive stars at a very young age can remain in the cluster and sink into the core through dynamical friction. Dynamical interactions, including collisions, of these massive remnants can also alter the core properties of the GCs in certain regimes (Mackey *et al.* 2007; Trenti *et al.* 2007; Trenti 2006). Another possibly important effect left out of this study for now is the role of primordial binaries (Fregeau & Rasio 2007; Fregeau *et al.* 2004). On the one hand, the presence of primordial binaries will increase the rate of collisions through resonant encounters (Fregeau *et al.* 2004), on the other hand, binaries will provide further support of a cluster against collapse and hence may prevent the core from reaching high enough densities for significant collisions.

Acknowledgements

This work was supported by NASA Grants NNG04G176G and NNG06GI62G.

References

Fekadu, N., Sandquist, E. L., & Bolte, M. 2007, *ApJ* 663, 277
Fregeau, J. M. & Rasio, F. A. 2007, *ApJ* 658, 1047
Fregeau, J. M., Cheung, P., Potegies Zwart, S. F., & Rasio, F. A. 2004, *MNRAS* 352, 1
Goodman, J. & Hernquist, L. 1991, *ApJ* 378, 637
Goodman, J. & Hut, P. 1989, *Nature* 339, 40
Gürkan, M. A., Fregeau, J. M., & Rasio, F. A. 2006, *ApJ* 640, L39
Hills, J. G. & Day, C. A. 1976, *Aplett* 17, 87
Hurley, J. R., Tout, C. A., & Pols, O. R. 2002, *MNRAS* 329, 897
Joshi, K. J., Nave, C. P., & Rasio, F. A. 2001, *ApJ* 550, 691
Joshi, K. J., Rasio, F. A., & Portegies Zwart, S. 2000, *ApJ* 540, 969
Lee, H. M. 1987, *ApJ* 319, 801
Leigh, N., Sills, A., & Knigge, C. 2007, *ApJ* 661, 210
Lombardi, Jr., J. C., Proulx, Z. F., Dooley, K. L., Theriault, E. M., Ivanova, N., & Rasio, F. A. 2006, *ApJ* 640, 441
Mackey, A. D., Wilkinson, M. I., Davies, M. B., & Gilmore, G. F. 2007, *MNRAS* 379, L40
Mapelli, M., Sigurdsson, S., Ferraro, F. R., Colpi, M., Possenti, A., & Lanzoni, B. 2006, *MNRAS* 373, 361
Merritt, D., Piatek, S., Portegies Zwart, S., & Hemsendorf, M. 2004, *ApJLett* 608, L25
Sills, A., Faber, J. A., Lombardi, Jr., J. C., Rasio, F. A., & Warren, A. R. 2001, *ApJ* 548, 323
Sills, A., Lombardi, Jr., J. C., Bailyn, C. D., Demarque, P., Rasio, F. A., & Shapiro, S. L. 1997, *ApJ* 487, 290
Trenti, M. 2006, *ArXiv Astrophysics e-prints* 12040
Trenti, M., Ardi, E., Mineshige, S., & Hut, P. 2007, *MNRAS* 374, 857
Vesperini, E. & Chernoff, D. F. 1994, *ApJ* 431, 231

Dynamical Evolution of Dense Stellar Systems
Proceedings IAU Symposium No. 246, 2007
E. Vesperini, M. Giersz & A. Sills, eds.

Multiple Stellar Populations in Globular Clusters: Collection of Information from the Horizontal Branch

Francesca D'Antona[1] and Vittoria Caloi[2]

[1]INAF - Osservatorio Astronomico di Roma, via Frascati 33, I-00040 Monte Porzio, Italy
email: dantona@oa-roma.inaf.it

[2]INAF - IASF, via Fosso del Cavaliere, I-00133 Roma, Italy
email: vittoria.caloi@iasf-roma.inaf.it

Abstract. The majority of the inhomogeneities in the chemical composition of Globular Cluster (GC) stars appear due to primordial enrichment by hot-CNO cycled material processed in stars belonging to a first stellar generation. Either massive AGB envelopes subject to hot bottom burning, or the envelopes of massive fastly rotating stars could be the progenitors. In both cases, the stars showing chemical anomalies must have also enhanced helium abundance, and we have proposed that this higher helium could be at the basis of the many different morphologies of GC horizontal branches (HB) for similar ages and metallicities. The helium variations have been beautifully confirmed by the splitting of the main sequence in the clusters ω Cen and NGC 2808, but this effect can show up only for somewhat extreme helium abundances. Therefore it is important to go on using the HB morphology to infer the number ratio of the first to the second generation in as many clusters as possible. We exemplify how it is possible to infer the presence of a He − rich stellar component in different clusters thanks to different HB features (gaps, RR Lyr periods and period distribution, ratio of blue to red stars, blue tails). In many clusters at least 50% of the stars belong to the second stellar generation, and in some cases we suspect that the stars might all belong to the second generation. We shortly examine the problem of the initial mass function required to achieve the observed number ratios and conclude that: 1) the initial cluster must have been much more massive than today's cluster, and 2) formation of the second stellar generation mainly in the central regions of the cluster may help in obtaining the desired values.

Keywords. globular clusters: general − globular clusters: formation − globular clusters: individual NGC 2808, NGC 6388, NGC 6441, M 3 − Stars: Horizontal Branch

1. Introduction

The observations of GC stars are still to be interpreted in a fully consistent frame. Nevertheless, a general consensus is emerging on the fact that most GCs can not be considered any longer "simple stellar populations", and that "self − enrichment" is a common feature among GCs. This consensus has been the consequence of three independent lines of evidence:

• Spectroscopic observations: the discovery of "chemical anomalies", such as the Na − O and Mg − Al anticorrelation, dates back to the seventies. The anomalies are now observed also at the turnoff (TO) and among the subgiants (e.g.,Gratton *et al.* 2001, Briley *et al.* 2002, 2004), so they must be attributed to some process of "self − enrichment" occurring at the first stages of the cluster life. There must have been a first epoch of star formation that gave origin to the "normal" (first generation) stars, with CNO and other abundances similar to the population II field stars of the same metallicity. Afterwards, there must have been some other epoch of star formation, including material heavily processed through the CNO cycle. This material either was entirely ejected by stars belonging to the first stellar generation, or it is a mixture of ejected and pristine matter of the initial star forming cloud. We can derive this conclusion as a consequence of the fact that there is no appreciable difference in the metallicity of the "normal" and chemically anomalous stars belonging to the same GC. (Needless to say, this statement *does not* hold for ω Cen, which must indeed be considered a small galaxy and not a typical GC. In the following, we will only examine "normal clusters", those which do not show signs of metal enrichment due to supernovae ejecta). This is an important fact that tells us, e.g., that it is highly improbable that the chemical anomalies are due to mixing of stars born in two different clouds, as there is no reason why the two clouds should have a unique metallicity. In addition, the clusters showing chemical anomalies have a huge variety of metallicities, making the suggestion of mixing of two different clouds even more improbable. The matter must have been processed through the hot CNO cycle, and not, or only marginally, through the helium burning phases, as the sum of CNO elements is the same in the "normal" and anomalous stars (Cohen & Meléndez 2005). The progenitors then may be either massive Asymptotic Giant Branch (AGB) star (Ventura *et al.* 2001, 2002) or fast rotating massive stars (Decressin *et al.* 2007).

• Interpretation of the Horizontal Branch morphology in terms of helium content variations among the GC stars. Whichever the progenitors, some helium enrichment must be present in the matter processed through hot CNO. D'Antona *et al.* (2002) recognized that this could have a strong effect on the Horizontal Branch (HB) morphology, and even help to explain some features (gaps, hot blue tails, second parameter effect) which had defied all reasonable alternative explanation. A wide variety of problems has been examined in the latest years: the very peculiar morphology of the HB in the massive cluster NGC 2808, interpreted in terms of varying helium among its stars; the second parameter effect in the clusters M 13 and M 3 (Caloi & D'Antona, 2005); the very peculiar features of the GCs NGC 6441 and NGC 6388 (Caloi & D'Antona, 2007a), which can all be modeled by assuming that quite a large fraction of stars have high helium. Hints that this was correct came from the analysis of the helium content in hot HB stars (Moehler *et al.* 2004).

• Photometric splitting of the main sequence in a few clusters. The necessity of a varying helium content in NGC 2808 was later confirmed by the first analysis of the cluster main sequence (D'Antona *et al.* 2005), that showed a tail of "blue" stars. This could only be interpreted as a very helium rich sequence. In fact, Carretta *et al.* (2006) had shown that the metallicity of oxygen poor and oxygen normal stars is the same. The recent new HST observations by Piotto *et al.* (2007) leave no doubt that there are at least three different populations in this cluster. This came after the first discovery of a peculiar blue main sequence in ω Cen (Bedin *et al.* 2004), interpreted again in terms of a very high helium content (Norris 2004, Piotto *et al.* 2005). The HB observations however have shown that the blue main sequences are only the tip of the iceberg of the self – enrichment. In most clusters the higher helium abundances remain confined below Y~ 0.30, and the presence of such stars will not be clearcut from main sequence observations (D'Antona *et al.* 2002, Salaris *et al.* 2006). Viceversa, if we wish to shed light on the entire process of formation of GCs we must have a rough idea of the total number of chemically anomalous stars.

Consequently, we decided to continue our investigation of the HB in as many clusters as possible, in the hypothesis that the HB morphology can be mainly interpreted in terms of a helium content distribution among the stars. We summarize here some of the results, as a basis to discuss the initial mass function of the first generation, required to produce the needed fraction of stars in the second generation.

2. Different clusters, different necessity for helium rich stars

We recall that the role of helium rich stars is different according to which is the basic HB morphology of the first generation stars (D'Antona *et al.* 2002).

2.1. *NGC 2808*

If the age, metallicity and mass loss are such that normal – helium stars populate a red clump, the stars with helium enhancement (which are less massive) will populate the bluer HB and the RR Lyr region. If there is a gap between the normal – helium stars and the *minimum* helium content of the second generation (a situation which does probably occur, if the processed matter comes from the massive AGBs), the case of NGC 2808 shows up: a red clump (first generation), almost no RR Lyr (due to the helium gap) and a blue part with larger helium content ($Y \sim 0.28$ according to D'Antona and Caloi 2004). In addition, the presence of a separate very high helium population -as derived from the main sequence- may explain the two blue tails of the HB (D'Antona *et al.* 2005). The cluster seems to be divided into 50% normal – helium stars, and 50% helium enriched stars, but remember that the very high helium ($Y \sim 0.40$) stars are only $\sim$15%.

2.2. *NGC 6441 and NGC 6388. And also 47 Tuc*

The case of these two high metallicity clusters is even more interesting: here the red clump extends for about a magnitude thickness, and any attempt to attribute this to differential reddening has failed (Raimondo *et al.* 2002). The RR Lyr have a very long period, unexplicable for the metallicity (Pritzl *et al.* 2000). And the HB extends even to the region of blue, hot stars. Caloi & D'Antona(2007a) show that this is another case study: the morphology requires not only some helium enrichment for the bluer side of the HB, as we could naively think, but extreme helium enrichment *even for the red clump stars!* In fact, the high helium – high metallicity helium core burning low mass stars make long loops from red to blue in the HB (Sweigart & Gross 1976). This is due to the fact that the higher mean molecular weight – leading to a high H – burning shell temperature – and the high metallicity – leading to a stronger CNO shell – both conspire towards the result that the H – shell energy source prevails with respect to the He – core burning. The consequent growth of the helium core leads evolution towards the blue. Therefore, if we must explain the luminous (long period) RR Lyr by stars having high helium, the same stars will also populate the red clump: this is exactly what we observe: if the helium content is not as large as $Y \sim 0.35$ *in the red clump*, the HB finds no satisfactory explanation. The percentage of helium enriched stars is in this case $\sim$60% for NGC 6441 and the same for NGC 6388. The main difference among the two clusters is that NGC 6388 seems to have a higher tail of very high helium ($Y>0.35$) stars, reaching $\sim$20%. If we analyze 47 Tuc, the prototype of metallic GCs, the red clump seems to imply higher helium only for $\sim$25% of the stars. Why a cluster almost as massive as the other two is so different – more normal – remains to be explained.

2.3. *M3 and the problem of RR Lyr*

The case of M3 is entirely different: it has a well populated red HB, variable region and blue side (R, V and B samples), with no blue tails, so it has always been taken as the prototype of HBs: its color distribution can be reproduced by assuming an average mass loss along the RGB, with a standard deviation $\sigma \sim 0.025 M_\odot$. Unfortunately, the RR Lyr period distribution *is not* so easily explained, as it is terribly peaked! Castellani *et al.* (2005) realized that the only way to reproduce this peak was to reduce the dispersion in mass loss rate: unfortunately, they also had to add a different average mass loss, with a different spread, to account for the blue side of the HB! Of course, if we assume that the

Table 1. Helium history of 8 clusters

NGC 2808		NGC 6441		NGC 6388		47 Tuc		M3		M 5		M53		M13	
Y	%	Y	%	Y	%	Y	%	Y	%	Y	%	Y	%	Y	%
0.24	50	0.25	38	.25	39	.25	75	.24	.50	.24	.40	.24	1.	.24	0.0
0.26–0.29	35	0.27–0.35	48	0.27–0.35	41	0.27–.32	25	.26	50	.26	60	>.24	0.	.28	1.0
~0.4	15	>0.35	14	>0.35	20										

blue side is populated by helium rich stars, the conundrum is solved, as shown by Caloi & D'Antona, 2007b. In our simulations, we can explain both the period distribution and the color distribution along the HB for the R, V and B regions. This analysis poses another problem: we find that the dispersion in mass loss along the red giant branch must be at most $\sigma \sim 0.003 M_\odot$ to be consistent with the period distribution. We now pose the question whether this small dispersion is peculiar to M3 or we have always been mislead by the HB morphology, when assuming dispersion in mass loss of some hundreths of $M_\odot$.

2.4. *Preliminary analysis of a few other clusters*

We list in Table 1 the results of the synthetic HB computations, including also the preliminary analysis for some other clusters. M5 is similar to M3, but has a larger fraction of higher helium stars (the peak of the number of stars is in the blue HB). M53 is a massive cluster, but its entirely blue HB could be explained by a single (normal) stellar population, maybe with a small tail of stars having slightly larger mass loss. We list in the table also M13, for which the analysis implying that only the second generation has survived in the cluster is taken from the relative location of the red giant bump, the turnoff and the HB (Caloi & D'Antona, 2005).

3. How did the GCs form?

Any model for the GC formation must be able to deal with this variety of results: there are clusters with no self – enrichment, and clusters which might have lost entirely their first generation. It is true that the most massive clusters have extreme helium enhancements, but also moderately massive clusters show considerable degrees of helium variations, and a *small* cluster like NGC 6397, which is apparently monoparametric, might be composed entirely of second generation stars: in fact the nitrogen abundance of all its stars is severely enhances, like in CNO processed material.

The typical situation is than 50% normal – helium stars and 50% enhanced – helium stars. It is almost obvious that the ejecta of a unique first stellar generation with a normal initial mass function (IMF) can not produce enough mass to give origin to such a large fraction of second generation stars (see, e.g., the case made by Bekki and Norris (2004), for the blue main sequence of ω Cen).

The only solution to this problem is that the starting initial mass from which the first generation is born was MUCH LARGER than todays first generation remnant mass (a factor 10 to 20 larger), so that the processed ejecta of the first generation provide enough mass to build up the second one. It is possible that a cooling flow collects the gas in the core, and that the second generation stars are preferentially born there (D'Ercole *et al.* , in preparation). This formation in the central cluster region may also by helped if the massive (or intermediate mass stars) progenitors were already segregated in the core when the second generation is born, as suggested by the observations of young clusters.

When the long term evolution leads to the loss of the external parts of clusters, these are populated preferentially by the first generation stars and the desired ratio of first to second generation can be achieved.

In some sense, the study of chemical anomalies leads us to the idea that *practically the GC forms in the second stellar generation*, and the remnant first generation is only the core fraction of the much less concentrated and wider system whose winds, collected in a cooling flow, gave birth to the peculiar stars! Should we then expect different velocity dispersions in first and second generation? Would the second generation at least in some cases- remain more concentrated than the first one? of course complete dynamical simulations of the fluid gas formation phase, coupled with N – body simulations to describe the interaction of the two stellar generations are needed to answer this question.

Acknowledgements

We would like to thank P. Ventura, A. D'Ercole and E. Vesperini for their support and collaboration through this adventure of building up a new view of globular clusters. We also thank the organizers of the IAU Symposium 246 for their successful effort to prepare a scientific and environmentally attractive meeting.

References

Bedin, L. R., Piotto, G., Anderson, J., Cassisi, S., King, I. R., Momany, Y., Carraro, G. 2004, *ApJ*, 605, L125
Bekki, K. & Norris, J. E. 2006, *ApJ Letters*, 637, L109
Briley, M. M., Cohen, J. G., & Stetson, P. B. 2002, *ApJ Letters*, 579, L17
Briley, M. M., Harbeck, D., Smith, G. H., & Grebel, E. K. 2004, *AJ*, 127, 1588
Caloi, V. & D'Antona, F. 2005, *A&A*, 121, 95
Caloi, V. & D'Antona, F. 2007, *A&A*, 463, 949
Caloi, V. & D'Antona, F. 2007, *ApJ*, in press
Carretta, E., Bragaglia, A., Gratton, R. G., Leone, F., Recio-Blanco, A., & Lucatello, S. 2006, *A&A*, 450, 523
Castellani, M., Castellani, V., & Cassisi, S. 2005, *A&A*, 437, 1017
Cohen, J. G. & Meléndez, J. 2005, *AJ*, 129, 303
D'Antona, F., Caloi, V., Montalbán, J., Ventura, P., & Gratton, R. 2002, *A&A* 395, 69
D'Antona, F. & Caloi, V. 2004, *ApJ*, 611, 871
D'Antona, F., Bellazzini, M., Caloi, V., Fusi Pecci, F., Galleti, S., & Rood, R. T. 2005, *ApJ*, 631, 868
Decressin, T., Meynet, G., Charbonnel, C., Prantzos, N., & Ekström, S. 2007, *A&A*, 464, 1029
Gratton, R. G. *et al.* 2001, *A&A*, 369, 87
Moehler S., Sweigart A.V., Landsman W.B., Hammer N.J., & Dreizler S. 2004, *A&A*, 415, 313
Norris, J. E. 2004, *ApJ Letters*, 612, L25
Piotto, G., *et al.* 2005, *ApJ*, 621, 777
Piotto, G., *et al.* 2007, *ApJ Letters*, 661, L53
Pritzl, B., Smith, H. A., Catelan, M., & Sweigart, A. V. 2000, *ApJ Letters*, 530, L41
Raimondo, G., Castellani, V., Cassisi, S., Brocato, E., & Piotto, G. 2002, *ApJ*, 569, 975
Salaris, M., Weiss, A., Ferguson, J. W., & Fusilier, D. J. 2006, *ApJ*, 645, 1131
Sweigart, A. V., & Gross, P. G. 1976, *ApJ Suppl. Series*, 32, 367
Ventura, P., D'Antona, F., Mazzitelli, I., & Gratton, R. 2001, *ApJ Letters*, 550, L65
Ventura, P., D'Antona, F., & Mazzitelli, I. 2002, *A&A*, 393, 215

Dynamical Evolution of Dense Stellar Systems
Proceedings IAU Symposium No. 246, 2007
E. Vesperini, M. Giersz & A. Sills eds.

Why Haven't Loose Globular Clusters Collapsed yet?

Guido De Marchi[1], Francesco Paresce[2] and Luigi Pulone[3]

[1] ESA, Space Science Department, 2200 AG Noordwijk, Netherlands
email: gdemarchi@rssd.esa.int
[2] INAF, Istituto di Astrofisica Spaziale e Fisica Cosmica, 40129 Bologna, Italy
email: paresce@iasfbo.inaf.it
[3] INAF, Observatory of Rome, 00040 Monte Porzio Catone, Italy
email: pulone@mporzio.astro.it

Abstract. We report on the discovery of a surprising observed correlation between the slope of the low-mass stellar global mass function (GMF) of globular clusters (GCs) and their central concentration parameter $c = \log(r_t/r_c)$, i.e. the logarithmic ratio of tidal and core radii. This result is based on the analysis of a sample of twenty Galactic GCs, with solid GMF measurements from deep HST or VLT data, representative of the entire population of Milky Way GCs. While all high-concentration clusters in the sample have a steep GMF, low-concentration clusters tend to have a flatter GMF implying that they have lost many stars via evaporation or tidal stripping. No GCs are found with a flat GMF and high central concentration. This finding appears counter-intuitive, since the same two-body relaxation mechanism that causes stars to evaporate and the cluster to eventually dissolve should also lead to higher central density and possibly core-collapse. Therefore, severely depleted GCs should be in a post core-collapse state, contrary to what is suggested by their low concentration. Several hypotheses can be put forth to explain the observed trend, none of which however seems completely satisfactory. It is likely that GCs with a flat GMF have a much denser and smaller core than suggested by their surface brightness profile and may well be undergoing collapse at present. It is, therefore, likely that the number of post core-collapse clusters in the Galaxy is much larger than thought so far.

Keywords. Stars: luminosity function, mass function – Galaxy: globular clusters: general

---

The dynamical evolution of globular clusters (GCs) is governed by the two-body relaxation process, whereby stars exchange energy via repeated distant encounters (see Spitzer 1987 and Elson, Hut & Ingaki 1987 for a review). Two-body encounters lead to the expansion of the outer regions of the cluster, while driving the stellar density in the central regions to increase dramatically towards an infinite value during the so-called core-collapse. The most visible effect of core-collapse is the appearance of a central cusp in the surface brightness profile of the cluster (Djorgovski & King 1986) and, correspondingly, an increase in the central concentration parameter $c = \log(r_t/r_c)$, i.e. the logarithmic ratio of tidal and core radii. Therefore, c has traditionally been seen as a gauge of the dynamical state of a cluster, with values of c in excess of ~ 2 indicating a post-core-collapse phase (Djorgovski & Meylan 1993; Trager, Djorgovski & King 1995).

Besides driving a cluster towards core-collapse, equipartition of energy through two-body relaxation also alters over time the mass distribution. More massive stars tend to transfer kinetic energy to lighter objects and sink towards the cluster centre, while less massive stars migrate outwards. The resulting mass segregation implies that the local stellar mass function (MF) within a cluster changes with time and place. Even if the stellar initial mass function (IMF) was the same everywhere when the cluster formed (a condition that does not seem to be true in some very young rich clusters where more massive stars are already more centrally concentrated; see e.g. Sirianni *et al.* 2002), after a few relaxation times there will be proportionately more low-mass stars in the cluster periphery and proportionately less in the core. The first tentative evidence of mass segregation in 47 Tuc (Da Costa 1982) has been fully confirmed by early HST observations in this and other clusters (see e.g. Paresce, De Marchi & Jedrzejewski 1995; De Marchi & Paresce 1995) and is now acknowledged in all observed GCs.

Finally, another important effect of the two-body relaxation process is that it drives the velocity distribution towards a Maxwellian and, therefore, an increasing number of stars in the tail of the velocity distribution will acquire enough energy to exceed the escape velocity and leave the cluster. This phenomenon, called evaporation, happens even in an isolated cluster, but its extent is greatly enhanced by the presence of the tidal field of the Galaxy, in a way that depends on the cluster's orbit. Evaporation, coupled with tidal truncation, is the leading cause of mass loss for most GCs after the first few billion years of their formation (see e.g. Gnedin & Ostriker 1997) and causes a preferential loss of low-mass stars, since these have typically higher velocities. The result is a selective depletion at the low-mass end of the cluster's stellar global MF (GMF). This effect, integrated over the orbit and time, implies a progressive departure of the GMF from the stellar IMF (Vesperini & Heggie 1997), namely a flattening at low masses that is now well established observationally (De Marchi *et al.* 1999; Andreuzzi *et al.* 2001; Koch *et al.* 2004; De Marchi, Pulone & Paresce 2006; De Marchi & Pulone 2007).

It would, therefore, appear natural to expect that, as clusters evolve dynamically, the increase in their central concentration should correspond to the flattening of their GMF, as both effects result from the same two-body relaxation process. To test this hypothesis, we have built a sample of 20 GCs for which reliable estimates exist of both c and the shape of the GMF (see De Marchi, Paresce & Pulone 2007 for details). The former comes from accurate surface photometry (Harris 1996), while the latter has been determined by us using high-quality HST and VLT photometry as briefly explained here below.

In order to obtain the GMF of a cluster, one would need to measure the MF of its entire stellar population, since MF measurements limited to a specific location need to be compensated for the effects of mass segregation. We have shown, however, that if the MF is measured at various locations inside the cluster (e.g. near the core, at the half-light radius and in the periphery), the GMF can effectively be derived by constraining a model MF to reproduce simultaneously three observables: the radial variation of the MF, the surface brightness profile and the velocity dispersion profile. Details on how this is done, using multi-mass Michie–King models, can be found in De Marchi *et al.* (2006) and references therein. Our analysis of clusters with an almost complete radial coverage of the MF (De Marchi *et al.* 2006; De Marchi & Pulone 2007) proves the long cherished belief that the local MF near the half-light radius is for all practical purposes indistinguishable from the GMF (Richer *et al.* 1991; De Marchi & Paresce 1995; De Marchi, Paresce & Pulone 2000).

In this work we have limited our analysis of the GMF to the mass range 0.3–0.8 $M_\odot$, in which a power-law distribution of the type $dN/dm \propto m^\alpha$ appears to adequately reproduce the observations. This choice of the mass range is dictated by the fact that below $\sim 0.3\,M_\odot$ the GMF of GCs departs from a simple power-law (see Paresce & De Marchi 2000 and De Marchi, Paresce & Portegies Zwart 2005) and, more importantly, because the number of clusters with reliable photometry at those masses is still limited.

In Fig. 1 we show the run of the GMF index α as a function of c for the clusters in our sample, as indicated by the labels. Besides the 20 GCs mentioned above (see De Marchi, Paresce & Pulone 2007 for details on the sample), we have also included in Fig. 1 the old open clusters NGC 188 and M 67 (NGC 2682), whose values of c and α (for stars in the range 0.6–1.2 $M_\odot$) were derived from the photometric studies of Stetson, McClure & VandenBerg (2004) and Fan *et al.* (1996), respectively.

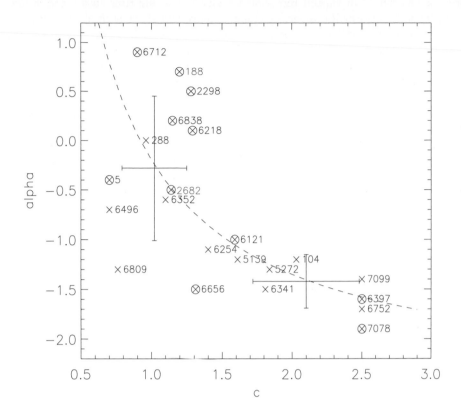

Figure 1. Observed trend between MF index α and the central concentration parameter c. Clusters are indicated by their NGC (or Pal) index number. Objects for which a GMF index is available are marked with a circled cross. For all others, the value of α is that of the MF measured near the half-light radius.

It is immediately obvious that the data do not follow the expected correlation or trend between increasing central concentration and flattening GMF mentioned above, as there are no high-concentration clusters with a shallow GMF. The median value of c (1.4) splits the cluster population roughly in two groups, one with lower and one with higher concentration. The mean GMF index of the first group is $\alpha = -0.3 \pm 0.7$, while the second has a much tighter distribution with $\alpha = -1.4 \pm 0.3$ (see large thick crosses in Fig. 1). The relationship $\alpha + 2.5 = 2.3/c$, shown as a dashed line in Fig. 1, is a simple yet satisfactory eye-ball fit to the distribution.

Fig. 1 suggests that a relatively low concentration is a necessary condition (although probably not also a sufficient one) for a depleted GMF. It appears, therefore, that mass loss, even severe, via evaporation and tidal truncation has not triggered core-collapse for low-concentration clusters (hence the title of this contribution). This finding is unexpected and counter-intuitive. Although no satisfactory explanation presently exists for the observed behavior, we briefly address here below some hypotheses. Some of the ideas put forth here are still preliminary, as they emerged from discussions during this Symposium.

IMF variations could explain the existence of clusters with a very depleted GMF and a loose core. Some clusters with shallow IMF have undergone severe stellar mass-loss and have therefore expanded considerably. This has led to a lower c and a shallower GMF because puffed-up systems of this type were more prone to tidal truncation. Most of these clusters have already disrupted but some survive for a long time in a state of low c and large α. This hypothesis, however, does not explain the absence of clusters with a dense core and a shallow GMF, as the most massive clusters with an originally shallow IMF should have long collapsed and still be visible in the upper-right portion of Fig. 1.

Alternatively, it is possible that the clusters with a depleted GMF and a loose core have in fact undergone core-collapse and have already recovered a normal radial density and surface brightness profile. Core re-expansion thanks to the energy released by hardening binaries has long been predicted (Hut 1985). However, the timescale for re-expansion, at least according to the predictions of Murphy, Cohn & Hut (1990), seems too long to account for the observed distribution. It is more likely that burning of primordial binaries may have halted the collapse altogether. Calculations by Trenti (2007; 2008) suggest that a binary fraction $\gtrsim 10\,\%$ in the core of loose clusters could be sufficient to avoid their collapse, without preventing further mass loss via evaporation or tidal stripping. This explanation seems particularly appealing in light of the discovery that many loose clusters appear to have a significant ($\gtrsim 6\,\%$) binary fraction in their cores (Sollima $et\ al.$ 2007). However, the problem remains that some of the most depleted objects in the upper-left quadrant of Fig. 1 have too narrow a main sequence in the colour–magnitude diagram to account for a binary fraction in excess of a few percent (Pulone & De Marchi, in preparation; see also Davis & Richer 2008).

Another explanation for the depleted clusters with a loose core is that proposed by Kroupa (2008) and Baumgardt (2008), in which the low concentration is the result of rapid gas expulsion from the cores of primordially segregated clusters in the early phases of their lives. Such a violent process would deplete the low-mass end of the GMF and could leave the GCs in an almost collisionless state, in which further dynamical evolution via two-body relaxation is prevented. The problem with this scenario, however, is how to explain why even the most depleted clusters such as NGC 2298 or NGC 6218 are in a condition of energy equipartition (De Marchi & Pulone 2007; De Marchi $et\ al.$ 2006), unless the observed mass stratification is the residue of primordial segregation.

The apparently simple dependence of α from c in Fig. 1 might also suggest that the observed distribution in practice represents an evolutionary sequence. In this scenario, the value of c at the time of cluster formation determines its evolution along two opposite directions of increasing and decreasing concentration. Clusters born with sufficiently high concentration ($c \gtrsim 1.5$) evolve towards core-collapse. Mass loss can be important via stellar evolution in the first $\sim 1\,\mathrm{Gyr}$, and to a lesser extent via evaporation or tidal stripping throughout the life of the cluster, but the GMF at any time does not depart significantly from the IMF. Clusters with $c \lesssim 1.5$ at birth also evolve towards core-collapse, but mass loss via stellar evolution and, most importantly, via relaxation and tidal stripping proceeds faster, particularly if their orbit has a short perigalactic distance or frequent disc crossings. Therefore, as the tidal boundary shrinks and the cluster loses preferentially low-mass stars, the GMF progressively flattens. This speeds up energy equipartition, but c still decreases, since the tidal radius shrinks more quickly than the luminous core radius (although the central density, particularly that of heavy remnants, is increasing). These clusters could eventually undergo core-collapse, but this might only affect a few stars in the core, thereby making it observationally hard to detect. The signature of core-collapse might only be present and should therefore be searched in the radial distribution of heavy remnants (Mark Gieles, private communication).

In summary, while no conclusive explanation still exists for the unexpected observed trend between central concentration and shape of the GMF, Fig. 1 should serve as a warning that the surface brightness profile and the central concentration parameter of GCs are not as reliable indicators of their dynamical state as we had so far assumed. In fact, if a central cusp in the surface brightness profile were the signature of a cluster's post core-collapse phase, it would be hard to explain why only about 20 % of the Galactic GCs show a cusp when the vast majority of them are an order of magnitude older than their half-mass relaxation time (Ivan King, private communication). Our current estimate of the fraction of post core-collapse clusters may therefore need a complete revision as a large number of them may be lurking in the Milky Way. A more reliable assessment of a cluster's dynamical state requires the study of the complete radial variation of its stellar MF and of the properties of its stellar population, particularly in the core.

Acknowledgements

We are grateful to Enrico Vesperini, Pavel Kroupa, Holger Baumgardt, Michele Trenti, Simon Portegies Zwart, Sidney van den Bergh, Oleg Gnedin, Mark Gieles and Evghenii Gaburov for helpful discussions and suggestions.

References

Andreuzzi, G., De Marchi, G., Ferraro, F., Paresce, F., & Pulone, L. 2001, *A&A*, 372, 851

Baumgardt, H. 2008, these proceedings

Da Costa, 1982, *AJ*, 87, 990

Davis, S. & Richer, H. 2008, these proceedings

De Marchi, G., Leibundgut, B., Paresce, F., & Pulone, L. 1999, *A&A* 343, 9L

De Marchi, G. & Paresce, F. 1995, *A&A*, 304, 202

De Marchi, G., Paresce, F., Portegies Zwart, S. 2005, in ASSL 327, The initial mass function 50 years later, Eds. E. Corbelli, F. Palla, H. Zinnecker (Dordrecht: Springer), 77

De Marchi, G., Paresce, F., & Pulone, L. 2000, *ApJ*, 530, 342

De Marchi, G., Paresce, F., & Pulone, L. 2007, *ApJ*, 656, L65

De Marchi, G. & Pulone, L. 2007, *A&A*, 467, 107

De Marchi, G., Pulone, L., & Paresce, F. 2006, *A&A*, 449, 161

Djorgovski, S. & King, R. 1986, *ApJ*, 305, L61

Djorgovski, S. & Meylan, G. 1993, in ASP Conf. Ser. 50, Structure and Dynamics of Globular Clusters, S. Djorgovski, G. Meylan (San Francisco: ASP), 325

Elson, R., Hut, P., & Ingaki, S. 1987, ARAA, 25, 565

Fan, X., *et al.* 1996, *AJ*, 112, 628

Gnedin, O. & Ostriker, J. 1997, *ApJ*, 474, 223

Harris, W. 1996, *AJ* 112, 1487

Hut, P. 1985, in IAU Symp. 113, Dynamics of star clusters, (Dordrecht: Reidel), 231

Koch, A., Grebel. E., Odenkirchen, M., Martinez–Delgado, D., & Caldwell, J. 2004, *AJ*, 128, 2274

Kroupa, P. 2008, these proceedings

Murphy, B., Cohn, H., & Hut, P. 1990, *MNRAS*, 245, 335

Paresce, F. & De Marchi, G. 2000, *ApJ*, 534, 870

Paresce, F., De Marchi, G., & Jedrzejewski, R. 1995, *ApJ*, 442, L57

Richer, H., Fahlman, G., Buonanno, R., Fusi Pecci, F., Searle, L., & Thompson, I. 1991 381, 147

Sirianni, M., Nota, A., De Marchi, G., Leitherer, C., & Clampin, M. 2002, *ApJ*, 579, 275

Sollima, A., Beccari, G., Ferraro, F., Fusi Pecci, F., & Sarajedini, A. 2007, *MNRAS*, 380, 781

Spitzer, L. 1987, Dynamical Evolution of Globular Clusters, (Princeton: Princeton Univ. Press)

Stetson, P., McClure, R., & VandenBerg, D. 2004, *PASP*, 116, 1012

Trenti, M. 2007, American Astronomical Society, DDA meeting #38, #2.01

Trenti, M. 2008, these proceedings

Trager, S., Djorgovski, S., & King, I. 1995, *AJ*, 109, 218

Vesperini, E. & Heggie, D. 1997, *MNRAS*, 289, 898

Dynamical Evolution of Dense Stellar Systems
Proceedings IAU Symposium No. 246, 2007
E. Vesperini, M. Giersz & A. Sills, eds.
© 2008 International Astronomical Union
doi:10.1017/S1743921308015524

Dynamical Evolution of Rotating Globular Clusters with Embedded Black Holes

J. Fiestas, O. Porth and R. Spurzem

Astronomisches Rechen-Institut, Zentrum für Astronomie Heidelberg,
Germany
email: fiestas, oporth, spurzem@ari.uni-heidelberg.de

Abstract. Evolution of self-gravitating rotating dense stellar systems (e.g. globular clusters) with embedded black holes is investigated. The interplay between velocity diffusion due to relaxation and black hole star accretion is followed together with cluster differential rotation using 2D+1 Fokker Planck numerical methods. The models can reproduce the Bahcall-Wolf $f \propto E^{1/4}$ ($\propto r^{-7/4}$) cusp inside the zone of influence of the black hole. Angular momentum transport and star accretion processes support the development of central rotation in relaxation time scales, before re-expansion and cluster dissolution due to mass loss in the tidal field of a parent galaxy. Gravogyro and gravothermal instabilities conduce the system to a faster evolution leading to shorter collapse times with respect to models without black hole.

Keywords. methods: numerical, gravitation, stellar dynamics, black hole physics, globular clusters: general

1. Introduction

The improvement of our knowledge and methods in the field of rotating dense stellar systems is extremely important for modelling systems like globular clusters and galactic nuclei, where a central star-accreting black hole comes into the game. Direct integration of orbits (N-Body method) has been applied to the problem. However, N-Body simulations only provide a very limited number of case studies, due to the enormous computing time needed even on the GRAPE computers. Moreover, in young dense clusters, supermassive stars may form through runaway merging of main-sequence stars via direct physical collisions, which may then collapse to form an IMBH.

2. Diffusion and loss-cone accretion

The Fokker-Planck approximation is applied for an axisymmetric system in flux conservation form, following :

$$\frac{df}{dt} = \frac{1}{p}\left(-\frac{\partial F_X}{\partial X} - \frac{\partial F_Y}{\partial Y}\right) \tag{2.1}$$

p is the phase volume per unit X (dimensionless energy) and Y (dimensionless angular momentum). The loss-cone limit is defined by the minimum angular momentum for an orbit of energy E:

$$J_z^{min}(E) = r_d\sqrt{2(E - GM_{\mathrm{BH}}/r_d)} \tag{2.2}$$

where r_d is the disruption radius of the BH, calculated following Frank & Rees (1976):

$$r_{\mathrm{d}} \propto r_*(M_{\mathrm{BH}}/m_*)^{1/3} \tag{2.3}$$

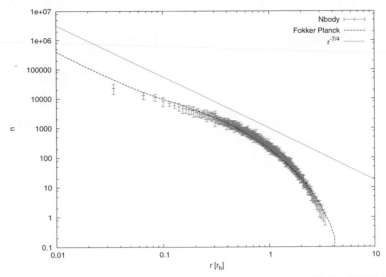

Figure 1. Comparison with an N-Body run using 10000 particles (Bahcall-Wolf cusp) at a time where $M_{\rm bh}/M_{\rm cl} = 0.1$.

r_* and m_* are the stellar radius and mass, respectively.

The contribution of $f(X, Y)$ to accretion, at each energy-angular momentum grid cell is given by:

$$\Delta f_{\rm acc} = P_a(Y)(f^{\rm old} + \Delta f) \qquad (2.4)$$

where $P_a(X, Y, Y^{\rm diff})$ is the probability of accretion by the BH and $\Delta f_{\rm acc}$ is the fraction of $f(X, Y)$ which goes into the loss-cone, due to diffusion in the inner/outer direction (Fiestas 2006).

The NBody6++ code has been modified in order to treat accretion of stars, which approach the BH inside its tidal disruption radius (Eq. 2.3)

3. Numerical results

As initial configurations, truncated King models with added bulk motion are used. Their adopted distribution function is

$$f(E, J_z) \propto \exp\left(-\beta \Omega_0 J_z\right) \cdot \left[\exp\left(-\beta(E - E_{\rm tid})\right) - 1\right], \qquad E < E_{\rm tid} \qquad (3.1)$$

and 0 otherwise. $\beta = 1/\sigma_c^2$ and Ω_0 is an angular velocity. The initial conditions of each model are fixed by the triple $(W_0, \omega_0, M_{\rm BH_i})$.

$M_{\rm BH}$ grows through accretion of low-J_z stars while central density increases and the BH-potential $(\sim GM_{\rm BH}/r)$ dominates the stellar distribution within its influence radius r_a.

The final steady-state, long-dashed line (FP) and crosses (N-Body) in Fig. 1, evolves towards a power-law of $\lambda = -1.75$, according to $n \propto r^\lambda$ (Bahcall & Wolf 1976; Lightman & Shapiro 1977; Marchant & Shapiro 1980). It forms inside r_a and is maintained in the post-collapse phase, while the evolution is driven through energy input from the central object. Very close to the center the density profile flattens due to the effective loss-cone accretion. Fig. 2 shows the evolution of density in the meridional plane (ρ, z). In the regions where BH star accretion dominates, the cusp forms a strong colour gradient towards the center. At the same time, the system loses mass through the outer

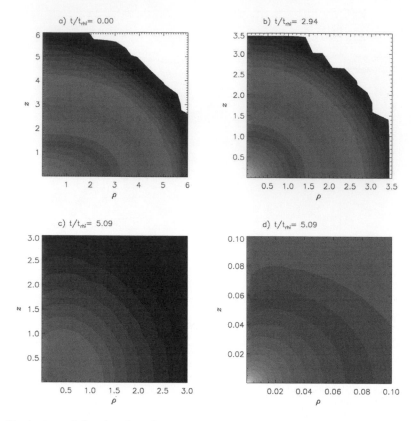

Figure 2. Evolution of density distribution in the meridional plane. Cylindrical coordinates (ρ, z) are used. Lighter zones represent higher isodensity contours. Note that scales are different in each plot and Fig. 2d shows a zoom of the central parts of Fig. 2c (insider r_a). The time is given in units of initial half-mass relaxation time (t_{rhi}).

tidal boundary. Particularly, at later times, the density of stars is higher in ρ-direction (equatorial plane, very close to the BH) than in z-direction (Fig. 2c,d).

M_{BH} stalls at $\sim 0.01\, M_{cl}$ at post-collapse time, and remains nearly constant afterwards, while BH mass accretion rate (dM_{BH}/dt) reaches a maximum at collapse time, due to the higher density of orbits in the core, and falls afterwards. Cluster mass (M_{cl}) loss in the tidal field of the parent galaxy is very strong during the re-expansion of the core.

BH models experience in a similar way, the onset of gravogyro instabilities (Hachisu 1979; Hachisu 1982), as angular momentum diffuses outwards, leading to an increase of central rotation. BH accretion of stars on orbits of low J_z sets off, an ordered motion of high-J_z bounded orbits around the central BH supports central rotation. At the same time, stars in the core are heated via the consumption of stars in bound, high energetic orbits in the cusp.

As seen in Fig. 3, angular velocity grows over time stronger in the inner Lagrangian-radii for the BH model, as a consequence of the faster dynamical evolution (gravogyro + gravothermal instabilities). Post collapse evolution is present only in the system harbouring a black hole. Thus, as the BH grows and rotation increases in its zone of influence $(r \sim r_a)$ angular momentum continues being transported out of the core. In the outer parts, rotation is continuously depleted.

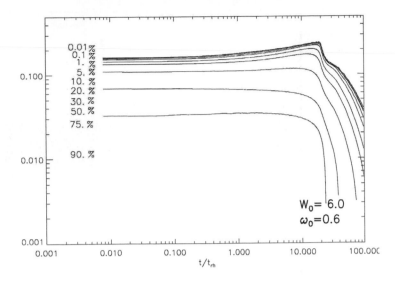

Figure 3. Evolution of angular velocity.

Fig. 4 shows the stellar distribution in the E-Jz plane, for a model with $M_{\mathrm{bh}}/M_{\mathrm{cl}} = 0.1$. In the non-rotating model ($W_0 = 3.0$) the symmetry of the stellar distribution makes a populated loss-cone from the beginning (Fig. 4a). As a consequence, the accretion rates are expected to be higher in the axisymmetric approximation ($J < J_{\mathrm{z,min}}$). In the E-J plane (Fig. 4c), the stellar distribution shows a less populated loss-cone, which would lead to a slower growth rate by using the criterion $J < J_{\mathrm{tot,min}}$ (which is implicitly used in the N-Body realisation). On the other side, the rotating model ($\omega_0 = 2.4$) shows an initial asymmetry in the distribution of stars (Fig. 4b), which is still present when $M_{\mathrm{bh}}/M_{\mathrm{cl}} = 0.1$. Nevertheless, the $J_{\mathrm{z,min}}$ approximation seems to overestimate the accretion rate, in comparison to the $J_{\mathrm{tot,min}}$ criterion (Fig. 4d). This effect is being currently investigated in order to accurate the Fokker-Planck approximation, in comparison to the direct N-Body method.

4. Conclusions and outlook

Although some constraints in the evolution of rotating clusters are still missing, like a mass spectrum or stellar evolution, as well as a more realistic criterium for galactic tidal mass loss (as observations suggest, e.g. Mackey & van den Berg 2005), the consistence of the general evolution of cluster structure in spherically symmetric systems embedding BHs and estimation of BH masses with observations makes clear that the models presented here can well reproduce the evolution of GC with embedded BHs, and that rotation constitutes an important constraint, which needs to be taken into account for the understanding of the formation and evolution of GCs, specially when it is high enough, at early times of evolution (e.g. in the young clusters of the LMC). Regarding the stellar spectrum, segregation of high mass stars is expected to drop the dispersion in the center (as reported by Kim, Lee & Spurzem 2004), leading possibly to a higher or at least more stable V_{rot}/σ in this region. Multi-mass models with BH are being currently developed and comparison to N-Body models are aimed to complement this calculations, using the highest particle number permitted at the time ($N \sim 10^6$) (Berczik *et al.* 2006).

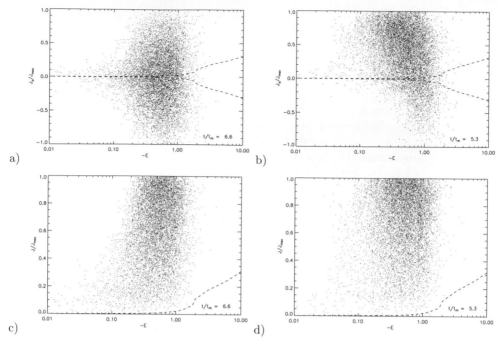

Figure 4. Number density of stars in the E-J plane.

References

Bahcall, J. N. & Wolf, R. A., 1976, *ApJ*, 209, 214

Berczik, P., Merritt, D., Spurzem, R., & Bischof, H.-P. 2006, *ApJ*, 642, 21

Fiestas, J. 2006, *Ph.D. thesis Univ. of Heidelberg*

Frank, J. & Rees, M. 1976, *MNRAS*, 176, 633

Hachisu, I. 1979, *PASJ*, 31, 523

Hachisu, I. 1982, *PASJ*, 34, 313

Kim, E., Lee, H. M., & Spurzem, R. 2004, *MNRAS*, 351, 220

Lightman, A. P. & Shapiro, S. L. 1977, *ApJ*, 211, 244

Mackey, A. & van den Bergh, S. 2005, *MNRAS*, 360, 631

Marchant, A. B. & Shapiro, S. L. 1980, *ApJ*, 239, 685

Dynamical Evolution of Dense Stellar Systems
Proceedings IAU Symposium No. 246, 2007
E. Vesperini, M. Giersz & A. Sills, eds.

© 2008 International Astronomical Union
doi:10.1017/S1743921308015536

Star Cluster Life-times: Dependence on Mass, Radius and Environment

Mark Gieles[1], Henny J. G. L. M. Lamers[2] and Holger Baumgardt[3]

[1]European Southern Observatory, Casilla 19001, Santiago 19, Chile
email: mgieles@eso.org

[2]Astronomical Institute, Utrecht University, Princetonplein 5,
3584 CC Utrecht, The Netherlands
email: lamers@astro.uu.nl

[3]Argelander Institut für Astronomie, Universität Bonn, Auf dem Hügel 71, Bonn, Germany
email: holger@astro.uni-bonn.de

Abstract. The dissolution time ($t_{\rm dis}$) of clusters in a tidal field does not scale with the "classical" expression for the relaxation time. First, the scaling with N, and hence cluster mass, is shallower due to the finite escape time of stars. Secondly, the cluster half-mass radius is of little importance. This is due to a balance between the relative tidal field strength and internal relaxation, which have an opposite effect on $t_{\rm dis}$, but of similar magnitude. When external perturbations, such as encounters with giant molecular clouds (GMC) are important, $t_{\rm dis}$ for an individual cluster depends strongly on radius. The mean dissolution time for a population of clusters, however, scales in the same way with mass as for the tidal field, due to the weak dependence of radius on mass. The environmental parameters that determine $t_{\rm dis}$ are the tidal field strength and the density of molecular gas. We compare the empirically derived $t_{\rm dis}$ of clusters in six galaxies to theoretical predictions and argue that encounters with GMCs are the dominant destruction mechanism. Finally, we discuss a number of pitfalls in the derivations of $t_{\rm dis}$ from observations, such as incompleteness, with the cluster system of the SMC as particular example.

Keywords. globular clusters: general, open clusters and associations: general, stellar dynamics, methods: n-body simulations

1. Theoretical predictions of cluster dissolution

1.1. *Dynamical evolution in a tidal field*

Simulations of star clusters dissolving in a tidal field have shown that the dissolution time ($t_{\rm dis}^{\rm tid}$) scales with the relaxation time ($t_{\rm rel}$) as $t_{\rm dis}^{\rm tid} \propto t_{\rm rel}^{0.75}$ (Baumgardt 2001; Baumgardt & Makino 2003). This non-linear dependence on $t_{\rm rel}$ is due to the finite escape time through one of the Lagrange points (Fukushige & Heggie 2000). The dependence on N, or cluster mass ($M_{\rm c}$), can be approximated as $t_{\rm dis}^{\rm tid} \propto M_{\rm c}^{0.62}$, which is accurate for $10^2 \lesssim N \lesssim 10^7$ (Lamers, Gieles & Portegies Zwart 2005). The half-mass radius ($r_{\rm h}$) of the cluster does not enter in the results, since it is assumed that clusters are initially "Roche lobe" filling, which implies $M_{\rm c} \propto r_{\rm h}^3$, i.e. a constant crossing time.

The assumption of Roche lobe filling clusters is computationally attractive since it avoids having $r_{\rm h}$ as an extra parameter. However, observations of (young) extra-galactic star clusters show that the dependence of $r_{\rm h}$ on $M_{\rm c}$ and galactocentric distance ($R_{\rm G}$) is considerably weaker ($r_{\rm h} \propto M^{0.1} R_{\rm G}^{0.1}$) than the Roche lobe filling relation ($r_{\rm h} \propto M^{1/3} R_{\rm G}^{2/3}$) (Larsen 2004; Scheepmaker, Haas, Gieles *et al.* 2007), implying that massive clusters at large $R_{\rm G}$ are initially underfilling their Roche lobe.

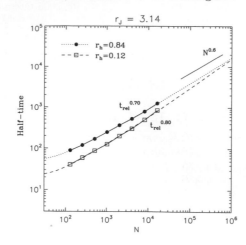

Figure 1. Half-mass time as found from N-body simulations of clusters dissolving in a tidal field. The filled circles represent clusters that initially fill their Roche lobe. The open squares are the results of runs where $r_{\rm h}$ was seven times smaller.

Gieles & Baumgardt (2007) simulated clusters with varying initial $r_{\rm h}$ in a tidal field to quantify the importance of $r_{\rm h}$. Fig. 1 shows the results of $t_{\rm dis}^{\rm tid}$ for two sets of clusters with different initial $r_{\rm h}$. The filled circles are for clusters that started tidally limited and the open squares are for runs where the initial $r_{\rm h}$ was a factor seven smaller. The difference in $t_{\rm dis}$ are within a factor two, while the "classical" expression of $t_{\rm rel}$ predicts a factor $7^{3/2} \simeq 20$. The reason that $t_{\rm dis}$ depends so little on $r_{\rm h}$ can be understood intuitively: for smaller clusters the tidal field is less important, but the dynamical evolution is faster. These effects happen to balance and result in almost no dependence on $r_{\rm h}$. *The crossing of the lines around $N \simeq 10^6$ implies that for globular clusters $t_{\rm dis}^{\rm tid}$ is completely independent of $r_{\rm h}$.*

This somewhat surprising result means that we can use the $r_{\rm h}$ independent results for $t_{\rm dis}$ of tidally limited clusters (Baumgardt & Makino 2003) as a general result for $t_{\rm dis}^{\rm tid}$ for clusters of different $r_{\rm h}$:

$$\frac{t_{\rm dis}^{\rm tid}}{\rm Gyr} = 1.0 \left(\frac{M_{\rm c}}{10^4\, M_\odot}\right)^{0.62} \frac{R_{\rm G}}{V_{\rm G}} \frac{220\,{\rm km\,s}^{-1}}{\rm kpc}. \tag{1.1}$$

From this it follows that a cluster with $M_{\rm c} = 10^4\, M_\odot$ in the solar neighbourhood would dissolve in approximately 8 Gyr due to tidal field. This is much longer than the empirically derived value of 1.3 Gyr (Lamers, Gieles, Bastian, *et al.* 2005), implying that there are additional disruptive effects that shorten the life-time of clusters.

1.2. *External perturbations: disruption by giant molecular clouds*

It has long been suspected that encounters with giant molecular clouds (GMCs) shorten the life-times of clusters (e.g. van den Bergh & McClure 1980). Gieles, Portegies Zwart, Baumgardt, *et al.* (2006) studied this effect using N-body simulations and found that $t_{\rm dis}$ due to GMC encounters ($t_{\rm dis}^{\rm GMC}$) can be expressed in cluster properties and average molecular gas density ($\rho_{\rm n}$) as

$$\frac{t_{\rm dis}^{\rm GMC}}{\rm Gyr} = 2.0 \left(\frac{0.03\, M_\odot\,{\rm pc}^{-3}}{\rho_{\rm n}}\right) \left(\frac{M_{\rm c}}{10^4\, M_\odot}\right) \left(\frac{3.75\,{\rm pc}}{r_{\rm h}}\right)^3. \tag{1.2}$$

Table 1. Columns 1–3: Estimates of tidal field strength, molecular gas densities and resulting predictions for t_4, the $t_{\rm dis}$ of a cluster with an initial $M_c = 10^4\,M_\odot$. Column 4: empirically derived values of t_4 are given, taken from: [1]Gieles *et al.* (2005); [2]Lamers, Gieles & Portegies Zwart (2005); [3]Lamers, Gieles, Bastian, *et al.* (2005); [4]Parmentier & de Grijs (2007); [5]Krienke & Hodge (2004); [6]Boutloukos & Lamers (2003).

Galaxy	Tidal field $R_{\rm G}/V_{\rm G}$ [Myr]	Molecular gas density $\rho_{\rm n}$ [$10^{-3}\,M_\odot\,{\rm pc}^{-3}$]	Predicted t_4 [Gyr]	Observed t_4 [Gyr]
M51[1]	10	450	0.13	0.1
M33[2]	15	25	1.4	0.6
Solar neighbourhood[3]	35	30	1.6	1.3
LMC[4]	30	–	<6.6	>1
NGC6822[5]	~35	–	<7.7	~4
SMC[6]	40	0.5	8.2	8

The scaling of $t_{\rm dis}^{\rm GMC}$ with cluster density $(M_c/r_{\rm h}^3)$ combined with the observed weak dependence of $r_{\rm h}$ on M_c, $r_{\rm h} \propto M_c^{0.13}$, results in a similar scaling of the mean $t_{\rm dis}^{\rm GMC}$ with M_c as found for $t_{\rm dis}^{\rm tid}$, i.e. $\propto M_c^{0.6}$ (1.1).

For the solar neighbourhood $(\rho_{\rm n} \simeq 0.03\,M_\odot\,{\rm pc}^{-3})$ $t_{\rm dis}^{\rm GMC} = 2$ Gyr, which combined with the tidal field (1.1) nicely explains the empirically derived $t_{\rm dis}$ of 1.3 Gyr and the observed age distribution of clusters in the solar neighbourhood (Lamers & Gieles 2006).

From (1.1) and (1.2) we see that the predicted $t_{\rm dis}$ scales with the tidal field strength $(R_{\rm G}/V_{\rm G})$ and the inverse of the molecular gas density $(1/\rho_{\rm n})$. In table 1 we give values for these parameters for six galaxies, combined with predictions for t_4. The values for $\rho_{\rm n}$ are taken from Gieles, Portegies Zwart, Baumgardt, *et al.* (2006) (and references therein), Heyer *et al.* (2004); Leroy, Bolatto, Stanimirovic *et al.* (2007) for the solar neighbourhood, M51, M33 and the SMC, respectively. In the next section we compare this to empirically derived values of $t_{\rm dis}$.

2. Comparison to observations

2.1. *Empirically derived $t_{\rm dis}$ values in different galaxies*

Under the assumption that $t_{\rm dis}$ scales with M_c, Boutloukos & Lamers (2003) (BL03) introduced an empirical disruption law: $t_{\rm dis} = t_4\,(M_c/10^4\,M_\odot)^\gamma$. The value of t_4 and γ can be derived from the age and mass distributions (see BL03 for details). BL03 found a mean γ of $\bar\gamma = 0.62$, agreeing nicely with (1.1) and (1.2), and values for t_4 ranging from $\sim$100 Myr to $\sim$8 Gyr. We summarise values of t_4 of clusters in six different galaxies taken from more recent literature in table 1.

Note that $R_{\rm G}/V_{\rm G}$ and $1/\rho_{\rm n}$ roughly increase with increasing t_4. The variation in $R_{\rm G}/V_{\rm G}$ is too small to explain the variation in t_4, which implies that in the galaxies with short t_4 the disruption is dominated by GMC encounters. *From Table 1 we see that the decreasing trend in the empirical t_4 can be explained by increasing gas density and increasing tidal field strength.*

2.2. *The clusters of the SMC*

A lot of attention has gone recently to the age distribution (dN/dt) of clusters in the SMC. Rafelski & Zaritsky (2005) (RZ05) found that dN/dt is roughly declining as t^{-1}, which Chandar *et al.* (2006) explain by mass independent cluster disruption† removing

---

† In fact the authors call their disruption model "infant mortality", but we prefer to reserve this term for the dissolution of clusters due to gas expulsion. In addition, 3 Gyr old clusters have survived 25% of a Hubble time, so they are not really infant anymore.

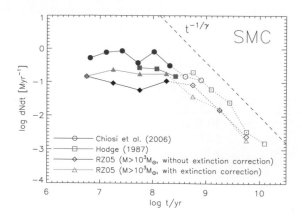

Figure 2. The age distribution (dN/dt) of star clusters in the SMC as found in different studies in literature. The data set of Rafelski & Zaritsky (2005) (RZ05) is very incomplete for low mass clusters at old ages (Gieles, Lamers & Portegies Zwart 2007), so a mass cut at $10^3\,M_\odot$ was applied. The general trend found in these studies is that dN/dt is flat up to an age of $0.3\text{--}1 \times 10^9$ yr and then it declines as $t^{-1.7}$. The dashed line is the predicted slope for dN/dt at old ages when $t_{\rm dis} \propto M_{\rm c}{}^\gamma$, with $\gamma = 0.62$.

90% of the clusters each age dex. Gieles, Lamers & Portegies Zwart (2007) showed that the decline is caused by incompleteness and that the dN/dt is flat in the first ~ 1 Gyr when using a mass limited sample. The dN/dt based on ages which are derived from extinction corrected colours starts declining a bit earlier than the one based on uncorrected colours (Fig. 2). However, the general shape is similar to that found by other authors: a flat part in the first 0.3–1.0 Gyr (recently reconfirmed by de Grijs & Goodwin 2007) and then a steep decline ($\propto t^{-1.7}$). When $t_{\rm dis} \propto M_{\rm c}{}^\gamma$, then the dN/dt at old ages declines as $t^{-1/\gamma}$ for both mass and magnitude limited samples (BL03). The decline of $t^{-1.7}$ implies $\gamma \simeq 0.6$, in agreement with the theoretical predictions (1.1 and 1.2).

2.3. *Selection effects and biases: a cautionary note*

Observed cluster samples are always heavily affected by the detection limit, causing the minimum observable cluster mass ($M_{\rm min}$) to increase with age, due to the fading of clusters. To illustrate this effect we create an artificial cluster population with a constant cluster formation rate (CFR) and with a power-law CIMF with index -2. In the left panels of Fig. 3 we show the ages and masses (bottom) and the corresponding dN/dt (top) when the sample is mass limited. The dN/dt is flat which is the result of the constant CFR we put it. In the right panel we remove the clusters which are fainter than $M_V = -4.5$. The mass of a cluster at the detection limit, $M_{\rm min}(t)$, increases with age as $M_{\rm min}(t) \propto 0.4\,M_V^{\rm SSP}(t)$, where $M_V^{\rm SSP}(t)$ is the evolution of M_V with age from an SSP model. For a power-law CIMF with index -2, the resulting dN/dt scales with $M_{\rm min}$ as $dN/dt \propto 1/M_{\rm min}(t)$ (BL03), which is shown in the top right panel of Fig. 3.

The detection limit is usually expressed in M_V. However, deriving cluster ages from broad band photometry requires the presence of blue filters such as U and B. We show the $M_{\rm min}(t)$ for a U-band detection limit of $M_U = -5$ as a dashed line in the age *vs.* mass diagram. The resulting dN/dt (shown as a dashed line in the top right panel) declines approximately as $dN/dt \propto t^{-1.1}$, i.e. steeper than the V-band prediction. *It is of vital importance to understand the effect of incompleteness in different filters before a disruption analyses can be done based on the slope of the dN/dt distribution.*

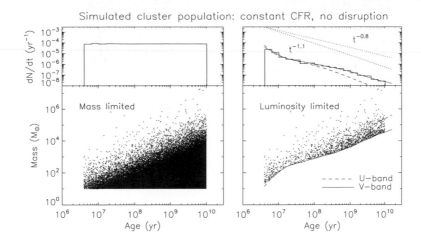

Figure 3. Simulated ages and masses of a cluster population that has formed with a constant cluster formation rate (CFR) and with a power-law CIMF ($N \propto M^{-2}$). In the left panels we show the result of mass limited sample, with $M_{\text{lim}} = 10\,M_{\odot}$. In the right panels we assume that the sample is magnitude limited, with $M_{V,\text{lim}} = -4.5$. The limiting mass due to a magnitude limit and the resulting prediction for dN/dt of a magnitude limited sample are shown as full lines (red). The prediction for a U-band limit ($M_U = -5$) is shown as dashed lines (blue). The dotted lines show power-law approximations for the predicted shapes of dN/dt.

References

Baumgardt, H. 2001, *MNRAS* 325, 1323

Baumgardt, H. & Makino, J. 2003, *MNRAS* 340, 227

Boutloukos, S. G. & Lamers, H. J. G. L. M. 2003, *MNRAS* 338, 717

Chandar, R., Fall, S. M., & Whitmore, B. C. 2006, *ApJ* (Letters) 650, L111

de Grijs, R. & Goodwin, S. P. 2007, *MNRAS* in press, astro-ph/0709.3781

Fukushige, T. & Heggie, D. C. 2000, *MNRAS* 318, 753

Gieles, M., Bastian, N., Lamers, H. J. G. L. M., & Mout, J. N. 2005, *A&A* 441, 949

Gieles, M. & Baumgardt, H. 2007, *MNRAS* to be submitted

Gieles, M., Lamers, H. J. G. L. M., & Portegies Zwart, S. F. 2007, *ApJ* 668, 268

Gieles, M., Portegies Zwart, S. F., Baumgardt, H., Athanassoula, E., Lamers, H. J. G. L. M., Sipior, M., & Leenaarts, J. 2006, *MNRAS* 371, 793

Heyer, M. H., Corbelli, E., Schneider, S. E., & Young, J. S. 2004, *ApJ* 602, 723

Krienke, K. & Hodge, P. 2004, *PASP* 116, 497

Lamers, H. J. G. L. M. & Gieles, M. 2006, *A&A* 455, L17

Lamers, H. J. G. L. M., Gieles, M., Bastian, N., Baumgardt, H., Kharchenko, N. V., & Portegies Zwart, S. F. 2005, *A&A* 441, 117

Lamers, H. J. G. L. M., Gieles, M., & Portegies Zwart S. F. 2005, *A&A* 429, 173

Larsen, S. S. 2004, *A&A* 416, 537

Leroy, A., Bolatto, A., Stanimirovic, S., Mizuno, N., Israel, F., & Bot C. 2007, *ApJ* 658, 1027

Rafelski, M. & Zaritsky, D. 2005, *AJ* 129, 2701

Scheepmaker, R. A., Haas, M. R., Gieles, M., Bastian, N., Larsen, S. S., & Lamers, H. J. G. L. M. 2007, *A&A* 469, 925

Tanikawa, A. & Fukushige, T. 2005 *PASJ* 57, 155

van den Bergh, S. & McClure, R. D. 1980, *A&A* 88, 360

Dynamical Evolution of Dense Stellar Systems
Proceedings IAU Symposium No. 246, 2007
E. Vesperini, M. Giersz & A. Sills, eds.

© 2008 International Astronomical Union
doi:10.1017/S1743921308015548

Black Holes and Core Expansion in Massive Star Clusters

A. D. Mackey[1], M. I. Wilkinson[2], M. B. Davies[3] and G. F. Gilmore[4]

[1]Institute for Astronomy, University of Edinburgh, Royal Observatory, Blackford Hill,
Edinburgh, EH9 3HJ, UK; email: dmy@roe.ac.uk
[2]Department of Physics & Astronomy, University of Leicester, University Road,
Leicester, LE1 7RH, UK
[3]Lund Observatory, Box 43, SE-221 00 Lund, Sweden
[4]Institute of Astronomy, University of Cambridge, Madingley Road,
Cambridge, CB3 0HA, UK

Abstract. Massive star clusters in the Magellanic Clouds are observed to follow a striking trend in size with age – older clusters exhibit a much greater spread in core radius than do younger clusters, which are generally compact. We present results from realistic N-body modelling of massive star clusters, aimed at investigating a dynamical origin for the radius-age trend. We find that stellar-mass black holes, formed as remnants of the most massive stars in a cluster, can constitute a dynamically important population. If retained, these objects rapidly form a dense core where interactions are common, resulting in the scattering of black holes into the cluster halo, and the ejection of black holes from the cluster. These processes heat the stellar component, resulting in prolonged core expansion of a magnitude matching the observations. Core expansion at early times does not result from the action of black holes, but can be reproduced by the effects of rapid mass-loss due to stellar evolution in a primordially mass segregated cluster.

Keywords. Stellar dynamics, N-body simulations, galaxies: star clusters, Magellanic Clouds

1. Introduction and Numerical Setup

For observational studies of star cluster evolution, the Galactic globular clusters, while close, are not ideal because they are exclusively ancient objects ($\tau \geqslant 10^{10}$ yr). We can therefore accurately assess the end-points of their evolution, but must infer the complete long-term development which brought them to these observed states. To directly observe cluster evolution, we must switch our attention to the Magellanic Clouds (LMC/SMC), which both possess extensive systems of star clusters with masses comparable to the Galactic globulars, but crucially *of all ages:* $10^6 \leqslant \tau \leqslant 10^{10}$ yr. These systems are of fundamental importance because they are the nearest places we can observe snapshots of all phases of massive star cluster development.

Elson, Freeman & Lauer (1989) discovered a striking relationship between core radius (r_c) and age for LMC clusters – the observed spread in r_c increases dramatically with increasing age. More recently, Mackey & Gilmore (2003a,b) used Hubble Space Telescope (HST) WFPC2 imaging of 63 massive Magellanic Cloud clusters to more clearly demonstrate the radius-age trend in the LMC and show, for the first time, that an indistinguishable radius-age trend also exists in the SMC. An additional 46 objects have since been observed with HST/ACS (Program #9891) to improve sampling of the radius-age plane. Structural measurements for all 107 clusters may be seen in Fig. 1.

The observed radius-age relationship provides strong evidence that our understanding of massive cluster evolution is incomplete, since standard quasi-equilibrium models do not predict long-term large-scale core expansion. Discerning the origin of the radius-age trend is therefore of considerable importance. Various observational evidence (summarised in Mackey *et al.* 2007b in preparation) suggests that the radius-age relation is driven by internal cluster processes, with any external or tidal effects second order (see also Wilkinson *et al.* 2003). In this contribution, we report on the results of direct, realistic N-body simulations designed to investigate an internal dynamical origin for the radius-age trend.

We have used the NBODY4 code (Aarseth 2003) in combination with a 32-chip GRAPE-6 special-purpose computer (Makino *et al.* 2003) to simulate massive star clusters over a Hubble time of evolution. We generate initial conditions with properties (masses, structures, densities, etc) as similar as possible to those observed for the youngest Magellanic Cloud clusters. Full details are provided in Mackey *et al.* (2007a) and Mackey *et al.* (2007b in preparation); here we simply note that our simulated clusters start with projected radial density profiles described by Elson, Fall & Freeman (1987; hereafter EFF) models with a power-law fall-off $\gamma = 3$. Initial stellar velocities are drawn from a Maxwellian distribution, where the velocity dispersion is calculated using the Jeans equations assuming an isotropic velocity distribution. We select the IMF of Kroupa (2001), with a stellar mass range 0.1–100 $M_\odot$, so that with $N \sim 10^5$ particles our cluster models possess total masses $\log M_{\mathrm{tot}} \sim 4.75$, consistent with massive young Magellanic Cloud objects. The models move on circular orbits of radius 6 kpc about a point-mass LMC with $M_g = 9 \times 10^9 M_\odot$.

For the models presented in this contribution, we are most interested in examining the dynamical effects of a retained population of stellar-mass black holes (BHs). We modified NBODY4 to control the production of BHs in supernova explosions. In the simulations described here, all stars initially above 20 $M_\odot$ produce BHs, with masses uniformly distributed in the range $8 \leqslant m_{\mathrm{BH}} \leqslant 12 M_\odot$. Our adopted IMF and total N lead to the formation of 198 BHs in all clusters. Natal BH velocity kicks are either much larger than the cluster escape velocity (BH retention fraction $f_{\mathrm{BH}} = 0$), or zero ($f_{\mathrm{BH}} = 1$).

Many young LMC and SMC clusters exhibit some degree of mass segregation. In order to include the effects of this in our models, we developed a method to generate clusters with primordial mass segregation in a self-consistent fashion; again, full details are in Mackey *et al.* (2007b in preparation). Here we consider models either with no primordial mass segregation, or a strong degree of primordial mass segregation chosen to match that observed in young Magellanic Cloud objects such as R136, NGC 330, 1805, and 1818.

To obtain structural measurements consistent with those for real clusters, we simulate observations of our N-body models. That is, we mimic the reduction procedures from which the HST r_c measurements were derived. We first convert the luminosity and effective temperature of each N-body star to magnitude and colour. Next, we impose bright and faint detection limits commensurate with the original observations, along with field-of-view limits appropriate for WFPC2 and ACS. We use the remaining stars to construct a projected radial profile and finally fit an EFF model to derive r_c.

2. Results

The parameter space of interest is spanned by non mass segregated clusters and those with significant primordial mass segregation. For each of these, we consider evolution with BH retention fractions $f_{\mathrm{BH}} = 0$ and 1. These four runs define the extremities of the parameter space, and hence are expected to cover the limits of cluster behaviour.

We first consider Runs 1 and 2, which have no primordial mass segregation, and $f_{\mathrm{BH}} = 0$ and 1, respectively. Their evolution is visible in the left panel of Fig. 1. Run 1 behaves exactly as expected for a classical massive star cluster. There is an early mass-loss phase ($\tau \leqslant 100$ Myr) due to the evolution of the most massive cluster stars. BHs are formed in

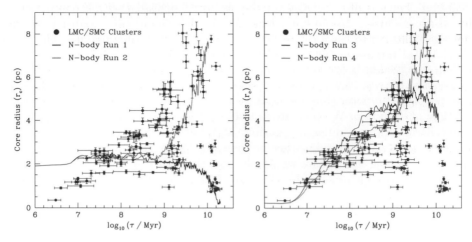

Figure 1. *Left:* Core radius evolution of Runs 1 and 2 (lower and upper lines, respectively).
Right: Core radius evolution of Runs 3 and 4 (lower and upper lines, respectively).

supernova explosions between 3.5–10 Myr; however, all receive large velocity kicks and
escape. The early mass-loss is not reflected in the evolution of r_c, presumably because
it is evenly distributed throughout the cluster. Subsequently, the general progression is
a slow contraction as two-body relaxation proceeds and mass segregation sets in. The
cluster core collapses near the end of the simulation.

Run 2 evolves similarly to a point, but, in striking contrast to Run 1, subsequently un-
dergoes dramatic and long-term core expansion. The only difference between this model
and Run 1 is that $f_{BH} = 1$ – i.e., 198 BHs are retained. Once early stellar evolution
is complete, the BHs are more massive than all other cluster members (of mean mass
$m_* \approx 0.5\,M_\odot$) and hence sink to the core on a time-scale of $\sim (m_*/m_{BH})\,t_{rh} \approx 100$ Myr.
By 200 Myr, the mass density of BHs at the cluster centre is similar to that of the stars;
by 400 Myr it is about three times larger. Soon after, the central BH subsystem becomes
unstable against further contraction and decouples from the stellar core in a runaway
collapse. At 490 Myr, the central density of the BH subsystem is $\sim$80 times that of the
stars. This is sufficient for the creation of stable BH binaries in three-body interactions –
the first is formed at $\sim$510 Myr, and by 800 Myr there are four.

Until this phase the evolution of Run 2 is observationally identical to that of Run 1.
Neither BH retention, nor the subsequent formation of a central BH subsystem leads to
differential evolution of r_c. Once formed, binary BHs undergo superelastic collisions with
other BHs in the core. The binaries become "harder", and the released binding energy
is carried off by the interacting BHs. This leads to BHs being *scattered* outside r_c, often
into the cluster halo, as well as to BHs being *ejected* from the cluster. Eventually a BH
binary is sufficiently hard that the recoil velocity imparted to it during a collision is
larger than the cluster escape velocity, and the binary is ejected. A BH scattered into
the halo gradually sinks back into the centre via dynamical friction, thus transferring its
newly-gained energy to the stellar component of the cluster. Most is deposited within
r_c, where the stellar density is greatest. Ejected BHs also transfer energy to the cluster,
since a mass m escaping from a cluster potential well of depth $|\Phi|$ does work $m|\Phi|$ on the
cluster. This mechanism is particularly effective in heating the stellar core, since BHs are
ejected from the very centre of the cluster, and the energy contributed to each part of the
cluster is proportional to the contribution which that part makes to the central potential.
Together these two processes result in significant prolonged core expansion, starting at

$\tau \approx 650$ Myr. The size of r_c behaves roughly as $\log \tau$, consistent with the upper envelope of the observed cluster distribution. However, in this model the expansion begins too late for the evolution to trace the upper envelope; rather, it runs parallel. Merritt *et al.* (2004) observed expansion due to similar processes in their simplified N-body models.

The number of stable BH binaries in the system peaks at 5 at $\tau \approx 890$ Myr, after which there are 0–5 at any given time. Single and binary BHs are continually ejected; however, empirically, both escape rates decrease with time such that $d^2 N_e/d\tau^2 \propto -1/\tau$. This is a result of the density of the central BH subsystem gradually decreasing with time due to the expansion of the cluster and the ejection of BHs. This in turn leads to a decreasing BH-BH encounter rate. By $\tau \approx 10$ Gyr, 65 single BHs and 2 binary BHs remain in the model. This is at odds with early studies (e.g., Kulkarni *et al.* 1993; Sigurdsson & Hernquist 1993) which predicted depletion of retained BH populations on timescales much less than cluster lifetimes. The decreasing BH encounter rate seen in our models prolongs the life of the BH subsystem for much longer than previously appreciated.

The velocity and mass distributions of stellar escapers are indistinguishable for Runs 1 and 2, implying that both lose stars solely due to relaxation processes. In Run 2 it is clear that stars interact closely with BH binaries only very rarely. Heating of the stellar component via such close interactions is hence negligible – the hardening of BH binaries is driven solely through interactions with other BHs.

Next, consider the evolution of Runs 3 and 4, which are primordially mass segregated versions of Runs 1 and 2, respectively (Fig. 1, right panel). In these models, early mass-loss due to stellar evolution is highly centrally concentrated – hence the amount of heating per unit mass lost is maximised, leading to dramatic early core expansion. Run 3 traces the observed upper envelope of clusters until several hundred Myr. Run 4 retains its BHs and hence loses less mass than Run 3 – this is reflected in its smaller r_c. After the early mass-loss phase is complete, core expansion stalls in both runs. Two-body relaxation gradually takes over in Run 3, leading to a slow contraction in r_c. At $\tau = 1$ Gyr, $t_{rh} \approx 4$ Gyr; hence this cluster is not near core collapse by the end of the simulation.

In Run 4, the BH population evolves similarly to that in Run 2. One might naively expect the earlier development of a compact BH subsystem in Run 4, because the BHs are already located in the core due to the primordial mass segregation. However, the centrally concentrated mass-loss acts against the accumulation of a dense BH core, and the first binary BH does not form until 570 Myr, a similar time to the non mass segregated model. The BH subsystem evolves more slowly than that in Run 2 – by $\tau = 10$ Gyr, there are still 95 single BHs and 2 binary BHs remaining in the cluster. As in Run 2, the evolution of the BH subsystem leads to expansion of r_c. This begins at $\tau \approx 800$ Myr and continues for the remainder of the simulation. As previously, r_c behaves roughly as $\log \tau$ during this phase. By $\tau \approx 10$ Gyr, Run 4 has $r_c \sim 11$ pc, comparable to that observed for the most extended old Magellanic Cloud clusters (e.g., Reticulum).

3. Discussion

Our four simulations cover the observed cluster distribution in radius-age space, thereby defining a dynamical origin for the radius-age trend. At ages less than a few hundred Myr, cluster cores expand due to centrally concentrated mass-loss from stellar evolution. At later times, expansion is induced via heating due to a BH population. Although we have assumed $f_{BH} = 1$, full retention is not necessary for cluster expansion. BH kicks of order $10 \leqslant v_{kick} \leqslant 20$ km s^{-1} result in $f_{BH} \sim 0.5$ in our models; evolution in such systems is intermediate between that of Runs 1 and 2, or 3 and 4 (Mackey *et al.* 2007b in preparation).

Our models require variations in BH population size between otherwise similar clusters. This is discussed fully by Mackey *et al.* 2007b in preparation; here, we simply note a few possibilities. First, the fraction of BH-forming stars in a cluster is small, so there will be sampling-noise variations between clusters. Further, any dispersion in stellar rotation may introduce mass-loss variations and further dispersion in BH numbers. Metallicity is also likely to be a key factor. Natal BH kicks are poorly constrained at present – typical estimates are $0 \leqslant v_{\text{kick}} \leqslant 200$ km s^{-1}. Stellar binarity may therefore play a significant role in retaining cluster BHs, as will the initial cluster mass and degree of primordial segregation.

Galactic globular clusters are typically an order of magnitude more massive than our N-body models. However, we expect the evolution described above to scale to such objects. The mass fraction of BHs formed in a cluster is dependent only on the IMF and minimum progenitor mass, neither of which should change with M_{tot}, while a larger M_{tot} implies a larger f_{BH} since it is easier to retain newly-formed BHs. The densities in our models are consistent with those observed for globular clusters; hence we expect the same processes to operate on similar time-scales, although BHs are likely to be more difficult to eject in more massive clusters – increasing the potential of each BH to heat the cluster via additional scattering-sinking cycles.

Core expansion due to early mass-loss and prolonged BH heating has strong implications for the observed properties of globular clusters as well as their survivability. Extended clusters are significantly more susceptible to tidal disruption, so it is important to account for expansion effects in studies of the evolution of the globular cluster mass function, for example. Core expansion due to BHs may offer a viable explanation for the origin of the luminous, unusually extended globular clusters found in M31, which are > 10 Gyr old metal-poor objects (Mackey *et al.* 2006). Our results imply that clusters possessing significant BH populations are, for most of their lives, low-density objects in which the timescale for close encounters between stars and BHs is very long. It is therefore unsurprising that no BH X-ray binaries are seen in the ~ 150 Galactic globulars.

Acknowledgements

ADM is supported by a Marie Curie Excellence Grant from the European Commission under contract MCEXT-CT-2005-025869. MIW acknowledges support from a Royal Society University Research Fellowship. MBD is a Royal Swedish Academy Research Fellow supported by a grant from the Knut and Alice Wallenberg Foundation.

References

Aarseth S. J. 2003, *Gravitational N-body Simulations.* Cambridge University Press, Cambridge
Elson R., Fall S. M., & Freeman K. C. 1987, *ApJ* 323, 54
Elson R., Freeman K. C., & Lauer T. R. 1989, *ApJ* 347, L69
Kroupa P. 2001, *MNRAS* 322, 231
Kulkarni S. R., Hut P., & McMillan S. 1993, *Nature* 364, 421
Mackey A. D. & Gilmore G. F. 2003a, *MNRAS* 338, 85
Mackey A. D. & Gilmore G. F. 2003b, *MNRAS* 338, 120
Mackey A. D., *et al.* 2006, *ApJ* 653, L105
Mackey A. D., Wilkinson M. I., Davies M. B., & Gilmore G. F. 2007a, *MNRAS* 379, L40
Makino J., Fukushige T., Koga M., & Namura K. 2003, *PASJ* 55, 1163
Merritt D., Piatek S., Portegies Zwart S., & Hemsendorf M. 2004, *ApJ* 608, L25
Sigurdsson S. & Hernquist L. 1993, *Nature* 364, 423
Wilkinson M. I., Hurley J. R., Mackey A. D., Gilmore G. F., & Tout C. A. 2003, *MNRAS* 343, 1025

Dynamical Evolution of Dense Stellar Systems
Proceedings IAU Symposium No. 246, 2007
E. Vesperini, M. Giersz & A. Sills, eds.

Dynamical Evolution of Mass-Segregated Clusters

Enrico Vesperini[1], Steve McMillan[1] and Simon Portegies Zwart[2]

[1]Department of Physics, Drexel University, Philadelphia, PA, USA

[2]Astronomical Institute "Anton Pannekoek" and Section Computational Science, University of Amsterdam

Abstract. We present the results of a survey of N-body simulations aimed at exploring the implications of primordial mass segregation on the dynamical evolution of star clusters. We show that, in a mass-segregated cluster, the effect of early mass loss due to stellar evolution is, in general, more destructive than for an unsegregated cluster with the same density profile and leads to shorter lifetimes, a faster initial evolution toward less concentrated structure and flattening of the stellar initial mass function.

Keywords. globular clusters: general, methods: n-body simulations, stellar dynamics

1. Introduction

Mass segregation, the tendency of more massive stars to preferentially populate the inner parts of a star cluster, is one of the consequences of two-body relaxation and of the evolution toward energy equipartition in stellar systems. The characteristic timescale for this process, T_{ms}, is, for a population of massive stars with mass m_h, $T_{ms} \sim \langle (m) \rangle / m_h) t_{relax}$ where t_{relax} is the cluster relaxation time and $\langle m \rangle$ the mean mass of the stars.

However, a number of young clusters with ages substantially less than the time needed to produce the observed segregation by standard two-body relaxation shows a significant level of mass segregation (e.g. Hillenbrand 1997, Hillenbrand & Hartmann 1998, Fischer *et al.* 1998, de Grijs *et al.* 2002, Sirianni *et al.* 2002, Gouliermis *et al.* 2004, Stolte *et al.* 2006). Several theoretical studies (e.g. Klessen 2001, Bonnell *et al.* 2001, Bonnell & Bate 2006 but see also Krumholz *et al.* 2005, Krumholz & Bonnell 2007) have suggested that massive stars would form preferentially in the center of star-forming regions and that the observed segregation in young clusters would be primordial.

Possible dynamical routes leading to early mass segregation in young star clusters have also been studied in McMillan, Vesperini & Portegies Zwart (2007 and this volume) and Vesperini, McMillan & Portegies Zwart (2008a *in prep.*).

Whatever the origin of the observed mass segregation in young clusters is, it is important to explore and understand the implications of initial mass segregation for the evolution of clusters. All the theoretical and numerical studies of the dynamical evolution of star clusters have adopted unsegregated initial conditions and nothing is known about the evolution of mass-segregated clusters. We have carried out a survey of N-body simulations to study the evolution of initially segregated clusters. We have explored the evolution of tidally truncated clusters located at different galactocentric distances and with different degrees of initial segregation and compared the evolution of the structure, stellar content and the lifetimes of segregated clusters with those of clusters with the same initial density profile but no initial mass segregation. We present here some of the results of our investigation; a complete description of all the simulations and results will appear in Vesperini, McMillan & Portegies Zwart (2008b, *in prep.*).

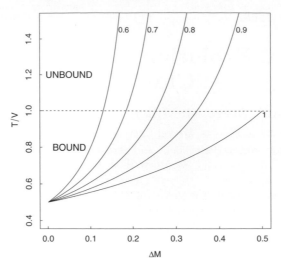

Figure 1. Virial ratio of a cluster with a Plummer density profile after the rapid mass loss of a ΔM fraction of its initial mass versus ΔM. The density profile of the mass lost is assumed to be that of a Plummer model, but with a scale radius a_{ML} equal to or less than the scale radius of the whole cluster, $a_{cluster}$. The number beside to each curve indicates the corresponding value of $a_{ML}/a_{cluster}$.

2. Segregated cluster dynamical evolution: analytical estimates

A number of studies (see e.g. Chernoff & Shapiro 1987, Chernoff & Weinberg 1990, Fukushige & Heggie 1995) have shown that early mass loss due to stellar evolution can have a significant impact on the evolution of clusters: this early mass loss causes a cluster to expand and, for a low-concentration cluster, leads to the cluster complete and quick dissolution (see e.g. Vesperini & Zepf 2003 for the possible implications of this early disruption on the properties of globular cluster systems). For initially segregated clusters, the mass lost due to the evolution of massive stars is removed preferentially from the cluster inner regions and the early expansion of the cluster is to be stronger and, potentially, more destructive than when the same amount of mass is lost in a non-segregated cluster.

Fig. 1 shows the results of a semi-analytical calculation illustrating the augmented destructive effect of stellar mass loss in a mass-segregated cluster. Specifically, in this figure we plot the virial ratio, T/V, of an isolated cluster with an initial Plummer density profile, $\rho = \frac{3M}{4\pi a^3}(1 + \frac{r^2}{a^2})^{-5/2}$, after the impulsive loss of a fraction ΔM of the total mass. In order to mimic the preferential mass loss from the inner regions of a mass-segregated cluster, we have assumed that the density profile of the mass lost also follows a Plummer model, but with a scale radius (a_{ML}) smaller than the scale radius of the cluster ($a_{cluster}$). The curves in Fig. 1 show T/V as a function of ΔM for different values of $a_{ML}/a_{cluster}$. For $a_{ML} = a_{cluster}$ one recovers the well-known result (Hills 1980) that a system becomes unbound for $\Delta M > 0.5$. For clusters preferentially losing mass from the inner regions, $a_{ML} < a_{cluster}$, we see that a significantly smaller amount of mass loss, innocuous for an unsegregated clusters, can lead to dissolution of a segregated cluster.

This simple semi-analytical calculation underscores the potential crucial implications of initial mass segregation for the evolution of star clusters.

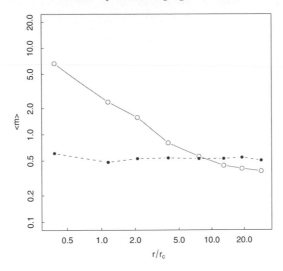

Figure 2. Initial radial profile of the mean mass of stars for the segregated (solid line) and the unsegregated clusters (dashed line) discussed in this paper.

3. Segregated cluster dynamical evolution: N-body simulations

In this section we discuss the results of a survey of N-body simulations aimed at comparing the evolution of initially segregated and unsegregated clusters. All the simulations have been carried out by the `starlab` package (Portegies Zwart *et al.* 2001; http://www.manybody.org) We have followed the evolution of tidally truncated clusters with the same density profile, a Kroupa (Kroupa, Tout & Gilmore 1993) stellar IMF, with and without initial mass segregation. Fig. 2 shows the initial radial profile of the star mean mass in the segregated and unsegregated clusters studied in our investigation.

We have performed a number of simulations comparing the evolution of segregated and unsegregated clusters at different galactocentric distances, R_g. For tidally truncated clusters with a fixed mass, the cluster global dynamical time and the amount of mass lost impulsively increase with the galactocentric distance, R_g (we have also carried out a number of simulations of isolated clusters and clusters underfilling their tidal radius and explored the evolution for different levels of initial segregation; the results of these simulations will be presented in Vesperini, McMillan & Portegies Zwart 2008b, *in prep.*)

Fig. 3 shows the dependence of the dissolution time, T_{diss}, (defined as the time when only 1 per cent of the initial mass is left in the cluster) on the clusters galactocentric distance for segregated and unsegregated clusters. All the unsegregated clusters survive the early mass loss due to stellar evolution and eventually dissolve as a result of the evaporation of stars due to two-body relaxation. On the other hand, early impulsive mass loss leads to the quick dissolution in a few dynamical times of all the initially segregated clusters. The only exception is the cluster closest to the galactic center for which the dynamical time is short enough that there is no significant impulsive mass loss; in this case the segregated cluster lifetime is similar to that of the corresponding unsegregated cluster.

The difference in the response to the same amount of impulsive mass loss in segregated and unsegregated clusters is further illustrated by the left panel of Fig. 4 which shows the time evolution of the core and half-mass radii for a segregated and an unsegregated cluster with the same initial density profile and located at the same galactocentric distance ($R_g = 18$ kpc). Both clusters initially lose the same fraction of their initial mass and

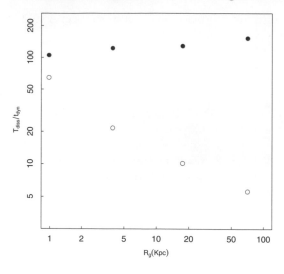

Figure 3. Ratio of the dissolution time, T_{diss}, to the dynamical time t_{dyn}, versus galactocentric distance for segregated (open dots) and unsegregated clusters (filled dots).

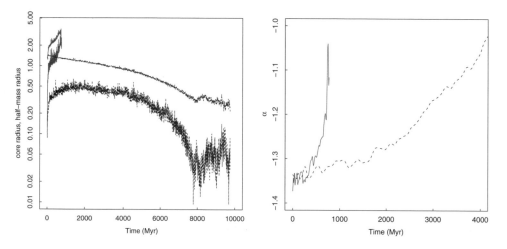

Figure 4. (Left panel) Time evolution of the core and the half-mass radii of a segregated (solid lines) and an unsegregated (dashed line) cluster. (Right panel) Time evolution of the slope of the mass function, α, for stars with $0.1 < m/m_\odot < 0.5$ for a segregated (solid lines) and an unsegregated (dashed line) cluster.

expand in response to this mass loss; however, as anticipated by the semi-analytical calculations discussed in the previous section, the preferential removal of mass from the innermost regions in a segregated cluster, leads to a stronger expansion and to the cluster quick dissolution.

The right panel of Fig. 4 shows the time evolution of the slope of the mass function for low-mass stars ($0.1 < m < 0.5$) and illustrates the implications of the different evolution of segregated and unsegregated clusters for the evolution of the clusters stellar content. Both segregated and unsegregated clusters preferentially lose low-mass stars but the timescale for the flattening of the mass function in the segregated cluster is significantly shorter than that of the unsegregated one.

4. Conclusions

We have presented the results of analytical calculations and numerical simulations aimed at exploring the differences in the evolution of initiallly segregated and unsegregated clusters. Our study shows that early mass loss due to stellar evolution leads to a stronger expansion of segregated clusters, a faster evolution toward a less concentrated structure and a faster dissolution. The evolution of the stellar content and, in particular, the preferential loss of low-mass stars also proceeds on a shorter timescale in segregated clusters.

A complete description of all the results of this investigation will be presented elsewhere (Vesperini, McMillan & Portegies Zwart 2008b, *in prep.*)

Acknowledgments. This work was supported in part by NASA grants NNG04GL50G and NNX07AG95G, NSF grant AST-0708299, and by the Royal Netherlands Academy of Arts and Sciences (KNAW).

References

Bonnell, I. A., Clarke, C. J., Bate, M. R., & Pringle, J. E. 2001, *MNRAS*, 324, 573

Bonnell, I. A. & Bate, M. R., 2006, *MNRAS*, 370, 488

Chernoff D. & Shapiro S., 1987, *ApJ*, 322, 113

Chernoff D. & Weinberg M., 1990, *ApJ*, 351, 121

de Grijs, R., Gilmore, G. F., Johnson, R. A., & Mackey, A. D. 2002, *MNRAS*, 331, 245

Fischer P., Pryor C., Murray S., Mateo M., & Richtler T. 1998, *AJ*, 331, 592

Fukushige T. & Heggie D., 1995, *MNRAS*, 276, 206

Gouliermis, D., Keller, S. C., Kontizas, M., Kontizas, E., & Bellas-Velidis, I., 2004 , *A&A*, 416, 137

Hillenbrand L. A., 1997, *AJ*, 113, 1733

Hillenbrand L. A. & Hartmann L. E., 1998, *ApJ*, 331, 540

Hills J., 1980, *ApJ*, 235, 968

Klessen R., 2001, *ApJ*, 556, 837

Kroupa P., Tout C., & Gilmore G., 1993, *MNRAS*, 262, 545

Krumholz M. R., Klein R. I. & McKee C. F., 2005, *Nature*, 438, 332

Krumholz M. R. & Bonnell I., 2007 astro-ph 0712.0828

McMillan S., Vesperini E., & Portegies Zwart S., 2007, *ApJ*, 655, L45

Portegies Zwart, S. F., McMillan, S. L. W., Hut, P., & Makino, J. 2001, *MNRAS*, 321, 199

Sirianni, M., Nota, A., De Marchi, G., Leitherer, C., & Clampin, M. 2002, *ApJ*, 579, 275

Stolte, A., Brandner, W., Brandl, B., & Zinnecker, H. 2006, *AJ*, 132, 253

Vesperini, E. & Zepf, S., 2003, *ApJ*, 587, L97

4. Conclusions

Dynamical Evolution of Dense Stellar Systems
Proceedings IAU Symposium No. 246, 2007
E. Vesperini, M. Giersz & A. Sills, eds.

© 2008 International Astronomical Union
doi:10.1017/S1743921308015561

N-body Simulations of Star Clusters

Peter Anders[1], Henny J. G. L. M. Lamers[1] and Holger Baumgardt[2]

[1]Sterrenkundig Instituut, Universiteit Utrecht, P.O. Box 80000, 3508 TA Utrecht,
The Netherlands; email: anders/lamers@astro.uu.nl

[2]Argelander-Institut für Astronomie, Auf dem Hügel 71, 53121 Bonn, Germany
email: holger@astro.uni-bonn.de

Abstract. Two aspects of our recent N-body studies of star clusters are presented:
1) What impact does mass segregation and selective mass loss have on integrated photometry?
2) How well do results compare from N-body simulations using NBODY4 and STARLAB/KIRA?

Keywords. methods: n-body simulations, methods: numerical, stars: mass function, galaxies:
star clusters

1. Selective mass loss and integrated photometry

"Mass segregation" describes the effect that high-mass stars are preferentially found in the center of clusters, while the outskirts are preferentially occupied by lower-mass stars.

(Primordial) mass segregation is studied observationally in a number of clusters. Some examples are: the Orion Nebula Cluster (Hillenbrand & Hartmann 1998), 6 LMC clusters (de Grijs *et al.* 2002).

Dynamical mass segregation is also found from N-body simulations (e.g. Inagaki & Saslaw 1985, Giersz & Heggie 1997), caused by two-body encounters and energy equipartition. A number of authors also point to the preferential mass loss of low-mass stars when the cluster evolves in a tidal field (e.g. Spitzer & Shull 1975, Giersz & Heggie 1997), as mass segregation populates the cluster outskirts preferentially with low-mass stars where they are most easily stripped from the cluster potential. The resulting changes in the overall stellar mass function inside the cluster are quantified by Baumgardt & Makino (2003).

In Lamers *et al.* (2006) we use the results from Baumgardt & Makino (2003) to incorporate the changing mass function slope in a simplified manner into the GALEV code (see Anders & Fritze-v. Alvensleben 2003 and references therein) to calculate its impact on the integrated photometry of star clusters.

Our main findings are:

• at 0 – 40 per cent of the cluster's lifetime: the cluster colours are comparable to standard models (i.e. without preferential loss of low-mass stars)

• at 40 – 80 per cent of the cluster's lifetime: a cluster appears too blue/young (compared to standard models) due to the loss of lower main-sequence stars

• at 80 – 100 per cent of the cluster's lifetime: a cluster appears too red/old (compared to standard models) due to the loss of main sequence turn-off stars

• when interpreting photometry of mass-segregated clusters, that have preferentially lost low-mass stars, with standard photometric models (without taking mass-segregation effects into account) the derived ages can be wrong by 0.3 – 0.5 dex

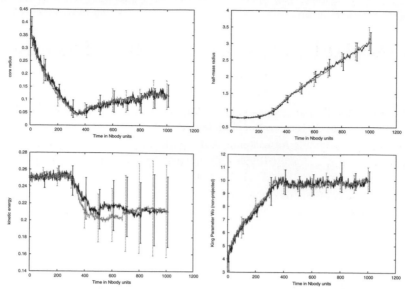

Figure 1. Comparison of simulations using STARLAB (dark lines) vs NBODY4 (light lines). The lines show the mean values; the error bars give the one-sigma spread from individual runs.

2. Benchmark test for N-body codes: Comparing NBODY4 with STARLAB/KIRA

Here we propose a benchmark test for comparison of N-body codes: 1024 equal-mass particles, Plummer sphere model, without primordial binaries, external tidal field or stellar/binary evolution.

We performed simulations using the same input files (10 individual runs) for NBODY4 (see e.g. Aarseth 1999) and STARLAB (see e.g. Portegies Zwart *et al.* 2001), following the setup from the proposed benchmark test. We find stochastic effects to be of importance for 10 runs, hence we supplemented the STARLAB results with 20 additional runs (equivalent NBODY4 runs are in preparation). Some results are shown in Fig. 1. For most parameters studied the results are virtually indistinguishable. Some questions remain for the kinetic energy and the parameter distributions of dynamically formed binaries. Any differences there could be originating from the different treatment of binaries, or due to stochastic effects. Further tests (e.g. with primordial binaries) are in preparation.

Acknowledgements

We thank the ISSI in Bern, Switzerland, where parts of this project were carried out. We thank Evghenii Gaburov and Simon Portegies Zwart for comments and suggestions.

References

Aarseth, S. 1999, *PASP*, 111, 1333
Anders, P. & Fritze-v. Alvensleben, U. 2003, *A&A*, 401, 1063
Baumgardt, H. & Makino, J. 2003, *MNRAS*, 340, 227
de Grijs, R., Gilmore, G. F., Johnson, R. A., & Mackey, A. D. 2002, *MNRAS*, 331, 245
Giersz, M. & Heggie, D. C. 1997, *MNRAS*, 286, 709
Hillenbrand, L. A. & Hartmann, L. W. 1998, *ApJ*, 492, 540
Inagaki, S. & Saslaw, W. C. 1985, *ApJ*, 292, 339
Lamers, H. J. G. L. M., Anders, P., & de Grijs, R. 2006, *A&A*, 452, 131
Portegies Zwart, S. F., McMillan, S. L. W., Hut, P., & Makino, J. 2001, *MNRAS*, 321, 199
Spitzer, L., Jr. & Shull, J. M. 1975, *ApJ*, 201, 773

Dynamical Evolution of Dense Stellar Systems
Proceedings IAU Symposium No. 246, 2007
E. Vesperini, M. Giersz & A. Sills, eds.

Numerical Modelling of the Tidal Tails of NGC 5466

M. Fellhauer[1], N. W. Evans[1], V. Belokurov[1], M. I. Wilkinson[1,2] and G. Gilmore[1]

[1]Institute of Astronomy, University of Cambridge, Madingley Road, Cambridge CB3 0HA, UK

[2]Department of Physics and Astronomy, University of Leicester, University Road, Leicester LE1 7RH, UK

Abstract. The study of sub-structures in the stellar halo of the Milky Way has made a lot of progress in recent years, especially since surveys like the Sloan Digital Sky Survey became available. In this paper we focus on the newly discovered tidal tails of the Galactic globular cluster NGC 5466. By means of numerical simulations we reproduce the tidal tails, which are the longest tails associated with a globular cluster known ($> 45°$) and hereby finding a possible progenitor of NGC 5466 and analyse its stability. We show that perigalactic passages are the dominant process in the slow dissolution of NGC 5466. Furthermore we use the position of the tails to verify the accuracy of the observationally determined proper motion. The proper motion has to be refined only slightly (within their stated error-margin) to match the location of the tidal tails.

Keywords. globular cluster: individual (NGC 5466), Galaxy: kinematics and dynamics, Galaxy: halo, methods: n-body simulations

1. Introduction

Recently, two different groups (Belokurov *et al.* 2006; Grillmair & Johnson 2006) claim to have detected tidal tails of various extents around the globular cluster NGC 5466. This is an old, metal-poor ([Fe/H] $= -2.22$) cluster, lying at Galactic coordinates $l = 42°15$, $b = 73°59$. In Belokurov *et al.* (2006), the observed tails of NGC 5466 are not as long as those of Pal 5, stretching about $2°$ or 500 pc in either direction. Grillmair & Johnson (2006) reported afterwards that they found evidence for a much larger extension of the tidal tails of NGC 5466. They claimed that the leading arm extends over ~ 30 degrees and the trailing arm extends at least 15 degrees, before it leaves the area covered by SDSS. This finding makes the tails of NGC 5466 even longer, but much fainter, than the tails of Pal 5. The aim of our paper is to confront these claims with theoretical expectation, as well as to study the survival of the cluster and the mechanisms of its mass-loss.

2. Setup

We use the particle-mesh code SUPERBOX (Fellhauer *et al.* 2000) to carry out our simulations. The Milky Way potential is added analytically consisting of a Hernquist bulge, a Miamoto-Nagai disc and a logarithmic halo with parameters described in Fellhauer *et al.* (2007). The globular cluster NGC 5466 is modelled as a Plummer sphere with Plummer radius of 10 pc and initial mass of 7×10^4 M$_\odot$.

3. Results

All our results can be found in detail in Fellhauer *et al.* (2007).

By running particle-mesh simulations using a million particles for the cluster we are able to trace its tidal tails in great detail. We find that the location of the near-field tails (around the cluster, as seen by Belokurov *et al.* 2006) do not conform with the proper motion given in the literature (Dinescu, Girard & Altena 1999). The proper motion has to be slightly (within the given error-margins) altered to $\mu_\alpha \cos\delta = -4.7$ mas yr^{-1} and $\mu_\delta = 0.42$ mas yr^{-1} to give the model tails the same orientation as the observed ones. We confirm the extend of the tidal tails and their density of about 25 $M_\odot$ deg^{-2} as claimed by Grillmair & Johnson (2006).

By comparing the mass-loss of the particle-mesh simulation with a direct N-body simulation, we show that only 1/3 of the mass-loss is caused by internal evolution and the major reason for the mass-loss is tidal shocking. We confirm the results of Gnedin, Lee & Ostriker (1999), that the initial NGC 5466 as well as the present cluster are able to survive for at least a Hubble-time and that the cluster resides in a regime where shocks are important. Calculating the disc-shock time-scale gives 110 Gyr for NGC 5466. In our simulations we found that without a disc present the mass-loss is reduced from 19 per cent down to 3 per cent over 10 Gyr of simulation. But a detailed inspection of the mass-loss at each pericentre- and disc-passage shows that it is not the actual disc-shock (i.e. the cluster travelling through the disc) but rather the general pericentre passage which is responsible for the bulk of the lost mass.

4. Conclusions

We have presented numerical simulations of the formation and evolution of the tidal tails of the globular cluster NGC 5466. We used direct N-body codes to argue that the evolution of the cluster is dominated by external effects rather than internal relaxation, and then grid-based codes to trace the faint tidal tails. This novel, hybrid approach is well-suited to map out the detailed morphology of the low-density tails of NGC 5466.

Naively, we might expect that a low mass cluster with observed and very lengthy tails on a disc crossing orbit would not be able to survive for too much longer. However, simulations by Dehnen *et al.* (2004) have already shown that the disrupting globular cluster Pal 5 has survived for at least many Gyr in a tidally-dominated and out-of-equilibrium state, although Pal 5 probably will be destroyed at the next disc crossing. Here, we have demonstrated that a progenitor cluster of NGC 5466, which is quite similar to the present cluster, could survive substantially longer, for at least a few Hubble times, with its extensive but tenuous tidal tails gradually wrapping around the whole Galaxy.

References

Belokurov, V., Evans, N. W., Irwin, M. J., Hewett, P. C., & Wilkinson, M. I. 2006, *ApJL* 637, L29

Dehnen, W., Odenkirchen, M., Grebel, E. K., & Rix, H.-W. 2004, *AJ* 127, 2753

Dinescu, D. I., Girard, T. M., & van Altena, W. F. 1999, *AJ* 117, 1792

Fellhauer, M., Kroupa, P., Baumgardt, H., Bien, R., Boily, C. M., Spurzem, R., & Wassmer, N. 2000, *New Astron.*, 5, 305

Fellhauer, M., Evans, N. W., Belokurov, V., Wilkinson, M. I., & Gilmore, G. 2007, *MNRAS* 380, 749

Gnedin, O. Y., Lee, H. M., & Ostriker, J. P. 1999, *ApJ* 522, 935

Grillmair, C. J. & Johnson, R. 2006, *ApJL* 639, L17

Dynamical Evolution of Dense Stellar Systems
Proceedings IAU Symposium No. 246, 2007
E. Vesperini, M. Giersz & A. Sills, eds.

© 2008 International Astronomical Union
doi:10.1017/S1743921308015585

Mass-Loss Timescale of Star Cluster in External Tidal Field

T. Fukushige[1] and A. Tanikawa[2]

[1]K&F Computing Research Co.
email: fukushig@kfcr.jp

[2]Department of General System Studies, College of Arts and Sciences, University of Tokyo
email: tanikawa@ea.c.u-tokyo.ac.jp

Abstract. We investigate evolution of star clusters in steady external tidal field by means of N-body simulations. We followed several sets of cluster models whose strength and Coriolis's contribution of the external tidal field are different. We found that the mass loss timescale due to the escape of stars, t_{mloss}, and its dependence on the two-body relaxation timescale, $t_{\mathrm{rh,i}}$, are determined by the strength of the tidal field. The logarithmic slope [$\equiv d\ln(t_{\mathrm{mloss}})/d\ln(t_{\mathrm{rh,i}})$] approaches unity for the cluster models in weaker tidal fields. We also found that stronger Coriolis force against others, produced by parent galaxy whose density profile is shallower, makes the mass loss timescale longer. This is due to the fact that a fraction of stars whose orbit are nearly regular increases as the Coriolis force becomes stronger.

Keywords. globular clusters: general, method: n-body simulations

1. Introduction

It has been clear that the mass-loss timescales of star clusters in an external tidal field are not proportional to the two-body relaxation times of the clusters (Heggie *et al.* 1998; Baumgardt 2001). This is due to escape time delay during which stars with the escape energy from the clusters (hereafter, "potential escapers") remain inside the clusters before finding exits (Fukushige & Heggie 2000). However, the dependence of the mass-loss timescales on the external tidal field have not been investigated. We investigated the dependence on the strength of the external tidal field (in detail in Tanikawa & Fukushige 2005), and the mass profiles of the parent galaxies of the clusters.

2. Method

We investigate the evolution of star clusters in the external tidal field by means of N-body simulations. We considered a cluster that moves in a circular orbit under a spherically symmetric galaxy potential. We used standard units, such $M = G = -4E = 1$, where G is gravitational constant, M is the initial total mass of the cluster, and E is the initial total energy within the cluster. The cluster has $W_0 = 3$ King's profile. We set the eight initial models whose $(r_{\mathrm{t,i}}/r_{\mathrm{kg}}, \kappa^2/\omega^2)$ are $(0.8, 1)$, $(1.0, 1)$, $(1.3, 1)$, $(2.2, 1)$, $(4.5, 1)$, $(1.0, 2)$, $(1.0, 2.5)$, and $(1.0, 3)$, where $r_{\mathrm{t,i}}$ is the initial tidal radius of the cluster, r_{kg} is the radius beyond which the density is zero in the King's model, κ is the epicyclic frequency of the cluster, and ω is the angular velocity of the cluster. The external tidal field becomes weaker with $r_{\mathrm{t,i}}/r_{\mathrm{kg}}$ increasing. Coriolis force becomes stronger and the mass profile of the parent galaxy becomes shallower with κ^2/ω^2 increasing.

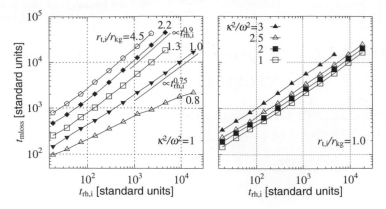

Figure 1. Mass-loss timescale of clusters as a function of initial half-mass relaxation time.

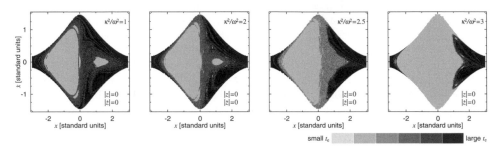

Figure 2. Escape time delay of the potential escapers as a function of their initial phase. The colors indicates the escape time delay, $t_e < 10^2$, $10^2 < t_e < 10^3$, $10^3 < t_e < 10^4$, $10^4 < t_e < 10^6$, and non-escaper, in order of the darkness.

3. Results

Fig. 1 shows the mass-loss timescale, t_{mloss}, of clusters as a function of the initial half-mass relaxation time, $t_{\mathrm{rh,i}}$. The mass-loss timescale is here defined as the time when 50 % of the initial total mass is lost. We found from the left panel of Fig. 1 that the mass-loss timescales, and the logarithmic slope, α [$\equiv d\ln(t_{\mathrm{mloss}})/d\ln(t_{\mathrm{rh,i}})$], depend on the tidal radii of the clusters. In the case of $r_{\mathrm{t,i}}/r_{\mathrm{kg}} = 1.0$, whose run parameters are set to be the same as that Baumgardt (2001) we could reproduce the asymptotic power, $\alpha \sim 0.75$. However, when the tidal radii were smaller the slope was smaller, and when it were larger, the slope was larger than $\alpha = 0.75$. The slope α in the case of larger tidal radii is seen to approach asymptotically to near unity, but roughly $\alpha \sim 0.9$.

We found from the right panel of Fig. 1 the mass-loss timescales become larger as κ^2/ω^2 become larger. Fig. 2 shows escape time delay of the potential escapers as a function of their initial phase. The larger mass-loss timescale in larger κ^2/ω^2 results from the fact that a fraction of stars whose orbit are nearly regular increases.

References

Baumgardt, H. 2001, *MNRAS* 325, 1323

Fukushige, T. & Heggie, D. C. 2000, *MNRAS* 318, 753

Heggie, D. C., Giersz, M., Spurzem, R., & Takahashi, K. 1998, in Highlights of Astronomy, 11A, ed. J. Andersen (Dordrecht: Kluwer Academic Publishers), 591

Tanikawa, A. & Fukushige, T. 2005, *PASJ* 57, 155

Dynamical Evolution of Dense Stellar Systems
Proceedings IAU Symposium No. 246, 2007
E. Vesperini, M. Giersz & A. Sills, eds.

Integrated Properties of Mass Segregated Star Clusters

E. Gaburov[1,2] and M. Gieles[3]

[1]Sterrenkundig Instituut "Anton Pannekoek", University of Amsterdam
egaburov@science.uva.nl
[2]Section Computational Science, University of Amsterdam
[3]European Southern Observatory, Casilla 19001, Santiago 19, Chile
email: mgieles@eso.org

Abstract. In this contribution we study integrated properties of dynamically segregated star clusters. The observed core radii of segregated clusters can be 50% smaller than the "true" core radius. In addition, the measured radius in the red filters is smaller than those measured in blue filters. However, these difference are small ($\lesssim 10\%$), making it observationally challenging to detect mass segregation in extra-galactic clusters based on such a comparison. Our results follow naturally from the fact that in nearly all filters most of the light comes from the most massive stars. Therefore, the observed surface brightness profile is dominated by stars of similar mass, which are centrally concentrated and have a similar spatial distribution.

Keywords. galaxies: star clusters

Mass segregation in star clusters is often observed from radial variations in stellar mass function (de Grijs *et al.* 2002a; de Grijs *et al.* 2002b; Kim *et al.* 2006). These methods are subject to biases and selection effects, such as incompleteness and blending and can, therefore, not be applied to clusters more distant than the Magellanic Clouds. Such extra-galactic star clusters can only be studied through their integrated properties (Bastian *et al.* 2007; McCrady, Graham & Vacca 2005). In this contribution we present simulated observations of integrated properties of dynamically segregated young star clusters.

We present a semi-analytical model of star clusters based on a simple analytical description of the mass function at different radii from the cluster centre (r). Based on both observations (Kim *et al.* 2006) and N-body simulations (Portegies Zwart *et al.* 2007; Gaburov, Gualandris & Portegies Zwart 2007) of young star clusters, we model the mass function in the following way. For $r < r_{\rm hm}$, with $r_{\rm hm}$ the half-mass radius, the mass function is

$$g(m, r < r_{\rm hm}) \propto \begin{cases} m^{\alpha_0}, & \text{if } m < \mu = 2\langle m \rangle \\ \mu^{\alpha_0} \left(\dfrac{m}{\mu} \right)^{\alpha(r)}, & \text{otherwise.} \end{cases},$$

whereas for $r > r_{\rm hm}$ it has power-law form with a constant slope $\alpha_{\rm hm}$, $g(m, r > r_{\rm hm}) \propto m^{\alpha_{\rm hm}}$. Here, $\langle m \rangle$ and α_0 are the mean mass and the slope of the initial mass function respectively, and $\alpha(r)$ is the index at the high-mass end which is a function of distance to the cluster centre. We choose $\alpha(r) = \alpha_{\rm inf} + (\alpha_c - \alpha_{\rm inf})(1 + (r/\epsilon)^\delta)$. The parameter α_c determines the degree of mass segregation, while parameters ϵ and δ define shape of $\alpha(r)$. The two other parameters, $\alpha_{\rm hm}$ and $\alpha_{\rm inf}$, are determined by constraints that the cluster integrated mass function results in IMF, and that the mean mass is continuous at $r_{\rm hm}$.

Given the mass function, we can calculate the mean stellar mass, $\mu(r)$, as well as the luminosity profile in different filters, $L_\lambda(r)$, as a function of r. To this end we use Padova isochrones (Girardi *et al.* 2002). For a given density profile, $\rho(r)$, we can obtain

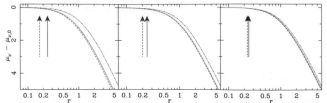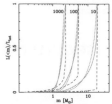

Figure 1. Normalised surface brightness profiles (left panel) of a star cluster as a function of its age. The dash-dot-dot-dot line is the surface brightness profile of a non-segregated cluster, whereas the solid, dotted and dashed lines display the profiles for mass segregated clusters in the V, U and K filters, respectively. The dashed and solid arrows show the core radius of segregated (U-filter) and non-segregated clusters respectively. The cumulative luminosity functions (right panel) are computed from a single stellar population with a power-law mass function. The numbers left of the lines show the age of the population in Myr.

spatial luminosity profile in a desired filter by using the following conversion $L_\lambda(r) = \rho(r) \cdot [\lambda_\lambda(r)/\mu(r)]$. As a test case, we choose $\rho(r)$ to be an EFF profile (Elson, Fall & Freeman 1987). The resulted surface brightness profiles are fitted to a projected EFF profile in order to obtain the core radius and power-law index.

In the case of non-segregated clusters, light distribution traces the density distribution, which is not the case for segregated star clusters. In the left panel of Fig. 1, we show the surface brightness profiles in different filters compared to surface brightness profile of non-segregated cluster. We notice that the observed core radius in segregated clusters is smaller by roughly 50% compared to the core radius of non-segregated cluster with the same density profile. Moreover, the former one is never larger than the latter one for ages $\lesssim 1$ Gyr. In addition, the difference between core radii in U- and K-filters is smaller than 10%, and it is larger for younger clusters.

Assuming a power-law mass function, we show the cumulative luminosity function in different filters in the right panel of Fig. 1. In older clusters most of the light comes from stars with similar masses which have light-to-mass ratio close to unity. Therefore, the light distribution approximates well the density distribution, and the core radii in different filters are nearly the same. In the case of young star clusters, however, the most massive stars dominate the K-filter, whereas stars with half the turn-off mass dominate the U-filter. Combined with the fact that light-to-mass ratio of these stars is notable large than unity, we expected that core radius is smaller than that of unsegregated star cluster. In addition, the core radius in K-filter is smaller than the core radius in U-filter. However, we do not expect this difference to be large, as the stars which differ in mass by a factor of two do not have significantly different spatial distribution.

Acknowledgements: We thank Nate Bastian and Søren Larsen for helpful discussions. This work is supported by NWO under grant #635.000.303.

References

Bastian, N., Konstantopoulos, I., Smith, L. J., Trancho, G., Westmoquette, M. S., & Gallagher, J. S. 2007, *MNRAS*, 379, 1333

Elson, R. A. W., Fall, S. M., & Freeman, K. C. 1987, *ApJ*, 323, 54

Gaburov, E, Gualandris, A., & Portegies Zwart, S. 2007, arXiv/0707.0406

Girardi, L., Bertelli, G., Bressan, A., Chiosi, C., Groenewegen, M. A. T., Marigo, P., Salasnich, B., & Weiss, A. 2002 *A&A*, 391, 195

de Grijs, R., Johnson, R. A., Gilmore, G. F., & Frayn, C. M., 2002a *MNRAS*, 331, 228

de Grijs, R., Gilmore, G. F., Johnson, R. A., & Mackey, A. D., 2002b *MNRAS*, 331, 245

Kim, S. S., Figer, D. F., Kudritzki, R. P., & Najarro, F. 2006, *ApJ*, 653, 113

McCrady, Na., Graham, J. R., & Vacca, W. D., 2005 *ApJ*, 621, 278

Portegies Zwart, S., Gaburov, E., Chen, H.-C. & Gürkan, M. A. 2007, *MNRAS*, 378, 29

Dynamical Evolution of Dense Stellar Systems
Proceedings IAU Symposium No. 246, 2007
E. Vesperini, M. Giersz & A. Sills, eds.

© 2008 International Astronomical Union
doi:10.1017/S1743921308015603

On the Efficiency of Field Star Capture by Star Clusters

S. Mieske[1] and H. Baumgardt[2]

[1]European Southern Observatory, Karl-Schwarzschild-Str.2, 85748 Garching b. München,
Germany
email: smieske@eso.org,mhilker@eso.org

[2]Argelander Institut für Astronomie, Auf dem Hügel 71, 53121 Bonn, Germany
email: holger@astro.uni-bonn.de

Abstract. An exciting recent finding regarding scaling relations among globular clusters is the so-called 'blue tilt': clusters of the blue sub-population follow a trend of redder colour with increasing luminosity. In this contribution we estimate by means of collisional N-body simulations to which extent this trend can be explained by field star capture occurring over a Hubble time. We investigate star clusters with 10^3 to 10^6 stars. We find that the ratio between captured field stars and total number of clusters stars is very low ($\lesssim 10^{-4}$), even for co-rotation of the star cluster in a cold disk. This holds for star clusters in the mass range of both open clusters and globular clusters. Therefore, field star capture is not a probable mechanism for creating the colour-magnitude trend of metal-poor globular clusters.

Keywords. globular clusters: general, open clusters and associations: general, stars: kinematics, galaxies: kinematics and dynamics

1. Introduction

An exciting recent finding regarding scaling relations among globular clusters is the so-called 'blue tilt': the colours of individual globular clusters (GCs) of the blue sub-population are correlated with their luminosities such that brighter globulars are redder (Harris *et al.* 2006, Mieske *et al.* 2006, Strader *et al.* 2006, Spitler *et al.* 2006, Cantiello *et al.* 2007). Various mechanisms have been discussed that may offer ways towards explaining the trend, like self-enrichment (Strader *et al.* 2006) or "sample contamination" by stripped nuclei of dwarf galaxies (Harris *et al.* 2006, Bekki *et al.* 2007). In Mieske *et al.* (2006) we indicate that also the capture of field stars in giant elliptical galaxies can in principle cause such colour-magnitude trends and its dependence on galactocentric distance as found by Mieske *et al.* (2006). This is because the field star population is much redder than the blue globular clusters.

2. Results and conclusions

Here, we present results of collisional N-body simulations made to quantify the amount of field star capture occurring over a Hubble time to star clusters with 10^3 to 10^6 stars (see also Mieske & Baumgardt 2007). In the simulations we follow the orbits of field stars passing through a star cluster and calculate the energy change that the field stars experience due to gravitational interaction with cluster stars during one passage through the cluster. The capture condition is that their total energy after the passage is smaller than the gravitational potential at the cluster's tidal radius (see Fig. 1). By folding this with the fly-by rates of field stars with an assumed space density as in the solar

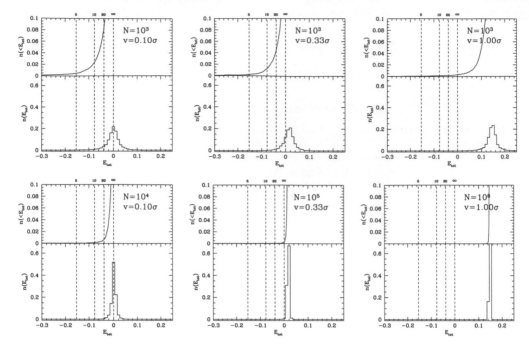

Figure 1. Distribution of total energy E_{tot} of field stars after interaction with the cluster stars. The three plots refer to different number of stars in the star clusters, and different assumed initial velocity v_{ini} of the field stars in units of the star cluster velocity dispersion σ. **Top row, from left to right:** star cluster with $N=10^3$ stars and field star initial velocities v_{ini} =0.1, 0.33 and 1.0 σ. **Bottom row, from left to right:** star clusters with $N=10^4$, 10^5 and 10^6 stars with the indicated v_{ini}. The lower part of each plot gives the energy histogram, the upper part gives the cumulative energy distribution. The requirement for field star capture is $E_{tot} < \frac{-GM_c}{r_{tid}}$, with r_{tid} being the tidal radius of the cluster in the gravitational field of the host galaxy. The vertical dashed lines indicate different assumed ratios of $\frac{r_{tid}}{r_h}$, where r_h is the cluster's half-mass radius. The *probability* of field star capture is non-negligible only for the case of a low mass cluster and low relative velocity. However, for the given example of $N=10^3$ stars, $v_{ini} = 0.1\sigma$ corresponds to 0.04 km/s, assuming an average cluster star mass of 0.5 solar masses. The extremely low number of field stars that have such low relative velocities makes the capture *rate* negligible.

neighbourhood and a range of velocity dispersions σ (15 to 485 km/s), we derive estimates on the mass fraction of captured field stars as a function of environment.

From this we find that integrated over a Hubble time, the ratio between captured field stars and total number of clusters stars is very low ($\lesssim 10^{-4}$), even for the smallest field star velocity dispersion $\sigma = 15$ km/s. This holds for star clusters in the mass range of both open clusters and globular clusters. **Field star capture is therefore not a probable mechanism for creating the 'blue tilt'.**

References

Bekki, K., Yahagi, H., & Forbes, D. A. 2007, *MNRAS*, 377, 215
Cantiello, M., Blakeslee, J. P., & Raimondo, G. 2007, astro-ph/0706.3943
Harris, W. E. *et al.* 2006, *ApJ*, 636, 90
Mieske, S. *et al.* 2006, *ApJ*, 653, 193
Mieske, S. & Baumgardt, H. 2007, *A&A* in press, arXiv:0709.1328
Spitler, L. R. *et al.* 2006, *AJ*, 132, 1593
Strader, J., Brodie, J. P., Spitler, L., & Beasley, M. A. 2006, *AJ*, 132, 2333

Part 4

Few-Body Systems

Dynamical Evolution of Dense Stellar Systems
Proceedings IAU Symposium No. 246, 2007
E. Vesperini, M. Giersz & A. Sills, eds.

© 2008 International Astronomical Union
doi:10.1017/S1743921308015615

Resonance, Chaos and Stability in the General Three-Body Problem

R. A. Mardling

School of Mathematical Sciences, Monash University, Victoria, 3800, Australia
email: mardling@sci.monash.edu.au

Abstract. Three-body stability is fundamental to astrophysical processes on all length and mass scales from planetary systems to clusters of galaxies, so it is vital we have a deep and thorough understanding of this centuries-old problem. Here we summarize an analytical method for determining the stability of arbitrary three-body hierarchies which makes use of the chaos theory concept of *resonance overlap*. For the first time the dependence on *all* orbital elements and masses can be given explicitly via simple analytical expressions which contain no empirical parameters. For clarity and brevity, analysis in this paper is restricted to coplanar systems including a description of a practical algorithm for use in N-body and other applications. A Fortran routine for arbitrarily inclined systems is available from the author, and animations of stable and unstable systems are available at www.maths.monash.edu.au/~ro/Capri.

Keywords. gravitation, instabilities, methods: n-body simulations, planetary systems, globular clusters: general, binaries (including multiple): close

1. Introduction

Most stable hierarchical triples are characterized by the following behaviour: (1) no energy exchange between the inner and outer orbits; (2) slow cyclic evolution of the eccentricities associated with angular momentum exchange between the orbits (except for coplanar systems in which the two inner bodies have the same mass); (3) apsidal advance of both orbits; and (4) nutation and precession of the orbital planes for inclined systems. The exceptions to (1) are those systems which are in resonance in which case a slow, cyclic exchange of energy between the orbits pertains.

In contrast, unstable hierarchies, defined here as those in which one body escapes the system (so-called *Lagrange* instability), by necessity involve substantial exchange of energy between the orbits. They therefore must involve orbital resonances, and the difference between a stable and an unstable resonant system is that the latter involves more than one resonance.† This, for example, explains why it is that Neptune and Pluto can exist in a stable resonance (in particular, the 3:2 resonance), while the overlap of two or more resonances explains the *absence* of orbits at some positions in the asteroid belt which are resonant with Jupiter (Murray & Dermott 2000).

It has been known since the 1960s (Walker & Ford 1969) that the overlap of neighbouring resonances is a diagnostic of chaotic behaviour (the so-called *resonance overlap stability criterion*, greatly expanded upon in Chirikov 1979). It is a consequence of the famous KAM theorem (Kolmogorov 1954), itself an outgrowth of Poincare's work on the restricted three-body problem early last century. The first application of this criterion to the restricted problem was performed by Wisdom (1980) who used a Hamiltonian

---

† More accurately, a stable system may involve more than one resonance if they are, in some sense, linearly superposed (Mardling 2008a, *in preparation*).

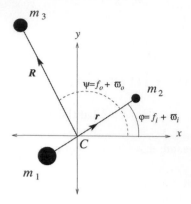

Figure 1. Coordinate system for an aligned coplanar system. The point C (the origin) corresponds to the centre of mass of m_1 and m_2.

formulation, setting the stage for the present application to the general problem (although here we are inspired by the simpler analysis outlined in Murray & Dermott 2000).

Thus the main task in determining the stability or otherwise of a given hierarchical configuration is that of identifying the widths of orbital mean-motion resonances and where they overlap. These resonances are nonlinear, and are defined by the pendulum-like libration (as opposed to circulation) of the so-called *resonance angles*, linear combinations of all the angles appearing in the problem which occur naturally in a Fourier expansion of the disturbing function, that is, the interaction term which couples the inner and outer orbits.

This paper is a summary of Mardling (2008a, *in preparation*).

2. The disturbing function

Using Jacobi coordinates (Fig. 1), the equations of motion for the inner and outer orbits of a hierarchical triple are

$$\mu_i \ddot{\mathbf{r}} + \frac{Gm_1 m_2}{r^2}\hat{\mathbf{r}} = \frac{\partial \mathcal{R}}{\partial \mathbf{r}} \tag{2.1}$$

and

$$\mu_o \ddot{\mathbf{R}} + \frac{Gm_{12} m_3}{R^2}\hat{\mathbf{R}} = \frac{\partial \mathcal{R}}{\partial \mathbf{R}} \tag{2.2}$$

respectively, where $\mu_i = m_1 m_2/m_{12}$ and $\mu_o = m_{12} m_3/m_{123}$ are the reduced masses associated with the inner and other orbits with $m_{12} = m_1 + m_2$ and $m_{123} = m_{12} + m_3$, and

$$\mathcal{R} = -\frac{Gm_{12} m_3}{R} + \frac{Gm_2 m_3}{|\mathbf{R} - \alpha_1 \mathbf{r}|} + \frac{Gm_1 m_3}{|\mathbf{R} + \alpha_2 \mathbf{r}|} \tag{2.3}$$

is the disturbing function which here has the dimensions of energy (in the study of the restricted three-body problem it has the dimensions of energy per unit mass). Here $\alpha_i = m_i/m_{12}$, $i = 1, 2$. Since the orbits interact via the disturbing function, it contains all the information about energy and angular momentum exchange and hence stability or otherwise of the system. We therefore focus attention exclusively on $\mathcal{R}$, our aim being to write it in a form which reveals in a simple way the explicit dependence of stability on all the orbital parameters. In particular these are $\{a_i, e_i, I_i, \varpi_i, \Omega_i, M_i\}$, that is, the inner semi major axis, eccentricity, inclination, longitudes of periastron and ascending nodes and mean anomaly respectively, with $\{a_o, e_o, I_o, \varpi_o, \Omega_o, M_o\}$ the corresponding elements for the outer orbit. This is best done using spherical harmonics (rather than Legendre polynomials, eg., Roy & Haddow 2003) because, together with the use of Euler angles, they allow one to explicitly separate the dependence on the inner and outer elements. Thus (2.3) becomes

$$\mathcal{R} = G\mu_i m_3 \sum_{l=2}^{\infty} \sum_{m=-l}^{l} \left(\frac{4\pi}{2l+1}\right) \mathcal{M}_l \left(\frac{r^l}{R^{l+1}}\right) Y_{lm}(\theta, \varphi) Y_{lm}^*(\Theta, \psi), \qquad (2.4)$$

where r is the distance between bodies 1 and 2, $\alpha_1 r$, θ and φ, and R, Θ and ψ are the spherical polar coordinates of bodies 2 and 3 respectively relative to a fixed (non-inertial) coordinate frame with origin at the centre of mass of bodies 1 and 2 (Fig. 1), Y_{lm} is a spherical harmonic defined as in Jackson (1975), and the mass factor $\mathcal{M}_l$ is given by

$$\mathcal{M}_l = \frac{m_1^{l-1} + (-1)^l m_2^{l-1}}{m_{12}^{l-1}} \qquad (2.5)$$

so that $\mathcal{M}_2 = 1$ for any masses while $\mathcal{M}_l = 0$ when l is odd and $m_1 = m_2$. Our intention is to expand $\mathcal{R}$ in a double Fourier series with basis frequencies ν_i and ν_o, the inner and outer orbital frequencies respectively. To clearly demonstrate how the formulation works, we assume that the system is coplanar and choose the coordinate system to be such that all three bodies lie in the $x - y$ plane. Then $\theta = \Theta = \pi/2$, $\varphi = f_i + \varpi_i$ and $\psi = f_o + \varpi_o$, where f_i and f_o are the inner and outer true anomalies, that is, the angular positions of bodies 2 and 3 relative to their periastron directions. The disturbing function then becomes

$$\mathcal{R} = G\mu_i m_3 \sum_{l=2}^{\infty} \sum_{m=-l,2}^{l} c_{lm}^2 \mathcal{M}_l e^{im(\varpi_i - \varpi_o)} \left(r^l e^{im f_i}\right) \left(\frac{e^{-im f_o}}{R^{l+1}}\right), \qquad (2.6)$$

where

$$c_{lm}^2 = \frac{4\pi}{2l+1} \left[Y_{lm}(\pi/2, 0)\right]^2 \qquad (2.7)$$

and the sum over m is in steps of two for the coplanar case. This paper will involve quadrupole $l = 2$, $m = -2, 2$ terms only with the relevant value of c_{lm}^2 being $c_{22}^2 = 3/8$. For uncoupled orbits, the quantities in brackets in (2.6) are periodic with period $2\pi/\nu_i$ and $2\pi/\nu_o$ respectively and so can be expanded in individual Fourier series such that

$$(r/a_i)^l e^{im f_i} = \sum_{n'=-\infty}^{\infty} s_{n'}^{(lm)}(e_i) e^{in' M_i}, \qquad (2.8)$$

and

$$\frac{e^{-im f_o}}{(R/a_o)^{l+1}} = \sum_{n=-\infty}^{\infty} F_n^{(lm)}(e_o) e^{-in M_o}, \qquad (2.9)$$

where the mean anomalies are related to the orbital frequencies by

$$M_i = \int \nu_i(t)\, dt + \epsilon_i - \varpi_i \quad \text{and} \quad M_o = \int \nu_o(t)\, dt + \epsilon_o - \varpi_o. \qquad (2.10)$$

Here ϵ_i and ϵ_o are the mean longitudes at epoch of the inner and outer orbits respectively (eg. Murray & Dermott 2000). This definition of the mean anomaly takes into account the fact that the semi major axes and hence the orbital frequencies vary with time when a system is in resonance (equations (3.2) and (3.3); Brouwer & Clements 1961 p. 286).

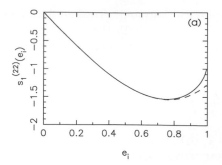 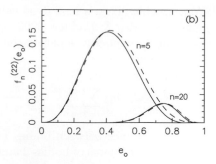

Figure 2. The eccentricity functions (a) $s_1^{(22)}(e_i)$ and (b) $f_n^{(22)}(e_o) \equiv F_n^{(22)}(e_o) \cdot (1 - e_o)^{l+1}$, $n = 5, 20$ (solid curves) together with their approximations (dashed curves) (2.13) and (2.14). The values of n were chosen to illustrate the general behaviour of $f_n^{(22)}(e_o)$ as n increases, and also to indicate that the approximation improves with increasing n.

Table 1. Data for scaling function (2.15).

	α_{22}	β_{22}	γ_{22}
$n \leqslant 9$	1.046	0.891	0.097
$n \geqslant 10$	0.448	0.134	2.4×10^{-4}

The Fourier coefficients

$$s_{n'}^{(lm)}(e_i) = \frac{1}{2\pi} \int_0^{2\pi} (r/a_i)^l e^{im f_i} e^{-in' M_i} \, dM_i \tag{2.11}$$

and

$$F_n^{(lm)}(e_o) = \frac{1}{2\pi} \int_0^{2\pi} \frac{e^{-im f_o}}{(R/a_o)^{l+1}} e^{in M_o} \, dM_o \tag{2.12}$$

give the dependence of the disturbing function on the inner and outer eccentricities and may be approximated respectively by Taylor series and by asymptotic expansions (Mardling 2008a, *in preparation*). To eighth-order in e_i, the relevant function of the inner eccentricity for this paper is

$$s_1^{(22)}(e_i) \simeq -3e_i + \frac{13}{8} e_i^3 + \frac{5}{192} e_i^5 - \frac{227}{3072} e_i^7. \tag{2.13}$$

In general, the leading term of such an expansion is $\mathcal{O}(e_i^{|m-n'|})$. This approximation is plotted in Fig. 2a (dashed curve) together with its numerically integrated "exact" solution (solid curve). The approximation diverges slightly for $e_i \gtrsim 0.8$. The asymptotic expression for $F_n^{(lm)}(e_o)$ is (Mardling 2008a, *in preparation*)

$$F_n^{(lm)}(e_o) \simeq s_{lmn} \cdot \frac{2^m/\sqrt{2\pi}}{(l+m-1)!!} \frac{(1 - e_o^2)^{(3m-l-1)/4}}{e_o^m} n^{(l+m-1)/2} e^{-n\xi(e_o)}, \tag{2.14}$$

where $\xi(e_o) = \mathrm{Cosh}^{-1}(1/e_o) - \sqrt{1 - e_o^2}$ and s_{lmn} is an empirical scaling factor designed to match the amplitudes of the exact and asymptotic expressions. It is given by

$$s_{lmn} = 1 - \alpha_{lm} n^{-\beta_{lm}} \exp(\gamma_{lm} n), \tag{2.15}$$

where the relevant fitting constants for this paper are given in Table 1. Note that $\lim_{e_o \to 0} F_n^{(lm)}(e_o) = 1$ when $n = 2$ and is zero otherwise. Note also that $\lim_{e_o \to 1} (1 - e_o)^{l+1} F_n^{(lm)}(e_o)$ is finite (Fig. 2) so that $\lim_{e_o \to 1} F_n^{(lm)}(e_o)$ is infinite, ensuring resonance overlap and hence instability for all configurations as $e_o \to 1$ (see next Section). Fig. 2b plots the numeri-cally integrated "exact" function $f_n^{(lm)}(e_o) \equiv (1 - e_o)^{l+1} F_n^{(lm)}(e_o)$ (solid curves) together with the scaled asymptotic version (dashed curves) for the cases $n = 5$ and 20. Since the scale factor is $\mathcal{O}(1)$ ($0.3 \lesssim s_{lmn} \lesssim 0.78$ for $2 \leqslant n \leqslant 1000$), the resonance widths (next Section) are not very sensitive to it and it can be omitted from the asymptotic expression (2.14). Note that (2.14) is closely related to Heggie's analysis of energy exchange during binary flybys (Heggie 1975).

Substituting (2.8) and (2.9) into (2.6) gives

$$\mathcal{R} = G\mu_i m_3 \sum_{l=2}^{\infty} \sum_{m=-l,2}^{l} \sum_{n'=-\infty}^{\infty} \sum_{n=-\infty}^{\infty} c_{lm}^2 \mathcal{M}_l \left(\frac{a_i^l}{a_o^{l+1}} \right) s_{n'}^{(lm)}(e_i) F_n^{(lm)}(e_o) \exp[i\phi_{mnn'}] \tag{2.16}$$

$$= 2\,G\mu_i m_3 \sum_{L0} \zeta_m c_{lm}^2 \mathcal{M}_l \left(\frac{a_i^l}{a_o^{l+1}} \right) s_{n'}^{(lm)}(e_i) F_n^{(lm)}(e_o) \cos\phi_{mnn'} \tag{2.17}$$

where

$$\phi_{mnn'} = n' M_i - n M_o + m(\varpi_i - \varpi_o) \tag{2.18}$$

is a *resonance angle*,

$$\sum_{L0} \equiv \sum_{l=2}^{\infty} \sum_{m=m_{min},2}^{l} \sum_{n'=-\infty}^{\infty} \sum_{n=-\infty}^{\infty}, \tag{2.19}$$

$$\zeta_m = \begin{cases} 1/2, & m = 0 \\ 1, & \text{otherwise} \end{cases} \quad \text{and} \quad m_{min} = \begin{cases} 0, & l \text{ even} \\ 1, & l \text{ odd.} \end{cases} \tag{2.20}$$

In going from (2.16) to (2.17) we have used the fact that $s_{n'}^{(lm)}$ and $F_n^{(lm)}$ are real so that $s_{n'}^{(lm)*} = s_{n'}^{(lm)}$ and $F_n^{(lm)*} = F_n^{(lm)}$ and consequently, $s_{-n'}^{(l-m)} = s_{n'}^{(lm)}$ and $F_{-n}^{(l-m)} = F_n^{(lm)}$, and have grouped together terms with the same value of $|m|$ (thus the factor $1/2$ in the definition of ζ_m).

3. Pendulum-like behaviour of the resonance angle

We now have the disturbing function expressed in such a way that the dependence on all the orbital elements is evident. In particular, all angles (orbital phases and longitudes of periasta) appear in various linear combinations in the resonance angles. For most con-figurations (in particular, non-resonant stable configurations), all resonance angles cycle rapidly through all angles with a frequency dominated by the inner orbital frequency. The main consequence of this is that no energy is exchanged *on average* between the inner and outer orbits. Energy is exchanged during outer periastron passage, but this is "returned" as the system moves away from periastron. This behaviour is demonstrated in an animation available at www.maths.monash.edu.au/~ro/Capri. However, for some configurations, one or more resonance angles *librate* between two fixed values, and this results in substantial permanent energy exchange between the orbits, except when a sys-tem is *exactly* in resonance which occurs when $\dot{\phi}_{mnn'} = 0$ for some $\{mnn'\}$. Permanent energy exchange in an unstable system is demonstrated at the above website.

A system is defined to be in resonance if a resonance angles librates, and is unstable if it resides in two overlapping resonances. Hence the task now is to identify the resonances and determine which configurations reside in two or more. This can be done by examining the behaviour of the resonance angles which, not surprisingly, satisfy the equation of motion of a pendulum. Referring to the definition of a resonance angle, (2.18), we label a resonance with the notation $[n : n'](m)$. In particular, resonances with $m = 0$ or $m = 2$ are referred to as *quadrupole* resonances while those with $m = 1$ or $m = 3$ are *octopole* resonances.

Starting with the definition (2.18) as well as (2.10) for the mean anomalies, we have

$$\ddot{\phi}_{mnn'} = n'\dot{\nu}_i - n\dot{\nu}_o + n'\ddot{\epsilon}_i - n\ddot{\epsilon}_o + (m - n')\ddot{\varpi}_i - (m - n)\ddot{\varpi}_o$$
$$\simeq n'\dot{\nu}_i - n\dot{\nu}_o. \tag{3.1}$$

Neglecting the second time derivatives of the longitudes is valid as long as the eccentricities are not vanishingly small. For systems near the stability boundary, this is only ever possible in the case of extreme mass ratios since eccentricity is always induced otherwise (Mardling 2008a, *in preparation*). To proceed we note that $\dot{\nu}_i/\nu_i = -\frac{3}{2}\dot{a}_i/a_i$ and similarly for $\dot{\nu}_o$, and use Lagrange's planetary equation for the rate of change of the semi major axis (Brouwer & Clements 1961):

$$\frac{1}{a_i}\frac{da_i}{dt} = \frac{2}{\mu_i \nu_i a_i^2}\frac{\partial \mathcal{R}}{\partial \lambda_i} = \frac{2}{\mu_i \nu_i a_i^2}\frac{\partial \mathcal{R}}{\partial M_i}$$
$$= -4\nu_i \left(\frac{m_3}{m_{12}}\right) \sum_{L0} n' \zeta_m c_{lm}^2 \mathcal{M}_l \left(\frac{a_i}{a_o}\right)^{l+1} s_{n'}^{(lm)}(e_i) F_n^{(lm)}(e_o) \sin(\phi_{mnn'}) \tag{3.2}$$

and

$$\frac{1}{a_o}\frac{da_o}{dt} = \frac{2}{\mu_o \nu_o a_o^2}\frac{\partial \mathcal{R}}{\partial \lambda_o} = \frac{2}{\mu_o \nu_o a_o^2}\frac{\partial \mathcal{R}}{\partial M_o}$$
$$= 4\nu_o \left(\frac{m_1 m_2}{m_{12}^2}\right) \sum_{L0} n \zeta_m c_{lm}^2 \mathcal{M}_l \left(\frac{a_i}{a_o}\right)^{l} s_{n'}^{(lm)}(e_i) F_n^{(lm)}(e_o) \sin(\phi_{mnn'}), \tag{3.3}$$

where $\lambda_{i,o} = M_{i,o} + \varpi_{i,o}$. The ratio of orbital frequencies is the most fundamental quantity in this problem. Putting

$$\sigma = \frac{\nu_i}{\nu_o} = \left[\left(\frac{m_{12}}{m_{123}}\right)\left(\frac{a_o}{a_i}\right)^3\right]^{1/2} \tag{3.4}$$

and using this to write a_i/a_o in terms of σ, then combining (3.1), (3.2) and (3.3), gives a pendulum equation for the evolution of the $[n : n'](m)$ resonance

$$\ddot{\phi}_{mnn'} = -n'^2 \nu_o^2 \mathcal{A}_{mnn'} \sin(\phi_{mnn'}), \tag{3.5}$$

where

$$\mathcal{A}_{mnn'} \equiv -6\zeta_m \sum_{l=l_{min},2}^{\infty} c_{lm}^2 s_{n'}^{(lm)}(e_i) F_n^{(lm)}(e_o) \left[M_i^{(l)}\sigma^{-(2l-4)/3} + M_o^{(l)}(n/n')^2\sigma^{-2l/3}\right]$$
$$\simeq -6\zeta_m c_{lm}^2 s_{n'}^{(lm)}(e_i) F_n^{(lm)}(e_o) \sigma^{-(2l-4)/3}\left[M_i^{(l)} + M_o^{(l)}\sigma^{2/3}\right], \tag{3.6}$$

and we have put $n/n' \simeq \sigma$ in the last step. Here we have assumed that the only term contributing to the variation of $\phi_{mnn'}$ is the one depending on $\phi_{mnn'}$. Except for low-order resonances with relatively low values of σ, it is also adequate to include only the lowest value of l in (3.6). Note that $l_{min} = 2$ if m is even and $l_{min} = 3$ if m is odd. The dependence on the masses is through the functions

$$M_i^{(l)} = \mathcal{M}_l \left(\frac{m_3}{m_{12}}\right)\left(\frac{m_{12}}{m_{123}}\right)^{(l+1)/3} \quad \text{and} \quad M_o^{(l)} = \mathcal{M}_l \left(\frac{m_1 m_2}{m_{12}^2}\right)\left(\frac{m_{12}}{m_{123}}\right)^{l/3}. \tag{3.7}$$

Taking $\mathcal{A}_{mnn'}$ to be approximately constant over one libration period, (3.5) can be integrated once to give

$$\tfrac{1}{2}\dot{\phi}^2_{mnn'} - n'^2\nu_o^2\mathcal{A}_{mnn'}\cos\phi_{mnn'} = \text{constant.} \tag{3.8}$$

From (3.5) we see that if $\mathcal{A}_{mnn'} > 0$ libration is about $\phi_{mnn'} = 0$, while for $\mathcal{A}_{mnn'} < 0$ it is about $\phi_{mnn'} = \pi$. The separatrix is the solution curve which contains the hyperbolic fixed point $(\phi_{mnn'}, \dot{\phi}_{mnn'}) = (\pi, 0)$ for $\mathcal{A}_{mnn'} > 0$ and $(0,0)$ for $\mathcal{A}_{mnn'} < 0$. For most systems of interest in this paper, $\mathcal{A}_{mnn'} > 0$. Therefore we define the quantity

$$\mathcal{E}_{mnn'} = \tfrac{1}{2}\dot{\phi}^2_{mnn'} - n'^2\nu_o^2\mathcal{A}_{mnn'}(1 + \cos\phi_{mnn'}), \tag{3.9}$$

so that the separatrix corresponds to $\mathcal{E}_{mnn'} = 0$ and is given by

$$\dot{\phi}_{mnn'} = \pm n'\nu_o\sqrt{2\mathcal{A}_{mnn'}(1 + \cos\phi_{mnn'})} = \pm 2\,n'\nu_o\sqrt{\mathcal{A}_{mnn'}}\cos\left(\frac{\phi_{mnn'}}{2}\right). \tag{3.10}$$

Solutions are libratory when $\mathcal{E}_{mnn'} < 0$ and circulatory when $\mathcal{E}_{mnn'} > 0$. From (2.18) we have that

$$\dot{\phi}_{mnn'} \simeq n'\nu_i - n\nu_o = n'\nu_o(\sigma - n/n') \equiv n'\nu_o\,\delta\sigma_{nn'}, \tag{3.11}$$

where we have neglected the contributions from the rates of change of the longitudes (since generally $\dot{\varpi}_{i,o} \ll \nu_o$). Thus we define the auxiliary quantity

$$\overline{\mathcal{E}}_{mnn'} = \tfrac{1}{2}(\delta\sigma_{nn'})^2 - \mathcal{A}_{mnn'}(1 + \cos\phi_{mnn'}), \tag{3.12}$$

which again indicates the libratory or circulatory nature of a system according to whether $\overline{\mathcal{E}}_{mnn'} < 0$ or $\overline{\mathcal{E}}_{mnn'} > 0$ respectively. Exact resonance corresponds to $(\phi_{mnn'}, \delta\sigma_{nn'}) = (0,0)$.

Since a system is defined to be in resonance when a resonance angle librates, from (3.10) and (3.11) the resonance half-width is defined to be

$$\Delta\sigma_{mnn'} = 2\sqrt{\mathcal{A}_{mnn'}}. \tag{3.13}$$

Since the expression for $\mathcal{A}_{mnn'}$, (3.6), depends on the orbital parameters through simple functions, it becomes easy to determine the stability or otherwise of any given system. However, formally a system potentially resides in infinitely many resonances so our next task is to determine which resonances govern the stability of a system.

4. The $[n:1](2)$ resonances

The first thing one must do to determine whether or not a system is in resonance is to check which exact resonances it is near. Since the rational numbers are dense on the number line, a system is always inside some resonance. However, as we have shown, resonance widths are proportional to $e^{-n\xi(e_o)}$ through the functions $F_n^{(lm)}(e_o)$, where $n \simeq n'\sigma$ so that for a given σ, $n \propto n'$. Thus the $[n:1](m)$ resonances always dominate. Moreover, the dependence of the resonance widths on n and $\sigma \simeq n$ together with their dependence on the mass ratios ensures that, for coplanar systems, the quadrupole $m = 2$ resonances are widest. Thus the $[n:1](2)$ resonances determine the stability of most coplanar configurations except planetary-like systems for which *both* m_2 and m_3 are less than around $0.01m_1$. In the latter case resonances with $n' > 1$, that is, resonances which sit between the $[n:1](2)$ resonances (see Fig. 3) are important (Mardling 2008a, *in preparation*) because the $[n:1](2)$ widths are limited by the mass ratio dependence in (3.6).

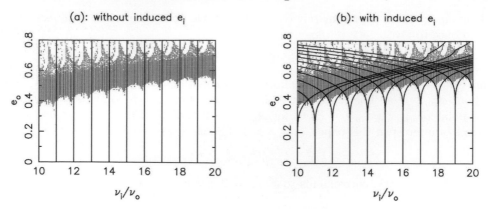

Figure 3. An illustration of the importance of including the induced inner eccentricity. Here the masses are equal and $e_i(0) = 0$. The curves are the resonance boundaries given by (3.6): the resonance widths are zero in (a) for $e_i = 0$. The resonance overlap stability criterion correctly predicts the stability boundary in (b) when $e_i = e_i^{(ind)}$ (eqn (5.1)).

5. Induced eccentricity and secular effects

Consider an equal mass coplanar configuration for which the initial inner eccentricity is zero. According to (3.13), (3.6) and (2.13) the resonance widths should be zero. Fig. 3a is a stability plot for equal mass configurations with initially circular inner binaries, for various initial period ratios and outer eccentricities. A dot corresponding to the initial conditions is plotted if a direct numerical integration of the three-body equations of motion results in an unstable system. Rather than integrating the system until one of the bodies escapes, two almost identical systems (the given system and its "ghost") are integrated in parallel and the difference in the inner semi major axes at outer apastron is monitored (because this variable is approximately constant for non-resonant systems). Taking advantage of the sensitivity of a chaotic system to initial conditions, this difference will grow in proportion to the initial difference between two systems (10^{-7} in the inner eccentricity) for a stable system, but will grow exponentially for an unstable system (Mardling 2001). The stability boundary should correspond to points where neighbouring resonances overlap. Clearly zero resonance width is incorrect! However, if one uses the inner eccentricity *induced* after the first outer periastron passage, resonance overlap correctly predicts the stability boundary (Fig. 3b).

If the initial inner eccentricity is $e_i(0)$, the inner eccentricity following outer periastron passage, $e_i^{(ind)}$, is given approximately by (Mardling 2008a, *in preparation*)

$$e_i^{(ind)} = \left[e_i(0)^2 - 2\beta_n e_i(0)\sin(\phi_{2n1}) + \beta_n^2 \right]^{1/2}, \tag{5.1}$$

where

$$\beta_n = \tfrac{9}{2}\pi(m_3/m_{123})f_n^{(22)}(e_o)/n. \tag{5.2}$$

5.1. *Octopole variations for coplanar systems*

For systems with $m_1 \neq m_2$, secular octopole contributions to the disturbing function (ie., terms with $n = n' = 0$) can cause the inner eccentricity to vary considerably on timescales of thousands of inner orbits (Murray & Dermott 2000; Mardling 2008b, *in preparation*). This is especially important for close planetary systems. While the outer eccentricity also varies, the main effect on the resonance widths comes from the variation of $s_1^{(22)}(e_i)$ which is a maximum at the maximum of the octopole cycle in e_i. Referring

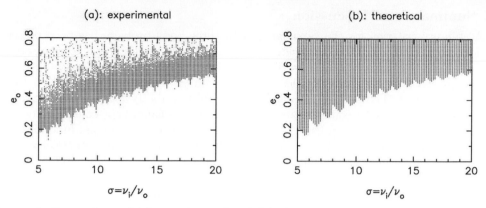

Figure 4. Comparison of (a) experimental and (b) theoretical data for equal mass coplanar systems with $e_i(0) = 0$. The structure at the top of (a) is a result of exchanges being deemed stable (the "ghost" orbit behaves almost identically before rapid escape).

to this maximum as $e_i^{(oct)}$, it is given approximately by (Mardling 2007)

$$e_i^{(oct)} = \begin{cases} (1+A)e_i^{(eq)}, & A \leqslant 1, \\ e_i(0) + 2e_i^{(eq)}, & A > 1, \end{cases} \tag{5.3}$$

where $A = |1 - e_i(0)/e_i^{(eq)}|$ and $e_i^{(eq)}$ is the "equilibrium" or "fixed point" eccentricity

$$e_i^{(eq)} = \frac{(5/4)(a_i/a_o)\, e_o/(1 - e_o^2)}{\left|1 - \sqrt{a_i/a_o}\,(m_2/m_3)/\sqrt{1 - e_o^2}\right|}. \tag{5.4}$$

6. A stability algorithm

Here we summarize the steps one can follow to implement the resonance overlap stability criterion for coplanar systems for which the $[n:1](2)$ resonances alone determine the stability. This scheme is valid for coplanar systems with *both* $m_2/m_1 > 0.01$ and $m_3/m_1 > 0.01$, <u>or</u>, for systems with *at least one of* $m_2/m_1 > 0.05$ or $m_3/m_1 > 0.05$.
(1) Identify which $[n:1](2)$ resonance the system is near and calculate the distance $\delta\sigma_n$ from that resonance: $\delta\sigma_n = \sigma - n$, where $n = \lfloor \sigma \rfloor$ (the nearest integer for which $n \leqslant \sigma$);
(2) Take the associated resonance angle to be zero rather than the definition (2.18) (see discussion below): $\phi_{2n1} = 0$;
(3) Calculate the induced eccentricity from (5.1) and (if $m_1 \neq m_2$) the maximum octopole eccentricity from (5.3). Determine $e_i = \max[e_i^{(ind)}, e_i^{(oct)}]$ for use in $s_1^{(22)}(e_i)$;
(4) Calculate $\mathcal{A}_{2n1}$ from (3.6);
(5) Calculate $\overline{\mathcal{E}}_{2n1}$ and $\overline{\mathcal{E}}_{2\,n+1\,1}$: deem the system unstable if $\overline{\mathcal{E}}_{2n1} < 0$ *and* $\overline{\mathcal{E}}_{2\,n+1\,1} < 0$.
Fig. 4 compares the experimental data shown in Fig. 3 with data generated using the algorithm above. A dot is plotted if a system is deemed to be unstable. The boundary structure is reproduced reasonably well, although the boundary itself should be slightly lower, a result of the fact that the resonance overlap criterion does not recognize the unstable nature of points near to but outside the separatrix.

A Fortran routine for arbitrarily inclined systems is available from the author.

7. Summary and Discussion

A stability criterion for the coplanar general three-body problem has been presented which involves *no empirical parameters*. While previous studies have concentrated on the coplanar circular restricted three-body problem, we can now see how stability works for all mass ratios and eccentricities through a simple transparent expression for the resonance widths. The $[n : 1](2)$ resonances determine stability for most configurations, while $[n : n'](2)$, $n' > 2$ resonances are important for two-planet-type configurations. The latter is a result of the dependence of the resonance widths on the mass functions (3.7).

We have also shown that it is vital to include the induced inner eccentricity as well as the maximum inner eccentricity achieved in an octupole cycle when $m_1 \neq m_2$. We note here that inclined systems also require a knowledge of the maximum eccentricity achieved in a *Kozai* cycle (Mardling 2008a, *in preparation*).

Given the simple functional form of (3.6), it is possible to determine which systems have similar stability properties. For example, given $e_i(0)$, e_o and σ, a system with $m_2/m_1 = 0.1$ and $m_3/m_1 = 0.35$ will have similar stability properties to a system with $m_2/m_1 = 20$ and $m_3/m_1 = 7.45$ (ie, $M_i^{(l)} + M_o^{(l)}\sigma^{2/3}$ is the same for both; note that m_3/m_{123} in (5.1) is similar for both systems).

An analysis of the success or otherwise of this formulation of the stability problem is given in Mardling (2008a, *in preparation*). Here we note that its main drawback is that the resonance overlap stability criterion doesn't recognize that systems outside but near the separatrix are often unstable (thus we take $\phi_{2n1} = 0$ in the algorithm). Nonetheless, it is possible to invent remedies for individual applications, many of which do not require such high resolution anyway. For example, the study of the evolution of triple systems formed through binary-binary collisions in N-body simulations requires that one knows unequivocally when such a system is stable, since a mistake has the potential to grind the whole simulation to a halt. The fact that the algorithm presented here will sometimes make the opposite mistake (deem an unstable triple stable) if the configuration is close to the stability boundary should not have much effect on the overall evolution of the cluster.

References

Brouwer, D. & Clements, G. M. 1961, *Methods of Celestial Mechanics*, New York and London: Academic Press

Chirikov, B. V. 1979, *Physics Reports*, 52, 263

Heggie, D. C. 1975, *MNRAS*, 173, 729

Jackson, J. D. 1975, New York: Wiley, 1975, 2nd ed.

Kolmogorov, A. N. 1954, *Dokl. Akad. Nauk. SSSR*— 98, 527

Mardling, R. A. 2001, in *Evolution of Binary and Multiple Star Systems*, eds Podsiadlowski *et al.* ASP Conference Series, 229, 101

Mardling, R. A. 2007, *MNRAS*, in press

Murray, C. D. & Dermott, S. F. 2000, *Solar System Dynamics*, Cambridge, UK: Cambridge University Press

Roy, A. & Haddow, M. 2003, *Cel. Mec. Dyn. Ast.*, 87, 411

Walker, G. H. & Ford, J. 1969, *Physical Review*, 188, 416

Wisdom, J. 1980, *AJ*, 85, 1122

Dynamical Evolution of Dense Stellar Systems
Proceedings IAU Symposium No. 246, 2007
E. Vesperini, M. Giersz & A. Sills, eds.

The Problem of Three Stars: Stability Limit

M. Valtonen[1], A. Mylläri[1], V. Orlov[2] and A. Rubinov[2]

[1]Department of Physics and Tuorla Observatory, University of Turku, 21500 Piikkiö, Finland
email: mavalto@utu.fi

[2]Sobolev Astronomical Institute, St. Petersburg State University, Russia
email: vor@astro.spbu.ru

Abstract. The problem of three stars arises in many connections in stellar dynamics: three-body scattering drives the evolution of star clusters, and bound triple systems form long-lasting intermediate structures in them. Here we address the question of stability of triple stars. For a given system the stability is easy to determine by numerical orbit calculation. However, we often have only statistical knowledge of some of the parameters of the system. Then one needs a more general analytical formula. Here we start with the analytical calculation of the single encounter between a binary and a single star by Heggie (1975). Using some of the later developments we get a useful expression for the energy change per encounter as a function of the pericenter distance, masses, and relative inclination of the orbit. Then we assume that the orbital energy evolves by random walk in energy space until the accumulated energy change leads to instability. In this way we arrive at a stability limit in pericenter distance of the outer orbit for different mass combinations, outer orbit eccentricities and inclinations. The result is compared with numerical orbit calculations.

Keywords. stellar dynamics, celestial mechanics

---

1. Introduction

Three-body scattering was studied comprehensibly in the pioneering work of Heggie (1975). Among the many results in this work was an expression for energy change when a single body passes by a binary. The calculation was carried out in greater detail by Roy & Haddow (2003) for a parabolic passage, and by Heggie (2006) for a hyperbolic passage. Valtonen & Karttunen (2006) calculate the same quantity for a passage in a low-eccentricity elliptic orbit.

In a bound triple system a number of passages takes place one after another. Even though each passage may change the orbits only slightly, a large number of them may lead to an accumulated energy change which eventually leads to the break-up of the triple system. The energy steps in this process may be in one direction only, or they may happen in both directions in the manner of random walk. In either case, the orbits evolve with time toward instability (Valtonen & Karttunen 2006).

In this paper we study the analytical expressions for the energy change in a single encounter, and simplify them a little for the purpose of the stability study. We then derive an analytical expression for the stability limit which improves the formula given by Valtonen & Karttunen (2006). This formula also improves the results given in previous works: Golubev (1967), Golubev (1968), Harrington (1977), Eggleton & Kiseleva (1995), Mardling & Aarseth (1999), Mardling & Aarseth (2001) (see Aarseth 2003 and Tokovinin 2004).

2. Energy change in a single encounter

Heggie (1975) and Roy & Haddow (2003) derive the expression for the relative energy change $\delta\varepsilon/\varepsilon$ in a single parabolic encounter. Heggie (2006) finds the corresponding formula for hyperbolic encounters and shows that it agrees with the result of Roy & Haddow (2003) at the parabolic limit. Let the binary have masses m_1 and m_2, with $M_{12} = m_1 + m_2$, semi-major axis a_i and eccentricity e_i. The third body orbit relative to the barycenter of the binary has pericenter distance q, with $Q = q/a_i$, and eccentricity e. The mass of the third body is m_3. Its orbital plane relative to the binary plane is described by the usual elements i, ω, and Ω. Then, as was derived in the papers cited above, by the first order perturbation theory

$$
\begin{aligned}
\frac{\delta\varepsilon}{\varepsilon} \simeq & -\frac{\sqrt{\pi}}{4}\frac{m_3}{M_{12}}Q^{-3}K^{5/2}e^{-(2/3)K}\left\{e_1\left[\sin(2\omega + nt_0)(\cos 2i - 1)\right.\right.\\
& - \sin(2\omega + nt_0)\cos(2i)\cos(2\Omega) - 3\sin(nt_0 + 2\omega)\cos(2\Omega)\\
& \left. - 4\sin(2\Omega)\cos(2\omega + nt_0)\cos i\right] + e_2(1 - e_i^2)\left[\sin(2\omega + nt_0)(1 - \cos 2i)\right. \qquad (2.1)\\
& - \sin(2\omega + nt_0)\cos(2i)\cos(2\Omega) - 3\sin(nt_0 + 2\omega)\cos(2\Omega)\\
& \left. - 4\cos(nt_0 + 2\omega)\sin(2\Omega)\cos i\right] + e_4\sqrt{1 - e_i^2}\left[-2\cos(2i)\cos(2\omega + nt_0)\sin(2\Omega)\right.\\
& \left.\left. - 6\cos(2\omega + nt_0)\sin(2\Omega) - 8\cos(2\Omega)\sin(2\omega + nt_0)\cos i\right]\right\}.
\end{aligned}
$$

Here n is the mean motion of the binary and t_0 is a reference time. The true anomaly of the binary $M = n(t - t_0)$. If we agree that at the pericenter $t = 0$, then the value of true anomaly at the pericenter is $M_0 \equiv 2\Phi_0 = -nt_0$. The quantity K is defined as

$$
K = \sqrt{2}\sqrt{\frac{M_{12}}{M_{12} + m_3}}\,Q^{3/2}.
$$

The functions e_1, e_2 and e_4 are

$$
\begin{aligned}
e_1 &= J_{-1}(e_i) - 2e_i J_0(e_i) + 2e_i J_2(e_i) - J_3(e_i),\\
e_2 &= J_{-1}(e_i) - J_3(e_i),\\
e_4 &= J_{-1}(e_i) - e_i J_0(e_i) - e_1 J_2(e_i) + J_3(e_i).
\end{aligned}
$$

Here $J_{-1}, \ldots, J_3$ are the Bessel functions.

Let us simplify (2.1) by assuming that e_i is small, and use only first order terms. Also we write the dependence on Q with the help of a scale distance Q_1 :

$$
Q_1 = 2.5\left(1 + \frac{m_3}{M_{12}}\right)^{1/3}
$$

whereby

$$
K = 5.59(Q/Q_1)^{3/2}
$$

and

$$
K^{5/2} = 73.9(Q/Q_1)^{15/4}.
$$

We also approximate the exponential factor by a power-law in the interval $1 \leqslant Q/Q_1 \leqslant 1.5$; thus

$$
(2.5)^{-3}(Q/Q_1)^{-3}K^{5/2}e^{-(2K/3)} \simeq 0.11(Q/Q_1)^{-7}.
$$

After all these modifications, the dominant term of (2.1) becomes

$$
\frac{\delta\varepsilon}{\varepsilon} \simeq e_i\frac{m_3}{M_{12}}\left(\frac{Q}{Q_1}\right)^{-7}\left(\frac{1 + \cos i}{2}\right)^2\sin 2(\Phi_0 - (\omega + \Omega)) \qquad (2.2)
$$

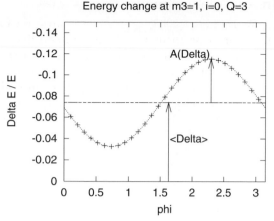

Figure 1. The relative energy change $\delta\varepsilon/\varepsilon$ in a single parabolic encounter between an equal mass circular binary and a single star. The energy change consists of constant shift (marked Delta) and of a sinusoidal variation with amplitude A(Delta).

Thus the energy change is a sinusoidally varying function of the initial phase angle (Φ_0) of the binary. For initially zero eccentricity binaries, a suitable value to insert in (2.2) is $e_i \simeq 0.05$ since the binary eccentricity does not remain zero during the encounter but changes to a non-zero value.

Valtonen & Karttunen (2006) derive a similar result by using a low-eccentricity outer orbit instead of a parabolic orbit. They find approximately for $e_i = 0$ that

$$\frac{\delta\varepsilon}{\varepsilon} \simeq 0.03 \, \frac{m_3}{M_{12}} \left(\frac{Q}{Q_1}\right)^{-7} \left(\frac{1+\cos i}{2}\right)^2 \cos 2(\Phi_0 - \Omega). \qquad (2.3)$$

Because of zero eccentricity, we have put $\omega = 0$, since the orientation of the major axis is of no consequence.

Numerical calculations show that the inclination function is in fact somewhat more complicated than in (2.2) and (2.3). We will find below an expression which is more suitable for the stability study. Note that the

$$\left(\frac{1+\cos i}{2}\right)^2$$

function cannot possibly be correct since it would make the energy change at retrograde orbit zero.

The functional forms of the relative energy change have been tested by single encounter parabolic orbit calculations. Fig. 1 shows that the $\delta\varepsilon/\varepsilon$ variation is sinusoidal. Fig. 2 shows the amplitude of the sinusoidal variation as a function of Q. It may be modelled as pure power-law at small Q, and by a combination of power-law and exponential at larger Q (dotted line). This simple sinusoidal behavior is valid only if $Q \geqslant Q_{st}$ where Q_{st} may be called the stability limit of a single encounter. Fig. 3 shows that Q_{st} depends strongly on the inclination. The stability limit also depends on the mass of the third body, as shown by Fig. 4. The limit may be defined in two ways: either from the Q−distance where exchanges between binary bodies and third bodies first start, or from the innermost point to which a power-law dependence on Q is valid. These stability limits are compared

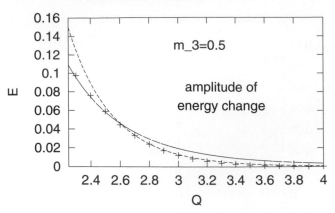

Figure 2. The amplitude of the sinusoidal (relative) energy change in a parabolic single star –
binary encounter, as a function of the pericenter distance Q (in units of the binary semi-major
axis). A pure power-law (solid line) fits at low Q while at higher Q a power-law times an
exponential gives a better fit.

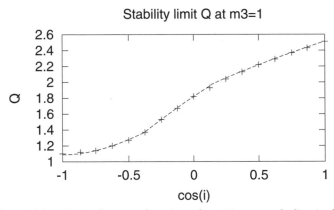

Figure 3. The stability limit Q_{st} as a function of $\cos i$ in a parabolic single – star – binary
encounter.

with the expression

$$2.1 \left(1 + \frac{m_3}{M_{12}}\right)^{1/3}$$

by a dashed line in Fig. 4. Note the connection with the scale distance Q_1 defined above.

The dependence of $\delta\varepsilon/\varepsilon$ on Q and m_3/M_{12} is shown in Fig. 5, while the dependence
on Q and i is displayed in Fig. 6. The lines show that simple analytical models describe
the data when $Q > Q_{st}$.

3. Evolution of the outer orbit

The energy of the outer orbit E_{out} is connected to the binary energy E_B by

$$\frac{E_B}{E_{out}} = \frac{m_1 m_2}{M_{12} m_3} \frac{a}{a_i} = \frac{m_1 m_2}{M_{12} m_3} \frac{Q}{1 - e}$$

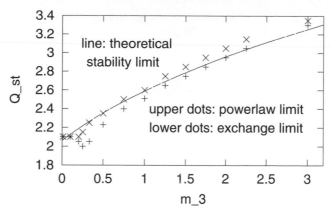

Figure 4. The stability limit Q_{st} as a function of third body mass, m_3/M_{12}, for $i = 0°$. The points may be defined either by the lower limit of power-law description in $\delta\varepsilon/\varepsilon \propto Q^{-m}$, where $m \simeq 7$, or by the limit where exchanges begin between the binary members and the third body.

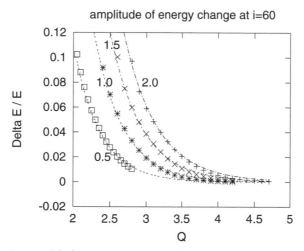

Figure 5. The dependence of $\delta\varepsilon/\varepsilon$ on pericentric distance Q at various mass values m_3/M_{12}.

where a is the semi-major axis of the outer orbit. If $m_1 = m_2$,

$$\frac{E_B}{E_{out}} \simeq \frac{M_{12}}{m_3} \frac{Q/Q_1}{1-e}.$$

Therefore the relative change in the binary energy

$$\frac{\triangle E_B}{E_B} = -\frac{\triangle E_{out}}{E_{out}} \frac{E_{out}}{E_B} \simeq -\frac{(1-e)}{Q/Q_1} \frac{m_3}{M_{12}} \frac{\triangle E_{out}}{E_{out}}.$$

Let us assume that the triple system breaks up when $\triangle E_{out} \simeq E_{out}$. It can happen after N steps of random walk. If the step size $\triangle E_{out}/E_{out} = x$, then

$$\sqrt{N}\, x = 1$$

or $N = x^{-2}$.

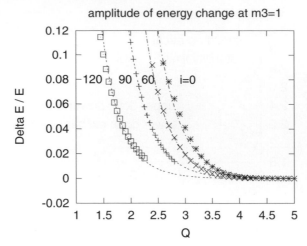

Figure 6. The dependence of $\delta\varepsilon/\varepsilon$ on pericentric distance Q at various inclination values.

If $x = 10^{-2}$, $N = 10^4$. In other words, if we put

$$\frac{\triangle E_{out}}{E_{out}} = 10^{-2},$$

the system should be stable over 10^4 revolutions of the outer orbit. Thus

$$\frac{\triangle E_{out}}{E_{out}} \simeq 10^{-2}\frac{(1-e)}{Q/Q_1}\frac{m_3}{M_{12}}.$$

Let us put this equal to the amplitude of $\delta\varepsilon/\varepsilon$:

$$10^{-2}\frac{(1-e)}{Q/Q_1}\frac{m_3}{M_{12}} = 0.03\frac{m_3}{M_{12}}\left(\frac{Q}{Q_1}\right)^{-7}\left(\frac{1+\cos i}{2}\right)^2.$$

Solve for Q :

$$Q \simeq 3\left(1+\frac{m_3}{M_{12}}\right)^{1/3}(1-e)^{-1/6}\left(\frac{1+\cos i}{2}\right)^{1/3}. \tag{3.1}$$

This value is referred to as the stability limit Q_{st}.

4. Stability experiments

Numerical orbit calculations have been carried for triple systems where the binary members are of equal mass. The value of e is varied in 5 steps from 0 to 0.9 and m_3/M_{12} in the range from 0.0005 to 5. Inclination values from zero to $180°$ have been studied. For the numerical integration of the equations of motion we used the code kindly provided by S. Mikkola that applies Wisdom-Holman method with time transformation in the extended phase-space (for the details see Mikkola 1997). The system is classified as stable if after $N = 10^4$ revolutions of the outer orbit there have been no escapes and no exchanges of binary members. The stability limit for fixed e, m_3/M_{12} and i is determined such that the triple systems are stable down to this value of Q for any value of ω and Ω. There may be stable systems even at smaller Q, but there is no guarantee that for $Q < Q_{st}$ the system is definitely stable.

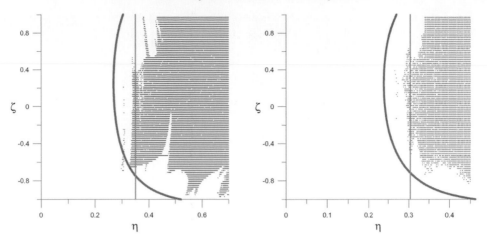

Figure 7. Results from numerical stability calculations for hierarchical triple systems for up to 10^4 revolutions of the outer orbit. The parameters are $\eta = a_i/a$ and $\xi = \cos i$. Every dot represents an unstable orbit. The vertical line follows the Mardling & Aarseth (2001) criterion while the curved line is from this work. There should be no dots to the left of the line if the boundary is correct. The case of $e = 0$, $m_3 = 0.1$ (left panel), $m_3 = 1$ (right panel) and $m_1 = m_2 = 1$.

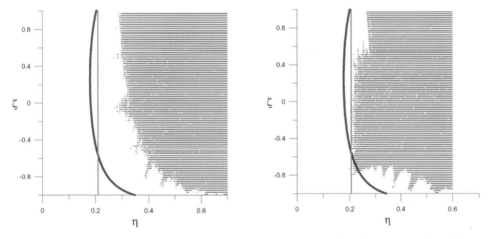

Figure 8. Same as Fig. 7, except that $e = 0.3$, $m_3 = 0.01$ (left), $m_3 = 0.1$ (right).

As we mentioned earlier, the functional forms present in (3.1) are not necessarily optimal for this problem. Therefore small variants have been tried. For the inclination, the functional form

$$\left(\frac{7}{4} + \frac{1}{2}\cos i - \cos^2 i \right)^{1/3}$$

have been found to improve the fit relative to the $[(1 + \cos i)/2]^{1/3}$ form. Then the whole formula becomes

$$Q_{st} = 3\,(1 + m_3/M_{12})^{1/3}\,(1 - e)^{-1/6}\left(\frac{7}{4} + \frac{1}{2}\cos i - \cos^2 i \right)^{1/3}.$$

The agreement between experiments and theory is shown in Figs. 7–11. The expression is not good for $0.025 < m_3/M_{12} < 0.17$ for retrograde orbits. In such cases one may use

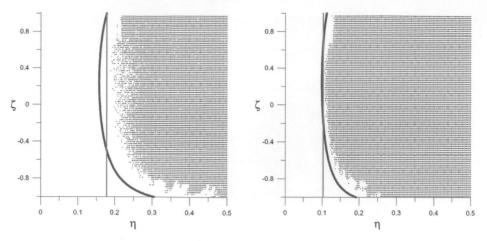

Figure 9. Same as Fig. 7, except that $e = 0.3$, $m_3 = 1$ (left), $m_3 = 10$ (right).

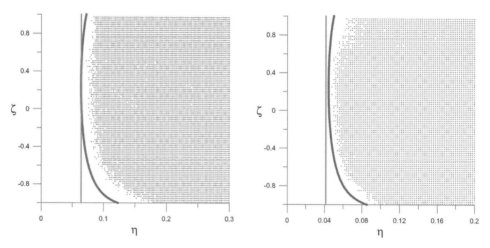

Figure 10. Same as Fig. 7, except that $e = 0.7$, $m_3 = 0.333$ (left), $m_3 = 5$ (right).

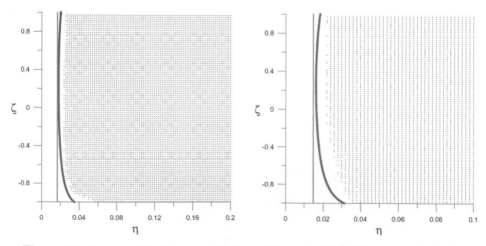

Figure 11. Same as Fig. 7, except that $e = 0.9$, $m_3 = 0.2$ (left), $m_3 = 1$ (right).

$\cos i = 0$ in the formula whenever $\cos i < 0$. Outside this mass range the formula works well, at least for $0 < m_3/M_{12} \leqslant 0.025$ and for $0.17 \leqslant m_3/m_B \leqslant 5$, where we have been able to test it. Typically there is about 5% safely margin in Q_{st}, i.e. the limit could be reduced by this much and only very few exceptions to the rule would arise.

5. Conclusions

As we mentioned in the introduction, several stability criteria have been proposed. In general, they are simpler formulae than what is presented in this paper. The Valtonen & Karttunen (2006) formula is an earlier and less accurate formulation of the same theory as is presented here. We have compared our data with Mardling & Aarseth (2001) criteria. In general it is good, but as can be seen in Fig. 7 it sometimes doesn't work in the middle inclinations. Zhuchkov, Orlov & Rubinov (2006) have shown that the stability criteria of Golubev, Harrington and Tokovinin are not always good. It would be an interesting project to compare the other criteria with ours and with numerical orbit calculations, but it will be left to another paper.

References

Aarseth, S. 2003, *Gravitational N-Body Simulations* Cambridge Univ. Press, Cambridge, p. 151.

Eggleton, P. & Kiseleva, L. 1995, *ApJ* 455, 640

Golubev, V. G. 1967, *Sov. Phys. Dokl.* 12, 529

Golubev, V. G. 1968, *Sov. Phys. Dokl.* 13, 373

Harrington, R. S. 1977, *Rev. Mexicana AyA* 3, 139

Heggie, D. C. 1975, *MNRAS* 173, 729

Heggie, D. C. 2006, in C. F. Flynn (ed.), *Few Body Problem: Theory and Computer Simulations*, Ann. Univ. Turku, Ser. 1 A, Vol. 358, p. 20

Mardling, R. & Aarseth, S. 1999, in B. A. Steves & A. E. Roy (ed.), *The Dynamics of Small Bodies in the Solar System Studies*, Kluwer, p. 385

Mardling, R. & Aarseth, S. 2001, *MNRAS* 321 398

Mikkola, S. 1997, *Cel. Mech. Dyn. Astr.*, 67, 145

Roy, A. & Haddow, M. 2003, *Cel. Mech. Dyn. Astr.*, 87, 411

Tokovinin, A. 2004, *Rev. Mexicana AyA* 21, 7

Valtonen, M. & Karttunen, H. 2006, *The Three-Body Problem*, Cambridge U. Press, Cambridge, Chapter 2

Zhuchkov, R., Orlov, V., & Rubinov, A. 2006, in C. F. Flynn (ed.), *Few Body Problem: Theory and Computer Simulations*, Ann. Univ. Turku, Ser. 1 A, Vol. 358, p. 79

Dynamical Evolution of Dense Stellar Systems
Proceedings IAU Symposium No. 246, 2007
E. Vesperini, M. Giersz & A. Sills, eds.

© 2008 International Astronomical Union
doi:10.1017/S1743921308015639

A Brief History of Regularisation

S. Mikkola

Tuorla Observatory, University of Turku, Väisäläntie 20, Piikkiö, Finland
email: seppo.mikkola@utu.fi

Abstract. The various methods for regularisation of the gravitational few-body problem, from the coordinate transformation by the Kustaanheimo-Stiefel method to the more recent methods of algorithmic regularisation, are reviewed. Numerical comparisons of the performance of the methods are presented and future research suggested.

Keywords. stellar dynamics, methods: n-body simulations, methods: numerical

1. Introduction

Close encounters of stars are common in dense stellar systems. Loss of precision was discovered by in the first attempts to simulate few-body systems (von Hoerner 1960, 1963) using the Newtonian form of equations of motion. This phenomenon is due to the $1/r^2$ singularity in the gravitational force at distance r. In addition to the force value variation the roundoff error is a significant source of computational difficulties.

Numerical studies became feasible after the publication of the two-body regularization method by Kustaanheimo & Stiefel (1965). Subsequently this KS-transformation became popular for treating dominant two-body interactions (*e.g.* Szebechely & Peters 1967; Aarseth 1972). Techniques to apply KS in multiparticle systems were developed by many authors (Aarseth & Zare 1974; Zare 1974; Heggie 1974; Mikkola 1985; Mikkola & Aarseth 1993). A major reference for these methods is the book by Aarseth (2003).

More recently an entirely new way for regularization was discovered in 1999 (Mikkola & Tanikawa 1999ab, Preto & Tremaine 1999). This, algorithmic regularization, does not require coordinate transformation but a time transformation, the leapfrog algorithm and extrapolation to zero step size (Bulirsch & Stoer 1966). Versions of few-body codes written using algorithmic regularization are more efficient than KS-based versions in case of very large mass ratios.

2. Events and methods in the history of regularization

• **Poincare's time transformation:** More than a century ago Poincare [cited in Siegel 1956, p.35] introduced a technique to transform the independent variable in a Hamiltonian system.

Let $H(\mathbf{p}, \mathbf{q}, t)$ be a Hamiltonian. One takes the time to be a canonical coordinate ($t = q_0$) by adding the momentum of time $p_0(= -E)$ to the Hamiltonian (here $E = H =$ energy = constant). Defining $dt = g(\mathbf{p}, \mathbf{q})ds$, where $g > 0$, one gets a new Hamiltonian

$$\Gamma = g(\mathbf{p}, \mathbf{q}) \left[H(\mathbf{p}, \mathbf{q}, q_0) + p_0 \right], \qquad (2.1)$$

which gives the equations of motion

$$\frac{dt}{ds} = \frac{\partial \Gamma}{\partial p_0} = g \tag{2.2}$$

$$\frac{dq}{ds} = \frac{\partial \Gamma}{\partial p}; \quad \frac{dp}{ds} = -\frac{\partial \Gamma}{\partial q} \tag{2.3}$$

In two-body regularisation one normally uses $g = r$ (where r is the distance). For multiparticle systems initially the product of distances were proposed, but later it was found that the inverse of potential $U = \sum_{i<j} m_i m_j / r_{ij}$ or the inverse of the Lagrangian $L = T + U$ (where $T = $ the kinetic energy) are more appropriate, e.g.

$$g = 1/U \quad \text{or} \quad g = 1/L. \tag{2.4}$$

- **Levi-Civita (1920)** published the regularisation in 2D.

$$x = Q_1^2 - Q_2^2; \; y = 2Q_1 Q_2, \quad \text{or} \quad x + iy = (Q_1 + iQ_2)^2 \tag{2.5}$$

which gives for the two-body Hamiltonian (with $g = r = \mathbf{Q}^2$)

$$\Gamma = r(\tfrac{1}{2}\mathbf{p}^2 - M/r - E_0) = \frac{1}{8}\mathbf{P}^2 - E_0\mathbf{Q}^2 - M, \tag{2.6}$$

which is the Hamiltonian of a harmonic oscillator.

- **Kustaanheimo & Stiefel (1965)** published their new regularization method which uses a transformation from 4D to 3D.

In matrix formulation the KS-transformation of coordinates $\mathbf{r}$ and momenta $\mathbf{p}$ may be written

$$\mathbf{r} = \widehat{\mathbf{Q}}\mathbf{Q}; \; \; \mathbf{p} = \widehat{\mathbf{Q}}\mathbf{P}/(2Q^2) \tag{2.7}$$

Here $\widehat{\mathbf{Q}}$ is the KS-matrix (e.g. Stiefel & Scheifele 1972 p. 24)

$$\widehat{\mathbf{Q}} = \begin{pmatrix} Q_1 & -Q_2 & -Q_3 & Q_4 \\ Q_2 & Q_1 & -Q_4 & -Q_3 \\ Q_3 & Q_4 & Q_1 & Q_2 \\ Q_4 & -Q_3 & Q_2 & -Q_1 \end{pmatrix}. \tag{2.8}$$

The time transformation

$$\frac{dt}{ds} = R = \mathbf{Q}^2 \tag{2.9}$$

gives the new Hamiltonian

$$\Gamma = R(H - E) = \frac{1}{8}\mathbf{P}^2 - M - E\mathbf{Q}^2, \tag{2.10}$$

where $H = \tfrac{1}{2}p^2 - M/r$ is the two-body Hamiltonian and $E = H(0)$ is the numerical value of the energy.

- **Aarseth & Zare (1974)** published a three-particle regularization method. The two shortest distances are regularised with the Kustaanheimo-Stiefel method and the arrangement is appropriately updated when necessary.

Taking the vectors $\mathbf{R}_1$ and $\mathbf{R}_2$ (see the figure) as new canonical coordinates, the generating function takes the form

$$S = \mathbf{W}_1 \cdot \mathbf{R}_1 + \mathbf{W}_2 \cdot \mathbf{R}_2 = \mathbf{W}_1 \cdot (\mathbf{r}_1 - \mathbf{r}_3) + \mathbf{W}_2 \cdot (\mathbf{r}_2 - \mathbf{r}_3) \tag{2.11}$$

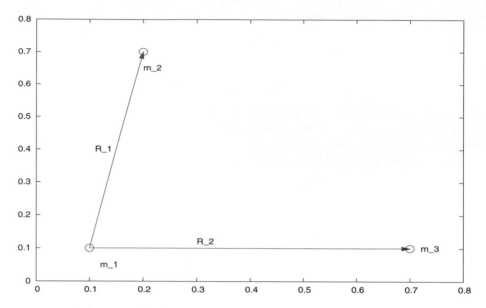

Figure 1. Regularised vectors in the Aarseth-Zare method.

and the physical momenta are $\mathbf{p}_1 = \partial S/\partial \mathbf{r}_1 = \mathbf{W}_1$, $\mathbf{p}_2 = \partial S/\partial \mathbf{r}_2 = \mathbf{W}_2$ and $\mathbf{p}_3 = \partial S/\partial \mathbf{r}_3 = -\mathbf{W}_1 - \mathbf{W}_2$.

KS-transformation and the time transformation $dt = R_1 R_2 ds = Q_1^2 Q_2^2 \, ds$, gives the Hamiltonian

$$\Gamma = \frac{1}{8\mu_{13}} Q_2^2 \mathbf{P}_1^2 + \frac{1}{8\mu_{23}} Q_1^2 \mathbf{P}_2^2 + \frac{1}{4m_3} \mathbf{P}_1^T \widehat{\mathbf{Q}}_1^T \widehat{\mathbf{Q}}_2 \mathbf{P}_2$$
$$- m_1 m_3 Q_2^2 - m_2 m_3 Q_1^2 - m_1 m_2 Q_1^2 Q_2^2 / R - E Q_1^2 Q_2^2, \qquad (2.12)$$

where $\mu_{k3} = m_k m_3/(m_k + m_3)$, and E is the total energy. The equations of motion i

$$\mathbf{P}_k' = -\frac{\partial \Gamma}{\partial \mathbf{Q}_k}; \quad \mathbf{Q}_k' = \frac{\partial \Gamma}{\partial \mathbf{P}_k}; \quad t' = Q_1^2 Q_2^2 \qquad (2.13)$$

are regular with respect to collisions between the body m_3 and any one of the two bodies m_1 and m_2 A switching of the reference body is carried out whenever the singular distance R becomes the smallest. In case of external perturbations the three-body energy E is no longer a constant, but must be obtained by integration.

- **Heggie (1974)** Discovered a **global N-body regularization** using KS-transformations. In this method each interparticle vector is a (formally) independent variable. The Hamiltonian is $H = \sum_{i=1}^{N} \frac{1}{2m_i} w_i^2 - \sum_{i=1}^{N-1} \sum_{j=i+1}^{N} \frac{m_i m_j}{r_{ij}}$. Heggie (1974) uses $\mathbf{r}_{ij} = \mathbf{r}_j - \mathbf{r}_i$ as new dependent variables. The generating function

$$S = \sum_{i=1}^{N-1} \sum_{j=i+1}^{N} \mathbf{w}_{ij} \cdot (\mathbf{r}_j - \mathbf{r}_i), \qquad (2.14)$$

defines the new momenta $\mathbf{w}_{ij}$. A proof for the correctness of such an increase of degrees of freedom is also given in Heggie's (1974) article.

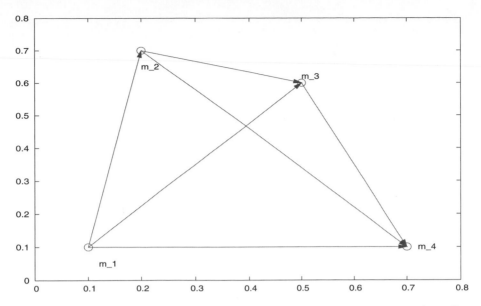

Figure 2. In the global regularization of Heggie, all the relative vectors are formally independent and regularised using KS.

After the introduction of the above transformation the old momenta are

$$\mathbf{w}_i = \sum_{j=i+1}^{N} \mathbf{w}_{ij} - \sum_{j=1}^{i-1} \mathbf{w}_{ji}. \tag{2.15}$$

Substituting these and applying the KS transformation to each conjugate pair $(\mathbf{w}_{ij}, \mathbf{r}_{ij})$. one obtains the globally regular Hamiltonian

$$\Gamma = g(\mathbf{r})[H(\mathbf{w}_{ij}, \mathbf{r}_{ij}) - E] \prod_{i=1}^{N-1} \tag{2.16}$$

$$= g(\mathbf{Q}) \left[H(\frac{\widehat{\mathbf{Q}}_{ij}\mathbf{P}_{ij}}{2Q_{ij}^2}, \widehat{\mathbf{Q}}_{ij}\mathbf{Q}_{ij}) - E \right]. \tag{2.17}$$

• **Mikkola (1985)** wrote a concise algorithm for N body integration using Heggie's global method. This was possible by using

$$g = 1/L, \tag{2.18}$$

instead of the originally (and customarily) used $g = \prod r_{ij}$.

• **Mikkola & Aarseth (1990), (1993)** developed what is known as the **chain method**.
Suppose a chain of vectors connecting N bodies has been selected. After re-labelling the bodies such that they are 1, 2, .., N along the chain, the generating function

$$S = \sum_{i=1}^{N-1} \mathbf{W}_k \cdot (\mathbf{q}_{k+1} - \mathbf{q}_k) \tag{2.19}$$

can be used to obtain the old momenta $\mathbf{p}_k = \partial S / \partial \mathbf{q}_k$ in terms of the new ones. In general,

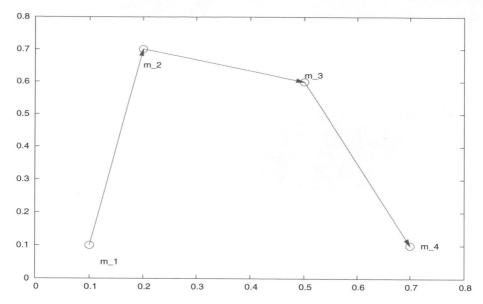

Figure 3. In the CHAIN method the chain vectors are regularised using KS. The chain also helps preventing serious roundoff errors.

$\mathbf{p}_k = \mathbf{W}_{k-1} - \mathbf{W}_k; \quad k = 2, \ldots, N-1$, but the first and last take the form $\mathbf{p}_1 = -\mathbf{W}_1$, $\mathbf{p}_N = \mathbf{W}_{N-1}$. By definition the corresponding chain vectors are given by $\mathbf{R}_k = \mathbf{q}_{k+1} - \mathbf{q}_k$. In the centre-of-mass system the Hamiltonian becomes

$$H = \sum_{k=1}^{N-1} \frac{1}{2}\left(\frac{1}{m_k} + \frac{1}{m_{k+1}}\right)\mathbf{W}_k^2 - \sum_{k=2}^{N} \frac{1}{m_k}\mathbf{W}_{k-1}\cdot\mathbf{W}_k - \sum_{k=1}^{N-1} \frac{m_k m_{k+1}}{R_k} - \sum_{1 \leqslant i \leqslant j-2} \frac{m_i m_j}{R_{ij}}, \quad (2.20)$$

where the non-chained distances are given by $R_{ij} = |\mathbf{r}_j - \mathbf{r}_i| = |\mathbf{q}_j - \mathbf{q}_i| = |\sum_{i \leqslant k' \leqslant j-1} \mathbf{R}_{k'}|$. Later different parts of this Hamiltonian, the kinetic energy, the chained and non-chained parts of the force function, respectively, will be denoted by T, U_c and U_{nc}.

Substitution of the KS-transformations $\mathbf{R}_k = \widehat{\mathbf{Q}}_k\mathbf{Q}_k$, $\mathbf{W}_k = \widehat{\mathbf{Q}}_k\mathbf{P}_k/(2Q_k^2)$ gives the Hamiltonian in terms of the regularising variables $\mathbf{Q}_k$, $\mathbf{P}_k$. With the time transformation $dt = g\,ds$, with $g = 1/(T+U)$, one obtains the regularised Hamiltonian $\Gamma = g(H-E) = (T - U - E)/(T + U)$ in the $(\mathbf{P}, \mathbf{Q}, s)$-system. The derivation of the equations of motion $\mathbf{P}_k' = -\partial\Gamma/\partial\mathbf{Q}_k$, $\mathbf{Q}_k' = \partial\Gamma/\partial\mathbf{P}_k$, where primes denote differentiation with respect to s, is now possible remembering that $\partial\widetilde{U}/\partial\mathbf{Q} = 2\widehat{\mathbf{Q}}^t\mathbf{F}$, where $\mathbf{F} = \partial\widetilde{U}/\partial\mathbf{R}$.

• **Algorithmic Regularisation (AR).**
(Mikkola & Tanikawa 1999ab, Preto & Tremaine 1999)
Case of two bodies: The Hamiltonian in extended phase space may be written

$$H = T + b - U = \mathbf{p}^2/2 + b - m/r, \quad (2.21)$$

where $\mathbf{p}$ is the momentum vector, r the distance and the momentum of time $b = m/r - \mathbf{p}^2/2 = \text{constant}$.

Instead of H one may use a new Hamiltonian (easy to prove)

$$\Lambda = \ln(T + b) - \ln(U), \quad (2.22)$$

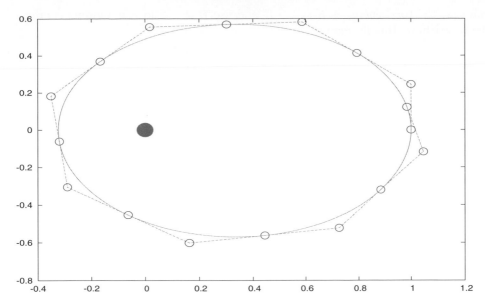

Figure 4. Logarithmic leapfrog trajectory for a two-body problem. The points outside the correct ellipse are the $\mathbf{r}_{\frac{1}{2}}$ points that are not to be considered physical coordinates, but only auxiliary quantities.

which gives equations of motion for coordinates t and $\mathbf{r}$

$$t' = 1/(\mathbf{p}^2/2 + b); \quad \mathbf{r}' = \mathbf{p}/(\mathbf{p}^2/2 + b) \tag{2.23}$$

end for momenta b and $\mathbf{p}$

$$b' = 0; \quad \mathbf{p}' = -\mathbf{r}/r^2. \tag{2.24}$$

One notices that a leapfrog algorithm is possible:

$$\delta t_0 = \tfrac{1}{2}h/(\mathbf{p}_0^2/2 + b); \quad t_{\frac{1}{2}} = t_0 + \delta t_0 \tag{2.25}$$

$$\mathbf{r}_{\frac{1}{2}} = \mathbf{r}_0 + \delta t_0 \mathbf{p}_0 \tag{2.26}$$

$$\mathbf{p}_1 = \mathbf{p}_0 - h\mathbf{r}_{\frac{1}{2}}/r_{\frac{1}{2}}^2 \tag{2.27}$$

$$\delta t_1 = \tfrac{1}{2}h/(\mathbf{p}_1^2/2 + b); \quad t_1 = t_{\frac{1}{2}} + \delta t_1 \tag{2.28}$$

$$\mathbf{r}_1 = \mathbf{r}_{\frac{1}{2}} + \delta t_1 \mathbf{p}_1. \tag{2.29}$$

A remarkable fact is that this simple, supposedly second order, algorithm gives a correct trajectory for the two-body motion. The only error is a phase error i.e. the time increment has a remainder of $O(h^3)$. The important fact is that even an exact collision orbit can be done by this algorithm. Consequently the **logarithmic Hamiltonian leapfrog can be used for regularization** without coordinate transformations.

• **N-Body formulation(s)**
Logarithmic Hamiltonian:
Let T and U be the kinetic and potential energies

$$T = \sum_{k=1}^{N} \frac{m_k}{2} \mathbf{v}_k^2; \quad U = \sum_{i<j\leqslant N} m_i m_j/r_{ij} + R(\mathbf{r}, t) \tag{2.30}$$

such that the total energy is $E = T - U$ and use $B = -E$ for the binding energy. The logarithmic Hamiltonian is

$$\Lambda = \ln(T + B) - \ln(U) \tag{2.31}$$

and the equations of motion

$$t' = \partial\Lambda/\partial B; \quad \mathbf{r}'_k = \partial\Lambda/\partial\mathbf{p}_k \tag{2.32}$$
$$B' = \partial\Lambda/\partial t; \quad \mathbf{p}'_k = \partial\Lambda/\partial\mathbf{r}_k, \tag{2.33}$$

allow the formation of the leapfrog algorithm. Define two subroutines: $\mathbf{X}(s)$ for moving coordinates over a step $=s$

$$\mathbf{X}(s): \quad \delta t = s/(T + B); \quad \mathbf{r}_k \to \mathbf{r}_k + \delta t \mathbf{v}_k, \quad t \to t + \delta t \tag{2.34}$$

and routine $\mathbf{V}(s)$ for velocity jumps

$$\mathbf{V}(s): \quad \delta\tau = s/U; \quad \mathbf{v}_k \to \mathbf{v}_k + \delta\tau\partial U/\partial\mathbf{r}_k; \quad B \to B + \delta\tau\partial U/\partial t \tag{2.35}$$

The leapfrog algorithm for n steps can then be symbolised as

$$\mathbf{X}(h/2)\left(\mathbf{V}(h)\mathbf{X}(h)\right)^{n-1}\mathbf{V}(h)\mathbf{X}(h/2). \tag{2.36}$$

This leads in principle to an algorithmically regularised method, results of which could be improved to high precision using the extrapolation method (Bulirsch & Stoer 1966.) Here the main problem that remains is roundoff. The cure for this is the use of the chain.

- **Time Transformed Leapfrog (TTL):**

Recently Mikkola & Aarseth (2002) suggested an alternative to the logH method. Here one introduces a new variable ω and a function $\Omega(\mathbf{r})$ such that $\omega(0) = \Omega(0)$ and the values of ω are obtained via the differential equation $\dot{\omega} = \dot{\Omega}(\mathbf{r}) = \partial\Omega/\partial\mathbf{r} \cdot \mathbf{v}$. This allows one to write the coordinate equations in the form

$$t' = 1/\omega \tag{2.37}$$
$$\mathbf{r}' = \mathbf{v}/\omega \tag{2.38}$$

and the velocity equations

$$\mathbf{v}' = \frac{\partial U}{\partial\mathbf{r}}/\Omega(\mathbf{r}) \tag{2.39}$$
$$\omega' = \frac{\partial\Omega}{\partial\mathbf{r}} \cdot \mathbf{v}/\Omega. \tag{2.40}$$

This formulation also allows a leapfrog construction and if $\Omega \sim 1/r_{min}$ when $r_{min} \to 0$, then the algorithm is asymptotically the same as the logH leapfrog. This fact makes it possible to regularise the close approaches of small bodies which do not affect considerably the value of the potential.

- **Algorithmic Regularization Chain**

One forms a chain of particles such that the shortest relative vectors are in the chain (Mikkola & Aarseth 1993). *It is necessary to stress that the main purpose of using the chain structure in this method is to reduce the effect of roundoff error.* In this the chain is effective.

Let us collect the chain coordinates

$$\mathbf{X}_k = \mathbf{r}_{i_k} - \mathbf{r}_{j_k} \tag{2.41}$$

in the vector

$$\mathbf{X} = (\mathbf{X}_1, \mathbf{X}_2, \ldots, \mathbf{X}_{N-1})$$

and let the corresponding velocities be

$$\mathbf{V} = (\mathbf{V}_1, \mathbf{V}_2, \ldots, \mathbf{V}_{N-1}).$$

Then the Newtonian equations of motion may be formally written

$$\dot{\mathbf{X}} = \mathbf{V} \tag{2.42}$$

$$\dot{\mathbf{V}} = \mathbf{A}(\mathbf{X}) + \mathbf{f}, \tag{2.43}$$

where $\mathbf{A}$ is the N-body acceleration and $\mathbf{f}$ is some external acceleration (e.g. due to other bodies).

One may use the two equivalent time transformations (Mikkola & Merritt 2006).

$$ds = [\alpha(T + B) + \beta\omega + \gamma]dt = [\alpha U + \beta\Omega + \gamma]dt, \tag{2.44}$$

where s is a new independent variable, B is the binding energy $B = -E$, α, β and γ are adjustable constants, Ω is an optional function of the coordinates $\Omega = \Omega(\mathbf{X})$. The initial value $\omega(0) = \Omega(0)$ and the differential equation

$$\dot{\omega} = \frac{\partial \Omega}{\partial \mathbf{X}} \cdot \mathbf{V}, \tag{2.45}$$

determines the value of ω (actually $\omega(t) = \Omega(t)$ along the exact solution).

It is possible to divide the equations of motion into two categories (when derivatives with respect to the new independent variable s are denoted by a prime).

Coordinate equations:

$$t' = 1/(\alpha(T + B) + \beta\omega + \gamma) \tag{2.46}$$

$$\mathbf{X}' = t'\, \mathbf{V} \tag{2.47}$$

Velocity equations:

$$\tilde{t}' = 1/(\alpha U + \beta\Omega + \gamma) \tag{2.48}$$

$$\mathbf{V}' = \tilde{t}'\, (\mathbf{A} + \mathbf{f}) \tag{2.49}$$

$$\omega' = \tilde{t}'\, \frac{\partial \Omega}{\partial \mathbf{X}} \cdot \mathbf{V} \tag{2.50}$$

$$B' = -\tilde{t}'\, \frac{\partial T}{\partial \mathbf{V}} \cdot f \tag{2.51}$$

In these equations the right hand sides do not depend on the variables at the left hand side. Consequently it is possible to construct a regular leapfrog algorithm for obtaining the solutions (Mikkola & Tanikawa 1999ab, Mikkola & Aarseth 2003, Preto & Tremaine 1999).

The time transformation thus introduced regularises the two-body collisions if one uses the simple leapfrog algorithm as a basic integrator (results of which can, and must, be improved using an extrapolation method (e.g. Bulirsch & Stoer 1966). [This is why the method is called Algorithmic Regularization.]

For the case of velocity dependent perturbation $\mathbf{f} = \mathbf{f}(\mathbf{X}, \mathbf{V})$, which occurs e.g if one introduces relativistic Post-Newtonian terms, related algorithms were discussed by (Mikkola & Merritt 2006).

In the presence of external perturbations the binding energy evolves according to

$$\dot{B} = -\frac{\partial T}{\partial \mathbf{V}} \cdot \mathbf{f} \tag{2.52}$$

The leapfrog for the chain vectors $\mathbf{X}$ and $\mathbf{V}$ can be written as the two mappings

$\mathbf{X}(s)$:

$$\delta t = s/(\alpha(T + B) + \beta\omega + \gamma) \tag{2.53}$$

$$t = t + \delta t \tag{2.54}$$

$$\mathbf{X} \to \mathbf{X} + \delta t \mathbf{V} \tag{2.55}$$

$$\tag{2.56}$$

$\mathbf{V}(s)$:

$$\widetilde{\delta t} = s/(\alpha U + \beta\Omega + \gamma) \tag{2.57}$$

$$\mathbf{V} \to \mathbf{V} + \widetilde{\delta t}(\mathbf{A} + \mathbf{f}) \tag{2.58}$$

$$B \to B + \widetilde{\delta t}\left\langle \frac{\partial T}{\partial \mathbf{V}} \right\rangle \cdot \mathbf{f} \tag{2.59}$$

$$\omega \to \omega + \widetilde{\delta t}\frac{\partial \Omega}{\partial \mathbf{X}} \cdot \langle \mathbf{V} \rangle, \tag{2.60}$$

where $\langle \frac{\partial T}{\partial \mathbf{V}} \rangle$ and $\langle \mathbf{V} \rangle$ are the averages over the advancement of $\mathbf{V}$.

The leapfrog with the above maps reads

$$\mathbf{X}(h/2)\left(\mathbf{V}(h)\mathbf{X}(h)\right)^{n-1}\mathbf{V}(h)\mathbf{X}(h/2), \tag{2.61}$$

for a macro-step of length $= nh$.

- **Alternatives for the time-transformation**

If one takes

$$\Omega = \sum_{i<j} \Omega_{ij}/r_{ij} \text{ and } \Omega_j = m_i m_j, \tag{2.62}$$

then $\alpha = 0$, $\beta = 1$, $\gamma = 0$ is mathematically equivalent to $\alpha = 1$, $\beta = \gamma = 0$ as was shown in Mikkola & Aarseth (2002). However, numerically these are not equivalent, but the logH alternative is much more stable. On the other hand, as noted above, it is desirable to get step size shortening (and thus regularization) also for encounters of small bodies and thus some function Ω should be used.

The increase the numerical stability for strong interactions of big bodies and also smooth the encounters of small bodies one may use $\alpha = 1$, $\beta \neq 0$ and

$$\Omega_{ij} = \begin{cases} = C_m; & \text{if } m_i * m_j < \epsilon C_m \\ = 0; & \text{otherwise} \end{cases}, \tag{2.63}$$

where $C_m = \sum_{i<j} m_i m_j/(N(N-1)/2)$ is the mean mass product and ϵ an adjustable parameter (usually $\epsilon \sim 10^{-3}$ or smaller, may be recommended).

3. Numerical Experiments

Here we discuss only results from a 7-body system with masses $m_1 = 1, m_2 = .1$ and 5 masses of $m_k = .001$ $k = 3, .., 7$. Experiments were made with a KS-CHAIN and an AR-CHAIN code. Integrations were carried out with and without relativistic perturbations and output was done with and without requiring exact output time (slower, because of iteration to exact time). It was also checked that the solutions obtained were similar. In all cases the accuracies were comparable ($|\delta E/L| < 10^{-12}$), but invariably the AR-CHAIN was faster by a factor of few. This does not necessarily mean that this is always the case, but seems to be typical for large mass ratios.

4. Discussion and Conclusion

The various alternatives to write regular N-Body algorithms (the KS and AR) were briefly reviewed. Numerical experiments have shown that the new AR method is often efficient even when the mass ratios are large (in this case the KS based algorithms are more difficult to use).

There is still a problem, related to the TTL (and AR-CHAIN) algorithm: the auxiliary quantity ω is obtained by summation of increments $\omega_{new} = \omega_{old} + \delta\omega$, where the increments $\delta\omega$ can be large and of varying sign. There is often considerable cancellation and loss of significant figures. This is one of the most important remaining problems in the AR-algorithm and needs to be investigated in future research.

References

Aarseth, S. J. 1972, in *Gravitational N-body Problem*, proceedings of IAU colloquium No. 10, ed. M. Lecar, Reidel, Dordrecht, pp.373–387.

Aarseth, S. J. 2003, *Gravitational N-Body Simulations*, Cambridge Univ. Press, Cambridge

Aarseth, S. J. & Zare, K. 1974, *Celestial Mechanics*, 10, 185

Bulirsch, R. & Stoer, J. 1966, *Numerical Mathematics*, 8, 1

Heggie, D. C. 1974, *Celestial Mechanics*, 10, 217

Kustaanheimo, P. & Stiefel, E. 1965, *J. Reine Angew. Math.*, 218, 204

Levi-Civita, T. 1920, *Acta Mathematica*, 42, 99

Mikkola, S. 1985, *MNRAS*, 215, 171

Mikkola, S. & Aarseth, S. J. 1990, *Celestial Mechanics and Dynamical Astronomy*, 47, 375

Mikkola, S. & Aarseth, S. J. 1993, *Celestial Mechanics and Dynamical Astronomy*, 57, 439

Mikkola, S. & Aarseth, S. 2002, *Celestial Mechanics and Dynamical Astronomy*, 84, 343

Mikkola, S. & Merritt, D. 2006, *MNRAS*, 372, 219

Mikkola, S. & Tanikawa, K. 1999a, *Celestial Mechanics and Dynamical Astronomy*, 74, 287

Mikkola, S. & Tanikawa, K. 1999b, *MNRAS*, 310, 745

Preto, M. & Tremaine, S. 1999, *AJ*, 118, 2532

Siegel. C. L. 1956, *Vorlesungen über Himmelsmechanik*, Springer, Berlin-Göttingen-Heidelberg.

Stiefel, E. L. & Scheifele, G. 1971, *Linear and Regular Celestial Mechanics*, Springer, Berlin.

Szebehely, V. & Peters, C. F. 1967, *AJ*, 72, 876.

von Hoerner, S. 1960, *Z. Astrophys.*, 50, 184

von Hoerner, S. 1963, *Z. Astrophys.*, 57 , 47

Zare, K. 1974, *Celestial Mechanics*, 10, 207

Dynamical Evolution of Dense Stellar Systems
Proceedings IAU Symposium No. 246, 2007
E. Vesperini, M. Giersz & A. Sills, eds.
© 2008 International Astronomical Union
doi:10.1017/S1743921308015640

Numerical Evolution of Single, Binary and Triple Stars

Peter P. Eggleton[1,2]

[1]Institute of Geophysics and Planetary Physics
[2]Physics and Allied Technologies Division
Lawrence Livermore National Laboratory, 7000 East Ave,
Livermore, CA94551, USA
email: ppe@igpp.ucllnl.org

Abstract. I discuss my stellar evolution code `Ev` in the context of simulations of large clusters of stars. It has long been able to handle single stars, and also binary stars up to a point. That point is far beyond what other codes are able to do, but well short of what is necessary for believable simulations. A recent version, `Ev(Twin)`, can in principle deal with the contact phase of binary evolution, but it is not yet clear what the physical interaction is that needs to be simulated.

An upgrade, which I hope will be only a few lines, should allow it to follow Kozai cycles with tidal friction, a process that strongly influences the orbital period of close pairs that reside within wide, non-coplanar triples. However, there are many substantial gaps in the physics of even single stars, let alone binaries or triples.

Keywords. stars: evolution, stars: binaries: general

---

1. Introduction

There is quite a lot that we seem to understand reasonably well about single and binary stars, but also quite a lot that we clearly don't understand and so we ought to try harder. For many people, a stellar evolution code is something you take off the shelf, turn into a module, and plug into some larger computational entity. For me, a stellar evolution code is a warm, living entity, a small child that needs constant nurturing and encouragement in the hope that it will grow to a comfortable and productive adulthood. It should develop the social skills to interact with other codes, but it should also develop its brain. This must be sufficiently supple that new ideas can be grafted in without major rearrangement.

The main difference between a code and a child is that a code does exactly what you tell it. This has disadvantages as well as advantages.

I call my code `Ev`. I run it usually in one of three operational modes that I call `Ev(Single)`, `Ev(Flip-Flop)` and `Ev(Twin)`. The first deals with single stars, the second with binaries under the assumption that Roche-Lobe overflow (RLOF) is the main interaction between them, and is reasonably (but not necessarily wholly) conservative both of total mass and of orbital angular momentum. `Flip-Flop` refers to the fact that it will advance the primary ('∗1') say 200 timesteps, then the secondary ('∗2') until it catches up, and then advances ∗1 another 200 timesteps, and so on. The reason for this is a small economy: very often when we follow ∗2 in a rather close binary we find that it fills its own Roche lobe *while ∗1 still does*. It evolves into contact, in other words, and there is no point in following ∗1 through to a supernova (SN) explosion if in practice ∗2 evolves into contact with ∗1 quite early in ∗1's evolution. Almost certainly the fact of

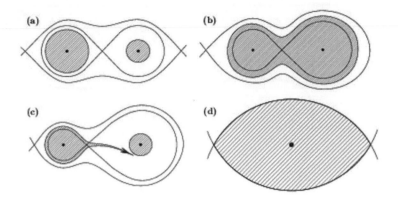

Figure 1. (a) A detached binary, also showing the critical inner (and outer) Lagrangian surface. (c) A semidetached system, with mass flowing from the loser to the gainer. The components have equal mass but different radii, perhaps as a result of differential stellar wind. (b) The gainer expands to fill its own Roche lobe, leading to contact. (d) Progressive mass transfer reduces one component to zero, and the other is then a rapidly rotating single star.

contact makes the later evolution of ∗1 obsolete – we need some different approach to compute the effect of contact. Nelson & Eggleton (2001) showed that contact is almost inevitable at a late stage in the evolution of short-period binaries; it is avoided only if the masses are fairly nearly equal, and the period is not too short. Fig. 1 is a cartoon of the progression of a binary from an initial detached configuration through a semidetached configuration, then into contact and ultimately (in many cases, I expect) into a merged single, rapidly-rotating star.

I believe that contact binaries are the proverbial elephant in the drawing room, which everyone prefers to ignore because they don't know what to do about it.

Ev(Twin) is intended to be the new process that will solve the contact problem, but it is still in development. It has in fact solved just one contact binary so far, which I will show shortly.

Even Ev(Single) contains some very powerful and original concepts, of which the main one is that equations for *all* of (a) the structure, (b) the composition, and (c) the adaptive mesh, are solved *simultaneously* and *implicitly*. Because of this concept, Ev(Flip-Flop) is enabled to be very powerful also. In order to deal with RLOF in a binary, it differs from the Single mode in one significant line only: the boundary condition $M_1 =$ given is replaced by one which says that $-dM_1/dt$ is proportional to the cube of $\ln R_1/R_{\mathrm{lobe},1}$, provided this is positive, when doing ∗1, and by $dM_2/dt = -dM_1/dt$ when doing ∗2. This is based on the approximination of Paczyński & Sienkiewicz (1972), itself based on a Bernoulli-type approximation to the fluid flow involved in RLOF.

Flip-Flop can incorporate what I call 'partially non-conservative' effects, such as stellar winds, magnetic braking, tidal friction, circularistion of eccentric orbits, and synchronism of non-corotating components, but *only for* ∗1. The wind from ∗2, for example, will depend on parameters relating to ∗2 such as its radius, luminosity and rotation rate. These will not be known at the time that one is advancing the evolution of ∗1.

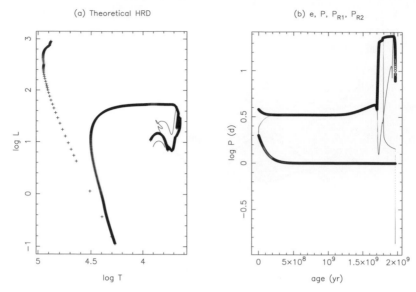

(a) Theoretical HRD   (b) e, P, P_{R1}, P_{R2}

Figure 2. Evolution of a system like Z Her, starting with $(1.8 + 1.68\,M_\odot; 3.8\,\mathrm{d}, e = 0.3)$, using Ev(Twin). (a) Theoretical HRD: thick line *1, thin line *2. The evolution terminated with a nova explosion on *1, following reverse RLOF. (b) Eccentricity (lower thick line), orbital period (upper thick line), and both rotational periods (thin lines) as functions of age. The rotational period of *1 ends shorter than that of *2.

Ev(Twin) is revolutionary in that it deals with this and several other problems by solving both of the components *simultaneously*. Actually it is not quite revolutionary: Robertson & Eggleton (1977) did this. But in 1977 big, powerful computers had only 50KB (KB!) of random-access memory, and it was quite a chore to squeeze in a matrix that was 2×2 times as large as the usual one. But the virtue of Twin is, or ought to be, that it can deal numerically with the contact phase. Contact involves what I would like to call a 'strong interaction', ie. the behaviour of one component is very strongly influenced by what the other component is doing at the same time. The obvious answer to this is to enlarge the concepts of 'simultaneous' and 'implicit' to the whole binary system.

Ev(Twin) can deal well with systems that are not in contact, but where non-conservative processes in both components are happening simultaneously. Fig. 2 shows the evolution of a system starting with parameters $(1.8 + 1.68\,M_\odot; 3.8\,\mathrm{d}, e = 0.3)$. It is strikingly different from what would go on in presumed conservative evolution, but space does not permit a detailed description.

The difficulty with contact binaries is to think of a reasonable mathematical/physical model of what goes on in contact. On the one hand mass can flow, in either direction in principle, between the outer layers of one component and the outer layers of the other, all the way down to the inner Lagrangian surface. On the other hand, heat can also flow. The direction of heat flow might change with depth, as might the direction of mass flow.

My best shot so far treats the heat flow by a model based on the fact that, in the Sun, it is well known (though not yet well understood) that there is differential rotation. A surface equatorial belt, about 30% of the depth of the Sun and about $\pm 30°$ in breadth, rotates about 10% faster than the mean. In a frame that rotates with the mean Sun, this belt is carrying a *colossal* flow of heat 'sideways'. Work it out: it's about $3000\,L_\odot$. Of course this makes no difference to the Sun, because the Sun is roughly axially symmetric. But in a contact binary consisting of say $1 + 0.5\,M_\odot$ components which share a single

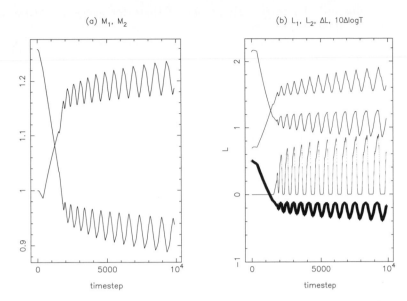

Figure 3. (a) M_1 and M_2 plotted against timestep number. In actual time, the oscillatory stretch is only about 2% of the entire evolution. (b) Evolution of L_1, L_2 (upper two curves), ΔL, and $10\Delta \log T$ (thick curve) against timestep number, in a binary which evolves from detached through semidetached to early contact. ΔL is the amount of luminosity transferred in contact, and so is zero during a detached or semidetached phase. $\Delta \log T$ is the difference in effective temperatures of the two components.

envelope that accounts for the outer 5–10% of each by radius, even 0.1% of this flux, flowing largely around the equatorial belt of one star and on around the equatorial belt of the other, and on further, would represent a large exchange of luminosity between the components, quite possibly equalising, more or less, the two surface temperatures as is normally observed in contact binaries.

Fig. 3 is an attempt (Yakut & Eggleton 2005) to model part of this with Ev(Twin). Fig 3a shows the two masses, and the upper two curves of Fig 3b the two luminosities. M_1 and L_1 started to decrease rapidly at RLOF, but shortly after the mass and luminosity ratio reversed the system came into contact. There followed a series of oscillations, but with a slight trend discernible: the mass ratio continued to depart from unity, on average. The oscillations are rather rapid, and demand small timesteps; the 11 oscillations occupy about 2% of the entire evolutionary sequence.

It is not clear how real these oscillations are, although similar oscillations were seen (using semi-analytical techniques) by Lucy (1976) and Flannery (1976), as well as by Robertson & Eggleton (1977). They appear to contradict the observational datum that the great majority of contact binaries have closely equal temperatures. The thick line of Fig 3b ($10\Delta \log T$) oscillates between -0.1 and -0.4; the first value is acceptable, but the second is not. However, there are a few arguments that suggest that the inconsistency may be more apparent than real:

(i) The model presented here has a mass ratio rather close to unity – of necessity, since I have had difficulty getting the code to work for more extreme mass ratios – and so is not representative of real contact binaries, whose mass ratios are usually in the range 2–10

(ii) There does exist a small population of close binaries that have similar periods to contact binaries, but rather more unequal temperatures. They are sometimes called 'near-contact binaries'. There is perhaps one of these for every 10–20 normal contact binaries, but the statistics are very poor

(iii) We might hypothesise that as the mass ratio gets more extreme, the time spent with the temperatures substantially different becomes a smaller fraction of the time spent with temperatures nearly equal.

Models in contact have so far turned out to be very expensive of computational time. But I hope to produce some more, with more extreme mass ratios, in due course. However it is possible that some important element of the physics is missing, and that when this is included the evolution will actually be simpler to calculate. Perhaps there is some kind of 'thermal inertia' which turns the oscillatory behaviour into monotonic behaviour; but most attempts to fudge that lead to the two components having permanently different temperatures, rather than permanently equal temperatures.

I believe, or at least hope, that it will not be very difficult to add in the equations that govern Kozai cycles. These are cycles of eccentricity, but *not* semimajor axis or period, that are induced in a binary that is part of a triple system where the outer orbit is inclined at more than 39° to the inner orbit. When tidal friction is included (which it already is) the inner orbit *does* suffer a reduction in period and semimajor axis: see the poster by Kisseleva-Eggleton & Eggleton in this conference. We believe this may be vital in producing the shortest-period systems.

I and my colleagues at Lawrence Livermore Laboratory hope to gain insight into the physics of contact binaries, tidal friction and several other evolutionary problems, by modeling stars in 3 dimensions with the code `Djehuty`; see Dearborn *et al.* (2006) and Eggleton *et al.* (2006).

2. Conclusion

When evolving a cluster, the evolution of the single stars (and of wide binaries, if any) can be well approximated by interpolating in a pre-computed grid of, say, 1000 single stars with a range of masses. But binaries have too many parameters to be pre-computed. A code like `Ev(Twin)` will be needed for parallelised evolution, but further development of the contact model is necessary.

Acknowledgements

This work performed under the auspices of the U.S. Department of Energy by Lawrence Livermore National Laboratory under Contract DE-AC52-07NA27344.

References

Dearborn, D. S. P., Eggleton, P. P., & Lattanzio, J. C. 2006, *ApJ*, 639, 405
Eggleton, P. P., Dearborn, D. S. P., & Lattanzio, J. C. 2006, *Science*, 314, 1580
Flannery, B. P. 1976, *ApJ*, 205, 217
Lucy, L. B. 1976, *ApJ*, 205, 208
Nelson, C. A. & Eggleton, P. P. 2001, *ApJ*, 552, 664
Paczyński, B. & Sienkiewicz, R. 1972, *A&A*, 22, 73
Robertson, J. A. & Eggleton, P. P. 1977, *MNRAS* 179, 359
Yakut, K. & Eggleton, P. P. 2005, *ApJ*, 629, 1055

Dynamical Evolution of Dense Stellar Systems
Proceedings IAU Symposium No. 246, 2007
E. Vesperini, M. Giersz & A. Sills, eds.

Full Ionisation in Binary-Binary Scattering

W. L. Sweatman

Institute of Information and Mathematical Sciences, Massey University at Albany, Private Bag
102 904, North Shore 0745, New Zealand
email: w.sweatman@massey.ac.nz

Abstract. Encounters between binary stars and single stars and between binary stars and other binary stars play a key role in the dynamics of dense stellar systems. In the simple model, in which stars are approximated by point masses, a number of theoretical and numerical results are known. In particular there exist relationships to help to describe the destruction process of binary stars (ionisation) through three-body encounters between binary and single stars. Here we extend these results to the four-body case involving encounters between pairs of binary stars that lead to a disruption of the binaries into single stars.

Keywords. binaries: general, scattering, stellar dynamics, celestial mechanics

1. Introduction

In dense stellar systems, interactions involving binaries are important for energy transfer. In addition, binaries can be created and destroyed. The most likely encounters are those between a single star and a binary star and those between a pair of binary stars. In the simplest model, which is used here, the stars are approximated by point masses. Early studies incorporating simulations of binary-binary interactions extend back to those of Mikkola 1983 and Hoffer 1983. Numerical simulations that support the theory developed in the present investigation made use of the CHAIN code developed by Mikkola & Aarseth (Mikkola & Aarseth 1993, 1996; Aarseth 2003).

2. Full Ionisation

In full ionisation all binaries involved in a scattering encounter are destroyed to create a number of separating single masses. The likelihood is measured by the total cross-section (σ). A key parameter is the total energy (h) in the centre of mass rest frame or the closely related relative velocity of the two approaching objects at infinity (v_∞). Full ionisation is impossible for negative total energy. As total energy increases from zero, the cross-section rapidly rises from zero, peaks, and then steadily declines. This evolution has been illustrated and described in the four-body binary-binary case (Hut 1992; Sweatman 2007) as well as for three bodies (Hut & Bahcall 1983; Hut 1983).

Considering binary-binary interactions with equal masses (m), here we summarise the low energy results of Sweatman (2007) and then present new results for high energy.

2.1. Full ionisation in binary-binary encounters at low positive total energy

Here, full ionisation is barely possible. The energy needs to be distributed quite precisely among the masses, with comparable inter-body distances for a period of expansion. The orbit must pass close to a central configuration (Heggie & Sweatman 1991; Heggie & Hut 1993, 2003) and the cross-section is found to be proportional to h^β where $\beta = \sum_{Re\lambda_i > 0, \lambda_i \neq 2/3} 3\,Re\,\lambda_i\,/2$, and the λ_i are related to the 'Siegel exponents' (Siegel &

Moser 1971). With sufficiently small energy, the dominant configuration gives

$$\sigma \propto h^\beta \quad \text{with} \quad \beta = -\frac{1}{2} + \frac{1}{2}\sqrt{23 - 4\sqrt{3} + 2\sqrt{469 - 236\sqrt{3}}} \approx 2.310. \qquad (2.1)$$

2.2. *Full ionisation in binary-binary encounters at high total energy*

Here, ionisation cross-section decreases due to the requirement for sufficient momentum transfer during the brief close encounter. The asymptotic relationship $\sigma \propto 1/v_\infty^2$ derived in the three-body case (Heggie 1975; Hut 1983), has also been found in the four-body case (Hut 1992; Sweatman 2007). In the present investigation, an approach, similar to that of Hut (1983), has been used to find the coefficient of proportionality in various four-body cases. More details will be included in a forthcoming publication. The results summarised below apply to two circular binaries with semi-major axes a and aQ, where Q, the ratio of the axes, is taken to be greater than or equal to 1.

When Q is between 1 and $17 + 12\sqrt{2} \approx 33.97$ the full ionisation cross-section is

$$\sigma = \left(\frac{40}{3} + \frac{(1 + 2\beta^2)Q^{\frac{1}{2}} + 2 + \beta^2}{12\,\beta^4 Q^{\frac{1}{2}}}(Q^{\frac{1}{2}} - \beta^2)^3 + \frac{Q - \beta^4}{2\,\beta^2} + \frac{Q^{\frac{1}{2}}}{2} \ln\left[\frac{\beta^4}{Q}\right] \right) \frac{\pi a G m}{v_\infty^2}, \qquad (2.2)$$

where $\beta = \sqrt{2} + 1$. So that if the binaries have equal binding energies ($Q = 1$), then

$$\sigma = \left(\frac{40}{3} - 6\sqrt{2} + 2\ln(\sqrt{2} + 1) \right) \frac{\pi a G m}{v_\infty^2}, \qquad (2.3)$$

approximately 0.9916 of the three-body ionisation cross-section with the same a and v_∞.

If Q is greater than $17 + 12\sqrt{2}$ then all interactions ionising the tight binary will also ionise the loose binary: in any orientation the maximum ionisation impact parameter of the tightly bound binary is less than that of the other binary. In this case the four-body full ionisation cross-section is exactly twice that for an interaction between a single star and the tight binary, the factor of two being due to the two stars in the loose binary:

$$\sigma = \frac{40}{3} \frac{\pi a G m}{v_\infty^2}. \qquad (2.4)$$

References

Aarseth, S. J. 2003, *Gravitational N-Body Simulations*, (Cambridge: CUP)

Heggie, D. C. 1975, *MNRAS* 173, 729

Heggie, D. C. & Hut, P. 1993, *ApJS* 85, 347

Heggie, D. C. & Hut, P. 2003, *The Gravitational Million-Body Problem*, (Cambridge: CUP)

Heggie, D. C. & Sweatman, W. L. 1991, *MNRAS* 250, 555

Hoffer, J. B. 1983, *AJ* 88, 1420

Hut, P. 1983, *ApJ* 268, 342

Hut, P. 1992, in: E. P. J. van der Heuvel & S. A. Rappaport (eds.), *X-ray Binaries and Recycled Pulsars*, (Dordrecht: Kluwer), p. 317

Hut, P. & Bahcall, J. N. 1983, *ApJ* 268, 319

Mikkola, S. 1983, *MNRAS* 203, 1107

Mikkola, S. & Aarseth, S. J. 1993, *Cel. Mech.* 57, 439

Mikkola, S. & Aarseth, S. J. 1996, *Cel. Mech.* 64, 197

Siegel, C. L. & Moser, J. K. 1971, *Lectures on Celestial Mechanics*, (Berlin: Springer-Verlag)

Sweatman, W. L. 2007, *MNRAS* 377, 459

Dynamical Evolution of Dense Stellar Systems
Proceedings IAU Symposium No. 246, 2007
E. Vesperini, M. Giersz & A. Sills, eds.

On the Calculation of Average Lifetimes for the 3-body Problem

David Urminsky

School of Mathematics and Maxwell Institute for Mathematical Sciences,
University of Edinburgh, UK
email: david.urminsky@ed.ac.uk

Abstract. Numerical solutions for the 3-body problem can be extremely sensitive to small errors. We consider how small errors in calculations can affect the lifetime of these systems. In particular, we show that numerical errors can shorten the average lifetime of a 3-body system. This is illustrated using the Sitnikov Problem as an example. To give a theoretical explanation, we construct an approximate Poincaré map for this problem and delineate the structure of the escape regions. We show that numerical errors can destroy escape regions and can cause orbits to migrate to a region in which escape is faster.

Keywords. stellar dynamics, time

The Sitnikov problem is the problem of the motion of a massless particle, m_3, on the axis of symmetry of an equal-mass binary (see Moser (1973)). The equation of motion for m_3 is given by

$$\ddot{z} = -z/\sqrt{z^2 + r^2}^{3}, \qquad (0.1)$$

where z is its position and r the distance from the centre of mass to one of the binary masses. We approximate r to first order in the eccentricity, e, by $r \approx \frac{1}{2}(1 - e\cos t)$. Taking the plane of motion of the binary ($z = 0$) as a Surface Of Section (SOS), consider a map, $\phi : (v_0, t_0) \rightarrow (v_1, t_1)$, which takes m_3 from one crossing of the SOS to the next. If m_3 is on the SOS at time t_0, ϕ is a map which brings $v_0 = \dot{z}(t_0)$ to time $t_1 > t_0$ where $v_1 = \dot{z}(t_1)$ and $z(t_1) = 0$. The map ϕ has an open domain D in which every point returns to the SOS. As time enters into the problem with period 2π, we can consider D in polar co-ordinates where the radial variable is v and the angular variable is given by t.

We integrate initial conditions in D forward in time until they satisfy an escape theorem outlined in Urminsky (2007). During the integration we save the time and velocity values for the last 5 crossings of the SOS. Fig. 1(a) shows the structure of these crossings. The crossings form distinct bands which m_3 visits in turn until finally visiting the upper crescent shaped region before leaving the SOS for the last time. The bands wrap around each other in a fractal like structure spiralling outwards towards the boundary of D.

Consider a radial segment, R, of initial conditions which traverses these bands and approaches the boundary of D. We integrate the initial conditions R forward in time using the Bulirsch-Stoer method for various relative tolerances. Fig. 1(c) shows the average lifetime over R for increasing relative tolerances. There is a clear decrease in the average lifetime as the relative tolerance increases. The plateau for small relative tolerances is due to the maximum time we integrate over and can be increased by integrating longer.

To study the structure of the bands in the SOS we turn to a symplectic approximate Poincaré map (Urminsky (2007)),

$$\Phi(t_n, E_n) = \begin{cases} t_{n+1} = t_n + 2\alpha(-E_n - a\cos t_n - b\sin t_n)^{-\frac{3}{2}} \\ E_{n+1} = E_n + a\cos t_n + b\sin t_n - a\cos t_{n+1} + b\sin t_{n+1} \end{cases}, \qquad (0.2)$$

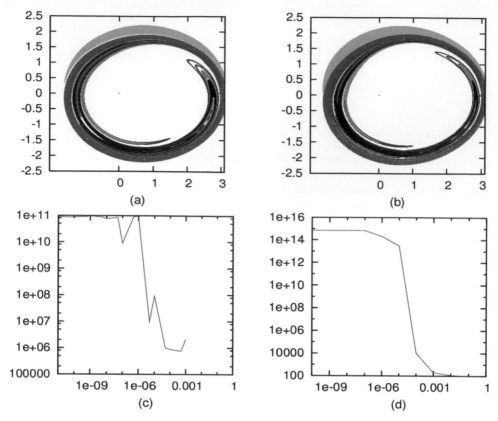

Figure 1. (a) The last five crossings m_3 makes with the SOS using equation (0.1) for $e = 0.61$. (b) The last five crossings m_3 makes with the SOS using (0.2) for $e = 0.61$. (c) Average lifetime vs. relative tolerance of the Bulirsch-Stoer method using equation (0.1). The maximum time of integration was 10^{11}. (d) Average lifetime vs. the magnitude of the noise using (0.2).

on an open domain $D^\dagger$ where t_i represents time at the ith crossing and E_i represents the energy of m_3. The constants a and b are approximately proportional to e and $\alpha = 2^{-3/2}\pi$. We repeat the experiment in Fig. 1(a) using Φ; the results are displayed in Fig. 1(b). We introduce uniformly distributed noise into (0.2) to mimic errors which would be present by numerically solving (0.1). Varying the magnitude of the noise over a radial segment of initial conditions we find a similar relationship between the average lifetime of orbits and the magnitude of the noise (Fig. 1(d)).

To explain this phenomenon we use Φ to determine the width of the bands as they wrap around the interior of $D^\dagger$. When the width of the bands are comparable to the amplitude of the noise, bands which represent previous crossing blend with successive crossings of the the SOS. This process causes orbits to migrate to bands which lead to quicker escape.

References

Moser, J. 1973, Stable and Random Motions in Dynamical Systems, Princeton University Press.

Urminsky, D. J. 2007, PhD thesis, University of Edinburgh. (in preparation)

Part 5

Binary Star Dynamics and its Interplay with Cluster Dynamical Evolution

Dynamical Evolution of Dense Stellar Systems
Proceedings IAU Symposium No. 246, 2007
E. Vesperini, M. Giersz & A. Sills, eds.

© 2008 International Astronomical Union
doi:10.1017/S1743921308015676

Binary Stars and Globular Cluster Dynamics

John M. Fregeau[1]

Northwestern University, Department of Physics and Astronomy, Evanston, IL 60208, USA.
email: fregeau@northwestern.edu
[1] Chandra Fellow

Abstract. In this brief proceedings article I summarize the review talk I gave at the IAU 246 meeting in Capri, Italy, glossing over the well-known results from the literature, but paying particular attention to new, previously unpublished material. This new material includes a careful comparison of the apparently contradictory results of two independent methods used to simulate the evolution of binary populations in dense stellar systems (the direct N-body method of Hurley, Aarseth, & Shara (2007) and the approximate Monte Carlo method of Ivanova *et al.* (2005)), that shows that the two methods may not actually yield contradictory results, and suggests future work to more directly compare the two methods.

Keywords. globular clusters: general, open clusters and associations: general, binaries: general, methods: n-body simulations, stellar dynamics

1. Preamble

As this conference is an occasion to celebrate Douglas Heggie's 60th birthday, I did the following to honor him. For each slide in my presentation for which Douglas had some impact – by directly working on a topic, by influencing the way people think about a topic, or by influencing my own personal thinking on a topic – I colored the slide background white. Interestingly, every single slide in my presentation had a white background.

2. Introduction

Globular clusters are observed to contain significant numbers of binary star systems – so many, in fact, that they must have born with binaries (Hut *et al.* 1992). Their presence in clusters is important for two complementary reasons. Through super-elastic dynamical scattering interactions, they act as an energy source which may postpone core collapse, and may be the dominant factor in setting the core radii of observed Galactic globulars. Similarly, the dense stellar environment and increased dynamical interaction rate in cluster cores is responsible for the high specific frequency of stellar "exotica" found in clusters, including low-mass X-ray binaries (LMXBs), cataclysmic variables (CVs), blue straggler stars (BSSs), and recycled millisecond pulsars (MSPs).

3. Evolution of Clusters

3.1. *The Negative Heat Capacity of Self-Gravitating Systems*

Imagine finding yourself piloting a spaceship in orbit about a planet. Your ship is equipped with rocket thrusters that can either fire in your direction of motion or opposite it. If you want to slow down, which way do you fire your thrusters?

The answer is that you fire your thrusters behind you – in your direction of motion. This causes work to be done on your spaceship, which increases your energy, expands your orbit about the planet, and slows you down. It's counterintuitive at first, since it's like depressing the accelerator pedal to slow down, but this behavior is typical of self-gravitating systems, and is a manifestation of their negative heat capacity – if you add energy to a system it cools down, if you take energy away from a system is heats up. A real-life example of this is air drag on an orbiting satellite, which will actually cause it to speed up.

A basic appreciation of the negative heat capacity of self-gravitating systems goes a long way in helping to understand the physics of the binary burning phase in clusters (the phase analogous to the main sequence in stars, in which clusters "burn" binaries instead of hydrogen to support their cores against collapse). Imagine a binary star system in a cluster encountering a single star, where the relative speed between the binary and the single star is smaller than the orbital speed in the binary system. When the three stars get close enough to interact strongly, the quickly moving binary members will tend to transfer some energy to the more slowly moving incoming single star (energy transfer from hot to cold). The result is that when the interaction is over one of the three stars (and it doesn't have to be the original single star) will leave with a higher relative velocity than the incoming single star initially had. Since the binary system gave up some energy to the single in the interaction it will become more tightly bound and thus have a larger orbital speed (energy was taken from it and it got hotter). The binary we have constructed in this thought experiment is a "hard" binary (since the orbital speed in the binary is larger than the encounter speed), which clearly becomes harder as a result of dynamical interactions (Heggie 1975)p. In general, a population of hard primordial binaries will act as an energy source that supports a cluster's core against collapse through dynamical scattering interactions (please see Heggie & Hut (2003) for a more detailed discussion).

It's easy to make an order of magnitude estimate of the importance of binaries in a cluster (the following discussion closely follows that in Heggie & Hut (2003)). Imagine a cluster with N objects, $f_b N$ of which are hard binaries. Denote the total cluster mechanical energy as E_{ext}, and the total binary binding energy as E_{int}. The binding energy of a binary with hardness x is then

$$E_b \equiv xkT \approx xE_{\mathrm{ext}}/N, \tag{3.1}$$

where kT represents thermal energy of motion. The total internal energy is then

$$E_{\mathrm{int}} \approx f_b x E_{\mathrm{ext}}. \tag{3.2}$$

Since a binary releases energy of order E_b through interactions, binaries are important when $f_b x \gtrsim 1$. For example, for binaries of hardness $x = 10$ (a reasonable value), a binary fraction of merely 10% can be enough to unbind a cluster completely. It would also appear that just one sufficiently hard binary could be dynamically very important. However, it should be noted that one key element has been left out of the discussion: interaction timescales. A very hard binary composed of stars that are roughly the average stellar mass in the cluster would have such a small semi-major axis as to make its interaction time so long that it is essentially dynamically irrelevant.

A more detailed analysis of energy generation due to binary burning can give a rough estimate of the equilibrium core radius in the binary burning phase, and can be compared with observations (the following discussion closely follows that in Goodman & Hut (1989)). In equilibrium the energy generated in the core via binary burning should equal the energy transported across the half-mass radius via two-body relaxation. The binary burning energy generation rate is

$$\dot{E}_{\mathrm{bin}} \approx n_c(n_c\sigma_{\mathrm{bin}}v_c)\left(\frac{4\pi r_c^3}{3}\right)\left(\frac{Gm^2}{2a}\right) \sim r_c^3 n_c^2 \frac{G^2m^3}{v_c}g(f_b, A_{\mathrm{bb}}, A_{\mathrm{bs}}), \tag{3.3}$$

where n_c is the core number density, the first term in parentheses is the n–σ–v estimate for the interaction rate of a binary, the second term in parentheses is the core volume, and the third term is the binding energy of a typical binary. The function $g(f_b, A_{bb}, A_{bs})$ is a dimensionless function of the binary fraction, the relative strengths of binary–binary and binary–single energy generation, and is of order unity. The two-body relaxation energy transport rate is

$$\dot{E}_{\mathrm{rel}} = \frac{|E|}{\alpha t_{\mathrm{rh}}} \approx \frac{1}{5\alpha} \frac{GM^2}{t_{\mathrm{rh}} r_h}, \tag{3.4}$$

where α is a constant, t_{rh} is the relaxation time at the half-mass radius, and r_h is the half-mass radius. Equating the two expressions yields

$$\frac{r_c}{r_h} \approx \frac{0.05}{\log_{10}(\gamma N)} g(f_b, A_{bb}, A_{bs}), \tag{3.5}$$

where the standard expression has been substituted in for the relaxation time, with $\log_{10}(\gamma N)$ the Coulomb logarithm. For $N = 10^6$ this expression yields $r_c/r_h \sim 0.02$ which is in rough agreement only with the $\sim 20\%$ of Galactic globulars that are observationally classified as core collapsed.

3.2. *Globular Cluster Core Radii*

Recently two independent and very different numerical methods for simulating the evolution of star clusters have been used to study the core radii of clusters in the binary burning phase. One is the direct N-body method, which utilizes very few approximations and thus treats the evolution of clusters on a dynamical (orbital) timescale. The other is the Monte Carlo method, which uses a number of assumptions in order to treat the evolution on a relaxation timescale. To accurately treat them, binary interactions are handled via direct few-body integration. Remarkably, the two methods agree quite well in the value of r_c/r_h predicted during the binary burning phase (Heggie, Trenti, & Hut 2006; Fregeau & Rasio 2007). Unfortunately, the value predicted by the simulations is at least an order of magnitude smaller than what's observed for the $\sim 80\%$ of clusters that are observed to be non-core collapsed. Since the longest phase of evolution for a cluster is the binary burning phase, it is expected that most clusters currently observed should be in this phase. The current state of the field thus represents a major discrepancy between theory and observations.

Several resolutions to the problem have been proposed. Hurley (2007), among others, has noted that there are in fact three different definitions of the core radius in popular use, with the observational definition possibly being larger than the standard dynamical definition used in some numerical codes by a factor of ~ 4. Another suggestion is that there are central intermediate-mass black holes (IMBHs) in most Galactic globular clusters, which act as an energy source that would increase core radii to roughly the value observed (Trenti 2006; Miocchi 2007; Trenti *et al.* 2007). It could also be that the "true" initial conditions for clusters are of much higher or lower stellar density than what has traditionally been assumed in simulations (Fregeau & Rasio 2007). Or it could be that stellar mass loss from enhanced stellar evolution of physical collision products could power the cores sufficiently (Chatterjee *et al.*, this volume). For young clusters, mass segregation of compact remnants (Merritt *et al.* 2004), or the evaporation of the sub-population of stellar-mass black holes (Mackey *et al.* 2007) could possibly explain the discrepancy.

3.3. *Globular Cluster Binary Fractions*

Observations show that globular clusters currently have core binary fractions ranging from a few % (NGC 6397, 47 Tuc, M4), to $\sim 30\%$ (Pal 13, E3, NGC 6752, NGC 288). The observational techniques used to determine the core binary fractions are varied, but include observation of a secondary main sequence, radial velocity studies, and searches for eclipsing binaries. All methods involve extrapolation of somewhat uncertain functions (e.g., the binary mass ratio distribution in the first method), although the first method is considered to be the most complete. When combined with observations of other cluster properties, measured core binary fractions enable detailed testing of cluster evolution models.

Two key processes govern the evolution of the binary fraction in clusters: binary stellar evolution, and stellar dynamical interactions. Two codes currently combine both processes to varying degrees of realism. The work of Hurley, Aarseth, & Shara (2007) uses full N-body calculations with binary stellar evolution. That of Ivanova *et al.* (2005) uses binary stellar evolution with a simplified dynamical model that assumes a two-zone cluster (core and halo) with constant core density, but performs direct few-body integration of binary scattering interactions. In perhaps oversimplified terms, Ivanova *et al.* (2005) find generally that the core binary fraction in clusters decreases with time, requiring clusters to have been born with large binary fractions to explain currently observed binary fractions. In similarly oversimplified terms, Hurley, Aarseth, & Shara (2007) generally find that the core binary fraction tends to increase with time. This is certainly a simplified comparison of the two apparently contradictory results, since there are several differences in initial conditions and assumptions used by the two different methods that need to be taken into account. The dynamics and evolution of binary populations in clusters – especially when coupled with a realistic treatment of binary stellar evolution – is a complex topic, however, and the important differences between the two methods and the clusters they model sometimes get lost in the discussion.

One of the most important parameters governing the evolution of binary fractions in clusters is the stellar density in the core. A larger density leads to a higher dynamical interaction rate, and it is these interactions that can affect stellar evolutionary processes by modifying orbital parameters of binaries, destroying binaries, exchanging members of binaries, and creating binaries via tidal capture. Fig. 1 shows the evolution of cluster evolution models in core number density and core binary fraction for the N-body simulations of Hurley, Aarseth, & Shara (2007) (solid arrows) and the Monte Carlo simulations of Ivanova *et al.* (2005) (dashed arrows). Note that the binary fraction plotted here is the *hard* binary fraction. The models of Hurley, Aarseth, & Shara (2007) use exclusively hard binaries, while those of Ivanova *et al.* (2005) start with a wide distribution of binaries that extends well into the soft regime. The globular clusters for which the core number density and binary fraction are known are plotted as open circles. The single triangle point represents a typical open cluster. In comparing the two models, a clear sign of fundamental disagreement would be if an N-body arrow and a Monte Carlo arrow begin in the same region of phase space but point in different directions. Only the two arrows at $n_c = 10^3 \, \mathrm{pc}^{-3}$ show this disagreement, pointing in opposite directions. However, the Monte Carlo method is less accurate at this low stellar density, since the assumption of constant core density breaks down here. Similarly, the N-body arrow appears to be horizontal since the initial and final densities are not known – only the average density is known for this model. In other words, the only two contradictory arrows in this diagram may not faithfully represent their respective methods. All other arrows are in very different areas of parameter space, and so unfortunately do not offer direct comparisons

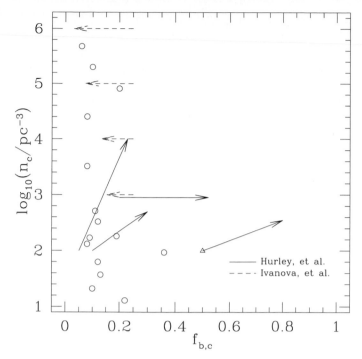

Figure 1. The evolution of cluster evolution models in core number density and core binary fraction for the N-body simulations of Hurley, Aarseth, & Shara (2007) (solid arrows) and the Monte Carlo simulations of Ivanova *et al.* (2005) (dashed arrows). Note that the binary fraction plotted here is the *hard* binary fraction. The models of Hurley, Aarseth, & Shara (2007) use exclusively hard binaries, while those of Ivanova *et al.* (2005) start with a wide distribution of binaries that extends well into the soft regime. The globular clusters for which the core number density and binary fraction are known are plotted as open circles. The single triangle point represents a typical open cluster.

between the two methods. N-body simulations are generally relegated to clusters with rather low initial densities ($\lesssim 10^3 \, \mathrm{pc}^{-3}$), while the Monte Carlo simulations are most accurate for rather higher densities ($\gtrsim 10^3 \, \mathrm{pc}^{-3}$). Thus it could very well be that the results of both methods represent the same underlying evolution. In order to fully compare the methods either many more Monte Carlo simulations should be performed with lower initial densities (although the assumption of constant core density breaks down as the density becomes lower), or N-body simulations should be performed for larger initial density and binary fraction (although this is currently quite computationally expensive).

A viable alternative to properly compare the two methods is to perform an N-body simulation at large initial core density for just a very short time to get a sense of the direction of evolution in phase space. It should be noted that the arrows in Fig. 1 are straight-line approximations to the true evolution. Thus it should first be tested whether the overall evolution of an N-body model in this parameter space has any relation to the initial, differential evolution.

4. Evolution of the Binary Population

Due to strong binary scattering interactions, globular clusters are home to large numbers of "exotic" stellar objects, including LMXBs, CVs, MSPs, and BSSs. The interactions can create and destroy classes of binaries directly through exchange and ionization, and indirectly by modifying orbital properties and thus affecting binary stellar evolution processes.

4.1. *The Interaction Frequency for X-Ray Sources*

It was realized over 30 years ago that globular clusters are overabundant per unit mass in X-ray binaries by orders of magnitude relative to the disk population (Clark 1975; Katz 1975). If X-ray binaries are formed in clusters mainly via binary scattering interactions, there should be a correlation between the interaction rate and the number of such binaries in each cluster (Verbunt & Hut 1987). Since the interaction rate is so heavily used in the literature, it is worthwhile to review its derivation.

Imagine a large volume of uniform density n_1 of an object labeled type 1, and similarly for type 2. The interaction rate for one member of species 2 with species 1 is

$$\frac{dN_2}{dt} = n_1 \sigma_{12} v_{12}, \tag{4.1}$$

where σ_{12} is the cross section for interaction of an object of species 1 with an object of species 2, and v_{12} is the relative speed between the two species. The total interaction rate per unit volume is then

$$\frac{dN_{\text{int}}}{dt\, dV} = n_1 n_2 \sigma_{12} v_{12}, \tag{4.2}$$

which is nicely symmetric under transformation between index 1 and 2. The total interaction rate for a cluster can be approximated by multiplying by the core volume, to give

$$\Gamma \equiv \frac{dN_{\text{int}}}{dt} \approx n_1 n_2 \sigma_{12} v_{12} \frac{4\pi r_c^3}{3} \propto \rho^2 r_c^3 / v_\sigma, \tag{4.3}$$

where r_c is the core radius, ρ is the core mass density, and v_σ is the core velocity dispersion. The last proportionality involves, among other things, substituting in the mass density and using the gravitational focusing dominated interaction cross section. Γ, in various incarnations, has been used for many years in analyses comparing with the observed numbers of X-ray sources in clusters. Recently, Pooley *et al.* (2003) have shown that the number of observed X-ray sources above 4×10^{30} erg/s in the 0.5–6 keV range displays a stronger correlation with cluster Γ than with any other cluster parameter.

The fact that the number of X-ray sources in clusters so strongly correlates with Γ is surprising given the number of approximations that go into deriving it. For example, the two interacting populations in eq. (4.3) are compact objects (neutron stars and white dwarfs) and stellar binaries, whose densities may differ dramatically. The use of ρ in place of n_i assumes a proportionality between the stellar density and the compact object density that is constant among *all* clusters. Furthermore, there is a factor of f_b, the binary fraction, that has been dropped, thereby implicitly assuming that it is constant among all clusters. There is also a factor of the binary semi-major axis that has been dropped from the variation, which may vary since the hard–soft binary boundary will vary among clusters. In addition, the recent dynamical history of the cluster may play an important role.

Improvements to the standard Γ analysis have recently been made. One of the difficulties present in the earlier analyses is that Γ strongly correlates with total cluster mass. This is unfortunate, since it makes it more difficult to distinguish dynamically-formed sources (whose number should correlate with Γ), from primordial sources (whose number should correlate with total cluster mass). An analysis of the CV populations in clusters using a normalized interaction rate, $\gamma \equiv \Gamma/M_{\mathrm{clus}}$, has shown that like LMXBs, CVs are predominantly formed via dynamical encounters (Pooley & Hut 2006).

5. Summary

In this proceedings article I have very briefly discussed the connection between binary stars and globular cluster dynamics, moving from basic physics to current research in the span of a few paragraphs. A thorough, easily readable, and fairly recent discussion of the material can be found in Heggie & Hut (2003).

The primary new material presented here is a comparison in phase space of the seemingly contradictory binary population evolution simulations of Ivanova *et al.* (2005) and Hurley, Aarseth, & Shara (2007), showing that they may in fact both represent the same underlying physics. In other words, new simulations must be performed to better compare the two very different methods.

Acknowledgements

For data, stimulating discussions, and general camaraderie, the author thanks Craig Heinke, Jarrod Hurley, Natasha Ivanova, Frederic Rasio, M. Atakan Gürkan, and Marc Freitag. JMF acknowledges support from Chandra theory grant TM6-7007X, as well as Chandra Postdoctoral Fellowship Award PF7-80047.

References

Clark, G. W. 1975, *ApJL*, 199, L143
Fregeau, J. M. & Rasio, F. A. 2007, *ApJ*, 658, 1047
Goodman, J. & Hut, P. 1989, *Nature*, 339, 40
Heggie, D. C. 1975, *MNRAS*, 173, 729
Heggie, D. & Hut, P. 2003, The Gravitational Million-Body Problem: A Multidisciplinary Approach to Star Cluster Dynamics, by Douglas Heggie and Piet Hut. Cambridge University Press, 2003, 372 pp.
Heggie, D. C., Trenti, M., & Hut, P. 2006, *MNRAS*, 368, 677
Hurley, J. R. 2007, *MNRAS*, 379, 93
Hurley, J. R., Aarseth, S. J., & Shara, M. M. 2007, *ApJ*, 665, 707
Hut, P., *et al.* 1992, *PASP*, 104, 981
Ivanova, N., Belczynski, K., Fregeau, J. M., & Rasio, F. A. 2005, *MNRAS*, 358, 572
Katz, J. I. 1975, *Nature*, 253, 698
Mackey, A. D., Wilkinson, M. I., Davies, M. B., & Gilmore, G. F. 2007, *MNRAS*, 379, L40
Merritt, D., Piatek, S., Portegies Zwart, S., & Hemsendorf, M. 2004, *ApJL*, 608, L25
Miocchi, P. 2007, *MNRAS*, 783
Pooley, D., *et al.* 2003, *ApJL*, 591, L131
Pooley, D., & Hut, P. 2006, *ApJL*, 646, L143
Trenti, M. 2006, ArXiv Astrophysics e-prints, arXiv:astro-ph/0612040
Trenti, M., Ardi, E., Mineshige, S., & Hut, P. 2007, *MNRAS*, 374, 857
Verbunt, F. & Hut, P. 1987, The Origin and Evolution of Neutron Stars, 125, 187

Dynamical Evolution of Dense Stellar Systems
Proceedings IAU Symposium No. 246, 2007
E. Vesperini, M. Giersz & A. Sills, eds.

Evolution of Compact Binary Populations in Globular Clusters: a Boltzmann Study

Sambaran Banerjee and Pranab Ghosh

Tata Institute of Fundamental Research, Mumbai 400005, India
email:sambaran@tifr.res.in, pranab@tifr.res.in

Abstract. We explore a Boltzmann scheme for studying the evolution of compact binary populations in globular clusters. We include processes of compact binary formation by tidal capture and exchange encounters, binary destruction by exchange and dissociation mechanisms and binary hardening by encounters, gravitational radiation and magnetic braking, as also the orbital evolution during mass transfer, following Roche lobe contact. From the evolution of compact-binary population, we investigate the dependence of the model number of X-ray binaries N_{XB} on two essential cluster properties, namely, the star-star and star-binary encounter-rate parameters Γ and γ (Verbunt parameters). We find that the values of N_{XB} and their expected scaling with the Verbunt parameters are in good agreement with results from recent X-ray observations of Galactic globular clusters.

Keywords. stellar dynamics, scattering, binaries: close, X-rays: binaries, globular clusters: general

1. Introduction

In this era of high-resolution X-ray observations with *Chandra* and *XMM-Newton*, studies of compact binaries in globular clusters (henceforth GC) have reached an unprecedented level of richness and detail, so that such observational studies can be compared with results obtained from theoretical modeling of binary dynamics in globular clusters (see Hut *et al.* (1992) for a review). In this study, we introduce a method for studying the evolution of compact binaries in GCs wherein we use a Boltzmann equation to trace the time evolution of such populations. We emphasize that the formalism we describe is not a Fokker-Planck description but the original Boltzmann one, which in principle is capable of handling both the *combined* small effects of a large number of frequent, weak, distant encounters *and* the *individual* large effects of a small number of rare, strong, close encounters.

The dynamical properties of a GC core with mean density ρ, velocity dispersion v_c and core radius r_c can be described by the two quantities $\Gamma \equiv (\rho^2/v_c)r_c^3$ and $\gamma \equiv \rho/v_c$ as pointed out by Verbunt (2002), which we shall refer to as *Verbunt parameters* hereafter. Γ is a measure the total two-body encounter rate within a GC core and γ measures the rate of encounter of a *single* binary with the surrounding stars (Verbunt 2002). If the GC core is assumed to be virialized ($v_c \propto \rho^{1/2}r_c$), the specification of these two quantities uniquely determines ρ, r_c and v_c.

A dynamically formed compact binary between a non-degenerate star and a compact star may in general be detached and becomes an X-ray binary (henceforth XB) after the non-degenerate companion fills its Roche-lobe through evolution of the binary. Evolution of such *pre X-ray binaries* (henceforth PXB) is governed by orbital angular momentum loss and stellar evolution of the companion. In this study, we focus on the evolution of compact binary population in both the XB and PXB phase using our Boltzmann scheme, with particular attention to (a) period distribution of XBs and (b) number of XBs, which we relate to observations.

2. Model of compact binary population evolution in globular clusters

We consider a binary population described by a number distribution $n(a, t)$, where a is the binary separation, interacting with a *unevolving uniform* background of stars representing the core of a globular cluster consisting equal mass stars of $m + f = 0.6 M_\odot$, a fraction k_X of compact stars of $m_X = 1.4 M_\odot$ and a fraction k_b of primordial binary fraction. $n(a, t)$ is defined such that $n(a, t) da$ is the total number of compact binaries in the core within the radius interval a to $a + da$.

2.1. A Boltzmann evolutionary scheme

We explore a Boltzmann evolutionary scheme, wherein the evolution of $n(a, t)$ is described by the *collisional Boltzmann equation* (Spitzer 1987):

$$\frac{\partial n}{\partial t} = R(a) - n D(a) - \frac{\partial n}{\partial a} f(a), \qquad (2.1)$$

where $R(a)$ is the total formation rate with the GC core, per unit a, of compact binaries with radius a, $D(a)$ is the destruction rate *per binary* of compact binaries of radius a and $f(a) \equiv da/dt$ is the total orbital evolution rate of the compact binaries (see Sec. 2.4). Eqn. (2.1) is the governing evolution equation a of compact binary population in an unevolving GC core (see Banerjee & Ghosh (2007) for derivation).

2.2. Compact binary formation processes

A compact binary can be formed by (a) tidal capture (tc) and (b) exchange encounter (ex1) as discussed below. If $r_{tc}(a)$ and $r_{ex1}(a)$ represents the rates of these processes respectively, then

$$R(a) = r_{tc}(a) + r_{ex1}(a), \qquad (2.2)$$

where a is the radius of the compact binary so formed.

In tidal capture formation, a compact star, during its close passage by an ordinary star, loses its kinetic energy by raising non-radial oscillations on the later by its tidal force so that they become bound, provided the first periastron separation r_p is shorter than a critical value (Fabian *et al.* 1975). After getting bound, the binary is usually highly eccentric, and circularizes within several periastron passages to the radius $a \approx 2 r_p$. We consider a simplified analytical approach involving the *impulsive approximation* (Spitzer 1987) which assumes that all the energy is deposited on the stellar surface instantly during the first periastron passage. It can be shown that (Banerjee & Ghosh 2007), for a Maxwellian velocity distribution, the rate function is nearly uniform in a for small a and falls off fairly sharply from about $a \approx 7 R_\odot$, as shown in Fig. 1 (left panel).

Compact binaries can also be formed by exchange encounter between a compact star and a primordial non-compact stellar binary. During a close encounter between the compact star and the stellar binary, the compact star being generally heavier, preferentially replaces one of the binary members to form a PXB. We use the well-known Heggie, Hut & McMillan (1996) exchange cross section to estimate the (Maxwellian averaged) ex1 exchange rate as a function of binary radius a. For primordial binaries, we take the widely-used radius distribution $f_b(a) \propto 1/a$ (*i.e.*, a uniform distribution in $\ln a$). In this case, the ex1 rate will be constant with a (see Banerjee & Ghosh (2007) for details) as shown in Fig. 1 (left panel).

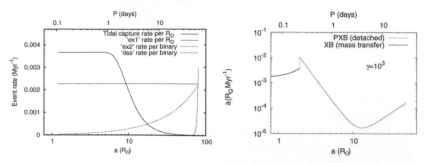

Figure 1. *Left:* A comparison of the dynamical rates. 'ex1' and 'ex2' rates have been multiplied by a factor of 50 and 60 respectively to make them visible in the same plot, while 'dss' rate have to be multiplied by a factor of $\sim 10^9$. *Right:* Hardening rate $\dot{a}$ of a compact binary as a function of the orbital radius a. For $a < 2R_\odot$ (solid line), mass transfer occurs representing the XB phase. Along abscissa, both orbital radius a and orbital period P scales are shown.

2.3. *Compact binary destruction processes*

A compact binary can be destroyed primarily by two processes, *viz.*(a) exchange encounter (ex2) and (b) dissociation (dss). Accordingly, the total destruction rate is:

$$D(a) = r_{ex2}(a) + r_{dss}(a) \tag{2.3}$$

In an exchange encounter (ex2) of a PXB with a compact star, the latter can replace the low-mass companion of the binary, forming a double compact-star binary. This, in effect, destroys the binary as an X-ray source as accretion is not possible in such a system, and it is essentially impossible for one of the compact stars in such a system to be exchanged again with an ordinary star in a subsequent exchange encounter, since $m_f = 0.6M_\odot$ is much lighter than $m_X = 1.4M_\odot$. As before, we estimate the Maxwellian averaged ex2 rate using the exchange cross section formula of Heggie, Hut & McMillan (1996), which is proportional to a as demonstrated in Fig. 1 (left panel).

As the compact binaries in the GC core are hard, they can only be dissociated by the small number of stars that constitute the high-velocity tail of the Maxwellian distribution. Thus dissociation constitutes a negligible channel for compact binary destruction (see Banerjee & Ghosh (2007) and references therein for details).

2.4. *Compact binary hardening processes*

As explained in detail in Banerjee & Ghosh (2006) (henceforth B06), the processes that harden binaries are of two types, *viz.*, (a) those which operate in isolated binaries, and are therefore always operational, and (b) those which operate only when the binary is inside a globular cluster. In the former category are the orbital angular momentum loss by *gravitational radiation* and *magnetic braking*, and in the latter category is that of *collisional hardening*. Collisional hardening refers to the process of preferential hardening due to repeated encounter by the background stars, according to Heggie's law (Heggie 1975). As discussed in detail in BG06, collisional hardening, which is proportional to a, dominates at larger orbital radii, while gravitational radiation and magnetic braking, which increase steeply with decreasing a, dominate at smaller orbital radii. It is these processes that harden a compact binary from its PXB phase, up to the point of Roche lobe contact ($a_L \approx 2R_\odot$), whence it turns on as an X-ray binary (XB) – either a CV or a LMXB, depending on the nature of the degenerate accretor. A typical form of hardening rate as a function of a is shown in Fig. 1 (right panel). At the minimum at $a \sim 14R_\odot$, gravitational radiation hardening ($\dot{J}_{\rm orb}/J_{\rm orb} \sim a^{-4}$) takes over from collisional hardening. Magnetic braking, having a steeper dependence on a (Verbunt-Zwaan scaling,

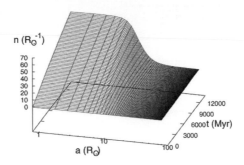

Figure 2. Three-dimensional surface $n(a,t)$ describing the model evolution of population-distribution function of compact binaries for GC parameters $\rho = 6.4 \times 10^4 \ M_\odot \ \mathrm{pc}^{-3}$, $r_c = 0.5 \ \mathrm{pc}$, $v_c = 11.6 \ \mathrm{km \ sec}^{-1}$ (roughly corresponding to 47 Tuc).

$\dot{J}_{\mathrm{orb}}/J_{\mathrm{orb}} \sim a^{-5}$), dominates at still shorter binary radius, during the mass transfer, which is shown with the thick line in Fig. 1 (right panel). It is important to note that during mass-transfer, the hardening rate remains nearly constant with a.

3. Results

A typical result from our computed evolution of the compact-binary distribution function $n(a,t)$ is shown in Fig. 2. The distribution function is seen to evolve such that the compact binary population grows predominantly, with a nearly uniform distribution function at shorter radii ($a < 10R_\odot$, say), and a sharp falloff longward.

The overall shape of the distribution function results from (a) the predominance of tidal capture formation rate for shorter binary radii (see Fig. 1), (b) inflow of binaries shortwards due to hardening and (c) higher ex2 destruction rate at larger radii. The uniformity in the distribution function for about $a < 10R_\odot$ mainly results from that in the tidal capture rate (Fig. 1).

The total number of X-ray binaries N_{XB} in a GC at any given time can be computed directly by integrating $n(a,t)$ over the range $a_{pm} \leqslant a \leqslant a_L$ representing mass transfer, where a_{pm} is the value of a corresponding to the period minimum $P \approx 80$ min, and a_L is the value of a at the first Roche lobe contact and onset of mass transfer. We determine N_{XB} for a representative evolutionary time of ~ 8 Gyr to study its dependence on the Verbunt parameters Γ and γ. Fig. 3 (left panel) shows the computed surface $N_{XB}(\gamma, \Gamma)$. As discussed in details in Banerjee & Ghosh (2007), the falloff of N_{XB} towards increasing γ from the fold is a signature of the increasing compact binary destruction rate with γ. Thus, the value of $\gamma(\approx 3 \times 10^3)$ corresponding to the fold seems to be a good estimate of the threshold γ above which the destruction processes dominate. However, the falloff towards decreasing γ is only an artifact of the assumption of virialization in evaluating the cluster parameters over the grid (Banerjee & Ghosh 2007). As can be seen in Fig. 3, most of the observed GC with significant numbers of XBs (filled squares), lie close to the fold of the $N_{XB}(\Gamma, \gamma)$ surface, indicating that the computed N_{XB} approximately follows the observed ones.

To further clarify these trends, we display in Fig. 3 (right panel) Γ/N_{XB} vs. γ, for a *particular value of* Γ. It has been shown in BG06 that the toy model of these authors leads to the scaling that Γ/N_{XB} is a function of γ *alone* (*i.e.*$\Gamma/N_{XB} \sim g(\gamma)$), which is a monotonically increasing function of γ. The close bunching of the $\Gamma/N_{XB} - \gamma$ curves, as

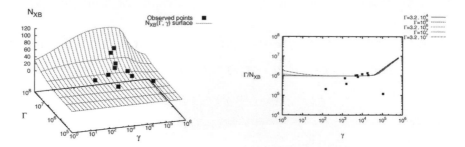

Figure 3. *Left:* Computed $N_{XB}(\Gamma, \gamma)$ surface. *Right:* Computed Γ/N_{XB} as a function of γ, showing scaling (see text). The computed curves for various values of Γ are closely bunched, as indicated. For both the figures, the overplotted filled squares are the positions of the Galactic globular clusters with significant numbers of X-ray sources from Pooley *et al.* (2003).

can be seen in Fig. 3, indicate that this scaling does carry over approximately to this more detailed study, thereby giving an indication of the basic ways in which the dynamical binary formation and destruction processes work. The above "universal" function $g(\gamma)$ of γ, except for a feature at low values of γ, is still a monotonically increasing one, reflecting the increasing strength of dynamical binary-destruction processes with increasing γ.

4. Concluding remarks

This work is an initiative of using Boltzmann equation to study the evolution of compact binaries in dense stellar systems. Not only we used simplified analytical models for dynamical formation and destruction of compact binaries, but also restricted ourselves only to systems like CVs and short period LMXBs, where the mass transfer occurs when the donor is in its main sequence, so that its stellar evolution is unimportant. To obtain a more realistic picture and consider other kinds of X-ray binaries, one should include more detailed treatment of tidal capture and consider the effects of stellar evolution in compact binary evolution. Such details can in principle be included in the Boltzmann scheme as the scheme itself is sufficiently generic. Such developments are in progress.

Acknowledgements

I am glad to thank the organizers of the "IAU Symposium 246" for having the opportunity to speak about this work during the event.

References

Banerjee, S. & Ghosh, P. 2006, *MNRAS*, 373, 1188. (BG06)
Banerjee, S. & Ghosh, P. 2007, accepted for publication in *ApJ*, arXiv:0708.1402.
Fabian, A. C., Pringle, J. E., & Rees, M. J. 1975, *MNRAS* (Short Communication), 172, 15P.
Heggie, D. 1975, *MNRAS*, 173, 729.
Heggie, D., Hut, P., & McMillan, S. L. W. 1996, *ApJ*, 467, 359.
Hut, P. *et. al.* 1992, *PASP*, 104, 981.
Pooley, D. *et al.* 2003, *ApJL*, 591, L131.
Portegies Zwart, S. F., Hut, P., McMillan, S. L. W., & Verbunt F. 1997, *A&A*, 328, 143.
Spitzer, L. Jr. 1987, *Dynamical Evolution of Globular Clusters*, Princeton Univ. Press.
Verbunt, F. 2002, in: *New horizons in globular cluster astronomy*, Proc. ASP conf. series.

Dynamical Evolution of Dense Stellar Systems
Proceedings IAU Symposium No. 246, 2007
E. Vesperini, M. Giersz, & A. Sills, eds.

© 2008 International Astronomical Union
doi:10.1017/S174392130801569X

Effects of Hardness of Primordial Binaries on Evolution of Star Clusters

A. Tanikawa[1] and T. Fukushige[2]

[1]Department of General System Studies, College of Arts and Sciences, University of Tokyo
email: tanikawa@ea.c.u-tokyo.ac.jp

[2]K&F Computing Research Co.
email: fukushig@kfcr.jp

Abstract. We performed N-body simulations of star clusters with primordial binaries using a new code, GORILLA. It is based on Makino and Aarseth (1992)'s integration scheme on GRAPE, and includes a special treatment for relatively isolated binaries. Using the new code, we investigated effects of hardness of primordial binaries on whole evolution of the clusters. We simulated seven $N = 16384$ equal-mass clusters containing 10% (in mass) primordial binaries whose binding energies are $1, 3, 10, 30, 100, 300$, and $1000kT$, respectively. Additionally, we also simulated a cluster without primordial binaries and that in which all binaries are replaced by stars with double mass, as references of soft and hard limits, respectively. We found that, in both soft ($\leqslant 3kT$) and hard ($\geqslant 1000kT$) limits, clusters experiences deep core collapse and shows gravothermal oscillations. On the other hands, in the intermediate hardness ($10 - 300kT$), the core collapses halt halfway due an energy releases of the primordial binaries.

Keywords. globular clusters: general, method: n-body simulations

1. Introduction

By means of previous numerical simulations, it has been clear that the presence of primordial binaries changes the dynamical evolution of star clusters. The clusters with many primordial binaries experience shallower core collapse than those without primordial binaries by one order of magnitude, and their gravothermal oscillations are delayed by several ten or hundreds half-mass relaxation time (Gao *et al.* 1991; Fregeau *et al.* 2003; Giersz & Spurzem 2003; Heggie *et al.* 2006). This is due to energy released by the primordial binaries that prevents the clusters from core collapsing.

Many previous works have numerically treated the dynamical evolution of star clusters with primordial binaries by means of N-body, Fokker-Planck, and Monte-Carlo simulations. However, they confined binding energy distribution of the primordial binaries to some ranges, and uniform distribution in logarithm of binding energy. There are no special reasons for these distributions. We systematically study the effect of the distributions on cluster evolutions. We set up the simplest binding energy distribution. All binding energies of primordial binaries are equal.

2. Method

We investigated the evolution of star clusters by means of N-body simulations. We used a new N-body simulation code, GORILLA, which we developed. GORILLA is neither based on NBODY (e.g. Aarseth 2003) nor kira (e.g Portegies Zwart *et al.* 2001), and applies for clusters of point-mass particles. It adopts the fourth-order Hermite integrator with individual timestep algorithm (Makino & Aarseth 1992, hereafter MA92),

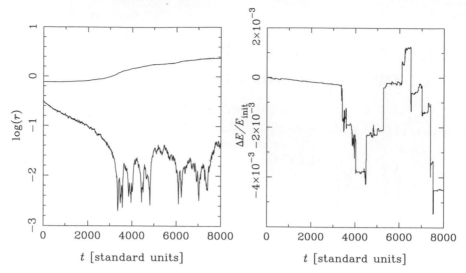

Figure 1. Time evolution of the core, r_c, and half-mass radius, r_h, (left) and the relative energy error from the initial time, $\Delta E/E_\mathrm{init}$, (right) of the cluster without primordial binaries.

and supports GRAPE-6/6A (Makino *et al.* 2003; Fukushige *et al.* 2005). Additionally, it is equipped with special treatments for close encounter between two or three particles.

In the special treatment, we approximate the internal motion of a binary as Kepler motion, by neglecting perturbations from other particles. The conditions of the binary are that the binding energy is more than $1kT$, where $3/2kT$ is the average kinetic energy of cluster stars, and the separation between the binary and the nearest particle are five times more than those between the binary components at the apocenter. If this binary and the nearest particle form a hierarchical triple system, in which the binary (hereafter "inner binary") and the nearest particle revolve around each other (hereafter "outer binary"), we approximate the internal motion between the center of mass of the inner binary, and the nearest particle as Kepler motion. The condition is that the binding energies of the inner and outer binary are more than $1kT$, the separation between the inner binary and the particle at the pericenter is five times more than that between the inner binary components at the apocenter, and the separation between the hierarchical triple system and the nearest particle is five times more than that between the inner binary and the particle at the apocenter.

We performed the test simulation of GORILLA. In the test simulation, we adopted a cluster with $N = 16384$ equal-mass and single stars, and the standard units. Fig. 1 shows the time evolution of the core and half-mass radii (left panel), and the relative energy error from the initial time (right panel) of the cluster. GORILLA can follow the evolution of the cluster from gravothermal core collapse to gravothermal oscillations. The energy error is less than 0.5% of the initial total energy.

By means of GORILLA, we performed N-body simulations of clusters with many primordial binaries. We also adopted the standard units. The initial distribution function is Plummer model. The number of particles is 16384, and they are equal-mass. These clusters contain 10 % primordial binaries in mass. The primordial binaries have equal binding energies in each model, and eccentricity with thermal distribution, $f(e) = 2e$, where e is eccentricity. We have 9 clusters with different binding energies, no binaries (in other words $0kT$), 1, 3, 10, 30, 100, 300, and $1000kT$, and double mass stars (in other

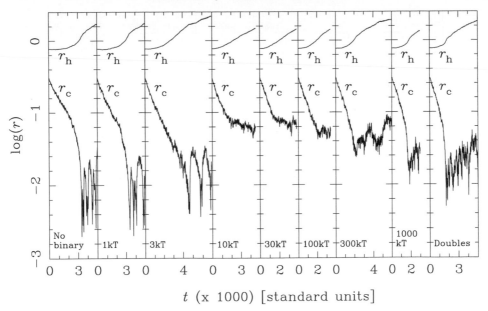

t (x 1000) [standard units]

Figure 2. Time evolution of the core, $r_{\rm c}$, and half-mass radii, $r_{\rm h}$, of the clusters without primordial binaries, with 1, 3, 10, 30, 100, 300, and $1000kT$ primordial binaries, and with double mass stars, from left panel to right panel.

words $\propto kT$). We call these clusters "No binary", $1 - 1000kT$, and "Doubles" clusters, respectively. The energy errors in all the simulations are $\sim 0.1 - 1$ %.

3. Results

Fig. 2 shows the time evolution of the core and half-mass radii of the clusters. The clusters with soft ($\leqslant 3kT$), and hard ($\geqslant 1000kT$) primordial binaries experienced deep core collapse, $r_{\rm c}/r_{\rm h} \sim 10^{-3} - 10^{-2}$, and gravothermal oscillations. The clusters with primordial binaries of intermediate hardness ($10 - 300kT$) experienced shallow core collapse, $r_{\rm c}/r_{\rm h} \sim 10^{-2} - 10^{-1}$, and steady core evolutions. We define core collapse time as the time when core collapses stop in the case of "No binary", 1, 3, and $1000kT$, and "doubles" clusters, and the time when core collapses slow down in the case of $10 - 300kT$ clusters. The core collapse time of $3kT$ is more delayed than that of $1kT$ cluster. $300kT$ cluster seems to show gravothermal oscillations. This cluster shows the intermediate behavior between 100 and $1000kT$.

The difference of the evolutions results from released energy by primordial binaries. Fig. 3 shows the total energies released by binaries from the initial times. In the case of $1kT$ cluster, the binaries release little energy before core collapse. In the case of $3kT$ cluster, the binaries release steadily halfway. This is due to the more delayed core collapse time than that of $1kT$ cluster. In the cases of $10 - 300kT$ clusters, the binaries release energy steadily from the beginning. Despite of large energy released in $1000kT$ cluster, the cluster experiences the deep core collapse. When a $1000kT$ binary encounters with other stars, this binary releases so much energy that they can escape from the cluster on crossing timescale. Therefore, the energy released by the $1000kT$ binaries is carried away by the binaries and stars involved in the encounters.

Fig. 4 show the time evolution of the number of binaries. Binaries of $1kT$ cluster decrease more rapidly than those in the cases of $10 - 1000kT$. This is due to little energy released by $1kT$ binaries.

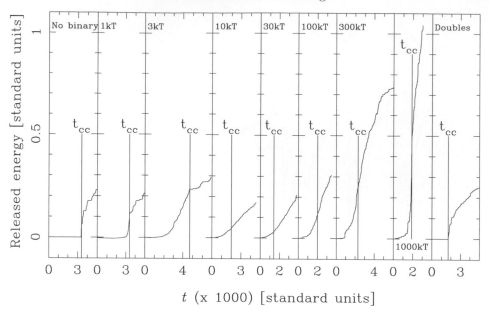

Figure 3. Time evolution of total energy released by binaries from initial time. Vertical lines show core collapse times, $t_{\rm cc}$.

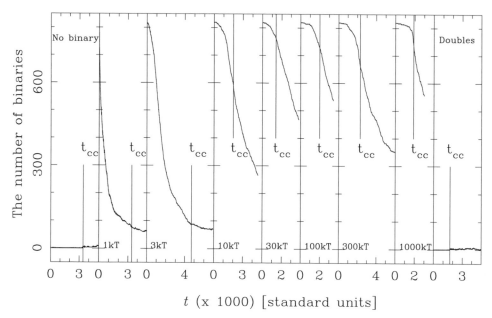

Figure 4. Time evolution of the number of binaries. Vertical lines show core collapse times, $t_{\rm cc}$.

4. Summary

We developed a new N-body simulation code, GORILLA. By means of GORILLA, we investigated the evolution of star clusters with 10 % primordial binaries in mass. We followed the evolution of nine sets of clusters that each have primordial binaries with equal binding energy.

The clusters experienced deep core collapses in soft ($\leqslant 3kT$) and hard ($\geqslant 1000kT$) limits, and shallow core collapses in the intermediate region of hardness ($10 - 300kT$).

The reasons are as follows. In $\leqslant 3kT$, little energy is released due to the rapid destruction. In $10-300kT$', energy is steadily released. In $\geqslant 1000kT$, released energy is carried away by the binary itself and stars involving the encounter. The fact shows that only primordial binaries with $10-300kT$ are effective sources of star clusters.

References

Aarseth, S. 2003, Gravitational N-body Simulations (Cambridge:Cambridge University Press)

CHW89] Cohn, H., Hut, P., & Wise, M. 1989, *ApJ*, 342, 814

Fregeau, J. M., Gurkan, M. A., Joshi, K. J., & Rasio, F. A. 2003, *ApJ*, 593, 772

Fukushige, T., Makino, J., & Kawai, A. 2005, *PASJ*, 57, 1009

Gao, B., Goodman, J., Cohn, H., & Murphy, B. 1991, *ApJ*, 370, 567

Giersz, M. & Spurzem, R. 2003, *MNRAS*, 343, 781

Heggie, D. C. 1975, *MNRAS*, 173, 729

Heggie, D. C. & Mathieu, R. D. 1986, in Lecture Notes in Physics Vol. 267, ed. P. Hut & S. McMillan (Berlin: Springer-Verlag), 233

Heggie, D. C., Trenti, M., & Hut, P. 2006, *ApJ*, 368, 677

Makino, J. & Aarseth, S. 1992, *PASJ*, 44, 141

Makino, J., Fukushige, T., Koga, M., & Narumi, K. 2003, *PASJ*, 55, 1163

Portegies Zwart, S. F., McMillan, S. L. W., Hut, P., & Makino, J. 2001 *MNRAS*, 321, 199

Vespereini, E. & Chernoff, D. F. 1994, *ApJ*, 431, 231

Dynamical Evolution of Dense Stellar Systems
Proceedings IAU Symposium No. 246, 2007
E. Vesperini, M. Giersz & A. Sills eds.

Dynamical Evolution of Star Clusters with Intermediate Mass Black Holes and Primordial Binaries

Michele Trenti[1]

[1] Space Telescope Science Institute,
3700 San Martin Dr., Baltimore, MD, 21218, U.S.
email: trenti@stsci.edu

Abstract. The evolution of a star cluster is strongly influenced by the presence of primordial binaries and of a central black hole, as dynamical interactions within the core prevents a deep core collapse under these conditions. We present the results from a large set of direct N-body simulations of star clusters that include an intermediate mass black hole, single and binary stars. We highlight the structural and dynamical differences for the various cases showing in particular that on a timescale of a few relaxation times the density profile of the star cluster does no longer depend on the details of the initial conditions but only on the efficiency of the energy generation due to gravitational encounters at the center of the system.

Keywords. stellar dynamics – globular clusters: general – methods: n-body simulations – binaries: general

1. Introduction

Dense stellar systems such as globular clusters are extremely fascinating astrophysical laboratories, where many physical processes – stellar dynamics, evolution and hydrodynamics – concur to shape their evolution. A realistic numerical modeling of these systems is outside the current hardware capabilities, even if the situation is likely to improve in the near future thanks to the GRAPE-DR (see Jun Makino's contribution in these proceedings). However globular clusters have appealing properties from the point of view of modeling, such as single old stellar population, quasi spherical symmetry, established dynamical equilibrium and quasi isotropy in the velocity dispersion tensor. Therefore even a simplified model has the hope to describe the key features in the evolution of these systems.

In this framework we decided to focus on the dynamical evolution driven by gravitational relaxation especially in presence of primordial binaries and intermediate mass black holes (IMBHs). Despite a growing evidence for the presence of a significant population of binaries in globular clusters and promising hints for IMBH signatures, these components are often neglected due to the dramatic increase in computational resources required in a simulation where the local dynamical timescale may be many orders of magnitude smaller than the global relaxation timescale (for example hard binaries have an orbital period of a few hours, while the half-mass relaxation time can be up to a few billion years). Thus we have started a simulation program aimed at systematically characterizing by means of direct N-body simulations the dynamical interplay of binaries and IMBHs in the center of dense stellar systems, initially using simple equal mass isolated models with up to 100% primordial binaries (see Heggie, Trenti & Hut 2006) and then progressively adding the influence of a tidal field (see Trenti, Heggie & Hut 2007), a central IMBH (see Trenti *et al.* 2007a, or primordial triples (see Trenti *et al.* 2007b).

To date we have completed about 1000 runs, exploring a wide parameter space for single stars and work is underway to extend the study to systems with a mass spectrum and stellar evolution.

One of the most important conclusions from the first part of this project and the one on which we focus these proceedings is that the collisional evolution of a star clusters does not depend on the details of the initial density profile one start with, but uniquely on the efficiency of the production of energy in the system. Therefore independently of the initial concentration, within a few to ten relaxation times star clusters with single stars only will reach a high concentration "core collapsed" configuration with a core to half mass radius $r_c/r_h \approx 0.01$. Instead in presence of a significant population of primordial binaries the core radius will be much larger $r_c/r_h \approx 0.05$ and in presence of a central IMBH the core may be even larger $r_c/r_h \gtrsim 0.1$. Given its insensitivity to the initial concentration, the core to half mass radius of well relaxed system does appear to be a powerful indicator of the dynamical properties of the system available even for globular clusters in nearby galaxies. Of course an accurate data-model comparison is required, as observations are based on light, while our core definition depends on mass (see Trenti, Heggie & Hut 2007 for a detailed discussion of different core radius definitions).

2. Cluster Evolution: The physical picture

Star clusters with up to a few hundred thousands stars have a two body relaxation time much shorter than the age of the universe, so that their internal long term evolution is driven by gravitational interactions.

When the system is made of single stars only, the classical mechanism of gravothermal instability is dominant (Spitzer 1987). Two-body encounters drive a heat flow from the central region of the star cluster, which behaves like a self gravitating system with negative specific heat, to the halo. This triggers a thermal "collapse" on the timescale of the heat flow process. The "collapse" phase lasts for several relaxation times until an efficient form of energy generation in the center can stop the process by providing an energy production rate equal to the energy loss rate by two body relaxation from the region of the core. This happens when one or more binaries are formed due to three body encounters in the dense central region, point at which a series of gravothermal oscillations with multiple core bounces can set in with an overall self-similar expansion of the cluster.

The presence of a more efficient central energy source, which may be in the form of primordial binaries, a central intermediate mass black hole, stellar mass black holes, mass loss from massive stars – possibly enhanced by physical collisions – changes significantly the standard gravothermal collapse picture. In all the cases where the energy release is higher than that due to dynamically formed pairs the equilibrium core size is larger due to a self-regulation mechanism. In fact, if the central energy production is too high the core is heated up and expands until a steady configuration is reached. This happens for example starting from a $W_0 = 11$ King model with 10% primordial binaries (see Trenti, Heggie & Hut 2007 and fig. 1). The presence of a tidal field does not influence much the self similar evolution of the density profile up to the half mass radius except in the latest stage of the life of the system, when more than 90% of the initial mass has been lost.

3. Results from direct N-body simulations

The initial conditions of our runs are tidally limited models with up to 19961 stars of equal mass. The initial density profile is that of a King model with different concentrations (W_0 parameter from 3,5,7 and 11). The runs with primordial binaries have a number

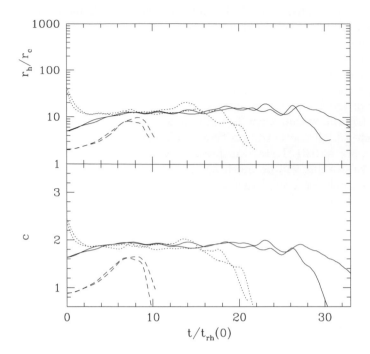

Figure 1. Evolution of the ratio of the half mass to core radius (upper panel) and of the concentration parameter $c = \log r_t/r_c$ for different King models with $f = 10\%, 20\%$. The solid line refers to simulations starting from $W_0 = 7$, the dotted line to $W_0 = 11$ and the dashed line to $W_0 = 3$; the number of particles used is 16384. The evolution quickly erases differences due to initial conditions, and during binary burning c and r_h/r_c evolve very similarly.

fraction of 10% and 20% (that is 20% and 40% in mass). In the runs with an IMBH, a central point particle has a mass of the order of a few percent of the total mass of the system (see Trenti *et al.* 2007a). The evolution of the system has been followed by direct integration until tidal dissolution of the system, using Aarseth's NBODY6 (Aarseth 2003), which has been slightly modified to guarantee and adequate treatment of the dynamical interactions around an IMBH (see Trenti *et al.* 2007a for full details).

The main result on the evolution of some representative models is shown in Fig. 1 for runs with primordial binaries and in Fig. 2 for runs with an IMBH. In Fig. 1 both the core to half mass ratio r_c/r_h and the tidal to core radius ratio $c = log(r_t/r_c)$ evolve within a few relaxation times toward a common value, which is then kept quasi constant until tidal effects becomes dominant in the final stage of the evolution. Interestingly simulations with 10% and 20% of primordial binaries have a very similar evolution of their density profile. This is because the efficiency of production through binary-single and binary-binary encounters saturates when mass segregation brings the binary population in the core above $\approx 50\%$, which happens starting with more than 10% of binaries (see also Vesperini & Chernoff 1994).

The evolution of r_c/r_h is similar to that observed for simulations with primordial binaries even when a IMBH is present at the center of the system (see Fig. 2). Again a universal profile is quickly reached and then the evolution proceeds in a self similar way. The BH however is a more efficient energy source, mainly due to an enhanced interaction rate between stars within its sphere of influence, so that a $W_0 = 7$ model initially expands its core rather than contracting it.

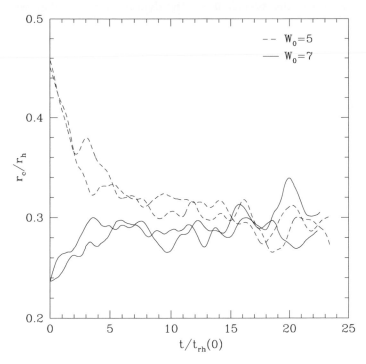

Figure 2. Core to half mass radius ratio for a series of simulations with $N = 8192$, $m_{BH} = 0.014$ and 10% primordial binaries. Different initial conditions (King profiles with $W_0 = 5, 7$) converge toward a common value for r_c/r_h.

4. Conclusion

In this proceeding contribution we highlight one important result from a project aimed at studying the dynamical evolution of star clusters with primordial binaries and intermediate mass black holes by means of direct N-body simulations (see Heggie, Trenti & Hut 2006, Trenti, Heggie & Hut 2007, Trenti *et al.* 2007a, Trenti *et al.* 2007b). We show that the density profile of a star clusters evolves toward a self-similar universal configuration on the two body relaxation time scale, with a concentration (measured for example in terms of the core to half mass radius ratio) which is sensitive only to the efficiency of the energy production in the system. Clusters with single stars only evolve toward a "core collapsed" state, as they need to reach very high central density to dynamically produce a few binaries that can eventually halt the gravothermal collapse and lead to a core bounce. Clusters with a significant population of primordial binaries and/or with a central IMBH never reach "core collapse" but rather settles in a configuration with $r_c/r_h \gtrsim 0.05$. With a detailed data-model comparison, the use of r_c/r_h could provide unique insight onto the dynamical status of observed old globular clusters not only in the Milky Way but also in nearby galaxies.

Acknowledgement

I would like to thank all my collaborators, Douglas Heggie, Piet Hut, Scott Ransom and Shin Mineshige not only for their fundamental contributions in our common projects but especially for sharing with me the interest in the fascinating world of globular clusters. Special thanks also to Enrico Vesperini for all his work in organizing the Symposium and for the many fruitful discussions about the dynamics of globulars we had together in Capri.

References

Aarseth, S., 2003, Gravitational N-body Simulations. Cambridge University Press

Heggie, D. C., Trenti, M., & Hut, P., 2006, *MNRAS*, 368, 677

Spitzer, L., Jr, 1987, Dynamical evolution of globular clusters, (Princeton University Press: Princeton)

Trenti, M., Heggie, D. C., & Hut, P., 2007, *MNRAS*, 374, 344

Trenti, M., Ardi, E., Mineshige, S., & Hut, P. 2007, *MNRAS*, 374, 857

Trenti, M., Ransom, S., Heggie, D. C., & Hut, P. 2007, *MNRAS*, submitted, astro-ph/0705.4223

Vesperini, E. & Chernoff, D. F., 1994, *ApJ*, 431, 231

Dynamical Evolution of Dense Stellar Systems
Proceedings IAU Symposium No. 246, 2007
E. Vesperini, M. Giersz & A. Sills, eds.

© 2008 International Astronomical Union
doi:10.1017/S1743921308015718

The Influence of Binary Stars on Post-Collapse Evolution

Rosemary Apple

School of Mathematics, University of Edinburgh, UK
email: r.apple@sms.ed.ac.uk

Abstract. The results in the N-body simulations in Giersz & Heggie (1996) show that although the masses segregate as expected during core collapse, after core collapse there is self-similar evolution with very little further evidence of mass segregation even though the system has not reached equipartition. Binary stars halt core collapse and it is possible that they also halt the tendency toward equipartition. To investigate this problem, we construct two models. One model is a two-component model which assumes that binary stars form in the region dominated by heavy stars. The other model is a single mass model which assumes that binary stars form only in the region of the core. In both models, when the binary heating term is included, we find the post-collapse evolution to be self-similar. The aim of our work is to combine these two models to form a two-component model which assumes that binary formation only occurs in the core.

Keywords. binaries: general, stellar dynamics, methods: analytical

1. Two-Component Model

In our first model, we consider a two-component system with individual masses $m_2 > m_1$. The total mass of the heavy stars and the total mass of the light stars are M_2 and M_1 respectively. We build our two-component model from equation (1) in Spitzer (1969). We assume that the heavy stars are uniformly distributed in the sphere characterized by the half-mass radius of the heavy stars, r_2. Similar assumptions are made with the light stars, where the half-mass radius is r_1. We use the binary heating term derived in Heggie & Hut (2003) and assume that binary stars form throughout the region of heavy stars. From these assumptions, we derive the equations for our dynamical system,

$$\dot{r}_1(r_1, r_2) = f(r_1, r_2) \frac{\phi(\frac{r_2}{r_1})}{t_{eq}(r_1, r_2)}, \tag{1.1}$$

$$\dot{r}_2(r_1, r_2) = \frac{g(r_1, r_2) f(r_1, r_2) \phi(\frac{r_2}{r_1})}{t_{eq}(r_1, r_2)} + \frac{9.688 G^{1/2} M_2^2 m_2^3 r_1^{27/2} r_2^{-1/2}}{\left(M_2 r_1^3 + M_1 r_2^3\right)^{7/2} \left(M_2 r_1^3 + 2M_1 r_2^3\right)}, \tag{1.2}$$

where, $f(r_1, r_2) = [r_1^4 r_2 M_1^2 M_2 (M_1 + M_2)]^{-1}$, $g(r_1, r_2) = \dfrac{M_1 r_2^2 (3 M_2 r_2^2 - M_1 r_1^2 - M_2 r_1^2)}{M_2 r_1 (M_2 r_1^3 + 2 M_1 r_2^3)}$,

$$\phi\left(\frac{r_2}{r_1}\right) = \frac{M_2}{M_1} \frac{m_2}{m_1} + \frac{m_2}{m_1} \left(\frac{r_2}{r_1}\right)^3 - \left(1 + \frac{M_2}{M_1}\right) \frac{r_2}{r_1}, \text{ and}$$

$$t_{eq}(r_1, r_2) = \frac{\sqrt{\pi} \left((M_1 + M_2) r_1^2 r_2 + M_2 r_1^3 + M_1 r_2^3\right)^{3/2}}{6\sqrt{6} M_1 m_2 \log\left(M_1/m_1\right) \left(r_1 r_2\right)^{3/2}}.$$

The second term on the right in Eqn. (1.2) is due to binary heating.

In Fig. 1(i) & (ii), we show the phase portraits for r_2 vs. r_1. We use an example in which the system is Spitzer stable, where $m_2/m_1 = 2$ and $M_2/M_1 = 0.1$. In the case where binary heating is excluded, we find two lines of equilibria (depicted by the dashed

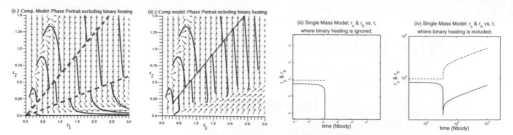

Figure 1. From left to right: Phase portraits for the two-component model (i) ignoring binary heating and (ii) including binary heating and the graphs of r_h (dashed) and r_c (solid) vs. t in the single mass model (iii) ignoring binary heating and (iv) including binary heating.

lines in Fig. 1(i)). The upper line is stable and the lower line is unstable. When we include binary stars, the system reaches what was the stable line of equilibria in the case without binary stars. When this line is reached, the system is in quasi-equilibrium.

2. Single Mass Model

The second model is a single mass model which assumes that binary stars only form in the core of the cluster. To model this system, we use approximations based on Lynden-Bell & Eggleton (1980) and again the binary heating term found in Heggie & Hut (2003). From this, we get the equations for our model

$$\dot{r}_h = \frac{765}{64\pi^2} G^{\frac{1}{2}} N^{-\frac{5}{2}} m^{\frac{1}{2}} r_c^{-4+\frac{1}{2}\alpha} r_h^{\frac{7}{2}-\frac{1}{2}\alpha}, \tag{2.1}$$

$$\dot{r}_c = \frac{1}{5-2\alpha} \left[-\frac{r_c}{t_{rc}} + \frac{765}{16\pi^2} G^{\frac{1}{2}} N^{-\frac{5}{2}} m^{\frac{1}{2}} r_c^{-8+\frac{5}{2}\alpha} r_h^{\frac{15}{2}-\frac{5}{2}\alpha} + (6-2\alpha)\frac{r_c}{r_h}\dot{r}_h \right], \tag{2.2}$$

where, $t_{rc} = G^{\frac{3}{2}} N^{\frac{1}{2}} m^{-\frac{1}{2}} r_c^{3-\frac{1}{2}a} r_h^{\frac{-3}{2}+\frac{1}{2}a} \left(3\log\left(.5N\left(r_c/r_h \right)^{3-a} \right) \right)^{-1}$.

In this model, all but the first term in equation (2.2) are caused by binary heating.

The evolution of the half-mass radius and the core radius over time is shown in Fig. 1(iii) & (iv). As expected, the core collapses when the binary heating term is ignored. When binary heating is included, the core radius collapses to a minimum value and then the post-collapse evolution is self-similar.

3. Further Work

We intend to combine these two models to create a new model which will include mass segregation and will assume that binary stars exist only in the core. From this model we can determine whether the influence of binary stars causes the post-collapse evolution to be self-similar.

References

Giersz, M. & Heggie, D. C. 1996, *MNRAS* 279, 1037

Heggie, D. C. & Hut, P. 2003, *Gravitational Million Body Problem* (Cambridge University Press, Cambridge)

Lynden-Bell, D. & Eggleton, P. P. 1980, *MNRAS* 483, 498

Spitzer, L. Jr. 1969, *ApJ* 158, 139

Dynamical Evolution of Dense Stellar Systems
Proceedings IAU Symposium No. 246, 2007
E. Vesperini, M. Giersz & A. Sills, eds.

© 2008 International Astronomical Union
doi:10.1017/S174392130801572X

The Binary Fraction of NGC 6397

D. Saul Davis[1], Harvey B. Richer[1], Jay Anderson[2] and James Brewer[1]

[1] Department of Physics & Astronomy, University of British Columbia,
6224 Agricultural Road, Vancouver, BC, V6T 1Z1, Canada
email: sdavis@astro.ubc.ca

[2] Department of Physics & Astronomy, Rice University,
6100 Main St., Huston, TX, 77005, USA

Abstract. The binary fraction, η, of a globular cluster (GC) is a key parameter in determining its dynamical evolution, as well as its content of rare stars, such as cataclysmic variables and blue stragglers. The precise value of η for a GC was historically difficult to constrain due to an inability to obtain reliable photometry for faint objects in dense stellar fields. However, today, the HST allows us to image the main sequence of the nearest GCs to their terminations. Using HST observations we constrain η for NGC 6397. While the necessary computing power is now available to realistically simulate entire GCs, large discrepancies in the assumed primordial binary fraction, η_p, of GCs still exist. Estimates range from 5% (Hurley *et al.* 2007) to 100% (Ivanova *et al.* 2005). The N-body models of Hurley *et al.* (2007) suggest that η beyond the half-mass radius remains close to η_p, while cluster evolution can increase the value in the core. We find η for NGC 6397 is $15.2 \pm 0.8\%$ in a field centered on the core, and $1.1 \pm 0.3\%$ in a field beyond the half mass radius. These findings suggests $\eta_p \sim 1\%$.

Keywords. binaries: general, globular clusters: individual (NGC 6397)

Observations: The data used in this project were obtained in 2005 by Richer *et al.* when 126 orbits were dedicated to imaging one ACS WFC field 5' South-East of the cluster core. WFPC2 parallels were obtained in the cluster core (Richer *et al.* 2006). The CMDs of the two fields are shown in Figs. 1C and 1D. Archival images (covering 60% of the ACS field and 80% of the WFPC field) were used to determine proper motions, shown in Figs. 1A and 1B.

The model: From isochrones (Dotter *et al.* 2006), and an empirically derived mass function and model for the photometric error, a model of the single-star sequence, F_s, was constructed (Fig. 1G). The only further assumption required to simulate the binary sequence, F_b (shown in Fig. 1H), is the distribution of mass ratios, q. We experimented with two distributions of q's: 1) a flat distribution, and 2) 55% of the binaries with a flat distribution, and 45% of the binaries with a twin (q⩾0.95) secondary (as suggested by Pinsonneault & Stanek 2006). The derived binary fraction was relatively insensitive to the assumed distribution of q. For a given η, F_s and F_b are added to make a cluster model in the following manner: $F(\eta) = (1 - \eta)F_s + \eta F_b$.

Results: For a given magnitude, m, and colour, c, the number of observed stars, n, and the value of the model, F, is determined. The probability, P, of observing n stars, is given by a Poisson distribution: $P(c, m|n) = F(c, m)^n e^{F(c,m)}/n!$. Following Romani & Weinberg (1991), the likelihood of the data given the model is the product of every bin: $L(\eta) = \prod_{i=1}^{N} P(c_i, m_i|n)$. The likelihood is calculated for all η, and the most probable, η_0, is determined. The allowed region (1-σ) is determined where $log\,[L(\eta)/L(\eta_0)] \geqslant -0.5$. Shown in Figs. 1E and 1F are the probability distributions for η in both the ACS field and the WFPC2 field. The allowed region is shown with hatching. The derived η in both fields is low: $15.2 \pm 0.8\%$ in the WFPC2 field and $1.1 \pm 0.3\%$ in the ACS field.

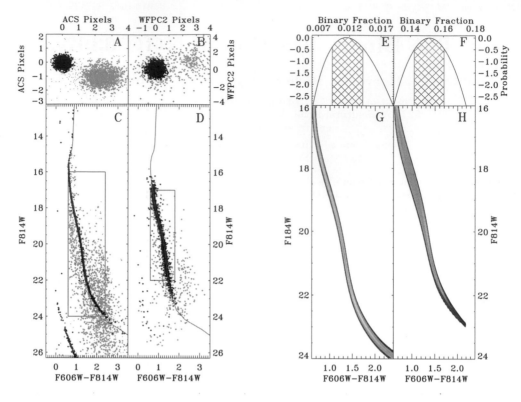

Figure 1. Panels A and B: The proper motions for the ACS and WFPC2 fields respectively, with cluster stars in black and field stars in gray. **Panels C and D:** The CMDs for the ACS and WFPC2 fields respectively. Again, stars with cluster proper motion are in black, while the field is in gray. Overlaid is the isochrone by Dotter *et al.* (2006). The color-magnitude space used for this study is shown with a box. **Panels E and F:** The derived binary fraction for the ACS and WFPC2 fields respectively. The 1-σ allowed region is designated with hatching. **Panels G and H:** The simulated probability for $\eta = 0$ and $\eta = 1$ for the ACS field respectively.

Conclusion: This work presents a very precise determination of η for a GC. Hurley *et al.* (2007) has shown that as a cluster evolves dynamically, η increases in the core while outside the half-mass radius it remains roughly constant. In this context, our results suggest that $\eta_{\mathrm{p}} \sim 1\text{--}2\%$ – significantly lower than typically assumed for N-body models.

References

Dotter, A. L., Chaboyer, B., Baron, E., Ferguson, J. W., Jevremovic, D., Lee, H., & Worthey, G. 2006, in Bulletin of the American Astronomical Society, Vol. 38, 958

Hurley, J. R., Aarseth, S. J., & Shara, M. M. 2007, *ApJ*, 665, 707

Ivanova, N., Belczynski, K., Fregeau, J. M., & Rasio, F. A. 2005, *MNRAS*, 358, 572

Pinsonneault, M. H. & Stanek, K. Z. 2006, *ApJL*, 639, L67

Richer, H. B., Anderson, J., Brewer, J., Davis, S., Fahlman, G. G., Hansen, B. M. S., Hurley, J., Kalirai, J. S., King, I. R., Reitzel, D., Rich, R. M., Shara, M. M., & Stetson, P. B. 2006, *Science*, 313, 936

Romani, R. W. & Weinberg, M. D. 1991, *ApJ*, 372, 487

Dynamical Evolution of Dense Stellar Systems
Proceedings IAU Symposium No. 246, 2007
E. Vesperini, M. Giersz & A. Sills, eds.

A Post-Newtonian Treatment of Relativistic Compact Object Binaries in Star Clusters

J. M. B. Downing and R. Spurzem

Astronomisches Rechen-Institut, Zentrum für Astronomie, Universität Heidelberg
email: downin@ari.uni-heidelberg.de

Abstract. Stellar mass compact object binaries are promising sources of gravitational radiation for the current generation of ground-based detectors, VIRGO and LIGO. Accurate templates for gravitational waveforms are needed in order to extract an event from the VIRGO/LIGO data stream. In the case of relativistic, compact object binaries accurate orbital parameters are necessary in order to produce such templates. Binary systems are affected by their stellar environment and thus the parameters of the binary population of a dense star cluster will be different from those of the field population. We propose to investigate the parameters of relativistic binary populations in dense star clusters using direct N-body simulations with a Post-Newtonian treatment of general relativity for the close binaries.

Keywords. stellar dynamics, relativity, gravitational waves, methods: n-body simulations, binaries: close

1. Introduction

Compact object binaries in the relativistic regime are usually assumed to be circularised by the emission of gravitational radiation and this assumption goes into the production of gravitational wave templates and estimates of relativistic merger event rates. It is known, however, that orbital parameters of compact object binaries can have significant effects on the gravitational radiation. In particular an eccentricity of $0.8 - 0.9$ can produce an enhancement of $10^2 - 10^3$ over a circular orbit with the same energy (Peters & Mathews 1963). Since this effect occurs primarily at periastron it is possible that eccentric binaries are capable of emitting significant gravitational radiation at larger separations than circularised binaries. Hard binaries with high eccentricities exist in dense star clusters and we propose to examine the relativistic segment of this population.

2. The NBODY6++ Code and the PN Formalism

To perform our direct N-body integration we use the code NBODY6++. This code is part of the series of NBODY codes (Aarseth 1999) modified for operation on parallel machines (Spurzem 1999). The code features a fourth-order Hermite predictor-corrector scheme, parallel operation, individual and block time stepping, an Ahmad-Cohen neighbour scheme (Ahmad & Cohen 1973), and Kustaanheimo and Stiefel (KS) regularisation for binaries (Kustaanheimo & Stiefel 1965). The KS regularisation rotates the Kepler problem through quaternion space into a simple harmonic oscillator. Other forces, such as Post-Newtonian corrections for relativistic motion, can then be applied as perturbations to the harmonic oscillator.

3. The Post-Newtonian Formalism

The Post-Newtonian formalism corrects the classical Newtonian acceleration with a series expansion in orders of v/c (Blanchet 2006):

$$G^{\alpha\beta}[g, \partial g, \partial^2 g] = \frac{8\pi G}{c^4} T^{\alpha\beta}[g] \quad \Longrightarrow \quad \vec{a} = \vec{a_N} + \frac{v}{c}\vec{a_1} + \frac{v^2}{c^2}\vec{a_2} + \frac{v^3}{c^3}\vec{a_3} + \frac{v^4}{c^4}\vec{a_4} + \frac{v^5}{c^5}\vec{a_5} + \dots$$

The first- and third-order accelerations vanish and the remaining terms are referred to by their *relative* order. The effect of the Post-Newtonian terms can be enhanced in a controlled manner by decreasing the velocity of light in the expansion with respect to the characteristic velocity in the N-body system. This allows us to produce more events per unit time for better statistics.

4. Current Work and Outlook

The Post-Newtonian expansion up to PN2.5 was previously incorporated into NBODY6 (single processor) by Kupi, Amaro-Seoane & Spurzem (2006). We have now incorporated the same expansion into NBODY6++ (multiple processor version) and are currently testing the code. We have some initial physical results such as number of mergers as a function of time (Fig. 1) but more work is needed before these can be meaningfully interpreted. One challenge is that there are no terms for the Post-Newtonian energies that are

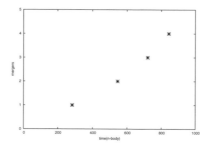

Figure 1. Number of relativistic merges per unit time for an 8k particle simulation with no primordial binaries. c is chosen so the the relativistic terms are enhanced by a factor of ~ 300.

conserved at each order, causing difficulty in monitoring the overall energy conservation of the code. Once this problem is solved we intend to start simulations with realistic numbers of primordial binaries using the Kyoto II initial conditions in order to produce data that can be used for the production of gravitational wave templates.

Acknowledgements

J.M.B. Downing would like to acknowledge the IMPRS-Astronomy, Heidelberg for funding his Ph.D. Thesis.

References

Aarseth, S. J. 1999, *PASP* 111, 1333
Ahmad, A. & Cohen, L. 1973, *J. Comput. Phys.* 12, 389
Blanchet, L. 2006, *Living Rev. Relativity* 9, 4,
Kupi, G., Amaro-Seoane, P., & Spurzem, R. 2006, *MNRAS* 371, L45
Kustaanheimo, P. & Stiefel, E. 1965, *J. Reine Angew. Math.* 218, 204
Peters, P. C. & Mathews, J. 1977, *Phys. Rev.* 131 no. 1, 435
Spurzem, F. 1999, *J. Comp. Appl. Maths.* 109, 407

Dynamical Evolution of Dense Stellar Systems
Proceedings IAU Symposium No. 246, 2007
E. Vesperini, M. Giersz & A. Sills, eds.

The Formation of Contact and Very Close Binaries

Peter P. Eggleton[1,2] and Ludmila Kisseleva-Eggleton[3]

[1]Institute of Geophysics and Planetary Physics
[2]Physics and Allied Technologies Division
Lawrence Livermore National Laboratory, 7000 East Ave,
Livermore, CA94551, USA
[3]San Francisco State University
email: ppe@igpp.ucllnl.org, peterluda1@juno.com

Abstract. We explore the possibility that all close binaries, i.e. those with periods < 3 days, including contact binaries, are produced from initially wider binaries by the action of a triple companion through the medium of Kozai Cycles with Tidal Friction (KCTF).

Keywords. stars: evolution, stars: binaries: general

1. Introduction

Pribulla & Rucinski (2006) have noted that in a reasonably complete sample of 88 northern contact binaries, 52 show evidence of a third body. Given the difficulty of recognising the presence of a third body except in favorable circumstances, this argues for the likelihood that *all* contact binaries are in triples, and hence that 'triplicity' is *necessary* for the formation of a contact binary.

Tokovinin *et al.* (2007) found that in a sample of spectroscopic binaries with $P < 30$ d, among the 41 with $P < 3$ d, 32 were triple; and making allowance, by a maximum-likelihood procedure, for incompleteness they concluded that the fraction of triples must be $\sim 96\%$. For spectroscopic binaries with $P > 12$ d the figure was much lower (34%).

Therefore there is a good case for the hypothesis that very close binaries form as a consequence of the presence of a third body. The most likely mechanism to do this is the combination of Kozai cycles with tidal friction (KCTF) – see Eggleton & Kisseleva-Eggleton (2006), Fabrycky & Tremaine (2007), and Kiseleva, Eggleton & Mikkola (1997). For contact binaries, the extra effects of magnetic braking and tidal friction will have to be included.

Fig. 1 shows an aspect of the behaviour of a reasonably close inner pair ($1.0 + 0.6\,M_\odot$, period $P_{\rm in}$) subjected to Kozai cycles by the presence of a third body in an orbit ($1.6 + 0.9\,M_\odot$, period $P_{\rm out}$) inclined at angle η to the inner orbit ($\cos\eta = 0.1$). Left of the boundary passing near the middle of the plot, and right of the boundary near the bottom right, there are no Kozai cycles. The symbols are explained in the legend. Plusses are systems where KCTF shortens the period to a few days in < 3 Gyr, but not so rapidly that tidal energy release rivals stellar luminosity (ie. the Kelvin-Helmhotz timescale).

We believe the systems marked by plusses are computed reasonably reliably, and all should arrive at periods of $\sim 2-10$ d. We expect 10% of all triples to be inclined as much (or more) as the systems shown; but a lower inclination like $\cos\eta = 0.3$ is much the same, with a smaller patch of circles. The circled systems need more detailed evolution, because the two close components will be swollen by the tidal-friction energy release,

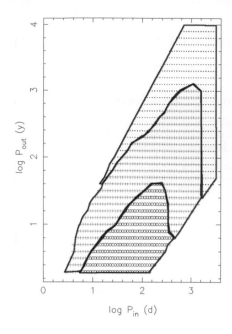

Figure 1. A sampling of systems with a range of both inner period (across) and outer period (up). Dots indicate Kozai cycling that is not significantly affected by tidal friction on a timescale $\lesssim 3$ Gyr, and circles indicate systems so rapidly modified by tidal friction that the luminosites of the two close components should be affected. Plusses are systems affected on times between the thermal timescale and 3 Gyr.

their quadrupole moments will be increased, and it is not clear whether this can stabilise or destabilise the KCTF evolution.

We expect to use our stellar-evolution code to follow some cases illustrated with circles. Our treatment of tidal friction, in the equilibrium-tide approximation, gives a specific form for the energy-generation rate as a function of radius in the star. We suspect that internal heating may affect the rate at which the inner system approaches short period and circularity, but may not have much effect on the parameters of the final system without affecting the stellar luminosities strongly, i.e the timescale of KCTF evolution is longer than the Kelvin-Helmholtz time, while shorter then ~ 3 Gyr.

Acknowledgements

This work was performed under the auspices of the U.S. Department of Energy by Lawrence Livermore National Laboratory under Contract DE-AC52-07NA27344.

References

Eggleton, P. P. & Kisseleva-Eggleton, L. 2006, *Ap&SS*, 304, 75
Fabrycky, D. & Tremaine, S. 2007, *ApJ* in press
Kiseleva, L. G., Eggleton, P. P., & Mikkola, S. 1998, *MNRAS*, 300, 292 (1998)
Pribulla, T. & Rucinski, S. M. 2006, *AJ*, 131, 2986
Tokovinin, A. A., Thomas, S., Sterzik, M., & Udry, S. 2006, *A&A*, 450, 681

Dynamical Evolution of Dense Stellar Systems
Proceedings IAU Symposium No. 246, 2007
E. Vesperini, M. Giersz & A. Sills, eds.

Binaries and the Dynamical Mass of Star Clusters

M. B. N. Kouwenhoven[1] and R. de Grijs[1,2]

[1] Department of Physics and Astronomy, University of Sheffield,
Hicks Building, Hounsfield Road, Sheffield S3 7RH, United Kingdom
email: t.kouwenhoven@sheffield.ac.uk

[2] National Astronomical Observatories, Chinese Academy of Sciences,
20A Datun Road, Chaoyang District, Beijing 100012, P.R. China
email: r.degrijs@sheffield.ac.uk

Abstract. The total mass of a distant star cluster is often derived from the virial theorem, using line-of-sight velocity dispersion measurements and half-light radii, under the implicit assumption that all stars are single (although it is *known* that most stars form part of binary systems). The components of binary stars exhibit orbital motion, which increases the measured velocity dispersion, resulting in a dynamical mass overestimation. In these proceedings we quantify the effect of neglecting the binary population on the derivation of the dynamical mass of a star cluster. We find that the presence of binaries plays an important role for clusters with total mass $M_{\rm cl} \lesssim 10^5$ $M_\odot$; the dynamical mass can be significantly overestimated (by a factor of two or more). For the more massive clusters, with $M_{\rm cl} \gtrsim 10^5$ $M_\odot$, binaries do not affect the dynamical mass estimation significantly, provided that the cluster is significantly compact (half-mass radius $\lesssim 5$ pc).

Keywords. star clusters, binaries: general, methods: numerical

1. Introduction

Young star clusters, with typical masses of $M_{\rm cl} = 10^{3-6}$ $M_\odot$, indicate recent or ongoing violent star formation, and are often triggered by mergers and close encounters between galaxies. Only a fraction of these young massive star clusters evolve into old globular clusters, while the majority ($60 - 90\%$) will dissolve into the field star population within about 30 Myr (e.g., de Grijs & Parmentier 2007). In order to understand the formation and fate of these clusters, it is important to study these in detail, and obtain good estimates of the mass, stellar content, dynamics, and binary population. The dynamical mass for a cluster in virial equilibrium, consisting of single, equal-mass stars, is given by:

$$M_{\rm dyn} = \eta \frac{R_{\rm hm} \sigma_{\rm los}^2}{G} \tag{1.1}$$

(Spitzer 1987), where $R_{\rm hm}$ is the (projected) half-mass radius, $\sigma_{\rm los}$ the measured line-of-sight velocity dispersion, and $\eta \approx 9.75$. For unresolved clusters, $\sigma_{\rm los}$ is usually derived from spectral-line analysis, neglecting the presence of binaries. However, observations have shown that the majority of stars form in binary or multiple systems (e.g., Duquennoy & Mayor 1991; Kouwenhoven *et al.* 2005, 2007; Kobulnicky, Freyer & Kiminki 2007). When binaries are present, $\sigma_{\rm los}$ does not only include the motion of the binaries (i.e., their centre-of-mass) in the cluster potential, but additionally the velocity component of the orbital motion. This results in an overestimation of the velocity dispersion, and hence of $M_{\rm dyn}$.

2. The effect of binaries on the dynamical mass M_{dyn} of a star cluster

The systematic error introduced by the single-star assumption depends on the properties of the star cluster and of the binary population. This effect is most easily seen in the extreme case when σ_{los} is dominated by the orbital motion of binaries (which is the case for most Galactic OB associations). In this binary-dominated case, the measured σ_{los} is independent of R_{hm} and the true cluster mass, M_{cl}. The inferred M_{dyn} from equation (1) is then proportional to R_{hm}. The dynamical mass overestimation is therefore $M_{\mathrm{dyn}}/M_{\mathrm{cl}} \propto R_{\mathrm{hm}}/M_{\mathrm{cl}}$. Sparse clusters are thus most sensitive to binaries; the systematic error in M_{dyn} could be as influential as that from the assumption of virial equilibrium. The cluster structure, stellar mass function and the presence of mass segregation also affect M_{dyn}, but these are of much less importance (Kouwenhoven & de Grijs, *in preparation*). Whether or not the binaries affect M_{dyn} additionally depends on the properties of the binary population. Obviously, the most important parameters are the binary fraction (which determines the relative weight between singles and binaries when measuring σ_{los}) and the semi-major axis, a, or period distribution (tight binaries have a larger orbital velocity component). As in a binary orbit $v_{\mathrm{orb}} \propto a^{-1/2}$, we have for the binary-dominated case $M_{\mathrm{dyn}} \propto \sigma_{\mathrm{los}}^2 \propto a^{-1}$. The distributions over eccentricity and mass ratio play a significantly smaller role than the binary fraction and semi-major axis distribution (Kouwenhoven & de Grijs, *in preparation*). Finally, we wish to stress that further systematic errors are introduced by observational selection effects. Firstly, σ_{los} is often derived from spectral lines of red giants; the velocities of these objects may or may not be representative for the cluster as a whole. Secondly, a measurement in the cluster centre, and near the tidal limit, will result in dynamical masses differing by $\sim 50\%$; caution should be exercised when interpreting the observational results.

3. When is binarity important?

The effect of binaries on the dynamical mass determination of a star cluster is important if the typical orbital velocity of a binary component is of order, or larger than, the velocity dispersion of the particles (single/binary) in the potential of the cluster. Our simulations indicate that, for example, the dynamical mass is overestimated by 70% for $\sigma_{\mathrm{los}} = 1\,\mathrm{km\,s^{-1}}$, 50% for $2\,\mathrm{km\,s^{-1}}$, 20% for $5\,\mathrm{km\,s^{-1}}$, and 5% for $10\,\mathrm{km\,s^{-1}}$. Due to spectral resolution and stellar atmospheric turbulence, most *measured* velocity dispersions are $\sigma_{\mathrm{los}} \geqslant 5\,\mathrm{km\,s^{-1}}$. Most of the known dynamical masses of massive star clusters are therefore only mildly affected by the presence of binaries. However, for low-mass star clusters, the spectroscopic velocity dispersion may result in an overestimation of the dynamical mass by a factor of two or more.

References

de Grijs, R. & Parmentier, G. 2007, *ChJA&A* 7, 155

Duquennoy, A. & Mayor, M. 1991, *A&A* 248, 485

Kobulnicky, H. A., Fryer, C. L., & Kiminki, D. C. 2006, *ApJ* in press (astro-ph/0605069)

Kouwenhoven, M. B. N., Brown, A. G A., Zinnecker, H., Kaper, L., & Portegies Zwart, S. F. 2005, *A&A* 430, 137

Kouwenhoven, M. B. N., Brown, A. G. A., Portegies Zwart, S. F., & Kaper, L. 2007, *A&A* in press (arXiv:0707.2746)

Spitzer, L. 1987, *Dynamical Evolution of Globular Clusters*, Princeton, NJ, Princeton University Press, 1987, p. 191

Dynamical Evolution of Dense Stellar Systems
Proceedings IAU Symposium No. 246, 2007
E. Vesperini, M. Giersz & A. Sills, eds.

© 2008 International Astronomical Union
doi:10.1017/S1743921308015767

Mass Transfer in Binary Systems: A Numerical Approach

C.-P. Lajoie and A. Sills

Department of Physics & Astronomy, McMaster University, Hamilton, ON L8S 4M1, Canada
email: lajoiec@mcmaster.ca, asills@mcmaster.ca

Abstract. We present preliminary work on the formation scenario of blue straggler stars by mass transfer in binary systems. More precisely, using Smoothed Particle Hydrodynamics (SPH), we want to model only the outer parts of the stars in order to get a much greater spatial resolution of the mass transfer flow itself. The inner boundary conditions are achieved using the so-called ghost particles and by replacing the inner mass by a central point mass. Stability of this central point mass is crucial, and it is shown that we get reasonable results. These simulations should give us indications on which layers of the donor star are actually transferred to the other star as well as how mass is transferred and how it settles on the accretor. This work is aimed at getting distinct observational signatures which would help identifying the dominant formation mechanism of blue straggler stars.

Keywords. binaries: close, blue stragglers, methods: numerical

1. Introduction

Globular clusters are dense stellar systems in which dynamical interactions between stars can significantly alter the stellar populations. It is thus important to understand how these interactions occur and what they result in if we want to better constrain the dynamical and formation histories of globular clusters. For example, blue stragglers, which are found above the main-sequence turn-off point in color-magnitude diagrams, are thought to be created either by collisions between two main-sequence stars or by mass transfer in primordial binaries. Observationally, however, it is rather difficult to distinguish between these two scenarios (see, e.g., Ferraro *et al.* 2006), so a better theoretical understanding of how these objects form may help identifying the formation of blue stragglers.

2. Numerical Approach

We address this problem by using our Smoothed Particle Hydrodynamics (SPH) code, which is based on the early version of Benz (1990) and, more recently, Bate, Bonnell & Price (1995). Briefly, this methods treats fluids elements as smoothed particles, which are subject to hydrodynamical forces such as pressure gradient and viscosity (Balsara 1995), as well as gravity (see Benz *et al.* 1990). Particles are smoothed over a volume defined by their smoothing length, which in turn is determined by the requirement that each particle must have a given number of neighbors. Thus, the total number of particles puts limits on the spatial resolution of the simulation.

To get a detailed description of the mass transfer flow between the two stars as well as how the matter settles onto the accretor, we need high spatial resolution. In order to reach such a high resolution without using extremely large number of particles, we intend to model only the outer parts of the stars, as shown in Fig. 1a. The boundary

271

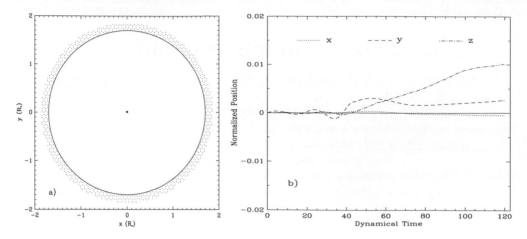

Figure 1. a) Initial configuration of our model star, with the SPH particles represented by small dots, the boundary shown as the solid line and the central point mass shown as the solid dot at the centre of the star. b) Normalized position of the central point mass as a function of time for the initial configuration shown in a). The solid line shows the equilibrium position of the point mass, which initially corresponds to the centre of the star.

conditions are achieved using the so-called ghost particles (see, e.g., Morris, Fox & Zhu 1997) and the SPH particles located inside this boundary are replaced by a point mass with the same total mass. We have tested the stability of the central point mass for an isolated star, and Fig. 1b shows that it drifts only slightly (i.e. one percent of the star's initial radius) over 120 dynamical times.

3. Future Work

The simultaneous implementation of boundaries for two stars is now underway, and following the encouraging results of § 2, we should start running simulations of main-sequence binaries in the near future. We think these simulations will be useful in identifying distinct observational signatures of blue stragglers formed by mass transfer and help identifying their main formation scenario.

Acknowledgements

This work was supported by NSERC Canada and made possible by the facilities of the Shared Hierarchical Academic Research Computing Network (SHARCNET: www.sharcnet.ca).

References

Balsara, D. S. 1995 *Jour. Comput. Physics* 121, 357
Bate, M. R., Bonnell, I. A., & Price, N. M. 1995, *MNRAS* 277, 362
Benz, W. 1990, in: Buchler J. R. (ed.), *The Numerical Modeling of Nonlinear Stellar Pulsations: Problems and Prospects*, Kluwer, Dordrecht, p. 269
Benz, W., Bowers, R. L., Cameron, A. G. W., & Press, W. 1990 *ApJ* 348, 647
Ferraro, F. R., Sabbi, E., Gratton, R., Piotto, G., Lanzoni, B., Carretta, E., Rood, R. T., Sills, A., Fusi Pecci, F., Moehler, S., Beccari, G., Lucatello, S., & Compagni, N. 2006, *ApJ* 647, L53
Morris, J. P., Fox, P. J., & Zhu, Y. 1997 *Jour. Comput. Physics* 136, 214

Dynamical Evolution of Dense Stellar Systems
Proceedings IAU Symposium No. 246, 2007
E. Vesperini, M. Giersz & A. Sills, eds.

© 2008 International Astronomical Union
doi:10.1017/S1743921308015779

Is our Sun a Singleton?

D. Malmberg[1], M. B. Davies[1], J. E. Chambers[2], F. De Angeli[3], R. P. Church[1], D. Mackey[4] and M. I. Wilkinson[5]

[1]Lund Observatory, Box 43, SE-221 00, Lund, Sweden
email: danielm@astro.lu.se

[2]Department of Terrestrial Magnetism, Carnegie Institution of Washington, 5241 Broad
Branch Road NW, Washington DC 20015, USA

[3]Institute of Astronomy, Madingley Road, Cambridge, CB3 OHA, UK

[4]Institute for Astronomy, University of Edinburgh, Royal Observatory, Blackford Hill,
Edinburgh, EH9 3HJ, UK

[5]Department of Physics and Astronomy, University of Leicester, Leicester, LE1 7RH, UK

Abstract. Most stars are formed in a cluster or association, where the number density of stars can be high. This means that a large fraction of initially-single stars will undergo close encounters with other stars and/or exchange into binaries. We describe how such close encounters and exchange encounters can affect the properties of a planetary system around a single star. We define a singleton as a single star which has never suffered close encounters with other stars or spent time within a binary system. It may be that planetary systems similar to our own solar system can only survive around singletons. Close encounters or the presence of a stellar companion will perturb the planetary system, often leaving planets on tighter and more eccentric orbits. Thus planetary systems which initially resembled our own solar system may later more closely resemble some of the observed exoplanet systems.

Keywords. binaries: general, open clusters and associations: general, planetary systems: general

---

Stars are most often formed in some sort of cluster or association. In such environments the number density of stars can be significantly higher than in the solar neighborhood. Thus, close encounters between stars might be common. If a single star with a planetary system suffers a close encounter with another star, the orbits of the planets might be changed. Sometimes this change can be enough for one or more planets to be ejected entirely from the system. Most likely this will happen long after the encounter, due to the strong planet-planet interactions induced by the close encounter. If one or more planets are ejected, the remaining planets will most often be left on tighter and more eccentric orbits. It is also possible that the close encounter does not cause the ejection of any planets and instead just stirs up the eccentricities of the planets somewhat.

If a single star instead encounters a binary system, it can be exchanged into it. When this occurs, the orientation of the orbital plane of the planets with respect to that of the companion star is completely random. This means that in about 70 per cent of the cases, the inclination between the two will be larger than 40°. When that happens, the Kozai Mechanism will operate (Kozai 1962). Given that the binary is not too wide, the Kozai Mechanism will cause the eccentricities of the planets to oscillate. If the planetary system contains multiple planets, this eccentricity pumping can cause strong planet-planet interactions, causing the orbits of the planets to change significantly and sometimes also ejecting one or more planets (Malmberg *et al.* 2007)

In our own solar system, the orbits of the planets are nearly circular. Furthermore, all the massive planets are found far out from the sun. This is however not the case for the observed exoplanets. Many of these systems contain one or more eccentric planets and

273

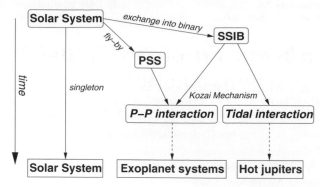

Figure 1. This flow-diagram outlines what can happen to a solar system orbiting an initially single star inside a stellar cluster. SSIB stands for Solar System In Binary, PSS stands for Perturbed Solar System and P-P interaction stands for Planet-Planet interaction.

also often massive planets on tight orbits. We propose that at least some of the observed exoplanet systems were once resembling the solar system, but were later altered into the planetary system we observe today, through interactions in stellar clusters. It might be that planetary systems like our own solar system can only exist around stars which formed single and has never experienced a close encounter with another star or been inside a stellar binary. We define such a star to be called a singleton. In Fig. 1 we give an outline of what can happen to a solar system like planetary system when inside a stellar cluster. Only around singletons will a planetary system remain solar-system-like until today, while if it orbits a star which suffer strong interactions with other stars it will today instead either be a Hot Jupiter system (see for example Fabrycky & Tremaine 2007; Wu *et al.* 2007) or contain planets on elliptical orbits (Takeda & Rasio 2005).

To explore how large the fraction of singletons is in the solar neighborhood (and thus to understand how common solar system like planetary system may be) we have numerically simulated a large range of stellar clusters, typical of those in which most of the stars in the solar neighborhood formed (Malmberg *et al.* 2007). From our simulations we estimate the singleton fraction for single stars with masses similar to that of the sun to be between 0.90 and 0.95. This means, that between 5 and 10 per cent of all planetary systems around solar mass stars can have been altered by dynamical interactions in stellar clusters, such as described above, into some of the observed exoplanet systems.

Acknowledgements

Melvyn B. Davies is a Royal Swedish Academy Research Fellow supported by a grant from the Knut and Alice Wallenberg Foundation. Ross P. Church is funded by a grant from the Swedish Institute. Dougal Mackey is supported by a Marie Curie Excellence Grant under contract MCEXT-CT-2005-025869. Mark Wilkinson acknowledges support from a Royal Society University Research Fellowship.

References

Kozai, Y. 1962, *AJ*, 67, 591

Malmberg, D., Davies, M. B., & Chambers, J. E. 2007, *MNRAS*, 377, L1

Malmberg, D., de Angeli, F., Davies, M. B., Church, R. P., Mackey, D., & Wilkinson, M. I. 2007, *MNRAS*, 378, 1207

Fabrycky, D. & Tremaine, S. 2007, arXiv:0705.4285

Takeda G. & Rasio F. A., 2005, *ApJ*, 627, 1001

Wu, Y., Murray, N. W., & Ramsahai, J. M. 2007, arXiv:0706.0732

Dynamical Evolution of Dense Stellar Systems
Proceedings IAU Symposium No. 246, 2007
E. Vesperini, M. Giersz & A. Sills, eds.

© 2008 International Astronomical Union
doi:10.1017/S1743921308015780

Getting a Kick out of the Stellar Disk(s) in the Galactic Center

H. B. Perets, G. Kupi and T. Alexander

Weizmann Institute of Science, POB 26, Rehovot 76100, Israel.

Abstract. Recent observations of the Galactic center revealed a nuclear disk of young OB stars, in addition to many similar outlying stars with higher eccentricities and/or high inclinations relative to the disk (some of them possibly belonging to a second disk). Binaries in such nuclear disks, if they exist in non-negligible fractions, could have a major role in the evolution of the disks through binary heating of this stellar system. We suggest that interactions with/in binaries may explain some (or all) of the observed outlying young stars in the Galactic center. Such stars could have been formed in a disk, and later on kicked out from it through binary related interactions, similar to ejection of high velocity runaway OB stars in young clusters throughout the galaxy.

Keywords. Galaxy: center, black hole physics, stars: kinematics, binaries: general

---

Recent observations have revealed the existence of many young OB stars in the galactic center (GC). Accurate measurements of the orbital parameters of these stars give strong evidence for the existence of a massive black hole (MBH) which govern the dynamics in the GC (Eisenhauer *et al.* 2005). Most of the young stars are observed to be OB stars in the central 0.5 pc around the MBH. Many of them are observed in a coherent stellar disk or two perpendicular disks configurations (Lu *et al.* 2006; Paumard *et al.* 2006). Others are observed to be have inclined and/or eccentric (> 0.5) orbits relative to the stellar disks (hereafter outliers). The inner 0.04 pc near the MBH contain only young B-stars, that possibly have a different origin (e.g. Levin 2007; Perets, Hopman & Alexander 2007).

It was suggested that the disk stars have been formed a few Myrs years ago in a fragmenting gaseous disk (Nayakshin & Cuadra 2005; Levin 2007). However, the origin of outliers from the disk is difficult to explain in this way. These stars are observed to have very similar stellar properties to the young disk stars (types, lifetimes), but have more inclined and/or eccentric orbits. Many suggestions have been made for the origin of these stars (Milosavljević & Loeb 2004; Paumard *et al.* 2006; Alexander, Begelman & Armitage 2007; Yu, Lu & Lin 2007, and references therein). Here we suggest a different process in which young stars in the GC stellar disks were kicked into high inclinations and/or eccentricities, in a similar way to OB runaway stars ejected from open clusters. Such a scenario could explain some of the puzzling orbital properties of the young stars in the GC.

A considerable fraction of the early OB stars in the solar neighborhood have large peculiar velocities ($40 \leqslant v_{pec} \leqslant 200$ km s^{-1}; e.g. Hoogerwerf, de Bruijne & de Zeeuw 2001) and are observed in isolated locations; these are the so-called runaway stars (Blaauw 1961). Two mechanisms are thought to eject OB runaway stars, both involve binarity (or higher multiplicity). In the binary supernova scenario (BSS; Blaauw 1961) a star is kicked at high velocity following a supernova explosion of its companion. In the dynamical ejection scenario (DES; Poveda, Ruiz & Allen 1967) runaway stars are ejected through gravitational interactions between stars in dense, compact clusters. The DES is more likely for the disk stars, given the short lifetime of the disk.

The binary properties of stars in the GC are unknown, but observations of eclipsing binaries in the GC suggests they are not fundamentally different from that observed in

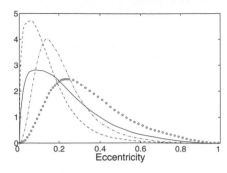 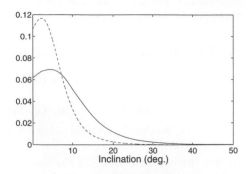

Figure 1. Left: Eccentricity distribution of outliers in the Galactic center. Monte-Carlo results assuming different models for the Maxwellian kick velocity distributions; isotropic, $v \sim 100$ km s^{-1} (solid) , isotropic, $v \sim 60$ km s^{-1} (dashed), planar (co-planar with the disk), $v \sim 100$ km s^{-1} (dotted) and planar, $v \sim 60$ km s^{-1} (dash-dotted). Right: Inclination distribution of outliers in the Galactic center. Monte-Carlo results assuming different models for the Maxwellian kick velocity distributions; isotropic, $v \sim 100$ km s^{-1} (solid), isotropic, $v \sim 60$ km s^{-1} (dashed).

the solar neighborhood (DePoyet *et al.* 2004; Martins *et al.* 2006; Rafelski *et al.* 2007). The conditions in the stellar disks in the GC are highly favorable for the DES given the high stellar densities ($> 10^6$ pc^{-3}), the low velocity dispersion and the masses of the stars in the stellar disk. It is thus quite likely that dynamical ejection in the disk is more frequent and efficient than in normal Galactic star forming regions. In order to escape the GC cusp, a disk star initially bounded to the MBH needs a kick velocity of a few$\times 10^2$ km s^{-1}. Most of the observed OB runaways in the galaxy do not reach such high velocities. Therefore runaways from the stellar disk would not escape, and would remain in the cusp. However, such kicks could considerably change the orbits of these stars. In Fig. 1 we show the inclination and eccentricity distribution of OB stars that were formed in a disk on circular orbits and were kicked out of the disk with velocities typical of OB runaways ($40 - 200$ km s^{-1}). Stars in the disk with near circular Keplerian orbits, would be kicked into more eccentric and inclined orbits (on average). Their orbits should also display some correlations between inclination, eccentricity and mass. Such "failed" runaways could possibly explain the outliers from and inside the disks in the GC, which are difficult to explain otherwise.

References

Alexander, R. D., Begelman, M. C., & Armitage, P. J. 2007, *ApJ* 654, 907
Blaauw, A. 1961, *Bull. Astro. Inst. Netherlands* 15, 505, 265
Depoy, D. L. *et al.* 2004, *ApJ* 617, 1127
Eisenhauer, F. *et al.* 2005, *ApJ* 628, 246
Hoogerwerf, R., de Bruijne, J. H. J., & de Zeeuw, P. T. 2001, *A&A* 365, 49
Levin, Y. 2007, *MNRAS* 374, 515
Lu, J. R. *et al.* 2006, *J. Phys.: Conf. Ser.* 54, 279
Martins, F. *et al.* 2006, *ApJL* 649, L103
Milosavljević, M. & Loeb, A. 2004, *ApJL* 604, L45
Nayakshin, S. & Cuadra, J. 2005, *A&A* 437, 437
Paumard, T. *et al.* 2006, *ApJ* 643, 1011
Perets, H. B., Hopman, C., & Alexander, T. 2007, *ApJ* 656, 709
Poveda, A., Ruiz, J., & Allen, C. 1967, *Bol. Obs. Tonantzintla y Tacubaya* 4, 86
Rafelski, M. *et al.* 2007, *ApJ* 659, 1241
Yu, Q., Lu, Y., & Lin, D. N. C. 2007, *ApJ* 666, 919

Dynamical Evolution of Dense Stellar Systems
Proceedings IAU Symposium No. 246, 2007
E. Vesperini, M. Giersz & A. Sills, eds.

© 2008 International Astronomical Union
doi:10.1017/S1743921308015792

A Search for Spectroscopic Binaries in the Globular Cluster M4

V. Sommariva[1], G. Piotto[2], M. Rejkuba[1], L. R. Bedin[3], D. C. Heggie[4], A. Milone[2], R. D. Mathieu[5] and A. Moretti[2]

[1] ESO, Karl-Schwarzschild-Str. 2, 85748 Garching, Germany email: vsommari@eso.org
[2] Astronomy Department, Padova University, vic. Osservatorio 2, 35122 Padova, Italy
[3] Space Telescope Science Institute, 3700 San Martin Drive, Baltimore, MD
[4] School of Mathematics, University of Edinburgh
[5] Astronomy Department, University of Wisconsin-Madison, Madison, WI

Abstract. We present preliminary results from an observational campaign aimed at the study of the binary fraction and binary radial distribution in Galactic globular clusters. In particular, we concentrate on the ongoing observational campaign for the search of spectroscopic binaries.

Keywords. binaries: spectroscopic, globular clusters

1. Introduction

Within the MODEST collaboration, we are conducting a multi-instrument observational campaign aimed at deriving the present day fraction and radial distribution of the binaries in the Galactic globular cluster (GC) M4. With the GRAPE-DR computer technology under development, N-body realistic modeling of GCs with up to 5×10^5 particles will be feasible. At that point, observational inputs, and the comparison of models with the observed parameters will become of fundamental importance to understand the evolution of a GC. Most of them are already available for the two closest clusters, M4 and NGC6397, except the fundamental information on their binary population. It is now the right time to start covering this gap. M4 is our first target for several reasons: 1) we expect to find a large fraction of binaries; 2) it has an extended and relatively uncrowded core, where the proportion of binaries should be highest; 3) it is nearby, which allows the study of binaries also below the turn-off (TO), and these are pristine tracers, as binaries containing giants, might be destroyed by internal mass transfer. We are following three complementary approaches, using the most up-to-date instrumentation: i) on ACS/HST images of the central $200'' \times 200''$, we are measuring the photometric binary fraction, mass distribution, and spatial distribution; ii) on adaptive Optics NACO/VLT observations of the inner core of M4 we search for binaries with massive companions on the basis of their wobble around the center of mass; iii) we are searching for spectroscopic binary systems through the radial velocity variations. Here, we present the first results from the multi-epoch high resolution spectroscopic campaign with FLAMES+GIRAFFE@VLT.

2. Observations and data reduction

We observed 2684 stars from the red giant branch tip to a couple of magnitudes below the main sequence turn-off (see Fig. 1, left panel). Our sample also cover most of the cluster extension, from the inner core, out to the cluster outskirts.

The first epoch data were taken with FLAMES+GIRAFFE@VLT in 2003. The second epoch was taken in 2006 for all targets, and for ∼484 stars we repeated the observations

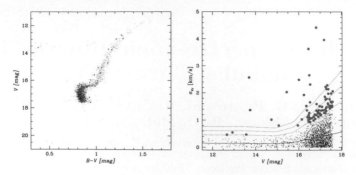

Figure 1. *Left panel:* The color-magnitude diagram of all the selected targets in M4 observed with FLAMES. The binary star candidates are indicated with large squares. *Right panel:* For all the stars in M4 which have been observed both in two epochs, we show the r.m.s. weighted by the fitting error to the cross-correlation function. The solid line is the average weighted r.m.s., which is the indication of the precision of the radial velocity measurement for non-variable stars. The dotted lines indicate 3, 4, and $5\sigma_{Vr}$ of the scatter around the mean as a function of magnitude. The candidates are indicated with large dots.

after few weeks. The second run is used to detect soft binary candidates with long periods from the cluster center to its outskirts, and the third run to identify hard binaries with short periods inside the cluster core. All stars were observed with HR9 setup centered at 525.8nm. This instrumental configuration allows the highest spectral resolution (R = 25800) and the best radial velocity accuracy (Royer *et al.* 2002). The data reduction was done with the GIRAFFE BaseLine Data Reduction Software (girBLDRS 1.13, Blecha *et al.* 2000). The radial velocities were calculated by cross-correlating the stars' spectra with a numerical mask constructed from the solar spectrum. The mean radial velocity for the sample is 70.1±0.3 km/s.

3. Results

To search for a first, preliminary sample of candidate spectroscopic binary systems we plotted the r.m.s (σ_{Vr}) in radial velocity as a function of magnitude. We paid attention to minimize all of the possible measurement errors (wavelength calibration, errors due to cosmic rays or bad pixels in the spectra, errors due to scattered light from cluster background and simultaneous lamp fibers). The stars presenting velocity differences more than $3\sigma_{Vr}$ from the average for their magnitude were considered binary star candidates. In this preliminary analysis we found 95 candidates, which would imply lower limit for the binary fraction f = $8.2 \pm 2\%$ (8 binaries out of 97 targets) for $r \leqslant r_c$ (r_c is the cluster core radius), and f = $3.6 \pm 2\%$ (87/2372) $r > r_c$. Interestingly enough, in the parallel project for the search of photometric binaries, we found similar binary fractions. The binary candidates are shown in Fig. 1 (left panel) in the color magnitude diagram and Fig. 1 (right panel) with larger dots. The present reduction allowed us to identify binaries with velocity variations greater than 0.2 km/s for the brightest stars, and 0.5 km/s for the fainter TO stars. For the binary candidates, we plan follow-up observations for the orbital solution.

References

Blecha, A., Cayatte, V., North, P., Royer, F. & Simond, G. 2000, *SPIE*, 4008, 467
Royer, F. *et al.*, 2002, *SPIE* 4847, 184

Part 6

Exotic Stellar Populations

Dynamical Evolution of Dense Stellar Systems
Proceedings IAU Symposium No. 246, 2007
E. Vesperini, M. Giersz & A. Sills, eds.

© 2008 International Astronomical Union
doi:10.1017/S1743921308015809

Blue Straggler Stars in Galactic Globular Clusters: Tracing the Effect of Dynamics on Stellar Evolution

Francesco R. Ferraro[1] and Barbara Lanzoni[1]

[1]Dipartimento di Astronomia, Università di Bologna,
Via Ranzani 1, 40127 Bologna, Italy
email: francesco.ferraro3@unibo.it – barbara.lanzoni@bo.astro.it

Abstract. In this contribution we review the main observational properties of Blue Stragglers Stars (BSS) in galactic GCs. A flower of results on the BSS frequency, radial distribution, and chemical composition are presented and discussed.

Keywords: globular clusters: general; stars: evolution; binaries: general; blue stragglers

1. Introduction

Ultra-dense cores of Galactic Globular Clusters (GCs) are very efficient "furnaces" for generating exotic objects, such as low-mass X-ray binaries, cataclysmic variables, millisecond pulsars, blue stragglers (BSS), etc. Most of these stars are thought to be the by-products of the evolution of binary systems, possibly originated and/or hardened by stellar interactions. Thus, studying the nature of these exotic objects and the properties of artificial sequences, as that of BSS, in the color-magnitude diagrams (CMDs) of GCs can serve as a powerful diagnostic of the dynamical evolution of clusters, and of its effects on the evolution of their stellar population and binary systems (see Bailyn 1995 and reference therein).

This topic has received strong impulse in the recent years. In this paper we review the main properties of the most known *exotic population* of GCs: the so-called BSS, that describe the very first sequence of exotic objects discovered in the CMD.

First discovered by Sandage (1953) in M3, BSS are commonly defined as stars brighter and bluer (hotter) than the main sequence (MS) turnoff along an apparent extension of the MS. Thus, they mimic a rejuvenated stellar population and their existence has been a puzzle for many years. Direct measurements (Shara *et al.* 1997) and indirect evidences show that BSS are more massive than the normal MS stars, pointing toward stellar mergers as possible explanation for their origin. Indeed, their formation mechanisms are not yet completely understood, and the leading explanations, at present, involve mass transfer (MT) between binary companions (McCrea 1964; Zinn & Searle 1976), possibly up to the complete coalescence of the binary system, or the merger of stars (whether or not in binaries) induced by collisions (COL; Hills & Day 1976). Thus, BSS represent the link between standard stellar evolution and the effects of cluster dynamics (see Bailyn 1995).

2. The UV approach to the study of BSS

The observational and interpretative scenario of BSS has significantly changed in the last 20 years. In fact, for almost 40 years since their discovery, BSS have been detected only in the outer regions of GCs or in relatively loose clusters, thus generating the idea

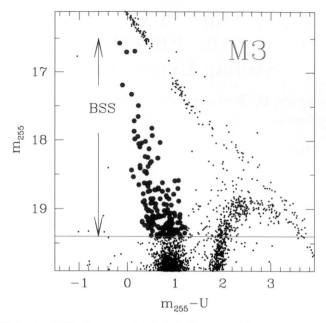

Figure 1. BSS in the UV: the case of M3. The horizontal line at $m_{255} = 19.4$ is the assumed limiting magnitude, corresponding to $\sim 5\sigma$ above the turnoff level (from F97).

that low-density environments were their *natural habitats*. However, this was just an observational bias, and, starting from the early '90, high resolution studies allowed to properly image and discover BSS also in the highly-crowded central regions of dense GCs (see the case of NGC 6397 by Auriere *et al.* 1990). In particular, the advent of the Hubble Space Telescope (HST) represented a really turning point in BSS studies, thanks to its unprecedented spatial resolution and imaging/spectroscopic capabilities in the ultraviolet (UV; see Paresce *et al.* 1991, Ferraro & Paresce 1993, Guhathakurta *et al.* 1994, etc).

In fact, the systematic study of BSS, especially in the central regions of high density clusters, still remains problematic in the optical bands, even if using HST. This is because the CMD of old stellar populations in the *classical* $(V, B - V)$ plane is dominated by the cool stellar component. Hence, the observation and the construction of complete samples of hot stars (as BSS, other by-products of binary system evolution, extreme blue horizontal branch stars, etc.) is "intrinsically" difficult in this plane. Moreover, BSS can be easily mimicked by photometric blends of red giant (RGB) stars in the optical CMDs. Instead, at UV wavelengths RGB stars are very faint, while BSS are among the brightest objects. In particular, BSS define a narrow, nearly vertical sequence spanning ~ 3 mag in the UV plane (see Fig. 1), thus being much more easily recognizable, while BSS-like blends are much less severe at these wavelengths because of the relative faintness of sub-giant and RGB stars. Indeed, the $(m_{255}, m_{255} - m_{336})$ plane is an ideal tool for selecting BSS even in the densest cores of GCs, and its systematic use allowed to put the BSS study into a more quantitative basis than ever before (e.g., Ferraro *et al.* 2003; see Fig. 2).

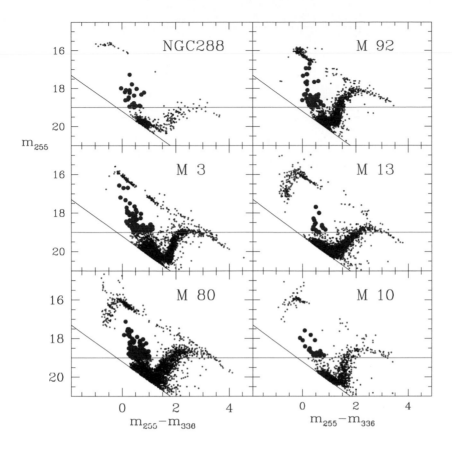

Figure 2. $(m_{255}, m_{255} - m_{336})$ CMDs for the six clusters discussed in F03. Horizontal and vertical shifts have been applied to all CMDs in order to match the main sequences of M3. The horizontal solid line corresponds to $m_{255} = 19$ in M3. The bright BSS candidates are marked as large filled circles.

3. BSS specific frequency

Based on these observations, the first catalogs of BSS have been published (e.g., Fusi Pecci *et al.* 1992; Ferraro, Fusi Pecci & Bellazzini 1995, hereafter FFB95), until the most recent collection of BSS which counts nearly 3000 candidates i 56 Galactic GCs (Piotto *et al.* 2004; the most recent results based on this data-set are discussed by Leigh *et al.* in this volume).

These works have significantly contributed to form the nowadays commonly accepted idea that BSS are a normal stellar population in GCs, since they are present in all properly observed clusters. However, according to Fusi Pecci *et al.* (1992), BSS in different environments could have different origins. In particular, BSS in loose GCs might be produced by the coalescence of primordial binaries, while in high density GCs (depending on survival-destruction rates for primordial binaries) BSS might arise mostly from stellar interactions, particularly those which involve binaries. While the suggested mechanisms of BSS formation could be at work in clusters with different densities (FFB95; Ferraro *et al.* 1999), there are evidences that they could also act simultaneously within the same

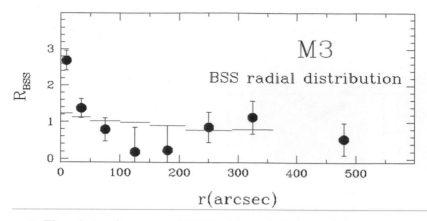

Figure 3. The relative frequency of BSS in M3 is plotted as a function of the radial distance from the cluster center. The horizontal lines show the relative frequency of the RGB stars used as a comparison population. For $r > 6'$ only the relative frequency of BSS has been computed using the Sandage (1953) candidates. (From F97)

cluster (as in the case of M3; see Ferraro *et al.* 1993, hereafter F93; Ferraro *et al.* 1997, hereafter F97).

A number of interesting results have been obtained from cluster-to-cluster comparisons. For this purpose we used the BSS specific frequency, defined as the number of BSS counted in a given region of the cluster, normalized to the number of "normal" cluster star in the same region, adopted as reference (generally we adopted the horizontal branch stars, hereafter HB). The BSS specific frequency has been found to largely vary from cluster to cluster: for the six GCs considered by Ferraro *et al.* (2003), the BSS frequency varies from 0.07 to 0.92, and does not seem to be correlated with central density, total mass, velocity dispersion, or any other obvious cluster property (see also Piotto *et al.* 2004). Even "twin" clusters as M3 and M13 harbor a quite different BSS populations: the specific frequency in M13 is the lowest ever measured in a GC (0.07), and it turns out to be 4 times lower than that measured in M3 (0.28). Which is the origin of this difference? The paucity of BSS in M13 suggests either that the primordial population of binaries in M13 was poor, or that most of them were destroyed. Alternatively, as suggested by F97, the mechanism producing BSS in the central region of M3 is more efficient than in M13, because the two systems are in different dynamical evolutionary phases.

In this respect, the most surprising result is that the largest BSS specific frequency has been found in two GCs which are at the extremes of central density values in our sample: NGC 288 and M80, with the lowest and the highest central density, respectively. This suggests that the two formation channels can have comparable efficiency in producing BSS in their respective typical environment.

4. The BSS radial distribution

M3 has played a fundamental role in the BSS history, because it is the GC not only where BSS have been first identified, but also where the BSS radial distribution has been studied for the first time over the entire cluster extension. In fact by combining UV HST observations of the cluster central region (F97) and extensive wide field ground-based observations (F93; Buonanno *et al.* 1994), F97 presented the BSS radial distribution

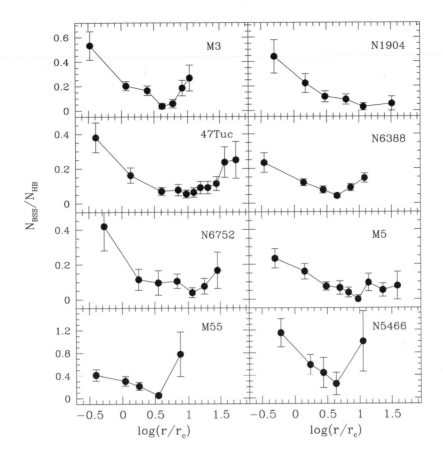

Figure 4. The BSS radial distribution observed in 8 GCs. In all cases, but NGC 1904, it is clearly bimodal, with a peak in the center, a dip at intermediate radii, and an upturn in the external regions.

of M3 all over its radial extent ($r \sim 6'$). The resulting distribution was completely unexpected: BSS appear to be more centrally concentrated than RGB stars in the central regions, and less concentrated in the cluster outskirts.

For further investigating such a surprising result, F97 divided the surveyed area in a number of concentric annuli, and counted the number of BSS and RGB stars normalized to the sampled luminosity in each annulus, accordingly to the following relations (F93):

$$R_{\mathrm{BSS}} = \frac{(N_{\mathrm{BSS}}/N_{\mathrm{BSS}}^{\mathrm{tot}})}{(L^{sample}/L_{tot}^{sample})}$$

and

$$R_{\mathrm{RGB}} = \frac{(N_{\mathrm{RGB}}/N_{\mathrm{RGB}}^{\mathrm{tot}})}{(L^{sample}/L_{tot}^{sample})},$$

respectively. The result is shown in Fig. 3, and clearly shows that the radial distribution of BSS in M3 is bimodal: it reaches maximum at the center of the cluster, shows a clear-cut dip in the intermediate region (at $100'' < r < 200''$), and rises again in the outer region (out to $r \sim 360''$).

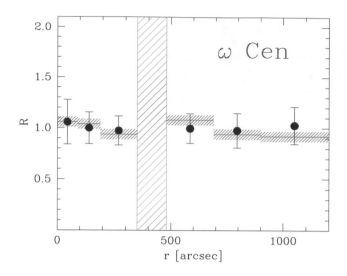

Figure 5. The double-normalized relative frequency R_{BSS} of the BSS (*filled circles*) in ω Cen. The shaded area marks the cluster region we excluded in order to avoid incompleteness problems. The horizontal lines show the relative frequency of the RGB stars used as reference population. (From Ferraro *et al.* 2006a).

While the bimodality detected in M3 was considered for years to be *peculiar*, the most recent results demonstrated that this is not the case. In fact, in the last years the same observational strategy adopted by F97 in M3 has been applied to a number of other clusters, with the aim of studying the BSS radial distribution over the entire cluster extension. Bimodal distributions with an external upturn have been detected in several cases (see Fig. 4): 47 Tuc (Ferraro *et al.* 2004), NGC 6752 (Sabbi *et al.* 2004), M55 (Zaggia *et al.* 1997; Lanzoni *et al.* 2007c), M5 (Warren *et al.* 2006; Lanzoni *et al.* 2007a), NGC 6388 (Dalessandro *et al.* 2007), NGC 5466 (Beccari *et al.* 2007, in preparation).

Originally, F97 argued that the bimodal distribution of BSS in M3 might be the signature of the two formation mechanisms acting simultaneously in the same cluster: the *external* BSS would arise from MT activity in primordial binaries, while the *central* BSS would be generated by stellar collisions leading to mergers. Sigurdsson *et al.* (1994) offered another explanation for the bimodal BSS distribution in M3. They suggested that all BSS were formed in the core and then ejected to the outer regions by the recoil from the interactions. Those BSS which get kicked out to a few core radii would rapidly drift back to the center of the cluster due to mass segregation, thus leading to the central BSS concentration and a paucity of BSS in the intermediate regions (around a few core radii). More energetic recoils would kick the BSS to larger distances and, since stars require much more time to drift back toward the core, these may account for the overabundance of BSS in the cluster outskirts. We are currently using Monte-Carlo dynamical simulations in order to discern between different possibilities (see also the contribution by Lanzoni in this book). Mapelli *et al.* (2004, 2006) and Lanzoni *et al.* (2007a) modeled the dynamical evolution of BSS in a number of clusters, by using a modified version

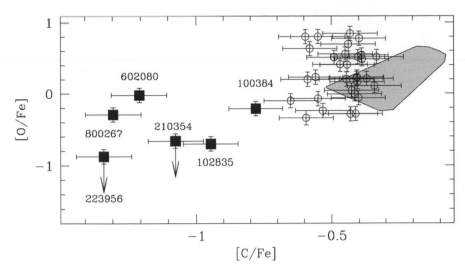

Figure 6. [O/Fe] ratio as a function of [C/Fe] for the BSS observed in 47 Tuc. Normal BSS are marked with *empty circles*, while CO-depleted BSS are marked with *filled squares* and their names are also reported. The gray regions correspond to the location of the 12 turnoff stars in 47 Tuc analyzed by Carretta *et al.* (2005).

of the code described by Sigurdsson & Phinney (1995). Their results demonstrate that the observed BSS bimodal distributions cannot be explained within a purely collisional scenario in which all BSS are generated in the core through stellar interactions. In fact, an accurate reproduction of the external upturn of the BSS radial distribution can be obtained only by requiring that a sizable ($\sim 20 - 40\%$) fraction of BSS is generated in the peripheral regions, where primordial binaries can evolve in isolation and experience mass transfer processes without suffering significant interactions with other cluster stars.

Even if the number of the surveyed clusters is low, the bimodal radial distribution first found in M3 and thought to be *peculiar* could instead be the *natural* one. However, generalizations cannot be made from a sample of a few clusters only, and such a statement needs to be characterized on a much more solid statistical base. Indeed, two exceptions are already known: NGC 1904, which does not present any external upturn (Lanzoni *et al.* 2007b), and ω Cen (Ferraro *et al.* 2006a), which shows a completely flat BSS radial distribution.

The case of ω Cen deserves specific comments: by using a proper combination of HST high-resolution data and wide-field ground-based observations sampling the entire radial extension of the cluster, Ferraro *et al.* (2006a) have detected the largest population of BSS ever observed in any stellar system: more than 300 candidates have been identified. At odds with all the GCs previously surveyed, BSS in ω Cen have been found not to be centrally segregated with respect to the other cluster stars (see Fig. 5). This is the cleanest evidence ever found that ω Cen is not fully relaxed, even in the central regions, and it suggests that the observed BSS are the progeny of primordial binaries, whose radial distribution was not yet significantly altered by stellar collisions and by the dynamical evolution of the cluster. Hence, most of these objects should have been produced essentially by MT processes, and *the population of BSS in ω Cen could represent the purest and largest population of non-collisional BSS ever observed.* Thus, ω Cen represents the best laboratory for studying the physical and chemical properties of MT-BSS.

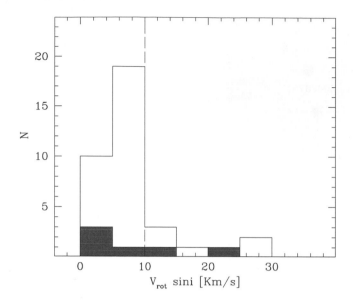

Figure 7. The rotation velocity distribution of normal BSS (*empty histogram*) is compared to the CO-depleted BSS distribution (*filled histogram*). The vertical *dashed line* marks the separation between slow and the 10 "fast" rotators (with $V_{rot} \sin i > 10$ km/s).

5. Searching for the chemical signature of the BSS formation process

Theoretical models still predict conflicting results on the expected properties of BSS generated by different production channels. For instance, Benz & Hills (1987) predict high rotational velocities for COL-BSS, whereas Leonard & Livio (1995) have shown that a substantial magnetic braking could occur, and the resulting BSS are *not* fast rotators. In the case of BSS formed through the MT production channel, rotational velocities larger than those of typical MS stars are predicted (Sarna & de Greve 1996). Concerning the chemical surface abundances, hydrodynamic simulations (Lombardi *et al.* 1995) have shown that very little mixing is expected to occur between the inner cores and the outer envelopes of the colliding stars. On the other hand, signatures of mixing with incomplete CN-burning products are expected at the surface of BSS formed via the MT channel, since the gas at the BSS surface is expected to come from deep regions of the donor star, where the CNO burning was occurring (Sarna & de Greve 1996).

Spectroscopic observations have recently begun to provide the first set of basic properties of BSS (effective temperature, mass, rotation velocity, etc.; see the recent work by De Marco *et al.* 2005). However, with the exception of a few bright BSS in the open cluster M67 (Mathys 1991; Shetrone & Sandquist 2000), an extensive survey of BSS surface abundance patterns is still lacking, particularly in GCs. In this context the advent of 8-meter class telescopes equipped with multiplexing capability spectrographs is giving a new impulse to the study of the BSS properties. By using FLAMES at the ESO VLT we are currently performing extensive surveys of surface abundance patterns for representative numbers of BSS in a sample of Galactic GCs. The first results of this search have lead to an exciting discovery: by measuring the surface abundance patterns of 43 BSS in 47 Tuc, we discovered a sub-population of BSS with a significant depletion of

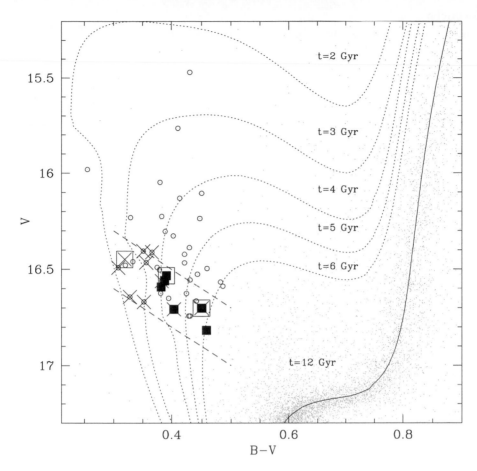

Figure 8. Zoomed CMD of 47 Tuc in the BSS region. Normal BSS are marked with *open circles*, while CO-depleted BSS are shown as *filled squares*. Isochrones of different ages (from 2 to 12 Gyr) from Cariulo *et al.* (2003) are overplotted for comparison. The three W UMa systems and the 10 BSS rotating with $V_{rot} \sin i > 10$ km/s are highlighted with *large empty squares* and *large crosses*, respectively.

Carbon (C) and Oxygen (O), with respect to the dominant population (see Fig. 6). This evidence is interpreted as the presence of CNO burning products on the BSS surface, coming from a deeply peeled parent star, as expected in the case of the MT formation channel. Thus, our discovery in 47 Tuc could be the first detection of a chemical signature clearly pointing to the MT formation process for BSS in a GC.

Indeed, the acquired data-set is a gold-mine of informations. In fact, our observations have shown that (1) only 10 BSS have been found to show rotational velocities larger than $v \sin i > 10$ Km/s, while most of the BSS are slow rotators at odds with what canonical models predict (see Fig. 7); (2) BSS with CO depletions and the few BSS with $v \sin i > 10$ Km/s appear "less evolved" than the others: they all lie within a narrow strip at the faint-end of the BSS luminosity distribution in the CMD (see Fig. 8); (3) some of them are WUma binary systems, suggesting that the evolution of these systems could be a viable channel for the formation of BSS in GCs.

References

Auriere, M., Lauzeral, C., & Ortolani, S. 1990, *Nature*, 344, 638

Bailyn, C. D. 1995, *ARA&A*, 33, 133

Bellazzini, M., *et al.* 2002, *AJ*, 123, 1509

Benz, W. & Hills, J. G. 1987, *ApJ*, 323, 614

Buonanno, R., *et al.* 1994, *A&A*, 290, 69

Cariulo, P., Degl'Innocenti, S., & Castellani, V., 2003, A&A, 412, 1121

Carretta, E., Gratton, R. G., Lucatello, S., Bragaglia, A., & Bonifacio, P. 2005, A&A, 433, 597

Dalessandro , E., *et al.* 2007, *ApJ*, submitted

De Marco, O., *et al.* 2005, *ApJ*, 632, 894

Ferraro, F. R. & Paresce, F., 1993, *AJ*, 106,154

Ferraro F. R., *et al.* 1993, *AJ*, 106, 2324 (F93)

Ferraro, F. R., Fusi Pecci, F., & Bellazzini, M. 1995, *A&A*, 294, 80 (FFB95)

Ferraro, F. R., *et al.* 1997, *A&A*, 324, 915 (F97)

Ferraro, F. R., Paltrinieri, B., Rood, R. T., & Dorman, B. 1999a, *ApJ*, 522, 983

Ferraro, F. R., *et al.* 2003, *ApJ*, 588, 464

Ferraro, F. R., *et al.* 2004, *ApJ*, 603, 127

Ferraro, F. R., *et al.* 2006a, *ApJ*, 638, 433

Ferraro, F. R., *et al.* 2006b, *ApJ* 647, L53

Fusi Pecci, F., *et al.* 1992, *AJ*, 104, 1831

Guhathakurta, P., *et al.* 1994, *AJ*, 108, 1786

Hills, J. G., & Day, C. A. 1976, *Astrophys. Lett.*, 17, 87

Lanzoni, B., *et al.* 2007a, *ApJ*, 663, 267

Lanzoni, B., *et al.* 2007b, *ApJ* 663, 1040

Lanzoni, B., *et al.* 2007c, *ApJ* in press

Leonard, P. J. T. & Livio, M. 1995, *ApJ*, 447L, 121

Lombardi, J. C. Jr., Rasio, F. A., & Shapiro, S. L. 1995, *ApJ*, 445, L117

Mapelli, M., *et al.* 2004, *ApJ*, 605, L29

Mapelli, M., *et al.* 2006, *MNRAS*, 373, 361

Mathys, G. 1991, *A&A*, 245, 467

McCrea, W. H. 1964, *MNRAS*, 128, 147

Paresce, F., *et al.* 1991, *Nature*, 352, 297

Piotto, G., *et al.* 2004, *ApJ*, 604, L109

Sabbi, E., Ferraro, F. R., Sills, A., & Rood, R. T. 2004, *ApJ*, 617, 1296

Sandage A. R. 1953, *AJ*, 58, 61

Sarna, M. J. & de Greve, J. P. 1996, *QJRAS*, 37, 11

Shara, M. M., Saffer, R. A., & Livio, M. 1997, *ApJ*, 489, L59

Shetrone, M. D. & Sandquist, E. L. 2000, *AJ*, 120, 1913

Sigurdsson, S., Davies, M. B., & Bolte, M. 1994, *ApJ*, 431, L115

Sigurdsson, S. & Phinney, E. S. 1995, *ApJS*, 99, 609

Warren, S. R., Sandquist, E. L., & Bolte, M. 2006, *ApJ*, 648, 1026

Zaggia, S. R., Piotto, G., & Capaccioli M. 1997, *A&A*, 327, 1004

Zinn, R. & Searle, L. 1976, *ApJ*, 209, 734

Dynamical Evolution of Dense Stellar Systems
Proceedings IAU Symposium No. 246, 2007
E. Vesperini, M. Giersz & A. Sills, eds.

© 2008 International Astronomical Union
doi:10.1017/S1743921308015810

Pulsars in Globular Clusters

Scott M. Ransom

NRAO, 520 Edgemont Rd., Charlottesville, VA, 22903 USA, email: sransom@nrao.edu

Abstract. Globular clusters produce orders of magnitude more millisecond pulsars per unit mass than the Galactic disk. Since the first cluster pulsar was uncovered 20 years ago, at least 138 have been identified – most of which are binary millisecond pulsars. Because their origins involve stellar encounters, many of the systems are exotic objects that would never be observed in the Galactic disk. Examples include pulsar-main sequence binaries, extremely rapid rotators (including the current record holder), and millisecond pulsars in highly eccentric orbits. These systems are allowing new probes of the interstellar medium, the equation of state of material at supra-nuclear density, the masses of neutron stars, and globular cluster dynamics.

Keywords. pulsars: general, globular clusters: general

1. Introduction

The first globular cluster (GC) pulsar was identified 20 years ago in the cluster M28 after intense efforts by an international team (Lyne *et al.* 1987). Since then at least 138 GC pulsars†, the vast majority of which are millisecond pulsars (MSPs), have been found. Finding these GC pulsars has required high-performance computing, sophisticated algorithms, state-of-the-art instrumentation, and deep observations with some of the largest radio telescopes in the world, primarily Parkes, Arecibo, and the Green Bank Telescope (GBT). The payoff has been an extraordinarily wide variety of science.

Low-Mass X-ray Binaries (LMXBs) have been known to be orders-of-magnitude more numerous per unit mass in GCs as compared to the Galactic disk since the mid-1970s (Katz 1975, Clark 1975). This overabundance is due to the production of compact binary systems containing primordially-produced neutron stars via stellar interactions within the high-density cluster cores. Since LMXBs are the progenitors of MSPs, this dynamics-driven production mechanism also applies to them, and it has made GCs (particularly the massive, dense, and nearby ones) lucrative targets for deep pulsar searches.

Camilo & Rasio (2005) produced an excellent review of the first 100 GC pulsars in 2005. This current review provides a significant update to Camilo & Rasio as it concentrates on the advances made (primarily with the GBT) within the past several years, including almost 40 additional pulsars and over 50 new timing solutions.

2. Basic Properties of GC Pulsars

There are currently 138 known pulsars in 25 different GCs. Over 100 GC pulsars have been found in the past 10 years, with almost 60 of these coming in the last 4 years from searches using the GBT (see Figure 1). The three clusters Terzan 5, 47 Tucanae, and M28 account for approximately half of these pulsars, with 33, 23, and 11 pulsars in each cluster respectively. Of the known pulsars, 80 are members of binary systems, 50 are isolated, and 8 are as yet undetermined.

† For an up-to-date catalog of known GC pulsars, see Paulo Freire's website at
http://www.naic.edu/~pfreire/GCpsr.html

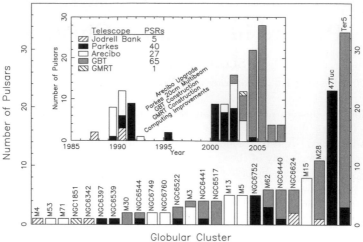

Figure 1. The number of pulsars per globular cluster and *(inset)* a timeline of their discovery.

Almost 90% of GC pulsars are true MSPs with spin periods $P_{spin} < 20$ ms (see Figure 2). Their spin properties seem consistent with the standard "recycling" scenario (e.g. Alpar *et al.*1982), with surface magnetic field strengths of $B \sim 10^{8-9}$ G and characteristic ages of $\tau_c \sim 10^{9-10}$ yrs. Most of the rest of the pulsars are partially recycled. However, there are several seemingly very out-of-place "normal" radio pulsars with $P_{spin} > 0.2$ s and $\tau_c \sim 10^7$ yrs as well (Lyne *et al.* 1996).

There are at least four distinct groups of binary GC pulsars (see Figure 2). The first two are similar to the binary MSPs found in the Galactic disk. The "Black Widows" have very low mass companions ($M_c \lesssim 0.04 \, M_\odot$) and orbital periods of several hours, while the "normal" low-mass binary MSPs (LMBPs) likely have Helium white dwarf (WD) companions of mass $M_c \sim 0.1 - 0.2 \, M_\odot$ and orbital periods of several to tens of days. The Black Widow systems are relatively much more common in GCs ($\sim$25% of the binaries) than in the Galactic disk, though ($\sim$4% of the binaries).

The other two groups of binary GC pulsars are possibly unique to clusters and their formation therefore likely depends on the high stellar densities and interactions found in the cores of GCs. Approximately 10% of GC binaries appear to have "main sequence"-like companions which show irregular eclipses, erratic timing, and often have hard X-ray and/or optical counterparts. The prototype system is J1740−5340 in NGC 6397 (D'Amico *et al.* 2001). Finally, $\sim$20% of the known GC binaries have highly eccentric orbits (with $e > 0.1$). The standard recycling scenario produces circular orbits due to tidal interactions during mass transfer, and so the large eccentricities are probably either induced during multiple stellar interactions with passing stars (e.g. Rasio & Heggie 1995) or produced directly during an exchange encounter with another star or binary. Over the past several years the numbers of pulsars in each of these two groups have grown dramatically. This is likely due to the fact that recent surveys have successfully probed many of the most massive and dense clusters in the Galaxy where these systems are preferentially produced (i.e. M28, M62, 47 Tucanae, Terzan 5, NGC 6440, and NGC 6441).

3. Searching for Cluster Pulsars

The signal-to-noise ratio from a radio observation towards a (perhaps unknown) pulsar is $\propto S_\nu \, A_e \, T_{sys}^{-1} \, \sqrt{B_\nu \, t_{obs}}$ (e.g. Lorimer & Kramer 2004). S_ν is the pulsar's flux density at

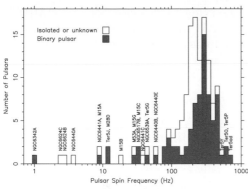 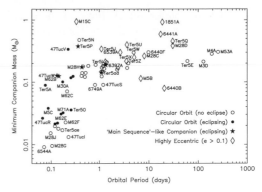

Figure 2. (**Left**) Spin frequency histogram of the 138 currently known GC pulsars. Eighty of the pulsars are confirmed members of binaries, 50 are isolated, and 8 are as yet undetermined. It is interesting to note that the binary MSPs seem to spin more rapidly on average than the isolated MSPs. Perhaps this is an indication that they are in general younger (i.e. more recently recycled) than the isolated MSPs. Such an explanation makes sense if all isolated MSPs originally come from binaries and therefore must destroy their companions over time. (**Right**) Orbital period P_{orb} vs. minimum companion mass ($M_{c,min}$; assuming a pulsar mass of $1.4\,M_\odot$) for the 70 binary GC pulsars with well-determined orbits. All 6 pulsars with "main sequence"-like companions and 14 of the known pulsars in highly eccentric ($e > 0.1$) orbits are labeled, as well as many more "normal" binary systems. Pulsars in "NGC" clusters are labeled without "NGC" to save space. The large grouping of pulsars with $P_{orb} < 1$ day and $M_{c,min} < 0.04\,M_\odot$ are the so-called "Black Widow" systems. The lack of pulsars in the lower right portion of the diagram is not due to selection effects, as those pulsars (if they existed) would be relatively easy to identify during searches or through timing observations. M15C is the only confirmed GC double neutron star system (Jacoby *et al.* 2006) while Ter5N is likely the only known GC pulsar with a Carbon-Oxygen WD companion (Ransom *et al.* 2005).

the observing frequency ν. A_e is the effective collecting area of the telescope. T_{sys} is the system temperature, which at the $\sim$GHz radio frequencies of interest is roughly the sum of the receiver temperature (typically $15-25\,$K), the $2.7\,$K cosmic microwave background, and the Galactic synchrotron background ($T_{Gal} \propto \nu^{-2.6}$ and typically a few to tens of K depending on sky position and observing frequency ν). B_ν is the radio bandwidth used for the observation and t_{obs} is the observation duration. In addition, signal-to-noise ratios improve when a pulsar has short duration pulsations compared to its pulse period (i.e. a small pulse duty cycle). If the pulses are smeared or broadened in time, perhaps by uncorrected orbital motion or interstellar medium effects, signal-to-noise ratios during searches can be reduced to effectively zero.

3.1. *The Problems with Cluster Distances*

MSPs are intrinsically very faint radio sources. Because of this, all of the wide-area Galactic pulsar surveys conducted to date have been severely sensitivity limited for MSPs (many have been instrumentation limited as well). Of the $\sim$60 known Galactic MSPs, $\sim$80% are within $2\,$kpc of the Sun. In contrast, the *nearest* GCs (M4 and NGC 6397) are just over $2\,$kpc from the Sun, and most, including the best targets for pulsar searches, are at distances of $5-15\,$kpc (Harris 1996).

In addition to the inverse square law problem, the large distances to GCs often (especially for the bulge clusters near the Galactic center) imply large column densities of the interstellar medium (ISM). The ionized ISM causes frequency-dependent dispersion of radio waves ($\propto \nu^{-2}$), scatter-broadening of the radio pulses ($\propto \nu^{-4.4}$), and for certain clusters dramatic fluctuations of observed pulse intensity due to diffractive scintillation (e.g. Camilo *et al.*2000). These ISM effects, as well as a substantially reduced Galactic synchrotron background, have pushed typical observing frequencies from ~400 MHz in the early 1990s up to 1.3−2 GHz in the past decade. At these frequencies, especially with much wider observing bandwidths available (hundreds of MHz), significant sensitivity gains have been realized despite the usually steep radio spectra of the pulsars themselves (flux densities $S_\nu \propto \nu^\alpha$ with $-3 \lesssim \alpha \lesssim -1$ and $<\alpha> \sim -1.8$; Maron *et al.*2000).

3.2. *Searching for Binaries*

Since most GC pulsars are in binaries†, orbital motion causes Doppler variations of the observed pulsation frequencies during an observation. If uncorrected, these variations can make even very bright MSPs undetectable. However, correcting for unknown orbital motion identically would be extremely computationally expensive and so current searches only account for linear changes in apparent spin frequencies (i.e. constant $\dot{f}$). These "acceleration" searches (e.g. Johnston & Kulkarni 1991) are valid when the orbital period is much longer than the observation duration ($P_{orb} \gtrsim 10\, t_{obs}$).

Acceleration searches add an extra dimension to the traditionally two-dimensional phase space of dispersion measure DM‡ and spin frequency f over which one must search for pulsars. Since the number of trials in the acceleration or $\dot{f}$ dimension is proportional to t_{obs}^2, the long observations used to improve the sensitivity of GC searches greatly increase their computational costs. Typically, searches for the first pulsar in a cluster are made using a large range of likely DMs but only a limited range of accelerations or $\dot{f}$. Once the first pulsar is found and the rough DM toward the cluster is known, a much smaller range of DMs is searched, but with a much larger range of possible accelerations for additional pulsars.

As an example, to properly search a single 7 hr GBT observation of Terzan 5 (where the DM is known to ~5%) with a full range of acceleration searches, requires approximately one CPU-*year* of processing on state-of-the-art CPUs. However, it is important to realize that without acceleration searches (or other advanced binary search techniques such as Dynamic Power Spectra; Chandler 2003), the majority of the binary GC MSPs that have been uncovered over the past decade would simply not have been found.

3.3. *A Renaissance in 2000*

The above paragraphs summarize why GC pulsar searches require long integrations (sensitivity) at GHz frequencies (minimize ISM effects), using the largest telescopes (collecting area), the best receivers (wide bandwidths and low system noise), and large amounts of high-performance computing (acceleration searches). The dramatic increases in the numbers of known GC pulsars beginning in 2000 (see Figure 1) resulted from significant improvements in each of these areas. First, new low-noise and wide-bandwidth ($B_\nu \sim 300$ MHz) observing systems centered near 1.4 GHz became available at Parkes and Arecibo. Second, the rise of affordable cluster-computing allowed acceleration searches to be conducted at investigator institutions rather than at special supercomputing sites. The first major success from these improvements (and a significant driver for further efforts) was the discovery of 9 new binary MSPs in 47 Tuc (Camilo *et al.* 2000).

The third and perhaps most important improvement was the completion of the GBT in 2001. With its state-of-the-art receivers, approximately three times greater A_e than Parkes, and the ability to observe over 80% of the celestial sphere, it is perfectly suited to make deep observations of GCs. By the end of 2003, a fantastic wide-bandwidth ($B_\nu \sim 600$ MHz) system centered near 2 GHz became available which provided 5−20 times

---

† The current binary fraction of ~60% is a lower limit since finding isolated pulsars is much easier than finding binaries.

‡ DM is the integrated electron column density along the line-of-sight to a pulsar.

more sensitivity for MSPs in certain GCs in the Galactic bulge than the 1.4 GHz system used at Parkes. The discovery of 30 new MSPs in Terzan 5 (Ransom *et al.* 2005), a cluster previously extensively searched at Parkes, including the fastest known MSP (J1748−2446ad aka Ter5ad; Hessels *et al.*2006), were some of the first results. Pulsar surveys of many additional GCs using the same system are ongoing.

3.4. *Future Cluster Pulsar Surveys*

Recent work on the luminosities L of GC pulsars (Hessels *et al.* 2007) has confirmed earlier results (e.g. Anderson 1992) suggesting that the luminosity distribution roughly follows a $d \log N = −d \log L$ relation. In addition, this work implies that we currently observe only the most luminous pulsars in each cluster. Together, these facts indicate that our current GC pulsar surveys are completely sensitivity limited such that even marginal improvements in search sensitivities will result in new pulsars[†]. The history of GC pulsar searches has directly demonstrated this fact many times.

However, there seems to be little likelihood of making very large improvements in GC pulsar search sensitivities over the next several years. Most of the variables in the signal-to-noise equation are already nearly optimal (e.g. T_{sys}, B_ν, t_{obs}, and ν). Dramatic improvements in sensitivities and therefore pulsar numbers will almost certainly require a new generation of larger telescopes (i.e. larger A_e) such as FAST or the SKA[‡].

4. Which clusters have pulsars?

Figure 1 shows the number of pulsars in each of the GCs with known pulsars. Currently there are 10 clusters with 5 or more pulsars and 3 clusters with 10 or more pulsars: M28 with 11, 47 Tucanae with 23, and Terzan 5 with 33. Camilo & Rasio (2005) pointed out that there are very few clear correlations between cluster parameters and the numbers of known pulsars. In fact, the only simple properties that seem to be related to the number of known pulsars are the total mass of the cluster (which likely influenced how many neutron stars were originally retained) and the distance D to the cluster (since all GC pulsar searches are currently sensitivity limited). However, even these indicators have exceptions. For example, ω Centauri has been searched extensively but unsuccessfully with the Parkes telescope, yet it is one of the nearest and most massive GCs in the Galactic system.

A more sophisticated indicator of which clusters may contain more LMXBs and therefore MSPs is the predicted stellar interaction rate Γ_c in the cores of the clusters, where LMXBs and MSPs are likely formed. Pooley *et al.* (2003) showed a strong correlation between the number of X-ray sources in a cluster and its Γ_c, which they expressed as $\Gamma_c \propto \rho_0^{1.5} r_c^2$, where ρ_0 is the central density and r_c is the core radius. We can attempt to adjust the indicator to account for our sensitivity issues by ranking clusters by $\Gamma_c D^{-2}$. Using this metric, we find that many of the clusters with numerous pulsars are near the top of the list, including 47 Tuc, Terzan 5, M62, NGC 6440, NGC 6441, NGC 6544, M28, and M15. Also near the top are several others which likely contain numerous pulsars but whose positions near or behind the Galactic center region (and therefore large amounts of ISM) make searches very difficult (e.g. NGC 6388 and Liller 1; Fruchter & Goss 2000).

It is important to realize, though, that because of limited amounts of telescope time, pulsar searchers have specifically *targeted* those clusters near the top of the $\Gamma_c D^{-2}$

[†] 47 Tuc is a possible exception to this rule as deep radio imaging (McConnell *et al.*2004) and X-ray observations (Grindlay *et al.* 2002) indicate that scintillation may have already allowed the identification of nearly all of the observable MSPs in the cluster.

[‡] FAST: `http://www.bao.ac.cn/LT/`, SKA: `http://www.skatelescope.org`

list first. Therefore, clusters further down that list may simply not have known pulsars because they haven't been searched to the same sensitivity levels as the clusters near the top of the list. Until we have a large number of GCs, independent of their $\Gamma_c\,D^{-2}$ values, searched to sensitivities comparable to the recent GBT 2 GHz surveys, it will be difficult to determine just how good of a predictor $\Gamma_c\,D^{-2}$ really is.

The recent 1.4 GHz survey of all 22 GCs within 50 kpc and visible with Arecibo (Hessels *et al.* 2007) is a good example of the type of surveys we need. The Arecibo survey found 11 new MSPs, the majority of which are in clusters with fairly average values of $\Gamma_c\,D^{-2}$. No pulsars were found (or have ever been found) in clusters with very low central luminosity densities, $\rho_0 < 10^3\,L_\odot\,pc^{-3}$. A similar unbiased survey of $\sim$60 GCs at 1.4 GHz using the Parkes telescope has uncovered 12 new pulsars (D'Amico *et al.*2001, Possenti *et al.*2001) in 6 GCs. Unfortunately, the limited sensitivity of that survey does not rule out even relatively bright MSPs in many of the clusters, thereby making it difficult to draw conclusions about cluster properties and their pulsar populations.

4.1. *What pulsars are in those clusters?*

We can compare the pulsar populations in the best studied clusters, such as Terzan 5 and 47 Tucanae, to attempt to determine if the properties of the clusters affect their pulsars. Two of the simplest things to compare are the spin-period distributions and the binary populations, both of which do show significant differences.

The spin-period distribution of the Terzan 5 pulsars is significantly flatter than that of the 47 Tuc pulsars with more faster and more slower pulsars. In fact, Ter 5 contains 5 of the 10 fastest known spinning pulsars (Hessels *et al.* 2006), while 47 Tuc contains only one in the top ten. Likewise, 47 Tuc has no MSPs slower than $\sim$8 ms, whereas Ter 5 has six. A Kolmogorov-Smirnov test suggests a $<$10% chance that the two period distributions were drawn from the same parent distribution. As for the binary populations, Ter 5 has only two known "Black-Widow" systems compared to five in 47 Tuc. Yet Ter 5 has six highly-eccentric binaries compared to none in 47 Tuc. Are these differences related to the current interaction rates in the cluster cores (Ter 5's is 2–3 times that of 47 Tuc's) or perhaps the "epochs" when the MSP production rates peaked?

The pulsars in M28, NGC 6440, and NGC 6441 are more similar to those in Ter 5 than to those in 47 Tuc, both in terms of their spin periods (i.e. flatter distributions) and their binary parameters (with a broader mix of different systems). It is interesting to note, though, that the 10 known pulsars in the very similar clusters NGC 6440 and NGC 6441 rotate on average significantly slower than those in either Ter 5 or 47 Tuc. Only one MSP is faster than 5 ms, and half are slower than 13 ms (Freire *et al.* 2007b).

5. Timing of Cluster Pulsars

While it is obviously essential to *find* the pulsars in GCs to do any science with them, without detailed follow-up observations, and in particular pulsar timing solutions, the amount of science one can do is severely limited. For most GC pulsars the extraordinary precision provided by MSP timing provides $\sim$0.1$''$ and often significantly better astrometric positions (crucial for multi-wavelength follow-up, e.g. Grindlay *et al.*2002), extremely precise Keplerian orbital parameters for binaries, and measurements of the pulsar's apparent spin period derivative. For some pulsars, particularly those in eccentric orbits, certain post-Keplerian orbital parameters can be measured (e.g. Lorimer & Kramer 2004) which determine or constrain the pulsar and/or companion star masses. Establishing timing solutions for as many GC pulsars as possible allows us to use them individually and collectively to probe a wide variety of basic physics and astrophysics.

Over the past two years, the number of GC pulsars with timing solutions (currently 107) has almost doubled, resulting in many interesting new results.

5.1. *Ensembles of pulsars*

The largest ensemble of pulsars with timing solutions in a GC are the 32 in Terzan 5 (only Ter5U, a weak eccentric binary, remains without a solution). However, there are five other GCs with five or more pulsars with timing solutions (47 Tuc, Freire *et al.*2003; M28, Bégin 2006; M15, Anderson 1992; NGC 6440, Freire *et al.*2007b; and NGC 6752, Corongiu et al. 2006). These pulsar ensembles can produce unique science.

Probes of Ionized ISM and Intra-Cluster Medium The precise DMs (errors $<0.1\,\mathrm{pc\,cm^{-3}}$) and timing positions for 32 pulsars in Ter 5 have recently allowed a unique probe of the ionized Galactic ISM between us and the cluster on parsec scales (Ransom 2007). A calculation of the DM structure function indicates that the fluctuations in the ISM on $0.2-2\,\mathrm{pc}$ scales roughly follow those predicted for Kolmogorov turbulence. Earlier work on 47 Tuc using the pulsar positions, accelerations (see Figure 3), and DMs provided the first definitive measurement of ionized gas within a GC (Freire *et al.* 2001).

Statistical Neutron Star Mass Measurement Using the projected offsets of pulsars from their cluster centers and a model for how relaxed components of different masses should be distributed within a cluster, it is possible to statistically measure the masses of the pulsar systems M_p (Heinke *et al.* 2003). Figure 3 shows the 107 GC pulsars with timing positions split into two different groups: 1) isolated pulsars or binary pulsars with very low-mass companions ($M_{c,min} \lesssim 0.05\,\mathrm{M_\odot}$) and 2) binary pulsars with more massive companions. Surprisingly, the "isolated" systems seem more centrally condensed than the (supposedly more massive) binary systems. Fits of the observed distributions give $q = M_p/M_\star \sim 1.5$ for the binaries and $q \sim 1.7$ for the "isolated" systems. Assuming that the dominant stars in the cluster cores have mass $M_\star = 0.9\,\mathrm{M_\odot}$ implies binary system masses of $M_{p,bin} \sim 1.35\,\mathrm{M_\odot}$ and *larger* masses $M_{p,iso} \sim 1.53\,\mathrm{M_\odot}$ for the "isolated" pulsars. The reason for this difference in mass segregation is currently unknown.

Cluster Proper Motions The very precise positions available from MSP timing allow individual pulsar proper motions given regular observations over $5-10$ years. The measurement of several pulsar proper motions from a single cluster allows a measurement of the proper motion of the GC itself. Currently this has been accomplished for three clusters (47 Tuc, Freire *et al.*2003; M15, Jacoby *et al.*2006; and NGC 6752, Corongiu *et al.*2006), and several more will likely be measured within the next couple of years. Such measurements are very important for determining the Galactic orbits of GCs and predicting the effects of tidal stripping and/or destruction. Measuring cluster proper motions is very difficult in the optical (using *HST*, for instance), especially for the Galactic bulge clusters which are distant and plagued by extinction.

Cluster Dynamics The projected positions of the pulsars with respect to the cluster centers as well as measurements of their period derivatives (which are usually dominated by gravitational accelerations within the GC; see Figure 3) provide a sensitive probe into the dynamics of the cluster and even constrain the mass-to-light ratio near the cluster center (Phinney 1992). These measurements can provide evidence for the presence (or absence) of black holes in the cores of the clusters (D'Amico *et al.* 2002). Cluster dynamics can also eject pulsars to the outskirts of the clusters or even entirely (Ivanova *et al.* 2007). A recently uncovered example of such a system is M28F, a bright (for a GC MSP) isolated pulsar located almost 3' from the center of M28. That offset is larger than for any other GC pulsar except NGC 6752A (Corongiu *et al.* 2006), where exotic ejection mechanisms have been invoked to explain its position (e.g. Colpi *et al.* 2003).

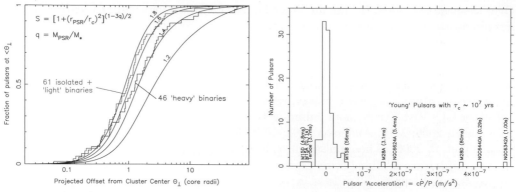

Figure 3. (Left) Radial distribution of "isolated" pulsars (including binaries with very low-mass companions, $M_c \lesssim 0.05\,M_\odot$) and heavier binary pulsar systems. The numbers on the black lines are $q = M_p/M_\star$ values where M_p is the pulsar *system* mass and $M_\star$ is the dominant stellar component mass in the GC cores ($M_\star \sim 0.9\,M_\odot$). Surprisingly, the "isolated" systems appear more centrally condensed (and therefore possibly more massive) than the binaries. **(Right)** Histogram of observed GC pulsar "accelerations" ($c\dot{P}_{obs}/P_{obs}$). The observed acceleration is the sum of the pulsar's intrinsic spin-down (i.e. $c\dot{P}_{int}/P_{int}$), an apparent acceleration due to proper motion, and gravitational accelerations from both the Galaxy and the GC (Phinney 1992, 1993). The proper motion and Galactic terms are typically small compared to the others. Since GC pulsars are usually recycled and have small intrinsic "accelerations" ($\sim 10^{-9}\,\mathrm{m\,s^{-2}}$), the GC gravitational term ($\sim |10^{-8}|\,\mathrm{m\,s^{-2}}$) typically dominates. This is apparent in the figure by a nearly symmetric clustering of pulsars around zero acceleration. Pulsars on the observer's side of a GC receive positive accelerations while those on the far side receive negative accelerations and appear to spin more rapidly with time. The five pulsars with observed accelerations $> 10^{-7}\,\mathrm{m\,s^{-2}}$ are anomalously young (characteristic ages $1 \times 10^7 \lesssim \tau_c \lesssim 3 \times 10^7$ years) with intrinsic accelerations much larger than the maximum possible gravitational accelerations from their GCs. Apparently GCs continue to produce such systems (e.g. Ivanova *et al.*2007). The three pulsars with the most negative accelerations provide unique probes of the central dynamics (and lower limits on the mass-to-light ratios) of their parent GCs (e.g. Anderson 1992).

5.2. *Individual Exotic Pulsars*

Ensemble studies are interesting, but many GC pulsars are worth studying individually.

Young and Slow Pulsars A handful of slow, "normal" ($\tau_c \sim 10^7$ yrs) pulsars have been known in GCs for some time: B1718−19 aka NGC 6342A (Lyne *et al.* 1993), B1820−30B aka NGC 6624B (Biggs *et al.* 1994), and B1745−20 aka NGC 6440A (Lyne *et al.* 1996). Recently, at least one more slow pulsar has been uncovered, NGC 6624C with $P_{spin} = 0.405\,\mathrm{s}$ (Chandler 2003), as well as M28D, an 80 ms binary pulsar that is definitely "young" (see Figure 3). The slow pulsars, which have likely not been through any recycling, must have formed relatively recently even though all of the massive stars would have gone supernova 10^{10} yrs ago. One possibility is that the pulsars formed via electron-capture supernovae, perhaps via accretion-induced collapse of a massive WD or merger-induced collapse of coalescing double WDs (Ivanova *et al.* 2007).

The fastest MSPs Ter5ad is the fastest MSP known, with a spin period of $P_{spin} = 1.396\,\mathrm{ms}$ (Hessels *et al.* 2006). Its discovery broke the 23-yr-old "speed" record established by the very first MSP known (Backer *et al.* 1982) and renewed hope for finding a sub-MSP (a pulsar with $P_{spin} < 1\,\mathrm{ms}$). A sub-MSP would provide the most direct and interesting constraints on the properties of matter at nuclear densities and would be of major significance to physics in general (Lattimer & Prakash 2007). Ter5 also hosts several very rapid rotators (Ter5O at 1.676 ms and Ter5P at 1.728 ms; Ransom *et al.* 2005), and it seems likely that the first sub-MSP (if they exist) might be found in a GC.

"Main-Sequence"—MSP Systems Several pulsars have been recently discovered (including 47TucW, Edmonds *et al.*2002, Bogdanov *et al.*2005; M62B, Possenti *et al.*2003; Ter5P, Ransom *et al.*2005; Ter5ad, Hessels *et al.*2006; and M28H, Bégin 2006) which appear to have bloated "main-sequence"-like companion stars much like the prototype system J1740−5340 in NGC 6397 (D'Amico *et al.*2001; Ferraro *et al.*2001). These pulsars are eclipsed for large fractions of their orbits and often show irregular eclipses. Timing positions usually associate them with hard X-ray point sources where the X-rays are likely generated via colliding MSP and companion winds. Several of the pulsars have been identified in the optical where they exhibit variability at the orbital period. In addition, at least some of them exhibit highly erratic orbital variability (resulting in several large amplitude orbital period derivatives) likely due to tidal interactions with the bloated companion stars. These systems could be the result of an exchange encounter between a main sequence star and a "normal" binary MSP system. Alternatively, perhaps the companions have recently recycled the pulsars and we are observing newly born MSPs (D'Amico *et al.*2001; Ferraro *et al.*2001). Multi-wavelength studies of these systems are difficult (e.g. Bogdanov *et al.*2005), but they allow constraints on MSP emission mechanisms, winds, evolutionary histories, and even tidal circularization theory.

Highly Eccentric Binaries At least 15 GC pulsar systems are members of eccentric binaries with $e > 0.1$, and most of those contain MSPs. In contrast, only a single eccentric binary MSP is known in the Galactic disk. Ten of these systems have been discovered since 2004: six in Terzan 5 (Ransom *et al.* 2005), two in M28 (Bégin 2006), and one each in NGC 1851 (Freire *et al.* 2007a) and NGC 6440 (Freire *et al.* 2007b). Given the angular reference that ellipses provide, pulsar timing easily measures the precession of the angle of periastron or $\dot{\omega}$. For compact companions, $\dot{\omega}$ is dominated by general relativistic effects, and its measurement provides the total mass of the binary system (Lorimer & Kramer 2004). Timing observations of four of these systems (Ter5I & J, Ransom *et al.*2005; NGC 6440B, Freire *et al.*2007b; and M5B, Freire *et al.*in prep.) indicate "massive" neutron stars ($>1.7\,M_\odot$) which constrain the equation-of-state of matter at nuclear densities (Lattimer & Prakash 2007). Such constraints are impossible to achieve in nuclear physics laboratories here on Earth. In addition, similar measurements for M28C, a 4.15 ms pulsar in an 8-day orbit with $e = 0.85$, indicate that the pulsar is *less* massive than $1.37\,M_\odot$. This is a fairly low mass for a neutron star which must have accreted a substantial amount of material during recycling and will likely constrain recycling models.

Other Exotica There is already one confirmed GC triple system (PSR B1620−26 in M4) which contains an MSP, a WD, and a planetary-mass component (Sigurdsson *et al.* 2003). Intriguingly, ongoing observations show very strange and seemingly systematic timing residuals from the "isolated" MSP NGC 6440C (Bégin 2006). One explanation for these residuals is the presence of one or more terrestrial-mass planets. Given the strange variety of systems already known in GCs, it is possible that one of the many currently uncharacterized systems could be another unique object.

6. Prospects for the Future

Given the wide variety of often unanticipated science that GC pulsars provide, and the fact that we probably currently observe only 10−20% of all the GC pulsars, the future of the field seems very bright. Search sensitivity improvements with current telescopes will likely uncover tens of new pulsars and next-generation facilities like FAST and the SKA promise to find hundreds. Among these pulsars may be more spectacular exotica, such as MSP−MSP or MSP−black hole binaries, and doubtless many surprising results.

Acknowledgments Thanks go to my collaborators on recent GC pulsar searches: Steve Bégin, Ryan Lynch, Jennifer Katz, Lucy Frey, Mike McCarty, Ben Sulman, Fernando Camilo, Vicky Kaspi, and especially Jason Hessels, Ingrid Stairs, and Paulo Freire.

References

Alpar, M. A., *et al.* 1982, *Nature*, 300, 728

Anderson, S. B. 1992, PhD thesis, California Institute of Technology

Backer, D. C., *et al.* 1982, *Nature*, 300, 615

Bégin, S. 2006, Master's thesis, University of British Columbia

Biggs, J. D., *et al.* 1994, *MNRAS*, 267, 125

Bogdanov, S., Grindlay, J. E., & van den Berg 2005, *ApJ*, 630, 1029

Camilo, F., *et al.* 2000, *ApJ*, 535, 975

Camilo, F. & Rasio, F. A. 2005, in ASP Conf. Ser., Vol. 328, Binary Radio Pulsars, ed. F. A. Rasio & I. H. Stairs, 147

Chandler, A. M. 2003, PhD thesis, California Institute of Technology

Clark, G. W. 1975, *ApJ*, 199, L143

Colpi, M., Mapelli, M. , & Possenti, A. 2003, *ApJ*, 599, 1260

Corongiu, A., *et al.* 2006, *ApJ*, 653, 1417

D'Amico, N., *et al.* 2001, *ApJ*, 561, L89

D'Amico, N., *et al.* 2002, *ApJ*, 570, L89

Edmonds, P. D., *et al.* 2002, *ApJ*, 579, 741

Ferraro, F. R., *et al.* 2001, *ApJ*, 561, L93

Freire, P. C., *et al.* 2001, *ApJ*, 557, L105

Freire, P. C., *et al.* 2003, *MNRAS*, 340, 1359

Freire, P. C. C., Ransom, S. M., & Gupta, Y. 2007a, *ApJ*, 662, 1177

Freire, P. C. C., *et al.* 2007b, *ApJ* submitted

Fruchter, A. S. & Goss, W. M. 2000, *ApJ*, 536, 865

Grindlay, J. E., *et al.* 2002, *ApJ*, 581, 470

Harris, W. E. 1996, *AJ*, 112, 1487, (http://www.physics.mcmaster.ca/resources/globular.html)

Heinke, C. O., *et al.* 2003, *ApJ*, 598, 501

Hessels, J. W. T., *et al.* 2006, *Science*, 311, 1901

Hessels, J. W. T., *et al.* 2007, *ApJ* in press, arXiv:0707.1602

Ivanova, N., *et al.* 2007, *MNRAS* submitted, arXiv:0706.4096

Jacoby, B. A., *et al.* 2006, *ApJ*, 644, L113

Katz, J. I. 1975, *Nature*, 253, 698

Lattimer, J. M. & Prakash, M. 2007, *Physics Reports*, 442, 109

Lorimer, D. R. & Kramer, M. 2004, Handbook of Pulsar Astronomy (Cambridge Univ. Press)

Lyne, A. G., *et al.* 1987, *Nature*, 328, 399

Lyne, A. G., *et al.* 1993, *Nature*, 361, 47

Lyne, A. G., Manchester, R. N., & D'Amico, N. 1996, *ApJ*, 460, L41

Maron, O., *et al.* 2000, *A&AS*, 147, 195

McConnell, D., *et al.*2004, *MNRAS*, 348, 1409

Phinney, E. S. 1992, *Phil. Trans. Roy. Soc. A*, 341, 39

Phinney, E. S. 1993, in ASP Conf. Ser., Vol. 50, Structure and Dynamics of Globular Clusters, ed. S. G. Djorgovski & G. Meylan, 141

Pooley, D., *et al.* 2003, *ApJ*, 591, L131

Possenti, A., *et al.* 2001, in Aspen Workshop: Compact Objects in Dense Star Clusters, astro-ph/0108343

Possenti, A., *et al.* 2003, *ApJ*, 599, 475

Ransom, S. M., *et al.* 2005, *Science*, 307, 892

Ransom, S. M. 2007, in ASP Conf. Ser., Vol. 365, SINS - Small Ionized and Neutral Structures in the Diffuse Interstellar Medium, ed. M. Haverkorn & W. M. Goss, 265

Sigurdsson, S., *et al.* 2003, *Science*, 301, 193

Dynamical Evolution of Dense Stellar Systems
Proceedings IAU Symposium No. 246, 2007
E. Vesperini, M. Giersz & A. Sills, eds.

© 2008 International Astronomical Union
doi:10.1017/S1743921308015822

Observational Evidence for the Origin of X-ray Sources in Globular Clusters

Frank Verbunt[1], Dave Pooley[2] and Cees Bassa[3]

[1] Astronomical Institute, Postbox 80.000, 3508 TA Utrecht, the Netherlands
email: verbunt@astro.uu.nl

[2] Dept. of Astronomy, University of Wisconsin-Madison Madison WI 53706-1582, U.S.A.
email: dave@astro.wisc.edu

[3] Physics Department, McGill University, Montreal, QC H3A 2T8 Canada
email: bassa@physics.mcgill.ca

Abstract. Low-mass X-ray binaries, recycled pulsars, cataclysmic variables and magnetically active binaries are observed as X-ray sources in globular clusters. We discuss the classification of these systems, and find that some presumed active binaries are brighter than expected. We discuss a new statistical method to determine from observations how the formation of X-ray sources depends on the number of stellar encounters and/or on the cluster mass. We show that cluster mass is not a proxy for the encounter number, and that optical identifications are essential in proving the presence of primordial binaries among the low-luminosity X-ray sources.

Keywords. X-rays: binaries, globular clusters: general, stellar dynamics

1. Introduction

The first celestial maps in X-rays, in the early 1970s, show that globular clusters harbour more X-ray sources than one would expect from their mass. As a solution to this puzzle it was suggested that these bright ($L_x \gtrsim 10^{36}$ erg/s) X-ray sources, binaries in which a neutron star captures mass from a companion star, are formed in close stellar encounters. A neutron star can be caught by a companion in a tidal capture, or it can take the place of a star in a pre-existing binary in an exchange encounter. Verbunt & Hut (1987) showed that the probability of a cluster to harbour a bright X-ray source indeed scales with the number of stellar encounters occurring in it; whereas a scaling with mass does not explain the observations.

With the *Einstein* satellite a dozen less luminous ($L_x \lesssim 10^{35}$ erg/s) X-ray sources were discovered in the early 1980. *ROSAT* enlarged this number to some 55, and now thanks to *Chandra* we know of hundreds of dim X-ray sources in globular clusters. The nature and origin of these dim sources is varied. Those containing neutron stars, i.e. the quiescent low-mass X-ray binaries in which a neutron star accretes mass from its companion at a low rate and the recycled or millisecond radio pulsars, have all formed in processes involving close stellar encounters. The magnetically active binaries, on the other hand, are most likely primordial binaries, with stars that are kept in rapid rotation via tidal interaction. Cataclysmic variables are binaries in which a white dwarf accretes matter from a companion. In globular clusters they may arise either via stellar encounters, or from primordial binaries through ordinary binary evolution – this is expected to depend on the mass and density of the globular cluster.

In this paper we describe the classification and identification of the dim sources in Section 2, and make some remarks on the theory of their formation in Section 3. In Section 4

301

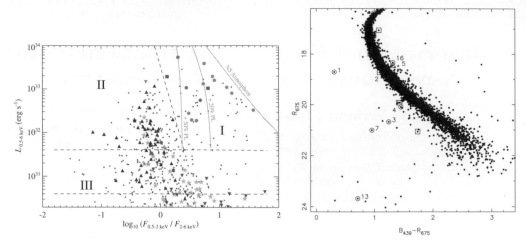

Figure 1. Left: X-ray hardness-luminosity diagram for dim sources in globular clusters. I: quiescent low-mass X-ray binaries, II: cataclysmic variables III: cataclysmic variables and magnetically active binaries. From Pooley & Hut (2006). Right: Colour-magnitude diagram of NGC 6752 on the basis of HST-WFPC2 data; objects within X-ray position error circles are marked. Left of the main sequence we find cataclysmic variables, above it active binaries. Updated from Pooley *et al.* (2002a).

we will discuss a new, and in our view more accurate, way to compare the numbers of these sources with theoretical predictions.

2. Classification and identification

Work on the dim sources is progressing along various lines. Grindlay and coworkers study one cluster, 47 Tuc, in great detail (Grindlay *et al.* 2001, Edmonds *et al.* 2003, Heinke *et al.* 2005). Webb and coworkers use XMM to obtain high-quality X-ray spectra (e.g. Webb *et al.* 2006, Servillat this meeting). Dim sources are also found in clusters in which individual sources are the main target, such as Terzan 1 and 5, and M 28 (Wijnands *et al.* 2002, Heinke *et al.* 2003, Becker *et al.* 2003). Lewin initiated a large program to observe clusters with very different central densities and core radii, and thereby to provide material for tests on the dependence on these properties of the numbers of dim sources. Further references to all this work may be found in the review by Verbunt & Lewin (2006); and in the remainder of this Section.

The first classification of the dim sources may be made on the basis of the **X-ray properties only** (Fig. 1). The brightest sources in the 0.5-2.5 keV band, at $L_x \gtrsim 10^{32}$ erg/s, tend to be quiescent low-mass X-ray binaries. To better use the Chandra range, one may also select the brightest sources in the 0.5-6.0 keV band, and select soft sources, with a high ratio of fluxes below and above e.g. 2 keV: $f_{0.5-2.0\text{keV}}/f_{2.0-6.0\text{keV}} \gtrsim 1$. Such sources also are mostly quiescent low-mass X-ray binaries. Between 10^{31} and 10^{32} erg/s most sources are cataclysmic variables, especially when they have hard spectra $f_{0.5-2.0\text{keV}}/f_{2.0-6.0\text{keV}} < 1$. The faintest sources include magnetically active binaries, often with soft X-ray spectra. For many faint sources the number of counts is too low to decide on the hardness of the spectrum.

The second step in classification can be made when **identification** with a source **at other wavelengths** is made. Positional coincidence of an X-ray source with the accurate radio position of a millisecond pulsar provides a reliable identification and classification. Positional coincidence with optical sources is only significant if the highest

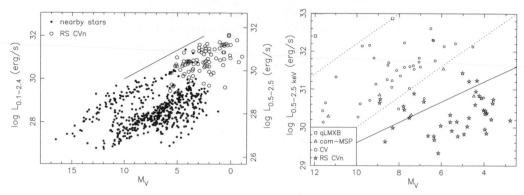

Figure 2. Left: X-ray luminosity as a function of absolute visual magnitude for nearby stars (*selected* from Hünsch *et al.* 1999, for details see Verbunt 2001) and for RS CVn systems (from Dempsey et al. 1993). The upper bound Eq. 2.3 is indicated with a solid line. We convert the X-ray fluxes in the 0.1-2.4 keV range (scale on the left) to the 0.5-2.5 keV range by multiplication with 0.4 (scale on the right). Right: X-ray luminosity as a function of absolute visual magnitude, for dim X-ray sources in globular clusters. The assumed separatrices Eqs. 2.1,2.2 are indicated with dotted lines, the upper bound Eq. 2.3 with a solid line. It is seen that some X-ray sources classified as active binaries in globular clusters are well above this bound.

possible astrometric accuracy is used to limit the number of possible counterparts (e.g. Bassa *et al.* 2004). The position of these possible counterparts in a colour-magnitude diagram is then used to select the probable counterparts. Cataclysmic variables are bluer than the main sequence stars, and magnetically active binaries may lie above the main sequence. Systems on the main sequence cannot be unambiguously classified: they may either be cataclysmic variables in which the optical flux is dominated by the donor star, or main-sequence binaries with unequal masses whose optical light is dominated by the brighter star. If a periodicity is found in the X-rays that corresponds to a period at another wavelength, e.g. the pulse period of a pulsar or the orbital period of a binary, identification and classification are secured simultaneously (e.g. Ferraro *et al.* 2001).

A very useful discriminant in X-ray astronomy in general is the X-ray to optical flux ratio. In the case of globular clusters we can use the known distance to determine the **optical to X-ray luminosity ratio** (Fig 2). On the basis of in particular the extensive data on 47 Tuc (Edmonds *et al.* 2003 and references therein) one finds that the lines of constant optical to X-ray luminosity ratio

$$\log L_{0.5-2.5\text{keV}}\,(\,\text{erg/s}) = 36.2 - 0.4 M_V \tag{2.1}$$

separates the quiescent low-mass X-ray binaries above it from the cataclysmic variables below. The line

$$\log L_{0.5-2.5\text{keV}}\,(\,\text{erg/s}) = 34.0 - 0.4 M_V \tag{2.2}$$

roughly separates the cataclysmic variables from the magnetically active binaries.

This latter separatrix leads to a surprise when one compares it with the X-ray luminosities of nearby stars and of known magnetically active binaries, i.e. RS CVn systems, near the Sun. For main-sequence stars in the solar neighbourhood, the X-ray luminosity increases with the rotation speed, up to an upper bound given approximately by

$$\log L_{0.5-2.5\text{keV}}\,(\,\text{erg/s}) = 32.3 - 0.27 M_V \tag{2.3}$$

as illustrated in Fig. 2 (left). This bound is lower than the separatrix given by Eq. 2.2, especially for brighter stars. This would imply that active binaries in globular clusters can have higher X-ray luminosities than similar binaries near the Sun. We suggest, however, that the classification must be reinvestigated, and that some of these objects are cataclysmic variables. The absence of the blue colour expected for a cataclysmic variable (see Fig. 1) then requires explanation – e.g. as a consequence of the non-simultaneous measurements at different colours combined with source variability.†

3. Some remarks on theory

Binaries in a globular clusters change due to their internal evolution and/or due to external encounters. To describe the current cluster binary population one must track the events for each primordial binary and for each binary that is newly formed via tidal capture, throughout the cluster. The first estimates of the formation of binaries with a neutron star necessarily made a number of drastic simplifications. The sum of all encounters (of a neutron star with a single star, or with a binary) was replaced with an integral over the cluster volume of the encounter rate per unit volume. Four assumptions followed: the number density n_1, n_2 of each participant in the encounter scales with the total mass density ρ, the relative velocity between the encounter participants scales with the velocity dispersion v, the interaction cross section A is dominated by gravitational focusing so that $A \propto 1/v^2$, and the encounter rate is dominated by the encounters in the dense cluster core. Hence one writes the cluster encounter rate Γ' as

$$\Gamma' = \int_V n_1 n_2 A v dV \propto \rho_o{}^2 r_c{}^3 / v \qquad (3.1)$$

where ρ_o is the central density and r_c the core radius. If one further eliminates the velocity dispersion through the virial theorem, $v \propto r_c \sqrt{\rho_o}$, one has

$$\Gamma' \propto \rho_o{}^{1.5} r_c{}^2 \equiv \Gamma \qquad (3.2)$$

where Γ is referred to as the collision number. With a life time τ the expected number of binaries of a given type is

$$N = \Gamma' \tau \propto \Gamma \tau \qquad (3.3)$$

A major advantage of these simple estimates is the clear connection between (the uncertainty in) the input and (the uncertainty of) the output. Thus, if n_1, n_2 are the number densities of neutron stars and of binaries, respectively, Eqs. 3.1-3.3 indicate that the uncertainty in the number N of neutron star binaries scales directly with the uncertainties in n_1 and n_2. Similarly, if we overestimate the life time τ of a binary by a factor 10, the estimated number N is overestimated by the same factor.

Thanks to a concerted effort by various groups fairly detailed computations of the happenings in globular clusters are now undertaken. This is a fortunate and necessary development, as many details cannot be understood from the simple scalings above. For example, the wide progenitors of cataclysmic variables are destroyed by close encounters before they evolve in a dense cluster core (Davies 1997), but evolve undisturbed into cataclysmic variables in the outer cluster regions, from where they can sink to the dense core to form a significant part of the current population there (Ivanova *et al.* 2006).

A disadvantage of complex computations is that they tend to hide the uncertainties. If the cross sections A are described in paper I of a series, and the life times τ in paper III, the large uncertainties in them tend to be less than obvious in paper V where the

---

† In his contribution to this meeting, Christian Knigge shows that the tentative counterpart of W24 in 47 Tuc, a possible active binary according to Edmonds *et al.* (2003, Sect.4.5), has blue FUV-U colours, which suggests that it is a cataclysmic variable. At $M_V = 2.6$ and $L_x = 8.7 \times 10^{30}$ erg/s, it actually lies below the line given by Eq. 2.3.

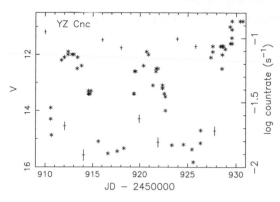

Figure 3. Optical (*, scale on left) and X-ray (+, scale on right) lightcurves of the dwarf nova YZ Cnc through several outburst cycles: the outburst in the optical luminosity is accompanied by a marked *drop* in the X-ray luminosity. After Verbunt *et al.* (1999)

final computations are described. The confidence expressed in summaries of the results of such computations is sometimes rather larger than warranted. An uncertainty in A or τ has an equally large effect in complex computations as in simple estimates. As a further illustration we discuss two other uncertainties.

The first relates to the question what happens when mass transfer from a giant to its binary companion is dynamically unstable. It is usually assumed that a spiral-in follows, in which the companion enters the envelope of the giant and expels it through friction. The outcome of this process is computed using conservation of energy, which implies a drastic shrinking of the orbit (Webbink 1984). However, the study of nearby binaries consisting of two white dwarfs shows that the mass ratios in them are close to unity (e.g. Maxted *et al.* 2002). Such binaries can only be explained if the consequences of dynamically unstable mass transfer are governed by conservation of angular momentum, rather than by the energy equation. If the mass leaving the binary has roughly the same specific angular momentum as the binary, the orbital period changes relatively little during the unstable mass transfer and concomitant mass loss from the binary (Van der Sluys *et al.* 2006). The standard prescription of dynamically unstable mass transfer hitherto implemented in globular cluster computations must be replaced.

The second uncertainty relates to the conversion of the mass transfer rate in a cataclysmic variable to the X-ray luminosity. This conversion does not affect the evolution of the binary but it is important for comparison with observations, as most cataclysmic variables in globular clusters are discovered as X-ray sources. It is generally assumed that the X-ray luminosity scales directly with the mass transfer rate: $L_x \propto \dot{M}$. Alas, reality is more complicated, and indeed in most cases the X-ray luminosity goes down when the mass transfer rate goes up. This is demonstrated unequivocally in dwarf novae whose X-ray luminosity drops precipitously during outbursts (Fig. 3), but there is evidence that it is true in the more stable nova-like variables as well (Verbunt *et al.* 1997). But exceptions are also known: the dwarf nova SS Cyg has higher X-ray flux during outburst than in quiescence (Ponman *et al.* 1995). Theoretical predictions of the numbers of cataclysmic variables in globular clusters that radiate detectable X-ray fluxes, are not believable when based on proportionality of X-ray flux and mass-transfer rate.

Observational evidence for the numbers of binaries of various types, and of the dependence of these numbers on cluster properties, may be collected and used to constrain the theories on formation and evolution of various types of binaries in globular clusters (e.g. Pooley *et al.* 2003, Heinke *et al.* 2006, Pooley & Hut 2006).

In the next Section we describe a new, and we hope more accurate, method of analysing source numbers: this method is based on direct application of Poisson statistics. This topic brings one of us, FV, to a brief Intermezzo.

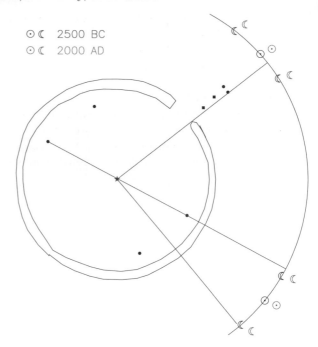

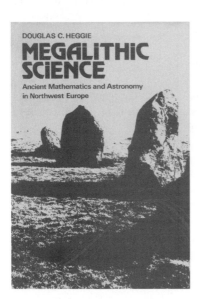

Figure 4. Left: the classic book by Douglas Heggie on Megalithic Science. Right: Sketch of Stonehenge with the circular bank & ditch (the *henge*), the rectangle of the four Station Stones within it, and the directions towards the extreme risings of Sun and Moon at the winter and summer solstitia of 2500 B.C. and 2000 A.D. The minute shift in 4500 yr is due to the small change in obliquity.

Intermezzo by FV: another side of Douglas Heggie

The first time that I had need of understanding of binomial and Poisson statistics was in the early 1980s, when I bought a book by Douglas Heggie (1981) called *Megalithic Science, Ancient Mathematics and Astronomy in Northwest Europe*. The first half of this book discusses the question whether the megalith builders used a standard measure of length (the answer is no), and the second half studies the question whether megalithic structures had astronomical orientations. Whereas Douglas is still rather modest about this book, it is still *the best introduction to the issues in megalithic astronomy* (McCluskey 1998, p.11). It, for the first time, explained the statistics of the study of astronomical orientations of megalithic structures. In my view it is not an exaggeration to state that archaeo-astronomy may be divided in a pre-Heggie era, in which the statistics is usually wrong, and a post-Heggie era, in which at least some people, following Douglas's prescription, do their statistics right. Among the latter I may specifically mention Ruggles (1999).

As an example of an application of binomial statistics I paraphrase the discussion on p.198 of the book, of the four Station Stones in Stonehenge. These Stones were placed, near 2500 B.C., in a rectangle with the short sides parallel to the main axis of Stonehenge (Fig. 4). This main axis points from the center of the rectangle through a gap in the bank-plus-ditch (the *henge*) surrounding it toward the location on the horizon of the most northern annual rising of the Sun, at summer solstice. The Moon ranges in a 19 yr cycle from about 5° north of the Sun to 5° south of the Sun, which leads to a range of its most northern and southern annual risings. The extremes of these ranges, together with the two solstitia, define 6 positions on the eastern horizon. Accepting a range of ±1.8° for each position, we find a probability that an arbitrary direction towards the eastern horizon hits one of these positions $p = 6 \times 2 \times 1.8/180 = 0.12$. The rectangle of the Station Stones defines three new directions towards the East: along the long side and along the diagonals. Two of these hit one of the 6 positions: extremes of the southernmost annual setting of the Moon in the 19 yr cycle. The probability of 1 chance hit in 3 trials with $p = 0.12$ is 28%, the probability of 2 chance hits 4%. Most likely one hit is due to chance, and one hit intentional.

Thanks to new research at Stonehenge, done after the publication of *Megalithic Science*, we now know that the central structure, circle and horse-shoe form, were put up simultaneous with the Station Stones, which implies that the view along the diagonals was blocked. The intentional hit is therefore the one along the long side. Since the long side of a rectangle is necessarily perpendicular to the short side, this implies that the location of Stonehenge was selected for its latitude.

4. Testing models against observations

For the denser clusters it is found that the X-ray sources lie well within the half-mass radius (e.g. NGC 6440, Fig.1 of Pooley *et al.* 2002b), whereas for (apparently) large clusters only the area within the half-mass radius is covered (e.g. 47 Tuc, Grindlay *et al.* 2001). Thus, the analysis generally deals with the sources within the half-mass radius. If N_h is the observed number of X-ray sources, and N_b the number of background (or foreground) sources not related to the cluster, the number of cluster sources within the half-mass radius is $N_c = N_h - N_b$. From a model we may obtain an estimate of the expected number of cluster sources μ_c, and we also may estimate the expected number of background sources μ_b.

In clusters with a small number of sources, $N_c \lesssim 10$, say, one cannot apply chi-squared statistics. One way of solving this is by adding such clusters together, to obtain sufficiently large numbers. This is done by Pooley & Hut (2006). In this process, the information on the individual clusters is lost.

To avoid this problem, we fit as follows. The probability of observing N when μ is expected according to a Poisson distribution is

$$P(N, \mu) = \frac{\mu^N}{N!} e^{-\mu} \tag{4.1}$$

An important aspect of the Poison function for our application is its asymmetry: $P(3, 0.1) \ll P(0, 3)$, i.e. the probability of observing 3 sources when 0.1 is expected is very much smaller than the probability of observing 0 sources when 3 are expected. We now consider

$$N_h = N_c + N_b = \mu_c + \mu_b \qquad \text{with} \qquad \mu_c \equiv a\Gamma + bM \tag{4.2}$$

Both N_c, N_b are realizations of Poisson distributions given by μ_c and μ_b respectively. μ_b is determined for each cluster separately, e.g. from the observed number of sources well outside the half-mass radius. For each value of N_c we take $N_b = N_h - N_c$, and then compute the combined probability $P(N_c, \mu_c)P(N_b, \mu_b)$. We then select the N_c, N_b pair with the highest combined probability (Fig. 5). The fitting procedure consists of varying a and b to maximize

$$P = \prod_j [P(N_c, \mu_c)P(N_b, \mu_b)]_j \tag{4.3}$$

where j indexes the clusters.

Table 1 lists the numbers that we use in our fitting. We note from the Table that mass cannot be used as a proxy for collision number. The cluster with the highest collision

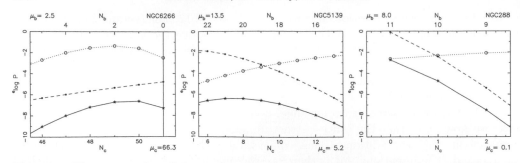

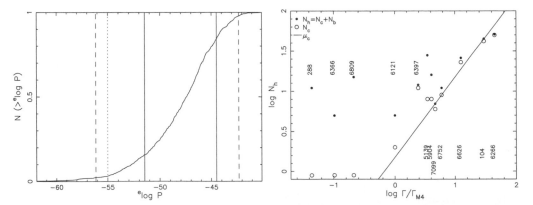

Figure 5. Three examples of the combined fit of the number of cluster sources N_c and background sources N_b for one cluster. The expected number of cluster sources μ_c according to the model $N_c = 1.5\Gamma$, and the expected number of background sources μ_b are indicated with each frame. For each realization $N_c \leqslant N_h$ the accompanying realization is $N_b \equiv N_h - N_c$. E.g. in NGC 6266 the observed number $N_h = 51$, allowing the combined realizations of N_c, N_b as 51,0 or 50,1 or 49,2 etc. The probability $P(N_c, \mu_c)$ is indicated with a dashed line; the probability $P(N_b, \mu_b)$ with a dotted line; and the combined probability – the product of these for each allowed pair – with a solid line. The cases shown are from left to right for low, roughly equal, and dominant background ($\mu_b \ll N_h$, $\mu_b \sim 0.5N_h$, $\mu_b \gg \mu_c$).

cluster	Γ	M	N_h	μ_b	N_s
NGC 6266	44.2	6.3	51	2.5	
NGC 104	29.7	7.7	45	4.0	
NGC 6626	12.6	2.5	26	2.5	
NGC 6752	6.0	1.6	11	2.5	
NGC 7099	4.7	1.2	7	1.5	
NGC 5904	4.1	4.4	16	5.5	
NGC 5139	3.5	17.2	28	13.5	
NGC 6397	2.4	0.6	12	0.5	
NGC 6121	$\equiv 1.0$	$\equiv 1.0$	5	2.0	
NGC 6809	0.2	1.4	15	7.0	3
NGC 6366	0.1	0.3	5	4.0	1
NGC 288	0.0	0.7	11	8.0	2

Table 1. Collision number Γ and mass M, normalized on the values for NGC 6121, for clusters studied with Chandra, together with the number N_h of sources within the half-mass radius, the expected contribution of background sources μ_b, and the number of secure members N_s. Most values N_h and μ_b are from Pooley et al. (2003); those for NGC 288, NGC 6366 and NGC 6809 are from estimated conversion of $L_{0.5-2.5\,keV}$ as given in Kong et al. (2006) and Bassa *et al.* (2007) to $L_{0.5-6.0\,keV}$. Collision numbers and masses are derived from the central density (in $L_\odot\,pc^{-3}$), core radii, and absolute magnitudes given in the Feb 2003 version of the Harris (1996) compilation, and they are scaled on the values for M 4 = NGC 6121. Thus we assume that the mass to light ratio is the same for all clusters.

Figure 6. Left: distribution of $^e\log P$ for 1000 random realizations of the best model $\mu_c = 1.5\Gamma$, with the 1- and 2-σ ranges indicates by solid and dashed lines. The $^e\log P$ value for the best fit is indicated with a dotted line. Right: The best model for $\mu_c = 1.5\Gamma$ is indicated with a solid line, the best number N_c with o (< 0 indicates $N_c = 0$), and the observed number $N_h = N_c + N_b$ with •.

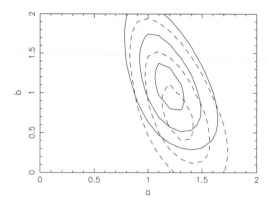

Figure 7. 1,2,3-σ contours for the best values for the model of Eq. 4.2. Dashed contours are for the fits which allow all observed sources to be background sources, solid contours for the fits where a minimum of secure cluster sources is imposed.

number has a ratio $\Gamma/M \simeq 7$ (normalized on M 4), the cluster with the highest mass has $\Gamma/M \simeq 0.2$. This suggests that mass and collision number can be well discriminated, and we confirm this below. The derived values for N_h and μ_b (for sources with $L_{0.5-6.0\text{keV}} > 4 \times 10^{30}$ erg/s) correlate with the derived values of Γ and M, since all are based on assumed values for the cluster distance d and interstellar absorption A_V. For the clusters in the Table the correlations thus introduced are small compared to the ranges of Γ and M, because the cluster distances are relatively well known, and we ignore them in what follows.

The models we fit all have the form given by Eq. 4.2. We first fit models with $b = 0$, i.e. assuming that the number of cluster sources depends only on the collision number. The best solution has $a = 1.5$ and is shown in Fig. 6. The total probability as defined in Eq. 4.3 is $P = e^{-55}$. To see whether this is acceptable, we have used a random generator to produce 1000 realizations by drawing random numbers from Poisson distributions with values μ_c as given by the best model and μ_b as listed in Table 1, and for each realization computed the total probability P. The cumulative distribution of $^e\log P$ is also shown in Fig. 6: the best solution is within the 95% range around the median value, and thus acceptable.

Our next fit allows non-zero values for a and b, i.e. allows a linear dependence both on mass and on collision number. The best solution now has $a = 1.3$ and $b = 0.7$; contours of 1,2,3 σ around these best values are shown in Fig. 7, where we assume that the distribution of $-2(^e\log P - ^e\log P_{\max})$ is given by a χ^2 statistic with one degree of freedom. ($P_{\max}$ is the total probability of the best solution.) We see that the solution with $b = 0$ is marginally acceptable, i.e. the evidence for the mass dependence is marginal. Because we find an acceptable solution for Γ as given by Eq. 3.2, there is no need for a different dependence on central density.

Our final fit uses more information, *viz.* the number of (almost) certain cluster members among the X-ray sources, as determined though optical identifications. The importance of this may be seen from Fig. 5 for NGC 288. With an estimated background $\mu_b = 8$ and, for a model depending on collision number only, $\mu_c \simeq 0.1$, the most probable solution is that all 11 sources within the half-mass radius are background sources (cf. Fig. 6). If we add the constraint that at least two sources included in N_h are cluster members (Kong *et al.* 2006) this solution is no longer possible, and the most probable solution now has $N_c = 2$ and $N_b = 9$. Because the collision number is so small, any cluster source must be a primordial binary, as indeed argued by Kong *et al.*(2006). The best solution for all clusters combined now has $a = 1.2$ and $b = 1.1$, and the solution with $b = 0$ now lies well outside the 3-σ contour, i.e. the dependence on mass M is significant (Fig. 7).

5. Conclusions

- mass M is *not* a proxy for collision number Γ
- the number of dim sources scales both with collision number Γ and with mass M
- scaling with mass only is not acceptable
- correct treatment of the background is important, esp. for faint sources
- to prove the mass-dependence optical identifications are essential

References

Bassa, C. *et al.* 2004, *ApJ*, 609, 755

Becker, W. *et al.* 2003, *ApJ*, 594, 798

Davies, M. 1997, *MNRAS*, 288, 117

Dempsey, R., Linsky, J., Fleming, T., & Schmitt, J. 1993, *ApJS*, 86, 599

Edmonds, P., Gilliland, R., Heinke, C., & Grindlay, J. 2003, *ApJ*, 596, 1177

Ferraro, F., Possenti, A., D'Amico, N., & Sabbi, E. 2001, *ApJ*, 561, L93

Grindlay, J., Heinke, C., Edmonds, P., & Murray, S. 2001, *Science*, 292, 2290

Harris, W. 1996, *AJ*, 112, 1487

Heggie, D. 1981, Megalithic Science, Ancient Mathematics and Astronomy in Northwest Europe (London: Thames and Hudson)

Heinke, C. *et al.* 2003, *ApJ*, 590, 809

Heinke, C. *et al.* 2005, *ApJ*, 625, 796

Hünsch, M., Schmitt, J., Sterzik, M., & Voges, W. 1999, *A&AS*, 135, 319

Ivanova, N. *et al.* 2006, *MNRAS*, 372, 1043

Kong, A. *et al.* 2006, *ApJ*, 647, 1065

Maxted, P., Marsh, T., & Moran, C. 2002, *MNRAS*, 332, 745

McCluskey, S. 1998, Astronomies and cultures in early medieval Europe (C.U.P.)

Ponman, T. *et al.* 1995, *MNRAS*, 276, 495

Pooley, D. & Hut, P. 2006, *ApJ*, 646, L143

Pooley, D. *et al.* 2003, *ApJ*, 591, L131

Pooley, D. *et al.* 2002a, *ApJ*, 569, 405

Pooley, D. *et al.* 2002b, *ApJ*, 573, 184

Ruggles, C. 1999, Astronomy in prehistoric Britain and Ireland (New Haven: Yale Univ. Press)

van der Sluys, M., Verbunt, F., & Pols, O. 2006, *A&A*, 460, 209

Verbunt, F. 2001, *A&A*, 368, 137

Verbunt, F., Bunk, W., Ritter, H., & Pfeffermann, E. 1997, *A&A*, 327, 602

Verbunt, F. & Hut, P. 1987, in The Origin and Evolution of Neutron Stars, IAU Symposium No. 125, ed. D. Helfand & J.-H. Huang (Dordrecht: Reidel), 187–197

Verbunt, F. & Lewin, W. 2006, in Compact stellar X-ray sources, ed. W. Lewin & M. van der Klis (Cambridge University Press), 341–379

Verbunt, F., Wheatley, P., & Mattei, J. 1999, *A&A*, 346, 146

Webb, N., Wheatley, P., & Barret, D. 2006, *A&A*, 445, 155

Webbink, R. 1984, *ApJ*, 277, 355

Wijnands, R., Heinke, C., & Grindlay, J. 2002, *ApJ*, 572, 1002

Dynamical Evolution of Dense Stellar Systems
Proceedings IAU Symposium No. 246, 2007
E. Vesperini, M. Giersz & A. Sills, eds.

© 2008 International Astronomical Union
doi:10.1017/S1743921308015834

Black Hole Motion as Catalyst of Orbital Resonances

C. M. Boily[1], T. Padmanabhan[2] and A. Paiement[3]

[1] Observatoire astronomique & CNRS UMR 7550, Université de Strasbourg I, F-67000
Strasbourg
email: cmb@astro.u-strasbg.fr

[2] I.U.C.A.A., Ganeshkhind Post Bag 4, Pune, India
email: paddy@iucaa.ernet.in

[3] E.N.S.P. de Strasbourg, Parc d'innovation, Bd. Sébastien Brant, F-67412 Ilkirch

Abstract. The motion of a black hole about the centre of gravity of its host galaxy induces a strong response from the surrounding stellar population. We consider the case of a harmonic potential and show that half of the stars on circular orbits in that potential shift to an orbit of lower energy, while the other half receive a positive boost. The black hole itself remains on an orbit of fixed amplitude and merely acts as a catalyst for the evolution of the stellar energy distribution function $f(E)$. We then consider orbits in the logarithmic potential and identify the response of stars near resonant energies. The kinematic signature of black hole motion imprints the stellar line-of-sight mean velocity to a magnitude $\simeq 13\%$ the local root mean-square velocity dispersion σ. The high velocity dispersion at the 5:2 resonance hints to an observable effect at a distance $\simeq 3$ times the hole's influence radius.

Keywords. stellar dynamics, gravitation, Galaxy: centre, galaxies: nuclei

1. Introduction

Black hole (BH) dynamics in galactic nuclei has attracted much attention for many years (e.g., Merritt 2006 for a recent review). The influence of a BH on its surrounding stars is felt first through the large velocity dispersion and rapid orbital motion of the inner-most stars ($\sigma \sim v_{1d} \lesssim 10^3$ km/s). This sets a scale $\lesssim GM_{bh}/\sigma^2$ ($\simeq 0.015 - 0.019$ pc for the Milky Way, henceforth MW) within which high-angle scattering or stellar stripping and disruption takes place. For the MW, low-impact parameter star-BH encounters are likely given the high density of $\rho \sim 10^7 M_\odot/\text{pc}^3$ within a radius of ≈ 10 pc (see e.g. Yu & Tremaine 2003; O'Leary & Loeb 2006; see also Freitag, Amaro-Seoane & Kalagora 2006 for a numerical approach to this phenomenon). Star-BH scattering occurring over a relaxation time (Preto, Merritt & Spurzem 2005; Binney & Tremaine 1987) leads to the formation of a Bahcall-Wolf stellar cusp of density $\rho_\star \sim r^{-\gamma}$ where γ falls in the range 3/2 to 7/4 (Bahcall & Wolf 1977). Genzel *et al.* (2003) modeled the kinematics of the inner few parsecs about Sgr A$\star$ with a mass profile $\rho_\star \sim r^{-1.4}$, which suggests of a strong interplay between the BH and the central stellar cusp. More recently, Schödel *et al.* (2007) presented a double power-law fit to the data, where the power index $\simeq 1.2$ inside a break radius r_{br}, and $\simeq 7/4$ outside, where $r_{br} \simeq 0.2$ pc. This is indicative of on-going evolution inside r_{br} not accounted for by the Bahcall-Wolf solution.

Most, if not all, studies of galactic nuclei dynamics assume a fixed BH (or BH binary) at the centre of coordinates. Genzel *et al.* (1997) had set a constraint of $\lesssim 10$ km/s for the speed of the BH relatively to the galactic plane, a constraint later refined to $\lesssim 2$ km/s (Backer & Sramek 1999; Reid & Brunthaler 2004). Stellar dynamics on scales of $\sim$ few pc surrounding Sgr A$\star$ is complex however, and the angular momentum distribution on that scale is a prime example of this complexity (Genzel *et al.* 2003). Reid *et al.* (2007) used maser emission maps to compute the mean velocity of 15 SiO emitters relatively to Sgr A⋆. They compute a mean (three-dimensional) velocity of up to 45 km/s from sampling a volume of $\simeq 1$ pc about the centre†. This raises the possibility that stars within the central stellar cusp experience significant streaming motion with respect to Sgr A⋆. The breaking radius $r_{br} \sim 0.2$ pc is suggestive of uncertain dynamics on that scale. Random, 'Brownian' BH motion may result from the expected high-deflection angle encounters (Merritt 2001, 2005; Merritt, Berczik & Laun 2007). Here we take another approach, and ask what net effect a BH set on a regular orbit will have on the stars. In doing so, we aim to fill an apparent gap in the modeling of BH dynamics in dense nuclei, by relaxing further the constraint that the hole be held fixed at the centre of coordinates. A full account will be found in Boily, Padmanabhan & Paiement (2007, MNRAS, in the press, hereafter BPP+07).

A rough calculation helps to get some orientation into the problem. Consider a point mass falling from rest from a radius R_o in the background potential of the MW stellar cusp. Let the radial mass profile of the cusp $\rho_\star(r) \propto r^{-3/2}$, consistent with MW kinematic data. If we define the BH radius of influence $\simeq 1$ pc to be the radius where the integrated mass $M_\star(< r) =$ the BH mass $\simeq 3$ to $4 \times 10^6 M_\odot$ (Genzel *et al.* 2003; Ghez *et al.* 2005), then R_o may be expressed in terms of the maximum BH speed in the MW potential as $[\max\{v\}/100 \,\mathrm{km/s}]^{4/5} = R_o/1 \,\mathrm{pc}$. For a maximum velocity in the range 10 to 40 km/s, this yields $R_o \simeq 0.3 - 0.5$ pc, or the same fraction of its radius of influence‡. We ask what impact this motion might have on the surrounding stars. To proceed further, let us focus on a circular stellar orbit outside R_o in the combined potential of the BH and an axisymmetric galaxy. When the BH is at rest at the centre of coordinates, the star draws a closed circular orbit of radius r and constant velocity v. We now set the BH on a radial path of amplitude R_o down the horizontal x-axis. Without loss of generality, let the angular frequency of the stellar orbit be $\omega_\star$, and that of the BH $\omega \geqslant \omega_\star$. The ratio $\omega/\omega_\star \geqslant 1$ is otherwise unbounded. The net force $\boldsymbol{F}$ acting on the star can always be expressed as the sum of a radial component $\boldsymbol{F}_r$ and a force parallel to the x-axis which we take to be of the form $F_x \cos(\omega t + \varphi)$; clearly the constant $F_x = 0$ when $R_o = 0$. The net mechanical work done on the star by the BH as the star completes one orbit is

$$\delta W = \int \boldsymbol{F} \cdot \boldsymbol{v} \mathrm{d}t = \int F_x v \sin(\omega_\star t) \cos(\omega t + \varphi) \,\mathrm{d}t \qquad (1.1)$$

where φ is the relative phase between the stellar and BH orbits. The result of integrating (1.1) is set in terms of the variable $\nu \equiv \omega/\omega_\star$ as

$$\frac{2\delta W}{v F_x} = \frac{1}{\nu + 1} \left[\cos(2\pi\nu + \varphi) - \cos(\varphi)\right]$$

$$+ \frac{1}{\nu - 1} \left[\cos(2\pi\nu - \varphi) - \cos(\varphi)\right] \qquad (1.2)$$

when $\nu > 1$, and $2\delta W/v F_x = 2\pi \sin(\varphi)$ when $\nu = 1$. Equation (1.2) encapsulates the essential physics, which is that δW changes sign when the phase φ shifts to $\varphi + \pi$. Thus whenever the stellar phase-space density is well sampled and all values of $\varphi : [0, 2\pi]$ are

---

† Statistical root-n noise $\sim 25\%$ remains large owing to the small number of sources but is inconsequential to the argument being developed here.

‡ These figures are robust to details of the stellar cusp mass profile, so for instance a flat density profile ($\gamma = 0$) would yield R_o in the range 0.3 to 0.6 pc.

realised with equal probability, half the stars receive mechanical energy ($\delta W > 0$) and half give off energy ($\delta W < 0$). In other words, stars in the first quadrant exchange energy with those in the third quadrant of a Cartesian coordinate system. (Similarly for those in the second and fourth quadrants.) By construction, the BH neither receives nor loses energy but merely acts as a *catalyst* for the redistribution of mechanical energy between the stars. Our goal, then, is to explore the consequences of this mechanism quantitatively for realistic stellar distribution functions.

We present a subset of results lifted from BPP+07 for the case of an BH orbiting in a logarithmic background potential. We focus on the effect of resonances and their likely detection at a distance equal to several times the BH radius of influence and show that streams that develop at the 5:2 resonance have larger Toomre parameters (hot streams).

2. Results

We set our problem in the framework of the logarithmic potential, which we write as

$$\Phi_g(\boldsymbol{r}) = -\tfrac{1}{2} v_o^2 \ln \left| \frac{R^2}{R_c^2} + 1 \right| \tag{2.1}$$

with v_o the constant circular velocity at large distances, and the radius R_c defines a volume inside of which the density is nearly constant. Thus when $r \ll R_c$ we have again harmonic motion of angular frequency $\omega = v_o/R_c$. If we let $q = 1$ and define $u \equiv r/R_c$, the integrated mass $M_g(u)$ reads

$$M_g(< u) = \frac{v_o^2 R_c}{G} \frac{u^3}{u^2 + 1} . \tag{2.2}$$

The mass $M_g(u \gg 1) \propto u$ diverges at large distances, however this is not a serious flaw since we consider only the region where $u \sim 1$. The mass $M_g(u = 1) = v_o^2 R_c/2G$ fixes a scale against which to compare the BH mass M_{bh}. Since the BH orbits within the harmonic core, we set

$$M_{bh} \equiv \tilde{m}_{bh} \frac{v_o^2 R_c}{2G} \tag{2.3}$$

with $0 < \tilde{m}_{bh} \leqslant 1$, and

$$\mathcal{M}(u) = 1 + \tilde{m}_{bh} \frac{1 + u^2}{u^3} \equiv \left(\frac{\omega_\star}{\omega} \right)^2$$

defines the position of orbital resonances when the frequencies are commensurate. The core radius offers a reference length to the problem. The position and velocity of the BH at any time follow from the equation of a harmonic oscillator, $\boldsymbol{R}(t) = R_o \sin(\omega t + \phi_o)\hat{\boldsymbol{x}}$, where the amplitude R_o defines the dimensionless number $u_o = R_o/R_c$ and $\hat{\boldsymbol{x}}$ is a directional unit vector. Our goal is to quantify the time-evolution of a large number of orbits in the combined logarithmic and BH potentials. If we pick parameters such that

$$m_\star \ll M_{bh} < M_g(\max\{u\})$$

then we may neglect the collective feedback of the stars on the BH and galactic potential and study only the response of individual orbits evolving in the time-dependent total potential. This approach will remain valid so long as the response of the stars are relatively modest. The time-evolution of orbits was done numerically using a standard integration scheme, see BPP+07 for details. As a specific case we chose dimension-less parameters $\tilde{m}_{bh} = u_o = 0.3$ (case 'C3' of BPP+07) with physical parameters $v_o = R_c = G = 1$. The radius of influence of the BH is then ≈ 0.58. We neglect the stars self-gravity. The stellar

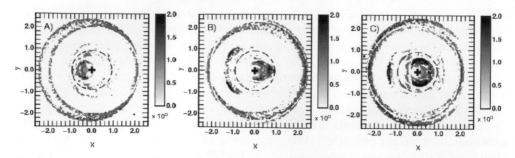

Figure 1. Maps of the Toomre number Q_J for a calculation with $\tilde{m}_{bh} = u_o = 0.3$ shown at times A) t = 15; B) t = 16 and C) t = 17. The dash circle marks the core length of the logarithmic potential, the cross is the origin of coordinates and the light dot marks the position of the BH. The shaded ring of radius $\simeq 2.2$ indicates large dispersion at the 5:2 resonance.

orbits were all co-planar with that of the BH but an extension to 3D only comforted our findings (see BPP+07).

2.1. *Resonances and hot streams*

Given the potential (2.1) it is straightforward to isolate for the radii where the circular orbital frequencies are in a commensurate ratio $m : n$ (see Table 2 of BPP+07). Inspection of the energy distribution function of the stars shows that BH motion induces highs and troughs when compared to the d.f. where the BH is fixed at the coordinate centre. The peaks seen in the d.f. match (roughly) the position of resonances, but, significantly, the d.f. is never steady because the potential varies continuously in time. A time-average of several snapshots, when all orbits are projected in space, shows that the largest resonances are still easily identifiable in the d.f., a result that would favour the detection of long-lasting hot streams (high velocity dispersion) at places where none is expected.

Our approach does not integrate the full response of the stars to their own density enhancements. These could become bound structures which would alter the dynamics globally. To inspect whether this could have an influence over the evolution of the velocity field, we computed the Toomre number

$$Q_J \equiv \frac{\sigma\Omega}{G\Sigma} = \frac{\sigma^2}{G\Sigma \mathrm{d}l}$$

on a mesh of 30×30 points in real space. We computed the dispersion σ with respect to the initial equilibrium flow; hence $\sigma = 0$ when the BH is at rest. The surface density Σ is calculated with an CIC algorithm. Stars are stable against self-gravitating local modes of fragmentation when $Q_J \gtrsim 1.7$ (Binney & Tremaine 1987). We applied a modified criterion for stability, because the disc is presumed initially stable against such modes, that is, when the BH is at the centre of coordinates. Because we subtracted from the local mean square velocity dispersion the value computed for the initial configuration, we set a conservative threshold for stability such that $Q_J > 1$. When that condition is satisfied, the BH contributes through its orbital motion more than 58% of the square velocity dispersion required to prevent local self-gravitating fragmentation modes from growing. Since the black hole already accounts for more than 50% to the gravity everywhere inside its radius of influence, it also provides the extra dispersion required to kill off all self-gravitating modes.

Fig. 1 maps out Q_J at three different times for the reference calculation; the dark shaded area have $Q_J > 1$ with an upper cutoff at 2, so white means instability on that figure. The outer dark ring at $u \simeq 2.2$ on the figure matches the position of the 5:2

resonance. Thus it is very likely that structures that would cross this area would be heated up and disrupted as a result of BH motion.

2.2. *Comparison with MW data*

The ring seen on Fig. 1 may have consequences for the streams of stars observed at the centre of the MW (Genzel *et al.* 2003). The dimensions of this ring, about three times the BH radius of influence, would correspond to a radius of 3 pc in the MW. This should be an element to incorporate into future modelling of the MW centre since actual resolution power already resolves sub-parsec scales.

The line-of-sight velocity is also of interest. This is derived from individual snapshots by projecting the orbits on a 1D mesh and averaging by number. The largest values of v_{1d} were obtained from a viewing angle parallel to the motion of the BH. Contrasting these values to the local root mean square velocity dispersion σ, we find a ratio of $<|v_{1d}|>/\sigma \approx 25\%$ at maximum value, which occurs inside the hole's radius of influence. Applying this to MW data, where the mean velocity dispersion rises to ~ 180 km/s inside 1 pc of Sgr A⋆ (Genzel *et al.* 1996), we obtain streaming velocities in the range ~ 40 km/s, a rough match to the values reported recently by Reid *et al.* (2007). The MW surface density profile shows a break at radius $r_{br} \sim 0.2$ pc (Schödel *et al.* 2007). Inside r_{br}, the volume density is fitted with a power-law index $\gamma \simeq 1.2$ which falls outside the range 3/2 to 7/4 of the Bahcall-Wolf solution. BH motion of an amplitude $R_o \sim r_{br}$ might cause such a break. The ratio $r_{br}/r_{bh} \sim 0.2$ is however lower than the value ≈ 0.5 adopted for the calculation discussed here. More detailed modelling in a galactic cusps is underway for a closer match to MW data.

Acknowledgments

CMB received a travel grant from IFAN which made possible a visit to IUCAA in 2007 where much of this work was carried out. We thank V. Debattista, T. Lauer and D. Merritt for comments.

References

Backer, D. C. & Sramek, R. A. 1999, *ApJ*, 524, 805
Bahcall, J. N. & Wolf, R. A 1977, *ApJ*, 216, 883
Binney, J. J. & Tremaine, S. D. 1987, *Galactic Dynamics*, Princeton: University Press
Boily, C. M., Padmanabhan, T., & Paiement A. 2007, *MNRAS*, astro-ph:0705.2756v2
Freitag, M., Amaro-Seoane, P., & Kalogera, V. 2006, *ApJ*, 649, 91
Genzel, R., Thatte, N., Krabbe, A., *et al.* 1996, *ApJ*, 472, 153
Genzel, R., Eckart, A., Ott, T., *et al.* 1997, *MNRAS*, 291, 219
Genzel, R., Schödel, R., Ott, T., *et al.* 2003, *ApJ*, 594, 812
Ghez, A. M., Salim, S., Hornstein, S. D., *et al.* 2005, *ApJ*, 620, 744
Merritt, D. 2001, *ApJ*, 556, 245
Merritt, D. 2005, *ApJ* 628, 673
Merritt, D. 2006, *RPPh*, 69, 2513
Merritt, D., Berczik, P., & Laun, F. 2007, *AJ*, 133, 553
O'Leary, R. M. & Loeb, A. 2006, submitted to *MNRAS*, astro-ph/0609046
Preto, M., Merritt, D., & Spurzem, R. 2004, *ApJ*, 613, L109
Reid, M. J. & Brunthaler, A. 2004, *ApJ* 616, 872
Reid, M. J., Menten, K. M., Trippe, S., *et al.* 2007, *ApJ*, 659, 378
Schödel, R., Eckart, A., Alexander, T., *et al.* 2007, astro-ph/0703178
Yu, Q. & Tremaine, S. D. 2003, *ApJ*, 599, 1129

Dynamical Evolution of Dense Stellar Systems
Proceedings IAU Symposium No. 246, 2007
E. Vesperini, M. Giersz & A. Sills, eds.

Neutron Stars in Globular Clusters

N. Ivanova[1], C. O. Heinke[2,3] and F. Rasio[3]

[1] CITA, University of Toronto, 60 St George St, Toronto, ON M5R 2N6, Canada
email: nata@cita.utoronto.ca

[2] Department of Astronomy, University of Virginia, 530 McCormick Road Charlottesville, VA 22904-4325, USA
email: cheinke@virginia.edu

[3] Physics and Astronomy Department, Northwestern University, 2145 Sheridan Rd, Evanston, IL 60208 USA
email: rasio@northwestern.edu

Abstract. Dynamical interactions that occur between objects in dense stellar systems are particularly important for the question of formation of X-ray binaries. We present results of numerical simulations of 70 globular clusters with different dynamical properties and a total stellar mass of $2 \times 10^7 M_\odot$. We find that in order to retain enough neutron stars to match observations we must assume that NSs can be formed via electron-capture supernovae. Our simulations explain the observed dependence of the number of LMXBs on "collision number" as well as the large scatter observed between different globular clusters. For millisecond pulsars, we obtain good agreement between our models and the numbers and characteristics of observed pulsars in the clusters Terzan 5 and 47 Tuc.

Keywords. binaries: close, globular clusters: general, X-rays: binaries

1. Introduction

In globular clusters (GCs), neutron stars (NSs) are seen as low-mass X-ray binaries (LMXBs), bright or quiescent, and as binary or single millisecond pulsars (MSPs). As the numbers per unit mass of LMXBs and MSPs in clusters greatly exceed their numbers in the Galaxy, their origin has been linked to stellar encounters that should occur frequently in an environment of high stellar density (Clark 1975). Galactic GCs are known to contain thirteen bright X-ray sources, and as many as ∼100-200 quiescent LMXBs (qLMXBs) are thought to exist in the Galactic GC system (Heinke *et al.* 2003). Millisecond pulsars, likely descendants of LMXBs (for a review see, e.g., Bhattacharya & van den Heuvel 1991), also are present in GCs in great numbers: about 140 GC millisecond radio pulsars have been detected by now†, with more than a dozen in several GCs – in 47 Tuc (Camilo *et al.* 2000), M28 (Stairs *et al.* 2006), and Terzan 5 (Ransom *et al.* 2005). As only a few X-ray binaries or few dozen MSPs are present per fairly massive (more than a million stars) and dense cluster, computationally this problem is very challenging. The target time for direct N-body methods to address the million-body problem is around 2020 (Hut 2006). In our studies, we use a modified encounter rate technique method, described in detail in Ivanova *et al.* (2005), and with the updates described in Ivanova *et al.* (2007).

† See http://www.naic.edu/~pfreire/GCpsr.html for an updated list.

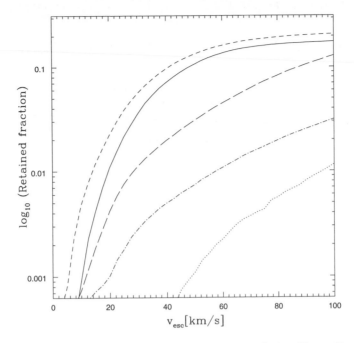

Figure 1. The retention fractions as a function of escape velocity (for stellar evolution unaffected by dynamics) for a Hobbs *et al.* (2005) kick distribution. Dotted and dash-dotted lines show the retention fractions for single and binary populations, core-collapse NSs only. Solid and short-dashed lines show the total retention fractions for single and binary populations, all NSs. For comparison, we show the total retention fraction of a binary population with the Arzoumanian, Chernoff & Cordes (2002) kick distribution (long-dashed line).

2. Production and retention

In our studies we adopt that a NS can be formed as a result of either a core-collapse (CC) supernova or an electron capture supernova (ECS). In the latter case, three possibilities are considered: ECS during normal stellar evolution, accretion induced collapse (AIC) and merger induced collapse. For the case of CC NSs, we adopt that a supernova was accompanied by a natal kick in accordance with the distribution by Hobbs *et al.* (2005). For ECS NSs, we adopt that the accompanying natal kick is 10 times smaller. We find that even considering a stellar population with 100% primordial binaries, the retention fraction of CC NSs is very small (Fig. 1) and the resulting number of retained NSs is just a few per typical dense globular cluster of $2 \times 10^5 \ M_\odot$. NSs formed via different ECSs channels are retained in reasonable numbers, providing about 200 retained NSs per typical GC, or more than a thousand in a cluster like 47 Tuc (similar numbers were found also in Kuranov & Postnov, 2006). Therefore, in contrast to the population of NSs in the Galaxy, the population of NSs in GCs is mainly low-mass NSs made by ECS.

3. X-ray binaries

In our simulations, we find that a typical GC can contain up to 2 LMXBs with a MS companion (most likely observed, at any particular time, as qLMXBs) and up to one LMXB with a WD companion (ultra-compact X-ray binaries, UCXBs). The scatter in the average number of observed LMXBs per cluster in independent simulations is very

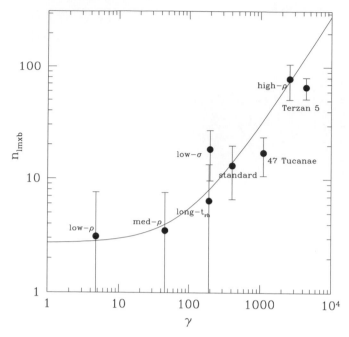

Figure 2. The collision numbers γ (Verbunt & Hut 1987) and numbers of LMXBs in simulated clusters. The solid line corresponds to $n_{LMXB} = (2.7 \pm 6) + 0.028\gamma$, γ is per $10^6 M_\odot$, as in Pooley & Hut (2006). The error bars correspond to the scatter in our simulations. The standard model has a central core density of 10^5 pc^{-3}, velocity dispersion 10 km/s and half-mass relaxation time 1 Gyr. The low-ρ, med-ρ, standard and high-ρ models have central core densities 10^3, 10^4, 10^5 and 10^6 pc^{-3}, respectively. Our low-σ model has velocity dispersion 5 km/s, while long-t_{rh} has half-mass relaxation time 3 Gyr.

large - e.g., for UCXBs, it can vary between 0.1 and 1.1. In the case of Terzan 5 and 47 Tucanae, the average number of LMXBs formed per Gyr, at the age of 11 Gyr, is ~ 5 for NS-MS LMXBs and ~ 8 for UCXBs. These numbers are in general agreement with the observations. Overall the numbers of NS that gain mass via mass transfer (MT) through 11 Gyr of cluster evolution are high: for our 47 Tuc model, about 40 NS-MS binaries and more than 70 UCXBs. As we observe fewer MSPs in these GCs, while the rate of LMXB formation in simulations is consistent with the observations, we conclude that not all NSs that gain mass via MT become currently active MSPs.

We analyzed how the specific number of LMXBs n_{LMXB} in our simulations depends on the specific collision frequency (see Fig. 2). For the case when only core density is varied, n_{LMXB} depends linearly on γ. Variation of other cluster dynamical properties leads to deviation from such a linear dependence, which may explain the scatter in γ in the observed GCs.

4. Millisecond pulsars

Suppose that all mass-gaining events in the life of a NS – mass transfer, physical collision with a red giant, common envelope hyper-accretion or merger – can lead to NS recycling. In this case we found that as many as 250 and 320 potential MSPs are made in our simulations of clusters like Terzan 5 or 47 Tuc, accordingly (the corresponding numbers of retained NSs are ~ 500 and ~ 1100). Although these numbers correlate

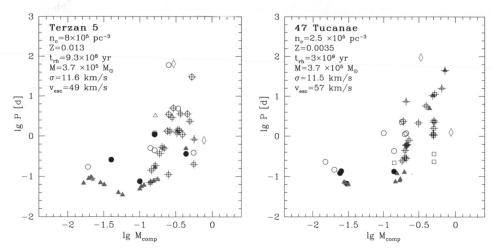

Figure 3. bMSPs in simulated models of 47 Tuc and Terzan 5 compared to observed bMSPs. The simulated populations correspond to several independent runs and represent a larger population than in the observed clusters. Observed bMSPs are shown with circles; triangles - bMSP formed via binary exchanges; stars - via tidal captures; squares - via physical collisions; diamond - primordial binaries. Cross signs mark eccentric bMSPs ($e \geqslant 0.05$) and solid symbols mark systems with a non-degenerate companion (in the case of simulations) or observed eclipsing systems.

well with the formation rate of LMXBs, it greatly exceeds the numbers of observed and inferred MSPs in both clusters, which are 33 (in Terzan 5; perhaps 60 total) and 22 (in 47 Tucanae; perhaps 30 total).

We analyzed the population of NSs that gained mass. We found that bMSPs formed from primordial binaries, where a common envelope event led to AIC, create a population of potential bMSPs with relatively heavy companions, in circular orbits with periods from one day to several hundreds of days. This population is not seen in either Terzan 5 or 47 Tuc. We considered primordial binaries that evolved through mass transfer from a giant donor after a NS was formed via AIC. Even though bMSPs that have similar periods, companion masses and eccentricities are present in Terzan 5, there are no such systems in 47 Tucanae. Also, bMSPs made from primordial binaries after AIC must inevitably be formed in low-dense clusters, but no bMSPs are observed there. These facts tell us that either AIC does not work, or the kicks in the case of AIC are stronger then we adopted, or a NS formed via AIC has such a strong magnetic field, that surface accretion does not occur.

A common understanding of MSP formation is that the NS is recycled through disk accretion, where a NS is spun up only if the accretion rate is not too low, $\dot{M} \geqslant 3 \times 10^{-3}\dot{M}_{\rm Edd}$, where $\dot{M}_{\rm Edd}$ is the Eddington limit (for a review see, e.g., Lamb & Yu 2005). In a UCXB, soon after the start of mass transfer, the accretion rate drops very quickly. After 1 Gyr, it is less than $10^{-4}\dot{M}_{\rm Edd}$. Such a MT leads to a spin-down of the previously spun-up NS, and no MSP is formed. Support for this statement is given by the fact that no UCXBs (those that have WD companions) are visible as MSPs (Lamb & Yu 2005).

The requirement of steady spin-up through disk accretion implies that not all physical collisions will lead necessarily to NS spin-up. In the case of a physical collision with a giant, the NS will retain a fraction of the giant envelope, with a mass of a few hundredths of $M_\odot$ (Lombardi *et al.* 2006). Immediately after the collision, this material has angular momentum and most likely will form a disk. We adopted therefore that in the case of a physical collision with a giant, an MSP can be formed, but, in the case of any other physical collision, the NS will not be recycled.

Considering all the exclusions described above, we form in our simulations at least 15 ± 7 MSPs for Terzan 5 and 25 ± 4 MSPs for 47 Tuc (for the formed population of bMSPs, see Fig. 3). The values for Terzan 5 are somewhat uncertain due to uncertainty in the properties of this heavily reddened cluster. The total number of NSs that gain mass in the simulations by one or other way are 250 and 320 in Terzan 5 or 47 Tuc, accordingly.

5. Conclusions

We studied the formation and evolution of NSs in globular clusters. We find that NS formation via different channels of ECS is very effective in GCs and provides most of the retained NSs. Having as many as a few hundred retained NSs per typical GC, or about 1100 per cluster like 47 Tuc, produces LMXBs in numbers comparable to observations. We note that if AIC does not lead to the formation of NSs, then the number of formed NSs is reduced only by $\sim 20\%$, but the number of appearing LMXBs is decreased by 2-3 times (per Gyr, at the cluster age of 11 Gyr), although it may still be consistent with the observations, given the large scatter in the simulations. We find that up to half of NSs could gain mass after their formation through mass transfer, hyper-accretion during a common envelope, or physical collision. It is likely that most of these mass-gaining events do not lead to NS spin-up, and that only a few per cent of all NSs appear eventually as MSPs, implying that there is a large underlying population of unseen NSs in GCs.

References

Arzoumanian, Z., Chernoff, D., F. & Cordes, J. M. 2002, *ApJ*, 568, 289

Bhattacharya, D. & van den Heuvel, E.P.J. 1991, *Phys. Rep.*, 203, 1

Camilo, F., Lorimer, D. R., Freire, P., Lyne, A. G. & Manchester R. N. 2000, *ApJ*, 535, 975

Clark, G. W. 1975, *ApJL*, 199, L143

Heinke, C O., Grindlay, J. E., Lugger, P. M., Cohn, H. N., Edmonds P. D., Lloyd D. A. & Cool, A. M., 2003, *ApJ*, 598, 501

Hobbs, G., Lorimer, D. R., Lyne, A. G. & Kramer, M, 2005, *MNRAS*, 360, 974

Hut, P. 2006, astro-ph/0610232

Ivanova, N., Belczynski, K., Fregeau, J. M. & Rasio, F. A., 2005, *MNRAS*, 358, 572

Ivanova, N., Heinke, C. O., Rasio, F. A., Taam, R E., Belczynski, K. & Fregeau, J. 2007, *MNRAS*, submitted

Kuranov, A. G. & Postnov, K. A. 2006, *Astron. Lett.*, 32, 393

Lamb, F. & Yu, W. 2005, *Binary Radio Pulsars*, 328, 299

Lombardi, J. C., Proulx, Z. F., Dooley, K. L., Theriault, E. M., Ivanova, N. & Rasio, F. A. 2006, *ApJ*, 640, 441

Pooley, D. & Hut P. 2006, *ApJL*, 646, L143

Ransom, S. M., Hessels, J. W. T., Stairs, I. H., Freire, P. C. C., Camilo, F., Kaspi, V. M. & Kaplan, D. L. 2005, *Science*, 307, 892

Stairs, I. H., *et al.* 2006, *BAAS*, 38, 1118

Verbunt, F. & Hut, P. 1987, in IAUS 125, *The Origin & Evolution of Neutron Stars*, p. 125

Dynamical Evolution of Dense Stellar Systems
Proceedings IAU Symposium No. 246, 2007
E. Vesperini, M. Giersz & A. Sills, eds.

© 2008 International Astronomical Union
doi:10.1017/S1743921308015858

Stellar Exotica in 47 Tucanae

C. Knigge[1], A. Dieball[1], J. Maíz-Apellániz[2], K. S. Long[3], D. R. Zurek[4] and Mike M. Shara[4]

[1] School of Physics & Astronomy, University of Southampton, UK
email:christian@astro.soton.ac.uk

[2] Institute de Astrofísica de Andalucía, Granada, Spain

[3] Space Telescope Science Institute, Baltimore, USA

[4] American Museum of Natural History, New York, USA

Abstract. We have used far-ultraviolet spectroscopy and broad-band photometry to identify and study dynamically-formed stellar exotica in the core of 47 Tucanane. Here, we present a subset of our main results, including: (i) the spectroscopic confirmation of three cataclysmic variables; (ii) the discovery of stripped sub-giant core in a binary system with a dark primary; (iii) the discovery of a Helium white dwarf; (iv) the discovery of a blue straggler with a white dwarf companion.

Keywords. globular clusters: individual (47 Tuc), ultraviolet: stars, stars: novae, cataclysmic variables, stars: white dwarfs, stars: blue stragglers, stellar dynamics

1. Introduction

Stellar collisions and near misses can create various exotic stellar populations in globular clusters. For example, direct stellar collisions can create blue stragglers (BSSs), whereas 2-body and 3-body encounters may produce cataclysmic variables (CVs) and X-ray binaries (XRBs). Some of these dynamically-formed populations can, in turn, affect the evolution of their host clusters. Stellar exotica are thus key tracers of the dynamical processes that drive cluster evolution.

Many of the most interesting dynamically-formed objects (such as all BSSs, CVs and XRBs) are characterized by spectral energy distribution that peak at much shorter wavelengths than ordinary clusters members. Surveys at X-ray (e.g. Grindlay *et al.* 2001; Pooley *et al.* 2002; Heinke *et al.* 2003) and ultraviolet (UV) wavelengths (e.g. Knigge *et al.* 2002, 2003, 2006; Dieball *et al.* 2005a, 2005b, 2007) are therefore particularly useful for finding and studying these objects. Here, we present a brief overview of results from a far-UV *spectroscopic* survey of the core of 47 Tuc that we have carried out with STIS onboard HST. The combination of our far-UV spectra with broadband optical photometry obtained with HST/ACS turns out to be extremely powerful in identifying and classifying the various types of stellar exotica in this cluster.

2. Cataclysmic Variables

Fig. 1 shows far-UV spectra and broad-band SEDs of three previously known CV candidates in our survey area. All of these are X-ray sources (Grindlay *et al.* 2001; Heinke *et al.* 2005) and exhibit variability at both optical (Edmonds *et al.* 2003a, 2003b) and far-UV wavelengths. All three exhibit clear C IV 1550 Å and He II 1640 Å emission lines, confirming their CV classification. The time-resolved far-UV spectroscopy of the brightest source – AKO 9 – was already studied in detail by Knigge *et al.* (2003). The

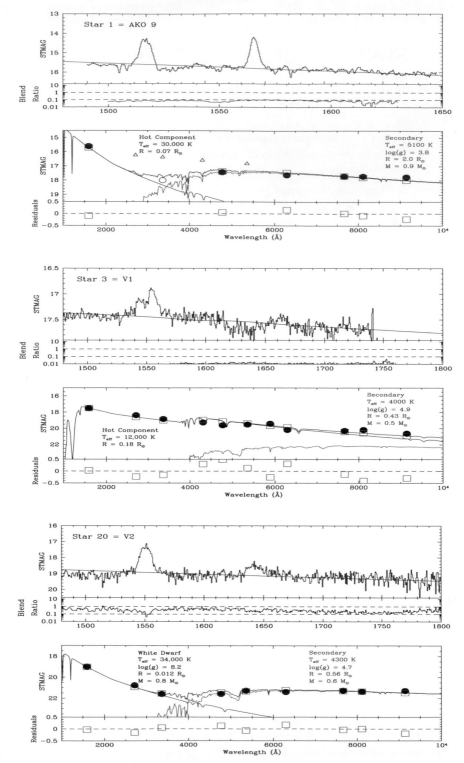

Figure 1. Far-UV spectra and broad-band SEDs of three previously suspected CVs in 47 Tuc. All three display obvious emission lines and the donor star is definitely detected in two of them.

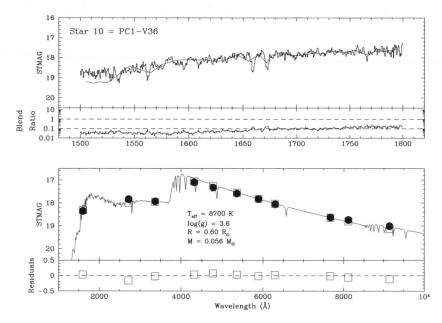

Figure 2. Far-UV spectrum and broad-band SED of PC1-V36. This object is probably an 8 hour binary system containing a dark, compact primary and a stripped sub-giant secondary.

secondary star is detected in two of the object; in both cases, the donor properties are consistent with orbital periods above the CV period gap (c.f. Knigge 2006). Only one other object in our spectroscopic database displays marginally convincing evidence for line emission.

3. A Stripped Sub-Giant in an 8-hour Binary

Fig. 2 shows the far-UV spectrum and SED of PC1-V36, a previously known variable star in 47 Tuc (Albrow *et al.* 2001). The optical light curve suggests that the object is a binary system with a likely orbital period of about 8 hrs in which a (nearly) Roche-lobe filling donor produces most of the light. Our new spectroscopy and photometry can be described by a single component SED, but implies very unusual parameters for the donor: $T_2 \simeq 8700$ K, $\log g \simeq 3.6$ and $R_2 \simeq 0.6$ R$_\odot$, suggesting a donor mass of only about $M_2 \simeq 0.055$ M$_\odot$. The implied mean density is completely consistent with that expected for a Roche-lobe filling star in an 8 hr binary system. We suspect the donor is the remnant of a sub-giant that has been stripped of most of its envelope, either due to mass-transfer in a pre-existing binary, or in the aftermath of the dynamical encounter that may have formed the binary. The primary is not detected and may thus itself be a compact stellar remnant (i.e. a massive white dwarf (WD) or a neutron star).

4. A Helium White Dwarf

Fig. 3 shows the far-UV spectrum and SED of a newly discovered low-mass Helium white dwarf. This is only the second directly detected He WD in 47 Tuc (Edmonds *et al.* 2001), and the first outside a milli-second pulsar system. He WDs are generally thought to be formed only in binary systems, where the progenitor can lose significant amounts of mass to a companion. We see no sign of such a companion in our data, which may suggest that it is again a dark, compact stellar remamnt.

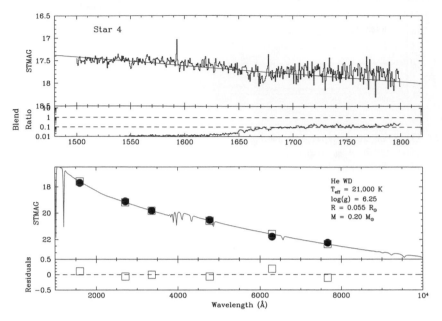

Figure 3. Far-UV spectrum and broad-band SED of a newly discovered Helium WD in 47 Tuc. There is no sign of a binary companion.

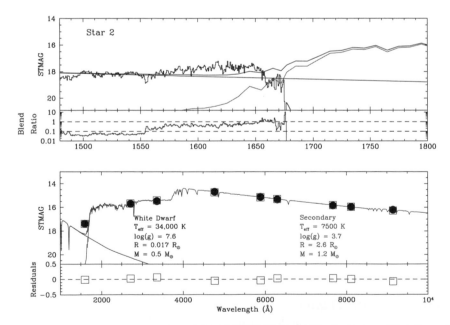

Figure 4. Far-UV spectrum and broad-band SED of a previously known blue straggler in 47 TUc. The spectrum shows a strong far-UV excess at the shortest wavelength, suggesting that the BSS is in a binary system with a hot WD.

5. A Blue Straggler with a White Dwarf Companion

Fig. 4 shows the far-UV spectrum and SED a BSSs in 47 Tuc (where the classification is based on its position in optical and far-UV CMDs). The far-UV spectrum, however, extends to much shorter wavelengths than would be expected for a "normal" BSS (with spectral type A or F). However, the far-UV excess can be accounted for self-consistently if the BSS is in a binary system with a hot WD. This, in turn, suggests that a specific formation scenario for this particular BSS, namely case B mass transfer from the WD progenitor.

References

Albrow, M. D., *et al.* 2001, *ApJ*, 559, 1060

Dieball, A., *et al.* 2007, *ApJ*, in press

Dieball, A., Knigge, C., Zurek, D. R., Shara, M. M. & Long, K. S. 2005a, *ApJ*, 625, 156

Dieball, A., *et al.* 2005b, *ApJ* (Letters), 634, L105

Edmonds, P. D., Gilliland, R. L., Heinke, C. O., Grindlay, J. E. & Camilo, F. 2001, *ApJ* (Letters), 557, L57

Edmonds, P. D., Gilliland, R. L., Heinke, C. O. & Grindlay, J. E. 2003a *ApJ*, 596, 1177

Edmonds, P. D., Gilliland, R. L., Heinke, C. O. & Grindlay, J. E. 2003b *ApJ*, 596, 1197

Grindlay, J. E., Heinke, C. O., Edmonds, P. D. & Murray, S. S. 2001, *Science*, 292, 2290

Heinke, C. O., *et al.* 2003, *ApJ*, 590, 809

Heinke, C. O., *et al.* 2005, *ApJ*, 625, 796

Knigge, C, Zurek, D. R., Shara, M. M. & Long, K. S. 2002, *ApJ*, 579, 752

Knigge, C., Zurek, D. R., Shara, M. M., Long, K. S. & Gilliland, R. L. 2003, *ApJ*, 599, 1320

Knigge, C., *et al.* 2006, *ApJ*, 641, 281

Knigge, C 2006, *MNRAS*, 373, 484

Pooley, D., *et al.* 2002, *ApJ*, 569, 405

Dynamical Evolution of Dense Stellar Systems
Proceedings IAU Symposium No. 246, 2007
E. Vesperini, M. Giersz & A. Sills, eds.

© 2008 International Astronomical Union
doi:10.1017/S174392130801586X

Observations and Simulations of the Blue Straggler Star Radial Distribution: Clues on the Formation Mechanisms

Barbara Lanzoni[1]

[1]Dipartimento di Astronomia, Università di Bologna,
Via Ranzani 1, 40127 Bologna, Italy
email: barbara.lanzoni@bo.astro.it

Abstract. By means of high-resolution and wide-field observations in the UV and optical bands we have derived the radial distribution of the Blue Stragglers Star (BSS) population in a number of galactic globular clusters. Monte-Carlo dynamical simulations have then been used to interpret the observed radial distributions in terms of percentage of collisional and mass-transfer BSS populating each cluster. I will present the main results thus obtained and an overall cluster–to–cluster comparison for the whole sample collected so far, mainly focusing on the clues that such an approach provides about the BSS formation mechanisms .

Keywords. globular clusters: general; stars: evolution; binaries: general; blue stragglers

1. Blue Straggler Stars and their radial distribution

In the color-magnitude diagrams (CMDs) of evolved stellar populations, like globular clusters (GCs), Blue straggler stars (BSS) populate a region which is brighter and bluer (hotter) than the Turn-Off (TO) point, along an extension of the Main Sequence. Thus, they appear as core hydrogen-burning objects, with masses larger than the normal cluster stars (Shara et al. 1997). The standard stellar evolution theory for single mass stars is unable to explain their existence, and a mechanism able to increase the initial stellar mass is required. Two main scenarios have been proposed for their formation: in the *collisional scenario* (Hills & Day 1976), BSS are the end-products of stellar mergers induced by collisions (COL-BSS), while in the *mass-transfer scenario* (McCrea 1964; Zinn & Searle 1976), BSS form by the mass-transfer activity between two companions in a binary system (MT-BSS), possibly up to the complete coalescence of the two stars. *Hence, understanding the origin of BSS provides valuable insight both on binary evolution processes, and on the complex interplay between dynamics and stellar evolution in dense stellar systems.*

The two formation mechanisms are likely to be at work simultaneously in every GC (Ferraro *et al.* 1993, 1997), with efficiencies that probably depend on the local density (Fusi Pecci *et al.* 1992; Ferraro *et al.* 1999a, 2003; Bellazzini et al. 2002). In fact, since stellar collisions are most probable in high-density environments, COL-BSS are expected to be formed preferentially in the cluster cores, while MT-BSS should mainly populate the cluster peripheries, where primordial binaries can more easily evolve in isolation. Since many GCs present high stellar crowding in their centers and large extensions in the sky, a combination of high-resolution and wide-field observations is needed to properly sample both the core and the outskirt environments. Moreover, the relatively high effective temperatures ($T_e \sim 7000 - 8000$ K) of BSS make the ultraviolet (UV) passbands ideal for studying the photometric properties of these stars (see the contribution by Ferraro & Lanzoni in this volume). For all these reasons, we are using a combination of high-resolution and wide-field observations, in the UV and optical bands, in a coordinated project aimed at understanding the formation mechanisms of these puzzling stars.

By adopting such kind of strategy for the study of M3 (Ferraro et al. 1997), 47 Tucanae (Ferraro *et al.* 2004), and NGC 6752 (Sabbi *et al.* 2004), BSS have been found to be more centrally concentrated than normal cluster stars (like red giant branch or horizontal branch stars; hereafter RGB and HB, respectively), consistently with the fact that they are more massive and that mass segregation process is at work in GCs. More surprisingly, the BSS population in these clusters also shows a bimodal radial distribution, i.e., with respect to HB or RGB stars, the BSS fraction is peaked in the center, decreases to a minimum at intermediate radii, and rises again in the outskirts. Such a bimodal radial distribution possibly contains precious information on the BSS formation mechanisms: in fact, the central peak might be due to COL-BSS formed in the core by stellar collisions, or to MT-BSS sunk to the center because of mass segregation and dynamical friction effects; the external rising branch might be produced by COL-BSS formed in the core and then kicked off by dynamical interactions, or by MT-BSS deriving from the normal evolution of binary systems in the cluster outskirts.

In order to disentangle among all these possibilities, we are using a modified version of the Monte-Carlo code originally described by Sigurdsson & Phinney (1995)) for simulating the BSS dynamical evolution within a background cluster, with the aim of reproducing the observed bimodal radial distributions. The background cluster is described as the multi-mass King model that best fits the observed projected density profile. Two populations of BSS are considered: all stars with (random) initial position r_i comprised within the cluster core radius ($r_i < r_c$) and with non-zero natal kick velocity are assumed to be COL-BSS; those with $r_i \gg r_c$ and zero kick velocity are MT-BSS. Their dynamical evolution within the cluster potential well is due to the effects of dynamical friction and distant encounters. By varying the initial percentage of the two kinds of BSS, different radial distributions within the background cluster are obtained at the end of the simulations. The one that best reproduces the observed radial distribution gives the relative amount of COL- and MT-BSS that most likely populate the cluster. By using these simulations, Mapelli *et al.* (2004, 2006) have shown that the external rising branch observed in M3, 47 Tuc and NGC 6752 cannot be due to COL-BSS generated in the core and kicked out by stellar interactions. Instead, the observed bimodality can be explained only by assuming that a non-negligible fraction ($\sim$ 40%) of the BSS population is made of MT-BSS. These results demonstrate that *detailed studies of the BSS radial distribution within GCs are powerful tools for better understanding the BSS formation mechanisms, and the effects of dynamical interactions on the (otherwise normal) stellar evolution.*

In the following I present recent results obtained for three further galactic GCs (namely M5, NGC 1904 and NGC 6388) studied with the strategy described above.

The case of M5 (Lanzoni *et al.* 2007a): We have combined HST (WFPC2 and ACS) observations of the cluster center, in the UV and optical bands, with wide-field optical observations from the ground (with ESO-WFI). The BSS sample has been selected primarily on the basis of the position of stars in the (m_{255}, $m_{255} - U$) plane, adopting a limiting magnitude $m_{255} = 18.35$ ($\sim$ one magnitude brighter than the TO) in order to avoid incompleteness bias and the possible contamination from TO and sub-giant branch stars. Once selected in the UV CMD, the bulk of the BSS in common with the optical-HST sample has been used to define the selection box and the limiting magnitude in the (B, $B - V$) plane. The resulting BSS population of M5 amounts to a total of 60 objects.

One BSS lies beyond the cluster tidal radius and might represent a very interesting case of a BSS evaporating from the cluster because of dynamical interactions. By dividing the sampled area in 8 concentric annuli, we have studied the radial distribution of the fraction of BSS with respect to HB stars ($N_{\mathrm{BSS}}/N_{\mathrm{HB}}$). It has been found to be bimodal: highly peaked within the cluster core radius, decreasing to a minimum at $r \simeq 10\,r_c$, and rising again outward. Dynamical simulations like those previously discussed suggest that the majority of BSS in M5 are collisional, with a content of MT-BSS ranging between 20% and 40% of the overall population (see Fig.1).

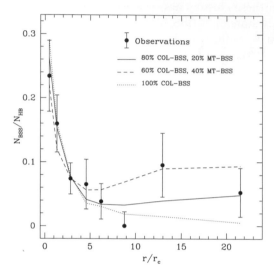

Figure 1. The BSS radial distribution of M5. The simulated distribution that best reproduces the observed one is shown as a solid line and is obtained by assuming 80% of COL-BSS and 20% of MT-BSS. The simulated distributions obtained by assuming 40% of MT-BSS (dashed line) and 100% COL-BSS (dotted line) are also shown for comparison.

The case of NGC 1904 (Lanzoni _et al._ 2007b): We have studied the BSS population of NGC 1904, over the entire cluster extension, by combining high-resolution images obtained with HST-WFPC2, and wide-field observations, both from the ground, with ESO-WFI, and from the space, with GALEX. By adopting selection criteria similar to those discussed for M5, we have found a total of 39 BSS (with $m_{218} \leqslant 19.5$ and $V \leqslant 18.9$). Approximately 38% of the entire BSS population is found within the cluster core, while only $\sim 13\%$ of HB stars are counted in the same region, thus indicating a significant overabundance of BSS in the center. The peak value is in good agreement with what is found in the case of M3, 47 Tucanae, NGC 6752 and M5 (Ferraro _et al._ 2004; Sabbi et al. 2004; Lanzoni _et al._ 2007a), but unlike these clusters, no significant upturn of the distribution at large radii has been detected in NGC 1904 (see second panel of Fig. 2). The Monte-Carlo simulations previously discussed suggest that only a negligible percentage (0–10%) of MT-BSS is needed to reproduce the observed BSS radial distribution in this cluster.

The case of NGC 6388 (Dalessandro _et al._ 2007): To study the BSS population of NGC 6388, we have combined high-resolution images, in the UV and optical bands, obtained with HST (ACS and WFPC2) and wide-field observations from the ground with

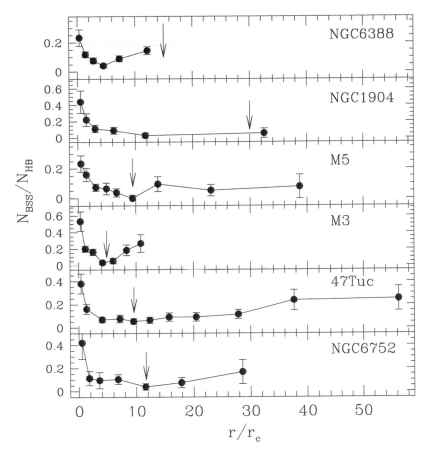

Figure 2. BSS radial distribution of NGC 1904, M5, M3, 47 Tucanae, and NGC 6752. The arrows indicate the position of the radius of avoidance of the clusters (see text), which nicely corresponds to the position of the minimum of the observed distributions, but in the case of NGC 6388.

ESO-WFI. Because of the high degree of Galaxy star contamination, we have limited the analysis to the inner $110''$ from the cluster center, entirely covered by the HST data. As reference population representative of normal cluster stars, we have considered the HB red clump. The resulting BSS radial distribution is clearly bimodal, with a minimum at $r \simeq 5\,r_c$ (see upper panel of Fig. 2). In order to infer the relative percentage of COL- and MT-BSS in this cluster, we are currently performing the Monte-Carlo dynamical simulations discussed above.

2. Discussion

The BSS radial distributions of all the GCs observed to date with the strategy described above are compared in Fig.2.

In all cases BSS appear to be highly concentrated in the cluster centers, and in all but one (NGC 1904) they show a bimodal radial distribution. Dynamical simulations have demonstrated that the peripheral rising branch of the bimodal distributions cannot be generated by COL-BSS formed in the core and kicked out by stellar interactions.

Instead, it can be properly reproduced only by assuming that a non-negligible fraction ($\sim 20-40\%$) of the BSS population is made of MT-BSS. The central peak is likely to be generated both by COL-BSS created inside the core by stellar collisions, and MT-BSS sunk to the center by mass segregation and dynamical friction processes.

These results seem to be confirmed by the fact that the observed position of the minimum nicely agrees with the estimated radius of avoidance of the systems (marked with arrows in Fig.2), i.e., the radius within which all the stars as massive as BSS (i.e., with $M \simeq 1.2\,M_\odot$) are expected to have already sunk to the core due to dynamical friction effects. Up to date, the only exception is represented by NGC 6388, which shows a radius of avoidance about 3 times larger than the position of the observed minimum. Such a result is currently under deeper investigation, and might somehow be connected with other peculiarities shown by this GC, like the long blue tail of its HB (unexpected for such a high-metallicity cluster; Rich *et al.* 1997), the double sub-giant branch sequence (see the contribution by Piotto in this volume), and the possible presence of an intermediate-mass black hole in its center (Lanzoni et al. 2007c).

In conclusion, the study of the BSS radial distribution contain precious information on the formation mechanisms of these puzzling stars and the cluster evolution processes. Extending this kind of investigation to a larger sample of GCs with different structural and dynamical characteristics is required for identifying the cluster properties that mainly affect the BSS formation mechanisms and their relative efficiency.

High-resolution spectroscopic follow-up studies are also crucial for distinguishing between the two types of BSS. In fact, anomalous chemical abundances are expected at the surface of BSS resulting from MT activity (Sarna & de Greve 1996), while they are not predicted in case of a collisional formation (Lombardi *et al.* 1995). The results found in the case of 47 Tucanae (Ferraro *et al.* 2006a; see also the contribution by Ferraro & Lanzoni in this volume) are encouraging, and we are extending this kind of study to other galactic GCs.

References

Bellazzini, M., *et al.* 2002, *AJ*, 123, 1509

Dalessandro , E., *et al.* 2007, *ApJ*, submitted

Ferraro, F. R., *et al.* 1997, *A&A*, 324, 915

Ferraro, F. R., Paltrinieri, B., Rood, R. T., & Dorman, B. 1999a, *ApJ*, 522, 983

Ferraro, F. R., *et al.* 2003, *ApJ*, 588, 464

Ferraro, F. R., *et al.* 2004, *ApJ*, 603, 127

Ferraro, F. R., *et al.* 2006a, ApJ 647, L53

Fusi Pecci, F., *et al.* 1992, *AJ*, 104, 1831

Hills, J. G., & Day, C. A. 1976, *Astrophys. Lett.*, 17, 87

Lanzoni, B., *et al.* 2007a, *ApJ*, 663, 267

Lanzoni, B., *et al.* 2007b, *ApJ* 663, 1040

Lanzoni, B., *et al.* 2007c, *ApJ* 668, L139

Lombardi, J. C. Jr., Rasio, F. A., & Shapiro, S. L. 1995, *ApJ*, 445, L117

Mapelli, M., *et al.* 2004, *ApJ*, 605, L29

Mapelli, M., *et al.* 2006, *MNRAS*, 373, 361

McCrea, W. H. 1964, *MNRAS*, 128, 147

Rich, R. M., *et al.* 1997, *ApJ*, 484, L25

Sabbi, E., Ferraro, F. R., Sills, A., & Rood, R. T. 2004, *ApJ*, 617, 1296

Sarna, M. J., & de Greve, J. P. 1996, *QJRAS*, 37, 11

Shara, M. M., Saffer, R. A., & Livio, M. 1997, *ApJ*, 489, L59

Sigurdsson, S., & Phinney, E. S. 1995, *ApJS* 99, 609

Zinn, R., & Searle, L. 1976, *ApJ*, 209, 734

Dynamical Evolution of Dense Stellar Systems
Proceedings IAU Symposium No. 246, 2007
E. Vesperini, M. Giersz & A. Sills, eds.

Where the Blue Stragglers Roam: Searching for a Link Between Formation and Environment

Nathan Leigh[1], Alison Sills[1] and Christian Knigge[2]

[1] Department of Physics and Astronomy, McMaster University, 1280 Main Street West,
Hamilton, ON, L8S 4M1, Canada
email: leighn@mcmaster.ca, asills@mcmaster.ca

[2] School of Physics and Astronomy, Southampton University, Highfield, Southampton, SO17
1BJ, UK
email: christian@astro.soton.ac.uk

Abstract. Current observational evidence seems to indicate that blue stragglers are a dynamically created population, though exactly how the mechanism(s) of formation operates remains a mystery. We search for links between blue straggler formation and environment by considering only those stars found within one core radius of the cluster center. In so doing, we aim to isolate a sample that is representative of an approximately uniform cluster environment where, ideally, a single blue straggler formation mechanism is predominantly operating. Normalized blue straggler frequencies are found and apart from new anticorrelations with the central velocity dispersion and the half-mass relaxation time, we find no other statistically significant trends.

Concerns regarding the method of normalization used to calculate relative blue straggler frequencies are discussed, specifically whether the previously observed anticorrelation with total cluster mass (see Piotto *et al.* 2004) is a consequence of the normalization process. A new correlation between the observed number of blue stragglers in the core and the number predicted from single-single collisions alone is presented. This new link between formation and environment represents the first direct evidence that the blue straggler phenomenon has, at least in part, a collisional origin.

Keywords. blue stragglers – globular clusters: general

1. Introduction

Blue stragglers are stars that are brighter and bluer (hotter) than the main-sequence (MS) turn-off. Stars having their same mass have evolved off the MS and begun to ascend the giant branch (GB). First discovered by Sandage (1953) in the cluster M3, blue straggler stars (BSSs) are the quintessential example of how stellar evolution alone cannot adequately explain the presence of every stellar sub-population in the color-magnitude diagram (CMD). Specifically, BSSs are suspected to be indicative of the complex interplay between stellar evolution and stellar dynamics that is thought to occur in various cluster environments (see for example Ferraro *et al.* 2004; Sills *et al.* 2005).

There is an overall consensus that blue stragglers are the products of stellar mergers between two (or more) low mass MS stars, either through direct stellar collisions or the coalescence of binary systems (see for example Leonard (1989); Stryker 1993; Bailyn 1995). There is evidence to suggest that both formation mechanisms do operate, however the preferred mode of creation appears to depend on the cluster environment (see for example Mapelli *et al.* 2006). Observations from the Hubble Space Telescope (HST) indicate that BSSs are centrally concentrated in globular clusters (for instance Ferraro *et al.* 1999), though they have been found to have a bimodal radial distribution in clusters like M55 (Zaggia *et al.* 1997), M3 (Ferraro *et al.* 1997), and 47 Tuc (NGC 104) (Ferraro *et al.* 2004). In other words, BSS populations seem to exhibit elevated numbers in the cores of GCs, followed by a "zone of avoidance" at a few core radii from the cluster center, and then another rise in numbers towards the outskirts. This bimodal trend has been speculated to have arisen as a result of two separate formation mechanisms dominating in the cluster center and the peripheral regions, with collisions dominating in the latter and mass transfer in the former.

Recently, Piotto *et al.* 2004 examined the CMDs of 56 different GCs, comparing the BSS frequency to cluster properties like total absolute luminosity (mass), central density and the collision rate. Their relative frequencies were found by normalizing the number of BSSs to the horizontal branch (HB) or the red giant branch (RGB). They found that the most massive clusters had the lowest frequency of BSSs, and that there was a weak anticorrelation between BSS frequency and the cluster collisional parameter. They also showed that the BSS luminosity functions (BSLFs) of the most massive clusters had brighter peaks and reached brighter luminosities than did those of less massive clusters.

2. Analysis

We analyzed the CMDs of 57 GCs from Piotto *et al.*'s 2002 database in order to search for links between blue straggler formation and environment (see Leigh *et al.* 2007). By considering only those stars found within one core radius of the cluster center where the density is the highest, we chose a uniform dynamical environment where a single BSS formation mechanism is primarily occurring - namely collisions (Ferraro *et al.* 2004; Mapelli *et al.* 2006).

First, we consistently defined each stellar sub-population, applying the same set of boundary conditions to each cluster CMD in our sample. As such, it was necessary to define two reference points, namely the center of the mass of points that populate the main-sequence turn-off (MSTO) and the middle of the HB in the F555-plane. Two such landmarks proved necessary since the separation in the F555-plane between the MSTO and the HB often varies appreciably from cluster to cluster. From there, it was a simple matter of fitting the same set of boundary criteria to each cluster in our sample.

After confirming that the previously observed anticorrelation with total cluster mass (Piotto *et al.* 2004) also exists in the purely core population, we searched for other trends between normalized blue straggler frequencies and various cluster parameters. Apart from anticorrelations with the half-mass relaxation time and the central velocity dispersion - both thought to stem from the fact that they are correlated with total cluster mass (see Spitzer 1987 and Djorgovski & Meylan 1994 respectively) - no new statistically significant trends were found with any other cluster parameters, including two separate collisional parameters (Piotto *et al.* 2004; Pooley & Hut 2006). This could reflect a complex interplay of influential parameters making it difficult to isolate any one connection, or that BSS formation does not really depend on the cluster environment.

We also generated cumulative BSLFs for every cluster in our sample, in addition to applying quadratic fits to each. Piotto *et al.* 2004 suggested that the resulting interior chemical profiles of collision products should differ from those of coalescence and that this could result in different observational signatures for each. If true and BSS formation does indeed depend on cluster mass, then one would expect the BSLF to similarly depend on cluster mass. However, the BSLFs of the most massive clusters showed no difference in peak magnitude with those off the least massive clusters. We also looked for trends between the various quadratic coefficients that accompanied our fits and the usual cluster parameters, but found none. These results suggest that either a single BSS formation mechanism is operating in all cluster cores, or that the products of collisions and coalescence are not observationally distinguishable, at least not long enough for the phenomenon to have much of a chance of being observed.

3. Normalization

As a result of finding a lack of trends between normalized BSS frequencies and the various cluster parameters considered, we became concerned that the normalization process could be obscuring trends. Indeed, the previously observed anticorrelation with total cluster mass appears to be, at least in part, a reflection of the correlation between the size of the stellar sub-population used for normalization - in our case the RGB - and total cluster mass. A plot of the total cluster V magnitude versus the logarithm of the inverse of the number of RGB stars shows that if one were to assume that each cluster in our sample has but a single blue straggler, normalizing the counts would still lead to a clear anticorrelation. We note that the effect is more pronounced in some clusters than others - that is, some of the BSS counts can fluctuate by a factor of quite a few (even getting close to a factor of ten at times) without greatly affecting their corresponding normalized BSS frequencies.

4. Results

A new correlation between observed core blue straggler counts and the number predicted from single-single collisions alone has been found, representing the first direct evidence that BSSs are, at least in part, formed via collisions. Fig. 1 shows a plot of the number of BSSs expected to populate the entire cluster core based on the application of a geometrical scaling factor (which proved necessary due to the varying degree of coverage exhibited by the HST field of view on a cluster-to-cluster basis) versus the number predicted from single-single collisions alone. Our predictions for the number of collisionally-produced BSSs comes from multiplying the average blue straggler lifetime τ_{BSS}, taken to be 1.5 Gyrs (see Sandquist *et al.* 1997), by the total number of single-single collisions expected to occur in the core based on the derived timescale of Leonard (1989). Note that upon applying a density cut at 10^4 $M_\odot$ pc^{-3}, a statistically significant correlation in the densest clusters becomes apparent. In other words, in clusters denser than this, there is good overall agreement between the observed number of BSSs and the number predicted from single-single collisions. In clusters less dense than this, however, relatively few BSSs are expected to be produced from single-single collisions and yet the observations show relatively populous blue straggler populations. This discrepancy implies that some other mechanism, possibly mass-transfer, is responsible for many BSSs in sparse cluster environments.

In dense cluster cores having a high single-binary (and binary-binary) encounter rate, binary systems that could undergo prolonged and unimpeded mass-transfer may continually be disrupted by single stars. This could account for the discrepancy seen in less dense cores wherein only a few BSSs are expected to be the products of single-single collisions since most binary systems undergoing mass-transfer in these clusters should remain uninterrupted. It seems that mass-transfer is responsible for many of the BSSs we see in low density cluster environments, specifically GC cores less dense than around 10^4 $M_\odot$ pc^{-3}. One has to wonder, however, why the agreement is so good in the densest clusters without having to factor in the inward migration of BSSs into the core from the cluster outskirts. If some of the core BSSs in those clusters less dense than 10^4 $M_\odot$ pc^{-3} are a

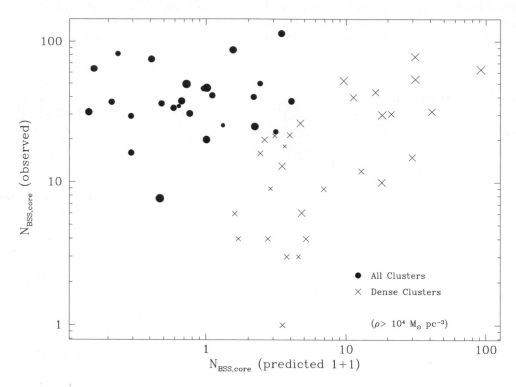

Figure 1. Plot of the number of core BSSs observed in Piotto et al's 2002 database after the geometrical scaling factor has been applied versus the number predicted in the core via single-single collisions alone. The filled circles represent clusters having a central density less than 10^4 $M_\odot$ pc^{-3}, whereas the crosses represent clusters denser than this - note the correlation in the densest clusters. The size of the circles and crosses scales with total cluster mass such that the largest ones correspond to the most massive clusters.

result of the inward migration of primordial binaries (PBs) from the cluster outskirts, then why is this effect not seen in the densest clusters where single-single collisions alone can account for the observations? Perhaps most PBs migrate into the core within the first few Gyrs of cluster evolution and have hence evolved away from the blue straggler phase long before they can be observed.

5. Conclusions

We showed that normalized blue straggler frequencies do not appear to be correlated with any global cluster parameters. Consequently, a direct link between blue straggler formation and environment remains elusive, at least when considering BSS frequencies. Even upon generating BSLFs for every cluster in our sample, in addition to applying quadratic fits to each, no trends with any cluster parameters were apparent.

We grew suspicious that normalizing the data was obscuring trends. After considering non-normalized BSS numbers, we found a statistically significant correlation between the number of observed BSSs in the core and the number predicted from single-single collisions in clusters having a core density greater than 10^4 $M_\odot$ pc^{-3}. In sparser cores, however, the agreement between our predictions and the observations is poor and we suspect that mass-transfer and coalescence are responsible for creating the majority of the BSSs therein. While more work still needs to be done, it certainly seems plausible that single-binary and binary-binary encounters could be inhibiting mass-transfer in dense cluster cores. With more numerous and better constrained estimates of GC core binary fractions, a great deal of light could be shed on how exactly these enigmatic stellar bodies are created.

References

Bailyn, C. D. 1995, *ARAA* 33, 133

De Angeli, F., Piotto, G., Cassisi, S., Busso, G., Racio-Blanco, A., Salaris, M., Aparicio, A., Rosenberg, A. 2005, *AJ* 130, 116

Djorgovski, S., & Meylan, G. 1994, *ApJ* 108, 1292

Ferraro, F. R. *et al.* 1997, *A&A* 324, 915

Ferraro, F. R., Paltrinieri, B., Rood, R.T., Dorman, B. 1999, *ApJ* 522, 983

Ferraro, F. R., Sills, A., Rood, R. T., Paltrinieri, B., & Buonanno, R. 2003, *ApJ* 588, 464

Ferraro, F. R., Beccari, G., Rood, R. T., Bellazzini, M., Sills, A., & Sabbi, E. 2004, *ApJ* 603, 127

Harris, W. E. 1996, *AJ* 112, 1487

Knigge, C., Zurek, D. R., Shara, M., M., Long, K. S., & Gilliland, R. L. 2003, *ApJ* 599, 1320

Knigge, C., Gilliland, R. L., Dieball, A., Zurek, D. R., Shara, M. M., & Long, K. S. 2006, *ApJ* 641, 281

Leigh, N. Sills, A. & Knigge, C. 2007, *ApJ* 661, 210

Leonard, P. J. T. 1989, *AJ*, 98, 217

Mapelli, M., Sigurdsson, S., Ferraro, F. R., Colpi, M., Possenti, A. & Lanzoni, B. 2006, *MNRAS*, 373, 361

Piotto, G. *et al.* 2002, *A&A* 391, 945

Piotto, G. *et al.* 2004, *ApJL* 604, L109

Pooley, D. & Hut, P. 2006, *ApJ* 646, 143

Pryor C. & Meylan, G. 1993, in: S.G. Djorgovski & G. Meylan (eds.) *ASP Conf. Series*, Structure and Dynamics of Globular Clusters (San Francisco: ASP), vol. 50, p. 357

Sandage, A. R. 1953, *AJ*, 58, 61

Sandquist, E. L., Bolte, M. & Hernquist, L. 1997, *ApJ*, 477, 335

Sills, A. R. & Bailyn, C. D. 1999, *ApJ*, 513, 428

Sills, A. R., Adams, T. & Davies, M. B. 2005, *MNRAS*, 358, 716

Spitzer, L. 1987, *Dynamical Evolution of Globular Clusters*, (Princeton: Princeton University Press)

Stryker, L. L. 1993, *PASP*, 105, 1081

Zaggia, S. R., Piotto, G. & Capaccioli, M. 1997, *A&A* 327, 1004

Zinn, R. & West, M. J. 1984, *ApJS* 55, 45

Dynamical Evolution of Dense Stellar Systems
Proceedings IAU Symposium No. 246, 2007
E. Vesperini, M. Giersz & A. Sills, eds.

© 2008 International Astronomical Union
doi:10.1017/S1743921308015883

An X-ray Emitting Black Hole in a Globular Cluster

T.J. Maccarone[1], G. Bergond[2], A. Kundu[3], K.L. Rhode[4,5,6], J.J. Salzer[4,5], I.C. Shih[3] and S.E. Zepf[3]

[1]School of Physics and Astronomy, University of Southampton, Southampton, UK, SO16 4ES
email: tjm@phys.soton.ac.uk

[2]Instituto de Astrofísica de Andalucía (IAA/CSIC), Camino Bajo de Huétor 50, 18008
Granada, Spain

[3]Department of Physics and Astronomy, Michigan State University, East Lansing, MI 48824,
USA

[4]Department of Astronomy, Indiana University, Bloomington, IN 47405, USA

[5]Department of Astronomy, Wesleyan University, Middletown, CT, 06459, USA

[6]Department of Astronomy, Yale University, New Haven CT, 06520, USA

Abstract. We present optical and X-ray data for the first object showing strong evidence for being a black hole in a globular cluster. We show the initial X-ray light curve and X-ray spectrum which led to the discovery that this is an extremely bright, highly variable source, and thus must be a black hole. We present the optical spectrum which unambiguously identifies the optical counterpart as a globular cluster, and which shows a strong, broad [O III] emission line, most likely coming from an outflow driven by the accreting source.

Keywords. X-ray: binaries, globular clusters: general, galaxies: individual (NGC 4472)

1. Introduction

Since the early days of X-ray astronomy, there has been considerable debate over whether globular clusters contained black holes. With the discovery of Type I X-ray bursts from all globular clusters in the Milky Way with bright X-ray sources (starting with Grindlay et al. 1976), and their subsequent explanation as episodes of thermonuclear burning on the surfaces of neutron stars (Woosley & Taam 1976; Swank *et al.* 1977), it became clear that there was no evidence for any accreting black holes in the Milky Way's globular cluster system.

Interpretations of the observations have been taken in two directions. One is simply that given only 13 bright X-ray sources in the Milky Way's globular cluster system, it is not so unlikely for them all to have neutron star accretors, especially in light of the fact that about 10 times as many neutron stars as black holes are expected to be produced for most stellar initial mass functions. The alternative is that dynamical effects are responsible for ejecting black holes from globular clusters. Severe mass segregation is likely to take place for globular cluster black holes, as they should be many times heavier than all the other stars in the cluster. This can lead to the formation of a "cluster within a cluster" where the heaviest stars (i.e. the black holes) feel negligible effects from the other stars in the cluster, which in turn leads to a cluster with a short evaporation timescale (Spitzer 1969). Numerical calculations have found that this evaporation can be accelerated further due to binary processes (e.g Portegies Zwart & McMillan 2000).

Early results from the Chandra X-ray Observatory gave new hope that globular cluster black holes might be detectable, by opening up the window of looking in other galaxies.

Previously, only ROSAT could resolve point sources in other galaxies, and its localization of sources was generally not good enough to allow for unique identification of optical counterparts. The first few years of Chandra observations revealed several extragalactic globular cluster X-ray sources brighter than the Eddington limit for a neutron star (e.g. Angelini *et al.* 2001; Kundu *et al.* 2002), but a globular cluster may contain multiple bright neutron stars (as does, for example M 15 in our own galaxy – White & Angelini 2001), and that the quality of X-ray spectra available from Chandra for even the brightest extragalactic sources is insufficient to make phenomenological determinations that a source has a black hole accretor. It was pointed out that only large amplitude variability could prove that we were seeing the emission from a single source, rather than multiple sources (Kalogera, King & Rasio 2003).

Furthermore, the optical catalogs used to identify globular cluster counterparts to X-ray sources have been predominantly photometric catalogs, sometimes made even without color selections being used to ensure that that the optical source in question truly is a globular cluster. Most studies done to date have focused on HST images of the central regions of elliptical galaxies with high specific frequencies of globular clusters. In these regions, and with the angular resolution of HST, contamination will be rare, especially if color cuts are used to ensure that the contribution of background quasars is minimized. In the halos of galaxies, the surface density of real globular clusters will drop, and contamination will be a more serious problem. In either case, when one is looking for conclusive proof that an object is a globular cluster black hole, spectroscopic confirmation that the object is a globular cluster is essential – the fractional contamination of the X-ray sources due to background AGN will be more serious at very high fluxes, corresponding to luminosities above 10^{39} ergs/sec than it will at lower levels consistent bright neutron star accretors. Furthermore, the optical to X-ray ratios for globular cluster black holes near the Eddington limit and background quasars are quite similar.

2. Discovery of a globular cluster black hole

In a recent paper, we found a source meeting all the strict criteria for identifying an X-ray emitting black hole in a globular cluster (Maccarone *et al.* 2007). XMMU J1229397+075333 has an X-ray luminosity of 4.5 $\times 10^{39}$ ergs/sec. It shows variability of a factor of 7 in count rate, in a time span of about 3 hours (see Fig. 1). It is located in a spectroscopically selected globular cluster. The object was first detected by ROSAT, and is included in the intermediate luminosity X-ray source catalog of Colbert & Ptak (2002), but as it is about 7 arcminutes from the center of NGC 4472, its host galaxy, good optical follow-up, combined with the positional accuracy obtained from archival Chandra data were necessary to ensure that it was not just a background active galactic nucleus.

3. Observational properties

This source has several unusual X-ray properties, all of which give some clues about its possible nature. Its X-ray spectrum is dominated by a very soft, quasi-thermal component. The best fitting models give inner disk radii of thousands of kilometers, with temperatures of about 0.2 keV. The change in X-ray count rate is predominately a change in the low energy X-ray emission, and is consistent with a change only in foreground absorption. This source was also observed in 1994 by ROSAT, and in 2000 by Chandra, in both cases as part of observations of the whole of NGC 4472. In both of these cases, the X-ray luminosity was within a factor of a few of 4×10^{39} ergs/sec, with bigger uncertainty

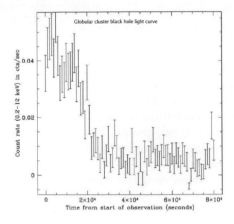

Figure 1. The XMM-Newton light curve of XMMU J1229397+07533.

in the luminosities than for XMM due to the lower count rates, narrower spectral energy range, and, in the case of ROSAT, much poorer spectral resolution, when compared with XMM. This suggests strongly that this source has been persistently bright for at least a 12 year period. The comparison of the different epochs' spectra and detailed spectral fitting can be found in Shih et al. (2007).

The optical spectrum also shows an unusual feature which may shed light on the nature of the accretor. Two optical spectra have been taken of this object, and in each case, a strong, broad [O III] emission line is seen (see Fig. 2 for one example). Unfortunately, the observations were aimed at making spectroscopic confirmation of the globular cluster nature of a large number of clusters, and at doing kinematic studies of the clusters. This means that the spectra have been taken with relatively high spectral resolution, but over a relatively narrow wavelength range, and this wavelength range includes only [O III]5007 Å among the emission lines which are commonly strong. Emission lines are rarely seen in globular clusters' integrated spectra, and are normally attributed to planetary nebulae when they are seen. The line is many times broader in velocity profile than lines from planetary nebulae. A detailed description of the line properties can be found in Zepf *et al.* (2007).

4. Interpretation: the nature of this object

The large inferred inner disk radius from the X-ray spectrum of this object implies that one of two things is most likely to be true. Either the object contains an intermediate mass black hole of a few hundred solar masses accreting at about a few percent of its Eddington rate, or it contains a stellar mass black hole accreting at a mildly super-Eddington rate. In the former case, the combination of observed temperature and luminosity is straightforward, as it requires only that the accretion disk extend in to the innermost stable circular orbit for the black hole.

The latter case, while not as intuitively obvious, is equally plausible physically. In the event of super-Eddington accretion, one can expect annuli where the luminosity is the local Eddington luminosity. This yields a total luminosity of the Eddington luminosity multiplied by the logarithm of the ratio of the mass accretion rate to the accretion rate needed to reach the Eddington luminosity (see e.g. Begelman, King & Pringle 2006). While the spectrum will not any more be exactly that predicted by the standard disk blackbody model, with moderate quality data like those which exist for extragalactic

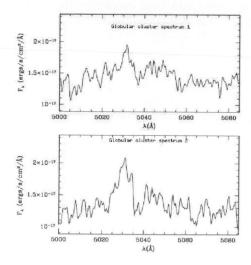

Figure 2. The optical spectrum of RZ2109, the globular cluster containing XMMU J1229397+07533. The emission line appears slightly redshifted from the position of [O III] at 5007 Å, with the redshift the same for the line as it is for the absorption lines in the stellar continuum. The two different epochs of VLT spectra are shown.

X-ray binaries, this model will provide an acceptable fit, and will yield an inner disk radius equal to the radius of the outermost annulus where the source is locally Eddington limited. Taking this into account, our data are well fit by a model in which the black hole mass is about 10 $M_{\odot}$, and the accretion rate is about 40 times what would be needed to produce an Eddington-luminosity source (see e.g. Begelman, King & Pringle 2006).

The variability can be explained by a puffed up warped region in the accretion disk which precesses, sometimes obscuring the central region of the accretion flow, while, most of the time, it is not blocking our line of sight to the source (see Shih *et al.* 2007 for a more detailed discussion of the source's X-ray variability). A precession period of about 100 days, combined with a radius of about 10^{11} cm for the location of the warp, gives a good match to the timescale of about 3 hours on which the obscuration takes place. One could alternatively consider self-obscuration by a variable disk wind, but it seems unlikely that the observed column density would change from a level which is negligible compared to the foreground column density of 1.6×10^{20} cm^{-2} to the level of approximately 3×10^{21} cm^{-2} needed to match the spectrum in the faint part of the light curve. Strongly warped accretion disks which are puffed up enough to allow self-obscuration are most likely to be seen for luminosities near the Eddington luminosity, so the interpretation of the source as a super-Eddington stellar mass black hole is favored in this scenario. Given the long inferred precession timescale, despite the high mass accretion rate which speeds the transfer of warp angular momentum, we expect such a system to have a relatively long orbital period of ~1 month. This is in good agreement with the finding that the object has been a bright X-ray source for over a decade, as long period systems show long outbursts followed by extremely long periods of quiescence (e.g. Portegies Zwart, Dewi & Maccarone 2005 and references within).

The optical spectrum is consistent with the idea that this system contains a super-Eddington stellar mass black hole, rather than an intermediate mass black hole (see Zepf *et al.* 2007 for a more detailed discussion of the cluster's optical spectrum). The broad [O III] emission line can be explained as the result of a bubble being inflated by strong disk winds from a stellar mass black hole exceeding the Eddington limit. This bubble then collides with the interstellar medium in the globular cluster, producing a shock. The shock velocity will approximately equal the physical width of the emission line. The shock velocity can be slowed down from the original outflow velocity to the observed velocity in a reasonable duration, given expected density of the interstellar medium in a globular cluster. Other scenarios for producing a broad line, especially from an intermediate mass black hole, run into problems – while [O III] can come from photoionization, and the line width could be due to virial motions near the black hole, it is difficult (albeit not impossible) to allow for enough material to be located close enough to the black hole to match both the line luminosity and the line width. Thus, an outflow would still be required in the case of an intermediate mass black hole, and outflows are generally quite weak for sources with luminosities well below their Eddington limits (Proga 2007).

5. Conclusions

We have shown for the first time clear evidence of a globular cluster black hole, on the basis of strong, highly variable X-ray emission from a source in a spectroscopically confirmed globular cluster. Based on the X-ray spectrum, the characteristic variability, and the [O III] emission in the optical spectrum, the black hole is most likely a stellar mass object accreting far faster than its Eddington rate. Given that it is difficult to develop a scenario in which a globular cluster could have both a stellar mass black hole in a binary and an intermediate mass black hole, this argues against the idea that all globular clusters contain intermediate mass black holes. This discovery also motivates future searches for quiescent stellar mass black holes in the Milky Way's globular clusters; these may be hiding among the X-ray sources currently classified as cataclysmic variable stars. Radio emission should be detectable only from quiescent stellar mass black holes, and should be the one feasible discriminant between the two classes of systems.

Acknowledgements

We are grateful to Tom Dwelly, Sebastian Jester, Elmar Körding, Phil Charles, Mike Eracleous, Steinn Sigurdsson, Robin Barnard, Andrew King, Guillaume Dubus and Mark Voit for useful discussions.

References

Angelini, L., Loewenstein, M., Mushotzky, R. F., 2001, *ApJL*, 557L, 35
Begelman, M. C., King, A. R., Pringle, J. E., 2006, *MNRAS*, 370, 399
Colbert, E. J. M. & Ptak, A. F., 2002, *ApJS* 143, 25
Grindlay, J., Gursky, H., Schnopper, H., Parsignault, D. R., Heise, J., Brinkman, A. C., & Schrijver, J. 1976, *ApJL*, 205, L127
Kalogera, V., King, A. R., Rasio, F. A., 2004, *ApJL*, 600, L17
Kundu, A., Maccarone, T. J., Zepf, S. E., 2002, *ApJL*, 574, L5
Maccarone, T. J., Kundu, A., Zepf, S. E., & Rhode, K. L. 2007, *Nature*, 445, 183
Portegies Zwart, S. F., Dewi, J., & Maccarone, T. 2004, *MNRAS*, 355, 413
Portegies Zwart, S. F., & McMillan, S. L. W. 2000, *ApJL*, 528, L17
Proga, D. 2007, *The Central Engine of Active Galactic Nuclei* in press
Spitzer, L. J. 1969, *ApJL*, 158, L139
Swank, J. H., Becker, R. H., Boldt, E. A., Holt, S. S., Pravdo, S. H., & Serlemitsos, P. J. 1977, *ApJL*, 212, L73
White, N. E. & Angelini, L., 2001 *ApJL*, 561, 101L
Woosley, S. E. & Taam, R. E. 1976, *Nature*, 263, 101

Dynamical Evolution of Dense Stellar Systems
Proceedings IAU Symposium No. 246, 2007
E. Vesperini, M. Giersz & A. Sills, eds.

© 2008 International Astronomical Union
doi:10.1017/S1743921308015895

Central Dynamics of Globular Clusters: the Case for a Black Hole in ω Centauri

Eva Noyola[1], Karl Gebhardt[2] and Marcel Bergmann[3]

[1]Max-Planck-Institut für extraterrestrische Physik, Giessenbachstrasse, 85748, Garching, Germany, email: noyola@mpe.mpg.de

[2]Department of Astronomy, University of Texas, Austin, TX, 78723

[3]NOAO Gemini Science Center, Casilla 603, La Serena, Chile

Abstract. The globular cluster ω Centauri is one of the largest and most massive members of the Galactic system. Its classification as a globular cluster has been challenged making it a candidate for being the stripped core of an accreted dwarf galaxy; this and the fact that it has one of the largest velocity dispersions for star clusters in our galaxy makes it an interesting candidate for harboring an intermediate mass black hole. We measure the surface brightness profile from integrated light on an *HST*/ACS image, and find a central power-law cusp of logarithmic slope -0.08. We also analyze Gemini GMOS-IFU kinematic data for a 5"x5" field centered on the nucleus of the cluster, as well as for a field 14" away. We detect a clear rise in the velocity dispersion from 18.6 kms^{-1} at 14" to 23 kms^{-1} in the center. Given the very large core in ω Cen (2.58'), an increase in the dispersion in the central 10" is difficult to attribute to stellar remnants, since it requires too many dark remnants and the implied configuration would dissolve quickly given the relaxation time in the core. However, the increase could be consistent with the existence of a central black hole. Assuming a constant M/L for the stars within the core, the dispersion profile from these data and data at larger radii implies a black hole mass of $4.0^{+0.75}_{-1.0} \times 10^4 M_\odot$. We have also run flattened, orbit-based models and find a similar mass. In addition, the no black hole case for the orbit model requires an extreme amount of radial anisotropy, which is difficult to preserve given the short relaxation time of the cluster.

Keywords. globular clusters: individual (Omega Centauri), Galaxy: kinematics and dynamics

1. Introduction

The globular cluster ω Centauri (NGC 5139) is the most massive member of the Galactic cluster system. It has a measured central velocity dispersion of 22 ± 4 kms^{-1} (Meylan *et al.* 1995). A rotating flattened model including proper motion and radial velocity datasets by van de Ven *et al.* (2006) calculate a total mass of $2.5 \times 10^6 M_\odot$. ω Cen has a complex stellar population with a broad metallicity distribution (Bedin *et al.* 2004). These result have led to the hypothesis that ω Cen is not a classical globular cluster, but instead is the nucleus of an accreted galaxy (Freeman 1993).

Two globular clusters have been suggested for harboring an intermediate mass black hole in their nucleus. One is the galactic cluster M15 (Gerssen *et al.* 2003) and the other is G1, a giant globular cluster around M31 (Gebhardt, Rich, & Ho 2005). M15 is assumed to be in a post-core collapse state. Its dynamical state has been debated between harboring a black hole or containing a large number of compact remnants in its center (Baumgardt *et al.* 2003a). Observational constraints between these two hypothesis remain inconclusive (van den Bosch *et al.* 2006). G1 on the other hand, has a core with characteristics closer to those of ω Cen, and observations support the black hole interpretation in this case. Baumgardt *et al.* (2003b) propose an alternative interpretation for G1 in which they match the observations with a model of two colliding globular clusters. The black hole

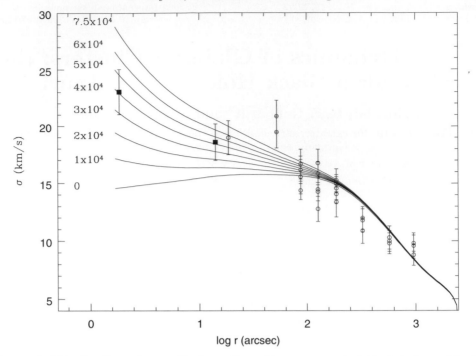

Figure 1. Velocity dispersion profile for ω Cen with various central black hole models. Filled squares are the dispersions and uncertainties from the GMOS-IFU and open circles are from individual radial velocity measurements. A set of isotropic spherical models of varying black hole masses is shown for comparison. The thick line is the no black hole model and the thin lines represent models with black holes as labeled

interpretation for G1 is strongly supported by radio (Ulvestad *et al.* 2007) and x-ray (Pooley & Rappaport 2006) detections centered on the nucleus.

2. Surface Brightness Profile

We measure the central part of the surface brightness profile for ω Cen using *HST* spatial resolution. We measure integrated light from an ACS F435W image (340 sec) applying the technique described in detail in Noyola & Gebhardt (2006). Since the image has a limited radial coverage, we use the Chebychev fit of Trager *et al.* (1995) for the surface brightness profile to cover the full radial extent of the cluster. The coordinates for our center are RA 13 : 26 : 46.043 and DEC −47 : 28 : 44.8 on the ACS dataset J6LP05WEQ using its WCS zeropoint.

The measured surface brightness profile shows a continuous rise toward the center with a logarithmic slope of -0.08 ± 0.03, which is in contrast to the common notion that ω Cen has a flat core. Baumgardt, Makino & Hut (2005) perform N-body models of star clusters with initial King profiles and containing a central black hole. They predict the formation of a shallow cusp of -0.1 to -0.3 logarithmic slope after $1.5 - 4$ relaxation times.

3. Kinematic Measurements

We obtained Gemini GMOS-South IFU nod-and-shuffle observations. The IFU has a field of view of 5'×5", comprised of 700 individual lenslets plus fiber elements, each of

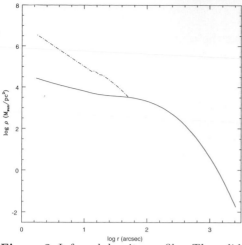

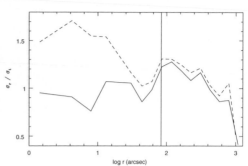

Figure 2. Inferred density profiles. The solid line is the deprojected density profile for the luminous component. The dashed line represents the required dark component to reproduce the observed kinematics.

Figure 3. Radial over tangential anisotropy vs. radius from orbit-based models. The solid line is for the best fit model containing a black hole. The dashed line is for a model with no black hole present. The vertical line marks the location of the core radius.

which covers approximately $0''.2$ on the sky. We use the R600 grating, yielding a resolving power R=5560, along with the Calcium Triplet filter. Two fields are observed, each for a total integration time of 900 sec on source and 900 sec on sky. The first of the two fields is located at the cluster center, and the second field is centered 14" away. Using the standard tasks from the IRAF-GEMINI package we sky subtract, flat-field, extract the spectra for each fiber, and apply a wavelength calibration.

We match the reconstructed image from the IFU fibers with the same region on the ACS image. Both fields contain $\sim$ 100 resolved stars. Using the photometric measurements of individual stars together with the reported seeing, we calculate how many stars contribute to each fiber. Excluding the fibers which are dominated by a single star we estimate that the integrated spectrum of the background unresolved light represent about 60 stars in both fields.

We focus on the Ca triplet region (8450Å-8700Å) for our analysis. To estimate the velocity dispersion we rely on the integrated light, and require template stars in this case. We use stars observed by Walcher *et al.* (2005), from VLT-UVES observations at around R=35000. We convolve the spectra to our measured resolving power. To extract the velocity dispersion from the integrated light we utilize the non-parametric, pixel-based technique as described in Gebhardt *et al.* (2000). We choose an initial velocity profile in bins, convolve it with the set of templates, and calculate residuals to the integrated spectrum. Monte Carlo simulations determine the uncertainties, and use the measured noise in the spectrum. We measure 23.0 ± 2.0 kms^{-1} for the central field and 18.6 ± 1.6 kms^{-1} for the outer one. Van den Ven *et al.*(2006) measure a line of sight velocity dispersion profile by combining various datasets. They use 2163 individual radial velocity measurements divided into polar apertures to obtain the final velocity dispersion profile. We use their dispersion estimates. Fig. 1 presents the velocity dispersion data.

4. Models

The presence of an intermediate black hole at the center of this cluster is one of the possibilities for explaining the observed rise in velocity dispersion. We have run two types of modeling in order to test this hypothesis

First, we create a series of models using the non-parametric method described in Gebhardt & Fischer (1995). We apply a reddening correction to the observed surface brightness profile, which we then deproject using Abel integrals assuming spheroidal symmetry. We apply a spline smoother to the surface brightness profile before deprojecting and thus obtain a luminosity density profile as discussed in Gebhardt et $al.$ (1996). By assuming a mass to light ratio of 2.7, we calculate a mass density profile, from which the potential and the velocity dispersion can be derived. We repeat the calculation adding a variety of central point masses ranging from 0 to $7.5 \times 10^4 M_\odot$ while keeping the global M/L value fixed. Fig. 1 shows the comparison between the different models and the measured dispersion profile. As it can be seen, an isotropic model with no black hole present predicts a slight decline in velocity dispersion toward the center, instead we observe a clear rise. The predicted central velocity for the no black hole model is 14.6 kms^{-1} which is well below any line of sight velocity dispersion measured inside 1'. The χ^2 curve implies a best-fit black hole mass of $4^{+0.75}_{-1} \times 10^4 M_\odot$.

An alternative to explain the observed rise in M/L toward the center is a concentration of dark stellar remnants. Using the observed velocity dispersion profile, we calculate the total enclosed mass and from this, the mass density profile. We then compare this with the enclosed mass implied by the luminosity density profile, assuming the same M/L as for the isotropic models. We estimate the density profile of the the implied extended dark component, assuming this was the cause of the velocity dispersion rise towards the center. Fig. 2 shows the estimated density profile for the dark and luminous components. If the velocity rise is due to an extended distribution of dark stellar remnants, the density profile of this dark component has to be very concentrated, with a logarithmic slope of ~ -2.0, resembling a cluster undergoing core collapse. The relaxation time for ω Cen implies a much slower dynamical evolution than the one necessary to reach such a configuration. Core-collapse models have shown that when a cluster has reached such a high degree of mass segregation, the observable core to half light radius gets very small, with values below 0.05 (Breeden, Cohn, & Hut 1994), while this ratio is 0.3 for ω Cen. Another possibility is that the observed rise in velocity dispersion is due to velocity anisotropy in the cluster. To explore this possibility, we construct orbit-based dynamical models. The models are based on the Schwarzschild formulation and are constructed as in Gebhardt et $al.$ (2000); Gehardt et $al.$ (2003). We use the same deprojection as described above, except we also include the observed flattening. Assuming an M/L ratio and a BH mass, the mass distribution of the cluster is obtained and from it, the potential can be computed. Using this potential, we generate about 10^4 representative orbits. The best match to the observed photometric and kinematical data provide the orbital weights for a given potential. The process is repeated for various M/L values and BH masses until the minimum χ^2 model is found. The best fit model requires a black hole mass around 3.5×10^4 $M_\odot$, close to the isotropic result. As expected, the difference in χ^2 compared to the no black hole case is smaller than in the isotropic case. However, the anisotropy profile is extremely different between the black hole and no black hole case. Fig. 3 shows that without the presence of a central black hole, a large degree of radial anisotropy $(\sigma_r/\sigma_t=1.5)$ is required inside 0.3 r_c. The models with and without a black hole are close to isotropic for $r > 28$", which is in agreement with the results of van den Ven et $al.$(2006). For a system as dense as ω Cen, such a degree of anisotropy is expected to be quickly erased through relaxation processes. Even with this extreme anisotropy profile, the black hole case provides a better match to the data.

5. Conclusions

The two pieces of observational evidence that ω Cen could harbor a central black hole come from the photometry and the kinematics. From the *HST* image of ω Cen, we measure a central logarithmic surface brightness slope of -0.08 ± 0.03. This value is very similar to that claimed by the N-body simulations of Baumgardt *et al.*(2005) that are most likely explained by a central black hole. Standard core-collapse does not lead to such a large core with a shallow central slope. The black hole tends to prevent core collapse while leaving an imprint of a shallow cusp. It will be important to run models tailored to ω Cen to see if one can cause and maintain a shallow cusp without invoking a central black hole. However, the main observational evidence for the central mass comes from the increase in the central velocity dispersion, where we detect a rise from 18.6 to 23 kms^{-1} from radii of 14 to 2.5". In fact, even excluding the Gemini data presented here, the previous ground-based data suggest a central mass concentration as well. The core of ω Cen is around 155" (about 2.5'), so the dispersion rise is seen well within the core.

References

Baumgardt, H., Heggie, D. C., Hut, P., & Makino, J. 2003a, *MNRAS*, 341, 247
Baumgardt, H., Makino, J., Hut, P., McMillan, S., & Portegies Zwart, S. 2003b, *ApJL*, 589, L25
Baumgardt, H., Makino, J., & Hut, P. 2005, *ApJ*, 620, 238
Bedin, L., *et al.* 2004, *ApJL*, 605, L125
Breeden, J., Cohn, H., & Hut, P. 1994, *ApJ*, 421, 195
Freeman, K. C. 1993, in ASP Conf. Ser. 48: The Globular Cluster-Galaxy Connection, 608
Gebhardt, K., & Fischer, P. 1995, *AJ*, 109, 209
Gebhardt, K., *et al.* 1996, *AJ*, 112, 105
Gebhardt, K., *et al.* 2000, *AJ*, 119, 1157
Gebhardt, K., *et al.* 2003, *ApJ*, 583, 92
Gebhardt, K., Rich, R. M., & Ho, L. 2005, *ApJ*, 634, 1093
Gerssen, J., van der Marel, R., Gebhardt, K., Guhathakurta, P., Peterson, R., Pryor, C. 2003, *AJ*, 125, 376
Meylan, G., Mayor, M., Duquennoy, A., Dubath, P. 1995, *A&A*, 303, 761
Noyola, E., & Gebhardt, K. 2006, *AJ*, 132, 447
Pooley, D., & Rappaport, S. 2006, *ApJL*, 644, L45
Trager, S., King, I, & Djorgovski, S. 1995, *AJ*, 109, 218
Ulvestad *et al.* 2007, *ApJL*, 661, L151
van den Bosch *et al.* 2006, *ApJ*, 641, 852
van de Ven *et al.* 2006, *A&A*, 445, 513
Walcher *et al.* 2005, *ApJ*, 618, 237

Dynamical Evolution of Dense Stellar Systems
Proceedings IAU Symposium No. 246, 2007
E. Vesperini, M. Giersz & A. Sills, eds.

Formation and Evolution of Black Holes in Galactic Nuclei and Star Clusters

R. Spurzem[1], P. Berczik[1], I. Berentzen[1], D. Merritt[2], M. Preto[1] and P. Amaro-Seoane[3]

[1] Astronomisches Rechen-Institut, Zentrum für Astronomie Univ. Heidelberg, Mönchhofstr. 12-14, 69120 Heidelberg, Germany
[2] Dept. of Physics, 85 Lomb Memorial Drive, Rochester Institute of Technology, Rochester, NY 14623-5604, USA
[3] Max Planck Institute for Gravitational Physics (Albert Einstein Institute), Am Mühlenberg 1, 14476 Golm, Germany

Abstract. We study the formation, growth, and co-evolution of single and multiple supermassive black holes (SMBHs) and compact objects like neutron stars, white dwarfs, and stellar mass black holes in galactic nuclei and star clusters, focusing on the role of stellar dynamics. In this paper we focus on one exemplary topic out of a wider range of work done, the study of orbital parameters of binary black holes in galactic nuclei (binding energy, eccentricity, relativistic coalescence) as a function of initial parameters. In some cases the classical evolution of black hole binaries in dense stellar systems drives them to surprisingly high eccentricities, which is very exciting for the emission of gravitational waves and relativistic orbit shrinkage. Such results are interesting to the emerging field of gravitational wave astronomy, in relation to a number of ground and space based instruments designed to measure gravitational waves from astrophysical sources (VIRGO, Geo600, LIGO, LISA). Our models self-consistently cover the entire range from Newtonian dynamics to the relativistic coalescence of SMBH binaries.

Keywords. methods: numerical, n-body simulations, gravitational waves, black hole physics, galaxies: star clusters, nuclei

1. Introduction

SMBH formation and their interactions with their host galactic nuclei is an important ingredient for our understanding of galaxy formation and evolution in a cosmological context, e.g. for predictions of cosmic star formation histories or of SMBH demographics (to predict events which emit gravitational waves). If galaxies merge in the course of their evolution, there should be either many binary or even multiple black holes, or we have to find out what happens to black hole multiples in galactic nuclei, e.g. whether they come close enough together to merge under emission of gravitational waves, or whether they eject each other in gravitational slingshot. For numerical simulations of the problem all models depend on an unknown scaling behaviour, because the simulated particle number is not yet realistic due to limited computing power (Milosavljević & Merritt 2001, Milosavljević & Merritt 2003, Makino & Funato 2004 and Berczik, Merritt & Spurzem 2005). Dynamical modelling of non-spherical dense stellar systems (with and without central black holes) is even less developed than in the spherical case. Here we present a set of numerical models of the formation and evolution of binary black holes in rotating galactic nuclei. Since we are interested in the dynamical evolution of SMBH binaries in their final phases of evolution (the last parsec problem) we somehow abstract from the foregoing complex dynamics of galactic mergers. We assume that after some violent dynamic relaxation a typical initial situation consists of a spherical or axisymmetric coherent stellar system (galactic nucleus), where fluctuations in density and potential due to the galaxy merger have decayed, which is reasonable on an astrophysically short time scale of a few ten million years. The SMBHs, which were situated in the centre of each of the previously merged galaxies, are located at the boundary of the dense stellar core, some few hundred parsec apart. This situation is well observable (Komossa *et al.* 2003).

According to the standard theory, the subsequent evolution of the black holes is divided in three intergradient stages (Begelman, Blandford & Rees 1980): 1. Dynamical friction causes an transfer of the black hole's kinetic energy to the surrounding field stars, the black holes spiral to the center where they form a binary. 2. While hardening, the effect of dynamical friction reduces and the evolution is dominated by superelastic scattering processes, that is the interaction with field stars closely encountering or intersecting the binary orbit, thereby increasing the binding energy. 3. Finally the black holes coalesce throughout the emission of gravitational radiation, potentially detectable by the planned space-based gravitational wave antennae LISA.

In this paper, the behavior of the orbital elements of a black hole binary in a dense stellar system is investigated in a self-consistent way from the beginning till the relativistic merger and its emission of gravitational waves. The evolution of the eccentricity has been discussed for some time e.g. Makino *et al.* (1993), Hemsendorf, Sigurdsson & Spurzem (2002), Milosavljević & Merritt (2001), Berczik, Merritt & Spurzem (2005) and Makino & Funato (2004). According to Peters & Mathews (1963) and Peters (1964) the timescale of coalescence due to the emission of gravitational radiation is given by

$$t_{gr} = \frac{5}{64} \frac{c^5 a_{gr}^4}{G^3 M_1 M_2 (M_1 + M_2) F(e)} \tag{1.1}$$

wherein M_1, M_2 denote the black hole masses, a_{gr} the characteristic separation for gravitational wave emission, G the gravitational constant, c the speed of light and

$$F(e) = (1 - e^2)^{-7/2} \left(1 + \frac{73}{24} e^2 + \frac{37}{96} e^4 \right) \tag{1.2}$$

a function with strong dependence on the eccentricity e. Thus the coalescence time can shrink by several orders of magnitude if the eccentricity is high enough, resulting in a strengthened burst of gravitational radiation. Highly eccentric black hole binaries would represent appropriate candidates for forthcoming verification of gravitational radiation through the planned mission of the Laser Interferometer Space Antenna mission LISA.

2. Numerical method, Initial Models

The simulations have been performed using NBODY6++, a parallelized version of Aarseth's NBODY6 (Aarseth 1999, Spurzem 1999 and Aarseth 2003). The code includes a Hermite integration scheme, KS-regularization (Kustaanheimo & Stiefel 1965) and the Ahmad-Cohen neighbour scheme (Ahmad & Cohen 1973). No softening of the interaction potential of any two bodies is introduced; this allows an accurate treatment of the effects due to superelastic scattering events, which play a crucial part in black hole binary evolution and require a precise calculation of the trajectories throughout the interaction. The code and its parallel performance has been described in detail in this series and elsewhere (Spurzem 1999 and Khalisi *et al.* 2003). The survey has been carried out for a total particle number of up to $N = 1\,000\,000$ including two massive black holes with $M_1 = M_2 = 0.01$ embedded in a dense stellar system of equal-mass particles $m_* \approx 1.0 \cdot 10^{-6}$.

The total mass of the system is normalized to unity. The initial stellar distribution was taken from generalized King models with and without rotation (Einsel & Spurzem 1999).

3. Simulations

3.1. *Newtonian Evolution of the binary black hole*

In the first evolutionary stage, each black hole individually suffers from dynamical friction with the surrounding low mass stars, which is the main process of losing energy. The role of dynamical friction decreases when a permanently bound state occurs, as the dynamical friction force acts preliminary on the motion of the now formed binary rather than on the individual black holes. Superelastic scattering events of field stars at the binary then become more and more important for the reduction of its energy. The process sustains an ongoing "hardening" of the binary (shrinking of semi-major axis and increase of energy) and also in many cases a high eccentricity. While the hardening rates are well understood (Sesana, Haardt & Madau 2007 and Quinlan 1996) and do not depend much on the initial parameters of the preceding galactic merger, this is not as clear for the eccentricity.

In a spherically symmetric system the binary hardening would stall after a few crossing times, because loss-cone orbits of stars, which come close to the central SMBH binary will be depleted, and replenishment takes place only on a much longer relaxation time. This effect is more dramatic for systems with large particle number, because the relaxation time increases strongly with N (Berczik, Merritt & Spurzem 2005 and Berczik *et al.* 2006). It has been claimed that in real galaxies with their very large particle numbers therefore the SMBH binary will not reach relativistic coalescence. This situation was relaxed from two sides, first by a careful analysis of loss-cone refilling time scales combining direct N-body and Fokker-Planck models (Merritt, Mikkola & Szell 2007), and by looking for a moderately rotating, axisymmetric galactic nucleus (Berczik *et al.* 2006), where the loss cone remains full even for large particle numbers. Since some degree of perturbation of a spherical model is quite natural for a remnant after galactic mergers, many of them might even be triaxial rather than axisymmetric, the stalling problem does not exist anymore.

4. Relativistic Dynamics of Black Holes in Galactic Nuclei

4.1. *Introduction*

Relativistic stellar dynamics is of paramount importance for the study of a number of subjects. For instance if we want to have a better understanding of what the constraints on alternatives to supermassive black holes are; in order to canvass the possibility of ruling out stellar clusters, one must do detailed analysis of the dynamics of relativistic clusters. Furthermore the dynamics of compact objects around SMBH and of multiple SMBH in galactic nuclei requires the inclusion of relativistic effects. Our current work deals with the evolution of two SMBHs, bound to each other, and looking at the phase when they get close enough to each other that relativistic corrections to Newtonian dynamics become important, which ultimately lead to gravitational radiation losses and coalescence.

Efforts to understand the dynamical evolution of a stellar cluster in which relativistic effects may be important have been already done by Lee (1987), Quinlan & Shapiro (1989), Quinlan & Shapiro (1990) and Lee (1993). In the earlier work $1\mathcal{PN}$ and $2\mathcal{PN}$ terms were neglected (Lee 1993) and the orbit-averaged formalism (Peters 1964) used. We describe here a method to deal with deviations from Newtonian dynamics more rigorously

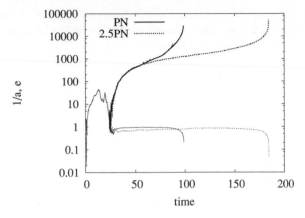

Figure 1. Effect of Post-Newtonian (PN) relativistic corrections on the dynamics of black hole binaries in galactic nuclei, plotted are inverse semi-major axis and eccentricity as a function of time. The solid line uses the full set of PN corrections, while the dotted line has been obtained by artificially only using the dissipative PN2.5 terms. Further details will be published elsewhere Berentzen *et al.* (2008, *in preparation*).

than in most existing literature (but compare Mikkola & Merritt (2007), Aarseth (2007), which are on the same level of PN accuracy). We modified the NBODY6++ code to allow for post-Newtonian ($\mathcal{PN}$) effects of two particles getting very close to each other, implementing in it the $1\mathcal{PN}$, $2\mathcal{PN}$ and $2.5\mathcal{PN}$ corrections fully from Soffel (1989).

4.2. *Method: Direct summation* NBODY *with Post-Newtonian corrections*

The version of direct summation NBODY method we employed for the calculations, NBODY6++, includes the *KS regularisation*. This means that when two particles are tightly bound to each other or the separation among them becomes smaller during a hyperbolic encounter, the couple becomes a candidate for regularisation in order to avoid problematical small individual time steps (Kustaanheimo & Stiefel 1965). We modified this scheme to allow for relativistic corrections to the Newtonian forces by expanding the acceleration in a series of powers of $1/c$ in the way given by Damour & Dereulle (1981), Soffel (1989) and Kupi, Amaro-Seoane & Spurzem (2006): We integrated our correcting terms as external forces into the *two-body KS regularisation* scheme.

4.3. *First results and Summary*

In Fig. 1 the impact of relativistic, Post-Newtonian dynamics to the separation of the binary black holes in our simulations is seen. The curve deviates from the Newtonian results when gravitational radiation losses set in and cause a sudden coalescence ($1/a \to \infty$) at a finite time. Gravitational radiation losses are supported by the high eccentricity of the SMBH binary. It is interesting to note that the inclusion or exclusion of the conservative 1 and 2PN terms changes the coalescence time considerably. Details of these results will be published elsewhere Berentzen *et al.* (2008, *in preparation*).

Once the SMBH binary starts to dramatically lose binding energy due to gravitational radiation its orbital period will drop from a few thousand years to less than a year very quickly (timescale much shorter than the dynamical time scale in the galactic center, which defines our time units). Then the SMBH binary will enter the LISA band, i.e. its gravitational radiation will be detectable by LISA. LISA, Laser Interferometer Space Antenna, is a system of three space probes with laser interferometers to measure gravitational waves, see e.g. http://lisa.esa.int/.

We have shown that supermassive black hole binaries in galactic nuclei can overcome the stalling barrier and will reach the relativistic coalescence phase in a timescale shorter than the age of the universe. A gravitational wave signal for the LISA satellite from these SMBH binaries is expected, in particular due to the high eccentricity of the SMBH binary when entering the relativistic coalescence phase. Our models cover self-consistently the transition from the Newtonian dynamics to the situation when relativistic, Post-Newtonian (PN) corrections start to influence the relative SMBH motion (Preto *et al.* (2008, *in preparation*)).

5. Acknowledgements

Computing time at NIC Jülich on the IBM Jump is acknowledged. Financial support comes partly from Volkswagenstiftung, German Science Foundation (DFG) via Schwerpunktprogramm 1177 (Project ID Sp 345/17-1) "Black Holes Witnesses of Cosmic History". It is a pleasure to acknowledge many enlightening discussions and support by Sverre Aarseth, very useful interactions about relativistic dynamics with A. Gopakumar and G. Schäfer.

References

Aarseth S. J. 1999, *PASP* 111, 1333
Aarseth, S. J. 2003, in: *Gravitational N-body simulations*, Cambridge University Press, Cambridge, p. 173
Aarseth, S. J. 2007, *MNRAS* 378, 285
Ahmad, A. & Cohen, L. 1973, *Journal of Computational Physics* 12, 349
Begelman, M. C., Blandford, R. D., & Rees, M. J. 1980, *Nature* 287, 307
Berczik, P., Merritt, D., & Spurzem, R. 2005, *ApJ* 633, 680
Berczik, P., Merritt, D., Spurzem, R., & Bischof, H.-P. 2006, *ApJ* (Letters) 642, L21
Damour, T. & Dereulle, N. 1987, *Phys. Rev.* 87, 81
Einsel, C. & Spurzem, R. 1999, *MNRAS* 302, 81
Hemsendorf, M., Sigurdsson, S., & Spurzem R. 2002, *ApJ* 581, 1256
Khalisi, E., Omarov, C. T., Spurzem, R., Giersz, M., & Lin, D. N. C. 2003, in: *High Performance Computing in Science and Engineering*, Springer Vlg., p. 71
Komossa, S., Burwitz, V., Hasinger, G., Predehl, P., Kaastra, J. S. & Ikebe, Y. 2003, *ApJ* (Letters) 582, L15
Kupi, G., Amaro-Seoane, P., & Spurzem, R. 2006, *MNRAS* 371, 45
Kustaanheimo, P. & Stiefel E. 1965, *Journal für die reine und angewandte Mathematik* 218, 204
Lee, H. M. 1987, *ApJ* 319, 801
Lee, M. H. 1993 *AJ* 418, 147
Makino, J., Fukushige, T., Okumura, S. K., & Ebisuzaki T. 1993, *PASJ* 45, 303
Makino, J. & Funato, Y. 2004, *ApJ* 602, 93
Mikkola, S. & Merritt, D. 2007, *arXiv:0709.3367*
Merritt, D., Mikkola, S., & Szell, A. 2007, *arXiv:0705.2745*
Milosavljević M. & Merritt D. 2001, *ApJ* 563, 34
Milosavljević, M. & Merritt, D. 2003, *ApJ* 596, 860
Peters, P. C. 1964, *Phys. Rev.* 136, 1224
Peters, P. C. & Mathews, J. 1963 *Phys. Rev.* 131, 435
Quinlan, G. D. 1996, *New Astron.* 1, 35
Quinlan, G. D. & Shapiro, S. L. 1989 *AJ* 343, 725
Quinlan, G. D. & Shapiro, S. L. 1990, *AJ* 356, 483
Sesana, A., Haardt, F., & Madau, P. 2007, *AJ* 660, 546
Soffel, M. H. 1989, in: *Relativity in Astrometry, Celestial Mechanics and Geodesy*, Springer-Verlag Berlin, Heidelberg, New York
Spurzem, R. 1999, *Journal of Computational and Applied Mathematics* 109, 407

Dynamical Evolution of Dense Stellar Systems
Proceedings IAU Symposium No. 246, 2007
E. Vesperini, M. Giersz & A. Sills eds.

© 2008 International Astronomical Union
doi:10.1017/S1743921308015913

The Imprints of IMBHs on the Structure of Globular Clusters: Monte-Carlo Simulations

Stefan Umbreit, John M. Fregeau and Frederic A. Rasio

Northwestern University, Dearborn Observatory
2131 Tech Drive, Evanston, Illinois, 60208, USA
email: s-umbreit@northwestern.edu

Abstract. We present the first results of a series of Monte-Carlo simulations investigating the imprints of a central black hole on the core structure of globular clusters. We investigate the three-dimensional and the projected density profile of the inner regions of idealized as well as more realistic globular cluster models, taking into account a stellar mass spectrum, stellar evolution and allowing for a larger, more realistic, number of stars than was previously possible with direct N-body methods. We compare our results to other N-body simulations published previously in the literature.

Keywords. stellar dynamics, methods: n-body simulations, globular clusters: general

1. Introduction

As recently as 10 years ago, it was generally believed that black holes (BHs) occur in two broad mass ranges: stellar ($M \sim 3 - 20 M_\odot$), which are produced by the core collapse of massive stars, and supermassive ($M \sim 10^6 - 10^{10} M_\odot$), which are believed to have formed in the center of galaxies at high redshift and grown in mass as the result of galaxy mergers (see e.g. Volonteri, Haardt & Madau 2003). However, the existence of BH with masses intermediate between those in the center of galaxies and stellar BHs could not be established by observations up until recently, although intermediate mass BHs (IMBHs) were predicted by theory more than 30 years ago; see, e.g., Wyller (1970). Indirect evidence for IMBHs has accumulated over time from observations of so-called ultraluminous X-ray sources (ULXs), objects with fluxes that exceed the angle-averaged flux of a stellar mass BHs accreting at the Eddington limit. An interesting result from observations of ULXs is that many if not most of them are associated with star clusters. It has long been speculated (e.g., Frank & Rees 1976) that the centers of globular clusters (GCs) may harbor BHs with masses $\sim 10^3 M_\odot$. If so, these BHs affect the distribution function of the stars, producing velocity and density cusps. A recent study by Noyola & Gebhardt (2006) obtained central surface brightness profiles for 38 Galactic GCs from HST WFPC2 images. They showed that half of the GCs in their sample have slopes for the inner 0.5" surface density brightness profiles that are inconsistent with simple isothermal cores, which may be indicative of an IMBH. However, it is challenging to explain the full range of slopes with current models. While analytical models can only explain the steepest slopes in their sample, recent N-body models of GCs containing IMBHs (Baumgardt *et al.* 2005), might explain some of the intermediate surface brightness slopes.

In our study we repeat some of the previous N-body simulations of GCs with central IMBH but using the Monte-Carlo (MC) method. This gives us the advantage to model the evolution of GCs with a larger and thus more realistic number of stars. We then compare the obtained surface brightness profiles with previous results in the literature.

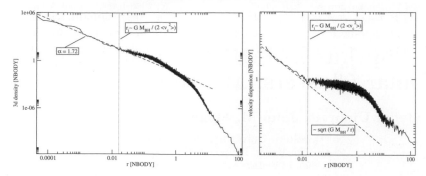

Figure 1. The imprints of an IMBH on the stellar distribution of a GC. On the left-hand side, the radial number density profile of an evolved single-mass GC is shown (solid) together with a power-law fit to its inner region (dashed line). The right hand side shows its velocity dispersion profile (solid) and the Keplerian velocity profile of the IMBH. The dotted line marks its radius of influence.

2. Imprints of IMBHs

The dynamical effect of an IMBH on the surrounding stellar system has first been described by Peebles (1972) who argued that the bound stars in the cusp around the BH must obey a shallow power-law density distribution to account for stellar consumption near the cluster center. Analyzing the Fokker-Planck equation in energy space for an isotropic stellar distribution, Bahcall & Wolf (1976) obtained a density profile with $n(r) \propto r^{-7/4}$, which is now commonly referred to as the Bahcall-Wolf cusp. The formation of this cusp has been confirmed subsequently by many different studies using different techniques and also, more recently, by direct N-body methods (Baumgardt *et al.* 2004). In Fig. 1 such a profile from one of our simulations is shown (for initial cluster parameters see Baumgardt *et al.* (2004) (run16)). As can be clearly seen, the density profile of the inner region of the evolved cluster can be very well fitted by a power-law and the power-law slope α we obtain is, with $\alpha = 1.72$, in good agreement with the value found by Bahcall & Wolf (1976). Also the extent of the cusp profile is given by the radius where the Keplerian velocity of a star around the central BH equals the velocity dispersion of the cluster core, the radius of influence of the IMBH.

However, Baumgardt *et al.* (2005) found that such a cusp in density might not be easily detectable in a real star cluster, as it should be much shallower and difficult to distinguish from a standard King profile. They find that this is mainly an effect of mass segregation and stellar evolution, where the more massive dark stellar remnants are concentrated towards the center while the lower-mass main sequence stars that contribute most of the light are much less centrally concentrated. In their simulations they found power-law surface brightness slopes ranging from $\alpha = -0.1$ to $\alpha = -0.3$. Based on these results they identified 9 candidate cluster from the sample of Galactic GCs of Noyola & Gebhardt (2006) that might contain IMBHs. However, the disadvantage of current N-body simulations is that for realistic cluster models, that take into account stellar evolution and a realistic mass spectrum, the number of stars is restricted to typically less than 2×10^5 as these simulations require a large amount of computing time. However, many GCs are known to be very massive, with masses reaching up to $1 \times 10^6 M_\odot$ resulting in a much larger number of stars one has to deal with when modelling these objects. In previous N-body simulations, such large-N clusters have been scaled down to low-N systems. Scaling down can be achieved in two ways (e.g. Baumgardt *et al.* 2005): either the mass of the central IMBH M_{BH} is kept constant and N is decreased, effectively decreasing the total cluster mass M_Cl, or the ratio M_{BH}/M_{Cl} is kept constant, while lowering both M_{BH} and M_{Cl}. As both M_{BH}/M_{Cl} and the ratio of M_{BH} to stellar mass are important parameters that influence the structure of a cluster, but cannot be held constant simultaneously when lowering N, it is clear that only with the real N a fully self-consistent simulation can be achieved. One such method that can evolve such large-N systems for a sufficiently long time is the MC method.

3. Monte-Carlo Method with IMBH

The MC method shares some important properties with direct N-body methods, which is why it is also regarded as a randomized N-body scheme (see e.g. Freitag & Benz 2001). Just as direct N-body methods it relies on a star-by-star description of the GC, which makes it particularly straightforward to add additional physical processes such as stellar evolution. Contrary to direct N-body methods, however, the stellar orbits are resolved on a relaxation time scale T_{rel}, which is much larger than the crossing time t_{cr}, the time scale on which direct N-body methods resolve those orbits. This change in orbital resolution is the reason why the MC method is able to evolve a GC much more efficiently than direct N-body methods. It achieves this efficiency, however, by making several simplifying assumptions: (i) the cluster potential has spherical symmetry (ii) the cluster is in dynamical equilibrium at all times (iii) the evolution is driven by diffusive 2-body relaxation. The specific implementation we use for our study is the MC code initially developed by Joshi *et al.* (2000) and further enhanced and improved by Fregau *et al.* (2003) and Fregau & Rasio (2007). The code is based on Hénon's algorithm for solving the Fokker-Planck equation. It incorporates treatments of mass spectra, stellar evolution, primordial binaries, and the influence of a galactic tidal field.

The effect of an IMBH on the stellar distribution is implemented similar to Freitag & Benz (2002). In this method the IMBH is treated as a fixed, central point mass while stars are tidally disrupted and accreted onto the IMBH whenever their periastron distances lie within the tidal radius, R_{disr}, of the IMBH. For a given star-IMBH distance, the velocity vectors that lead to such orbits form a so called loss-cone and stars are removed from the system and their masses are added to the BH as soon as their velocity vectors enter this region. However, as the star's removal happens on an orbital time-scale one would need to use time-steps as short as the orbital period of the star in order to treat the loss-cone effects in the most accurate fashion. This would, however, slow down the whole calculation considerably. Instead, during one MC time-step a star's orbital evolution is followed by simulating the random-walk of its velocity vector, which approximates the effect of relaxation on the much shorter orbital time-scale. After each random-walk step the star is checked for entry into the loss-cone. For further details see Freitag & Benz (2002). Comparison with N-body calculations have shown that in order to achieve acceptable agreement between the two methods, the MC time-step must be chosen rather small relative to the local relaxation time, with $dt \leqslant 0.01 T_{rel}(r)$ (Freitag *et al.* 2006). While choosing such a small time-step was still feasible in the code of Freitag & Benz (2002), to enforce such a criterion for all stars in our simulation would lead to a dramatic slow-down of our code and notable spurious relaxation. The reason is that our code uses a shared time step scheme, with the time-step chosen to be the smallest value of all $dt_i = f\, T_{rel}(r_i)$, where f is some constant fraction and the subscript i refers to the individual star. In Freitag & Benz (2002) each star is evolved separately according to its local relaxation time, allowing for larger time steps for stars farther out in the cluster where the relaxation times are longer. In order to reduce the effect of spurious relaxation we are forced to choose a larger f, typically around $f = 0.1$. This has the consequence that the time-step criterion is only strictly fulfilled for stars typically

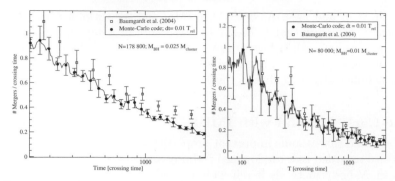

Figure 2. Comparison of the stellar merger rate per crossing time with simulations of Baumgardt *et al.* (2004). In both cases the evolution of single-mass clusters were calculated and the tidal radius of the IMBH was fixed to 1×10^{-7} in N-body units. Full circles with error bars are results from our MC runs at selected times, while the solid line goes through all obtained points.

outside of $0.1r_h$, where r_h is the half-mass radius of the cluster. To arrive at the correct merger rate of stars with the IMBH, despite the larger time step for the stars in the inner region, we apply the following procedure: (i) for each star i with $dt > 0.01T_{rel}(r_i)$ we take $n = dt/(0.01T_{rel})$ sub-steps. (ii) during each of these sub-steps we carry out the random-walk procedure as in Freitag & Benz (2002) (iii) after each sub-step we calculate the star's angular-momentum J according to the new velocity vector (iv) we generate a new radial position according to the new J. By updating J after each sub-step we approximately account for the star's orbital diffusion in J space during a full MC step, while neglecting any changes in orbital energy. This is at least for stars with low J legitimate (Shapiro & Marchant 1978), while for the other stars the error might not be significant as the orbital energy diffusion is still slower than the J diffusion (Frank & Rees 1976). A further assumption is that the cluster potential in the inner cluster region does not change significantly during a full MC step, which constrains the size of the full MC step.

4. Comparison to N-body Simulations

In Fig. 2 the rates of stellar mergers with the central BH per crossing time from two of our single-mass cluster simulations are compared to the corresponding results of Baumgardt *et al.* (2004). The initial cluster parameters were the same as in their study (run16 and run2). As can bee seen, the differences between our MC and the N-body results are within the respective error bars and thus in reasonable agreement with each other. However, the merger rates in the left panel of Fig. 2 seem to be consistently lower than in the N-body calculations. This might indicate that the agreement gets worse for other M_{BH}/M_{Cl} than we considered here ($0.25\% - 1\%$) and a different choice of time-step parameters for our MC code might be necessary in those cases. On the other hand, the differences might also be caused by differences in the initial relaxation phase before the cluster reaches an equilibrium state. This phase cannot be adequately modeled with a MC code because the code assumes dynamical equilibrium. Further comparisons to N-body simulations for different M_{BH}/M_{Cl} and N are necessary to test the validity of our method.

5. Realistic Cluster Models

In order to compare our simulations to observed GCs additional physical processes need to be included. Here we consider two clusters containing 1.3×10^5 and 2.6×10^5

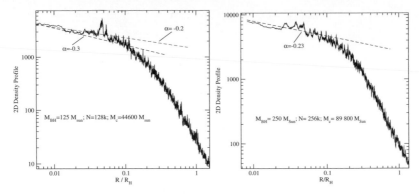

Figure 3. Two-dimensional density profile of bright stars for two clusters with different numbers of stars and BH to cluster mass ratios. The dashed line in the right panel is a power-law fit to the inner region of the cluster, while the two dashed lines in the left panel are for orientation only.

stars with a Kroupa mass function (Kroupa 2001), and follow the evolution of the single stars with the code of Belczynski et al. (2002) (for all other parameter see Baumgardt *et al.* 2005). Fig. 3 shows the two-dimensional density profiles of bright stars for the two clusters at an age of 12Gyr. The profile in the left panel can directly be compared to the corresponding result of Baumgardt *et al.*(2005) as N is the same. As was expected from the discussion in §2, the profile shows only a very shallow cusp with a power-law slope α between -0.2 and -0.3, consistent with the N-body results. The right panel shows the resulting profile for a cluster that is also similar but twice as massive and, consequently, has twice as many stars as in the N-body simulation. Here we still get a very similar profile with $\alpha = -0.23$ which is very close to the average slope of $\alpha = -0.25$ found in Baumgardt *et al.* (2005). Therefore, based on these very preliminary results, there seems to be no significant difference in cusp slopes for larger N clusters compared to small-N ones, but the parameter space must certainly be explored much further in order to confirm this finding.

References

Bahcall, J. N., & Wolf, R. A. 1976, *ApJ*, 209, 214
Baumgardt, H., Makino, J. & Hut, P. 2005, *ApJ*, 620, 238
Baumgardt, H., Makino, J. & Ebisuzaki, T. 2004, *ApJ*, 613, 1133
Belczynski, K., Bulik, T. & Kluźniak, W. ł. 2002, *ApJL*, 567, L63
Frank, J., & Rees, M. J. 1976, *MNRAS*, 176, 633
Fregeau, J. M., & Rasio, F. A. 2007, *ApJ*, 658, 1047
Fregeau, J. M., Gürkan, M. A., Joshi, K. J., & Rasio, F. A. 2003, *ApJ*, 593, 772
Freitag, M. & Benz, W. 2001, *A&A*, 375, 711
Freitag, M. & Benz, W. 2002, *A&A*, 394, 345
Freitag, M., Amaro-Seoane, P., & Kalogera, V. 2006, *ApJ*, 649, 91
Joshi, K. J., Rasio, F. A., & Portegies Zwart, S. 2000, *ApJ*, 540, 969
Kroupa, P. 2001, *MNRAS*, 322, 231
Noyola, E. & Gebhardt, K. 2006, *AJ*, 132, 447
Peebles, P. J. E. 1972, *ApJ*, 178, 371
Shapiro, S. L. & Marchant, A. B. 1978, *ApJ*, 225, 603
Volonteri, M., Haardt, F., & Madau, P. 2003, *ApJ*, 582, 559
Wyller, A. A. 1970, *ApJ*, 160, 443

Dynamical Evolution of Dense Stellar Systems
Proceedings IAU Symposium No. 246, 2007
E. Vesperini, M. Giersz, & A. Sills, eds.

© 2008 International Astronomical Union
doi:10.1017/S1743921308015925

The Formation and Evolution of Very Massive Stars in Dense Stellar Systems

Houria Belkus[1], Joris Van Bever[2] and Dany Vanbeveren[1,3]

[1] Astrophysical Institute, Vrije Universiteit Brussel, Pleinlaan 2, 1050 Brussel, Belgium
email: hbelkus@vub.ac.be, dvbevere@vub.ac.be
[2] Institute of Computational Astrophysics, Saint Mary's University,
923 Roby Street, Halifax, Nova Scotia, Canada B3H 3C3
email: vanbever@ap.smu.ca
[3] Mathematics department, Groep T, Vesaliusstraat 13, 3000 Leuven, Belgium

Abstract. The early evolution of dense stellar systems is governed by massive single star and binary evolution. Core collapse of dense massive star clusters can lead to the formation of very massive objects through stellar collisions ($M \geqslant 1000 \, M_\odot$). Stellar wind mass loss determines the evolution and final fate of these objects, and determines whether they form black holes (with stellar or intermediate mass) or explode as pair instability supernovae, leaving no remnant. We present a computationally inexpensive evolutionary scheme for very massive stars that can readily be implemented in an N-body code. Using our new N-body code 'Youngbody' which includes a detailed treatment of massive stars as well as this new scheme for very massive stars, we discuss the formation of intermediate mass and stellar mass black holes in young starburst regions. A more detailed account of these results can be found in Belkus, Van Bever & Vanbeveren (2007).

Keywords. stars: evolution, galaxies: starburst, stars: mass loss, stellar dynamics

---

Our calculations of the evolution of very massive stars are based on the results of Nadyozhin & Razinkova (2005) who constructed stellar structure models for these objects using the similarity theory of stellar structure. Their models correspond to chemically homogeneous stars, having Thompson scattering as the only opacity source throughout. This provides an accurate treatment, since extremely massive stars are almost completely convective during their evolution, and the opacity differs significantly from the Thompson scattering value only in a thin layer near the surface of the star. Our stellar parameters should therefore be reliable, with the possible exception of the effective temperature.

Kudritzki (2002) studied line-driven stellar winds of very massive stars and calculated mass loss rates of very massive O-type stars as a function of metallicity, in the range $6.3 \leqslant \log(L/L_\odot) \leqslant 7.03$ and $40 \, \text{kK} \leqslant T_{\text{eff}} \leqslant 60 \, \text{kK}$. The mass loss rates are smallest for the highest T_{eff}. We use his interpolation formula for the $40 \, \text{kK}$ models to compute the mass loss rates of our evolutionary models, meaning that we obtain an upper limit to the mass of a model star at all times. Note that for a given luminosity, we ensure that the mass loss rate never exceeds the maximum for line-driven winds, as given by Owocki (2004).

Fig. 1 shows the masses of very massive stars (with initial central hydrogen abundance $X_{c,0} = 0.68$) at the end of the core helium burning stage for 3 different metallicities. It is seen that only very massive stars at sufficiently low metallicities are expected to produce pair-instability supernovae and direct collapse black holes (compared to the limiting masses from Heger *et al.* (2003)). At high Z, the stellar wind mass loss rates

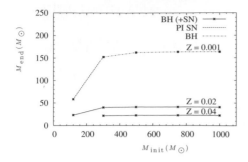

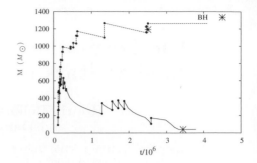

Figure 1. Masses of massive and very massive stars ($X_{c,0} = 0.68$) at the end of the core helium burning stage for 3 different metallicities (see labels). Note the almost constant final mass of very massive stars, which is due to the properties of the $\dot{M} - L$ relationship that is used. Also, the dashed part of the $Z = 0.001$ curve indicates stars that are expected to explode as pair instability supernovae, leaving no remnant.

Figure 2. Evolution of the mass of the runaway merger in a King ($W_0 = 9$) N-body model containing 3000 massive single stars ($M \geqslant 10\,M_\odot$) and with a half mass radius of 0.5 pc. The doted line represents a model in which any star more massive than 120 M0 is given a constant mass loss rate of $10^{-4}\,M_\odot/yr$. The full line represents a model in which those same stars are treated with the mass loss rates of Kudritzki (2002). The star symbols denote the moment at which the runaway merger collapses into a black hole.

reduce the stellar mass to such an extent that only black holes due to accompanying supernova and fallback result.

We implemented this evolution scheme in our direct N-body code 'Youngbody' and computed models of young dense starburst regions showing the creation of so-called runaway mergers. Fig. 2 shows a typical example and indicates that the reduction of mass by stellar wind is able to compete with the growth of the runaway star due to stellar collisions, at least in the later stages. In this model the core collapse stage is over before the merger ends its life and therefore mass loss is able to reduce the stellar mass sufficiently to prevent the formation of an intermediate mass black hole. This suggests that Ultra Luminous X-ray sources (ULXs) may not be IMBH accreting at rates close to the Eddington rate, but could be stellar mass BHs ($\approx 50 - 100\,M_\odot$) accreting at Super Eddington rates (Soria 2007).

References

Belkus, H., Van Bever, J. & Vanbeveren, D. 2007, *ApJ* 659, 1576
Heger, A., Fryer, C. L., Woosley, S. E., Langer, N. & Hartmann, D. H. 2003, *ApJ* 591, 288
Kudritzki, R. P. 2002, *ApJ* 577, 389
Nadyozhin, D. K. and Razinkova, T. L. 2005, *Astron. Lett.* 31, 695
Owocki, S. P., Gayley, K. G. and Shaviv, N. J. 2004, *ApJ* 616, 525
Soria, R. 2007, *Ap&SS* (in press)

Dynamical Evolution of Dense Stellar Systems
Proceedings IAU Symposium No. 246, 2007
E. Vesperini, M. Giersz & A. Sills, eds.

© 2008 International Astronomical Union
doi:10.1017/S1743921308015937

On the Dynamical Capture of a MSP by an IMBH in a Globular Cluster

B. Devecchi[1], M. Colpi[1], M. Mapelli[2] and A. Possenti[3]

[1] Department of Physics, University of Milano Bicocca, ,Italy
email: bernadetta.devecchi@mib.infn.it, monica.colpi@mib.infn.it

[2] Institute for Theoretical Physics, University of Zürich, Winterthurerstrasse 190, CH-8057
Zürich, Switzerland
email: mapelli@physik.unizh.ch

[3] INAF, Osservatorio Astronomico di Cagliari, Poggio dei Pini, Strada 54, 22 I-09012
Capoterra, Italy
email: possenti@ca.astro.it

Abstract.
Globular clusters (GCs) are rich of millisecond pulsars (MSPs) and might also host single or binary intermediate–mass black holes (IMBHs). We simulate 3- and 4-body encounters in order to test the possibility that an IMBH captures a MSP. The newly formed system could be revealed from the timing signal of the MSP, providing an unambiguous measure of the BH mass. In current surveys, the number of expected [IMBH,MSP] binaries in the Milky Way is ~ 0.1. If next-generation radio telescopes (e.g. SKA) will detect ~ 10 times more MSPs in GCs, we expect to observe at least one [IMBII,MSP] binary.

Keywords. pulsars: general, globular clusters: general

1. Introduction

Recent observations suggest that intermediate-mass black holes (IMBHs hereon) might be hosted in some globular cluster (GCs) (Gebhardt *et al.* 2005). There is strong evidence that GCs also host a high number of millisecond pulsars (MSP, Ransom *et al.* 2005). We study the possibility for an IMBH to capture a MSP by dynamical interactions.

2. Simulations

In order to determine the properties of [IMBH,MSP] binaries in Galactic GCs, we run 3- and 4-body simulations between a single or binary MSP interacting with a single or binary IMBH. The companion of the MSP, when present, is assumed to be a white dwarf (WD). The initial distributions adopted for the orbital parameters of the [MSP,WD] binaries are those typical of the Galactic population of binary MSPs (Camilo & Rasio 2005). We divide the entire population of MSP binaries into three families:

- class I (short period binaries): systems whose orbital period P is less than 1 day;
- class II (long period binaries): systems with 1 <P (days)<10;
- outliers: systems whit P> 10 days.

The companion of the IMBH is either a stellar mass BH or a star (for the initial distribution of the eccentricities and semi-major axes of the IMBH binaries see Devecchi *et al.* 2007).

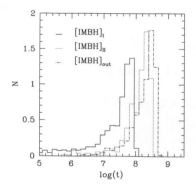

Figure 1. Lifetimes for [IMBH,MSP] binaries formed after the interaction of the single IMBH with [MSP,co] binaries belonging to class I ([IMBH]$_I$), class II ([IMBH]$_{II}$) and outliers ([IMBH]$_{out}$).

3. Results

From our simulations we determine the cross-sections Σ_X for the formation of [IMBH,MSP] binaries. We find that the presence of a secondary BH inhibits the formation of an [IMBH,MSP] binary. The highest value of Σ_X is found for a single IMBH interacting with outliers.

The formation rate for [IMBH,MSP] binaries for a particular channel X is related to the cross-section by:

$$\Gamma_X \sim n_{MSP} \omega_X \langle v_\infty \rangle \Sigma_X \qquad (3.1)$$

where n_{MSP} is the number density of MSPs in the Galactic GCs' cores, $\langle v_\infty \rangle$ is the mean relative velocity and ω_X corresponds to the probability of channel X. Even in the most favourable case (i.e. the single IMBH interacting with a binary MSP) the formation time-scales are of the order of the Hubble time.

Once these systems are formed, the probability of detecting an [IMBH,MSP] binary is also related to its lifetime, and thus depends on its orbital parameters. We infer the distributions of the semi-major axis and eccentricity of the [IMBH,MSP] from our simulations. Typical orbital separations are a few AU, in agreement with analytical models both for the single IMBH (Pfhal 2005) and for the [IMBH,star] cases (Devecchi *et al.* 2007). The eccentricities are very high, particularly for the single IMBH. The lifetimes of the newly formed binaries are calculated accounting for hardening by stellar encounters and by gravitational waves. Because of their high eccentricities, the binaries have typical lifetimes of $\sim 10^8$ yrs. This relatively short lifetime, combined with the low formation rate, makes [IMBH,MSP] binaries extremely rare: their expected number is ~ 0.1 for the Galactic population of GCs. If next-generation radio telescopes (e.g. SKA) will detect ~ 10 times more MSPs in GCs, we expect to observe at least one [IMBH,MSP].

References

Camilo, F. & Rasio, F. A. 2005, in Rasio F. A., Stairs 1200 I.H., eds, ASP Conf. Ser. Vol. 328, Binary Radio 1201 Pulsars.Astron.Soc.Pac., San Francisco, p.147
Devecchi, B., Colpi, M., Mapelli, M., & Possenti, A. 2007, *MNRAS* 380, 691
Gebhardt, K., Rich, R., & Ho, L. C. 2005, *ApJ* 634, 1093
Pfahl, E. 2005, *ApJ* 626, 849
Ransom S. M., Hessels J. W. T., Stairs I. H., Freire P. C. C., Camilo F., Kaspi V. M., & Kaplan D. L. 2005, *Science* 307,892

Dynamical Evolution of Dense Stellar Systems
Proceedings IAU Symposium No. 246, 2007
E. Vesperini, M. Giersz & A. Sills, eds.

Unveiling the Core of M 15 in the Far-Ultraviolet

A. Dieball[1], C. Knigge[1], D. R. Zurek[2], M. M. Shara[2], K. S. Long[3], P. A. Charles[4] and D. Hannikainen[5]

[1]Department of Physics and Astronomy, University Southampton, SO17 1BJ, UK
email: andrea@astro.soton.ac.uk

[2]Department of Astrophysics, American Museum of Natural History, New York, NY 10024

[3]Space Telescope Science Institute, Baltimore, MD 21218

[4]South African Astronomical Observatory, PO Box 9, Observatory, 7935, South Africa

[5]University of Helsinki, P.O. Box 14, SF-00014 Helsinki, Finland

Abstract. We present an analysis of our deep far- (FUV) and near-ultraviolet (NUV) photometry of the core region of the dense globular cluster M 15. Our FUV-NUV colour-magnitude diagram (CMD) is the deepest one presented for a globular cluster so far, and shows all hot stellar populations expected in a globular cluster, such as horizontal branch stars, blue stragglers, white dwarfs, cataclysmic variables and even main sequence stars. The main sequence turn-off is clearly visible and the main sequence stars form a prominent track that extends at least two magnitudes below the main sequence turn-off. We compare and discuss the radial distribution of the various stellar populations that show up in the FUV. We search for variability amongst our FUV sources and tentatively classify our variable candidates based on an analysis of the UV colours and variability properties. We find that RR Lyraes, Cepheids, and SX Phoenicis exhibit massive variability amplitudes in this waveband (several mags).

Keywords. globular clusters: individual (M 15), stars: cataclysmic variables, white dwarfs, stars: variables: other, ultraviolet: stars, blue stragglers, binaries: close

Stellar densities in the cores of globular clusters (GCs) are extremely high, so that dynamical encounters between the cluster stars are inevitable, leading to a variety of exotic stellar populations like blue stragglers (BSs), cataclysmic variables (CVs), low-mass X-ray binaries (LMXBs) and other close binary systems. These exotica all show a spectral energy distribution bluer than ordinary main sequence stars and red giants. As a consequence, crowding is considerably reduced, making FUV observations the ideal tool to study these dynamically-formed stellar populations in the dense cluster cores.

Results:

Our study confirms that FUV observations are particularly well suited in studying hot and especially dynamically formed stellar populations. We detected prominent populations of CV, BS and WD candidates and HB stars. CV and BS candidates are the most centrally concentrated stellar populations, which might either be an effect of mass segregation or reflect the preferred birthplace in the dense cluster core of such dynamically-formed objects. We found in total 41 variable sources. RR Lyrae, Cepheids and SX Phoenicis stars are easily detected as they show large FUV amplitudes. This is expected since the pulsation amplitude increases towards the FUV, but this is the first time they are actually observed for SX Phoenicis stars.

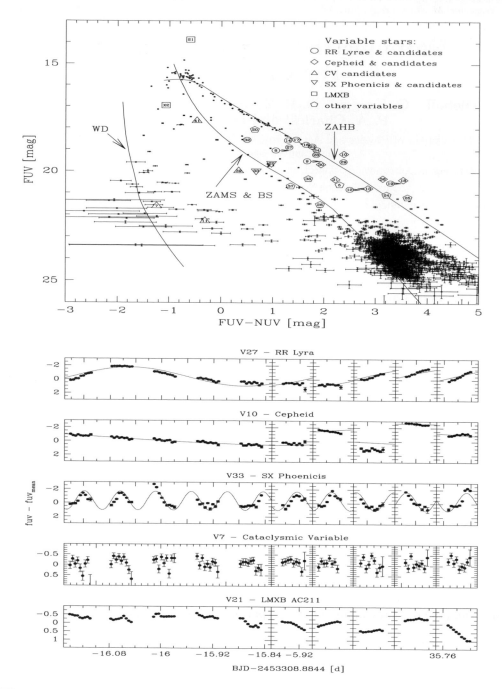

Figure 1. Top: Several stellar populations show up in our FUV-NUV CMD, such as the bright HB stars, BS and WD candidates. CVs and detached WD-MS binaries are located between the WD cooling sequence and the MS. The clump of MS stars and red giants reaches ≈ 2 mag below the cluster's turn-off. As such, this is the deepest FUV-NUV CMD presented for a GC so far. **Bottom:** We found in total 41 variable sources and suggest a classification based on their light curves and location in the CMD. We find four previously known RR Lyrae and 13 further candidates, one known Cepheid and six further candidates, one known and one probable SX Phoenicis star, six CVs, and the well known LMXB AC 211.

Proceedings Dynamical Evolution of Dense Stellar Systems
Proceedings IAU Symposium No. 246, 2007
E. Vesperini, M. Giersz & A. Sills, eds.

Building Blue Stragglers with Stellar Collisions

Evert Glebbeek and Onno R. Pols

Sterrekundig Instituut Utrecht, P.O. Box 80000, 3508 TA Utrecht, The Netherlands

Abstract. The evolution of stellar collision products in cluster simulations has usually been modelled using simplified prescriptions. Such prescriptions either replace the collision product with an (evolved) main sequence star, or assume that the collision product was completely mixed during the collision.

It is known from hydrodynamical simulations of stellar collisions that collision products are not completely mixed, however. We have calculated the evolution of stellar collision products and find that they are brighter than normal main sequence stars of the same mass, but not as blue as models that assume that the collision product was fully mixed during the collision.

Keywords. Stars: blue stragglers, evolution, formation

1. Introduction

The aim of the MODEST collaboration (Hut *et al.* (2003)) is to model and understand dense stellar systems, which requires a good understanding of what happens when two single stars or binary systems undergo a close encounter. A possible outcome of such an encounter is a collision followed by the merging of two or more stars. This is a possible formation channel for blue straggler stars (*e.g.* Sills *et al.* (1997)). Understanding the formation and evolution of blue stragglers is important for understanding the Hertzsprung-Russell diagram of clusters.

2. Method

We have developed a version of the Eggleton stellar evolution code (Eggleton (1971), Pols et al. (1995)) that can import the output of a collision calculation and calculate the subsequent evolution of the remnant, in principle without human intervention.

We have used this code to calculate detailed evolution models of collision remnants from the N-body simulation of M67 by Hurley et al. (2005) and compared these with normal main sequence stars of the same mass as well as homogeneous models with the same average abundances. The post-merger profiles were calculated with the parametrised code of Lombardi et al. (2002).

The collisions in the N-body simulation span a range of total masses $M = M_1 + M_2$, mass ratio $q = M_2/M_1$ and time of collision t. We have also calculated a grid of models spanning four collision times ($t = 2800, 3100, 3400$ and 3700Myr), four mass ratios ($q = 0.4, 0.6, 0.8$ and 1.0) and six total masses ($M = 1.5, 1.6, 1.7, 1.8, 1.9$ and 2.0), which covers the parameter space of collisions found in the N-body simulation.

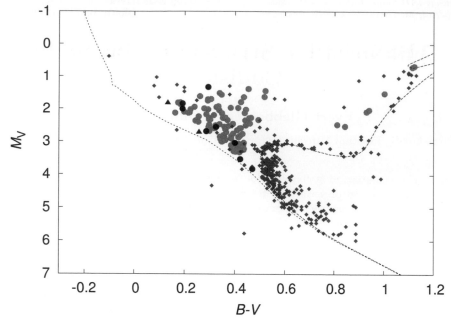

Figure 1. Colour-magnitude diagram of the open cluster M67 (♦). Overplotted are the locations of our models at 4Gyr, the age of M67. The black (●, ▲) symbols are collisions from the M67 simulation. Two of these are double collisions, which are indicated by ▲. The grey (●) symbols are from our larger grid.

3. Results and Conclusions

Compared to normal stars, collision products are helium enhanced. Most of the helium enhancement is in the interior and does not affect the opacity of the envelope. The helium enhancement does increase the mean molecular weight and therefore the luminosity of the star. This decreases the remaining lifetime of collision products compared to normal stars of the same mass and can be important for the predicted number of blue stragglers from cluster simulations. The increased luminosity changes the distribution of blue stragglers in the colour-magnitude diagram, moving it above the extension of the main sequence.

The evolution track of a fully mixed model can be significantly bluer than a self-consistently calculated evolution track of a merger remnant. Fully mixed models are closer to the zero age main sequence.

Our grid of models covers most of the observed blue straggler region of M67 (Fig. 1). A better coverage of the blue part of the region can be achieved by increasing the upper mass limit in the grid. The brightest observed blue straggler falls outside the region of our grid because it requires at least a double collision to explain.

References

Eggleton P. P., 1971, *MNRAS*, 151, 351
Hurley J. R., Pols O. R., Aarseth S. J., & Tout C. A., 2005, *MNRAS*, 363, 293
Hut P. *et al.* 2003, New Astronomy, 8, 337
Lombardi Jr. J. C., Warren J. S., Rasio F. A., Sills A., & Warren A. R., 2002, *ApJ*, 568, 939
Pols O. R., Tout C. A., Eggleton P. P., & Han Z., 1995, *MNRAS*, 274, 964
Sills A., Lombardi Jr. J. C., Bailyn C. D., Demarque P., Rasio F. A., & Shapiro S. L., 1997, *ApJ*, 487, 290

Dynamical Evolution of Dense Stellar Systems
Proceedings IAU Symposium No. 246, 2007
E. Vesperini, M. Giersz & A. Sills, eds.

© 2008 International Astronomical Union
doi:10.1017/S1743921308015962

On the Origin of Hyperfast Neutron Stars

V.V. Gvaramadze[1], A. Gualandris[2,3]
and S. Portegies Zwart[3]

[1]Sternberg Astronomical Institute, Moscow State University, Universitetskij Pr. 13, Moscow
119992, Russia
email: vgvaram@mx.iki.rssi.ru

[2]Center for Computational Relativity and Gravitation, Rochester Institute of Technology, 78
Lomb Memorial Drive, Rochester, NY 14623, USA
email: alessiag@astro.rit.edu

[3] Astronomical Institute 'Anton Pannekoek' and Section Computational Science, Amsterdam
University, Kruislaan 403, 1098 SJ, Amsterdam, the Netherlands
email: spz@science.uva.nl

Abstract. We propose an explanation for the origin of hyperfast neutron stars (e.g. PSR
B1508+55, PSR B2224+65, RX J0822−4300) based on the hypothesis that they could be the
remnants of a *symmetric* supernova explosion of a high-velocity massive star (or its helium core)
which attained its peculiar velocity (similar to that of the neutron star) in the course of a strong
three- or four-body dynamical encounter in the core of a young massive star cluster. This hypothesis implies that the dense cores of star clusters (located either in the Galactic disk or near
the Galactic centre) could also produce the so-called hypervelocity stars – ordinary stars moving
with a speed of $\sim 1\,000\,\mathrm{km\,s^{-1}}$.

Keywords. Stars: neutron, pulsars: general, pulsars: individual (B1508+55), galaxies: star clusters, methods: n-body simulations

1. Introduction

Recent proper motion and parallax measurements for the pulsar PSR B1508+55 (Chat-
terjee *et al.* 2005) gave the first example of a high velocity ($1\,083^{+103}_{-90}\,\mathrm{km\,s^{-1}}$) *directly*
measured for a neutron star (NS). A possible way to account for extremely high veloci-
ties of NSs† is to assume that they are due to a natal kick or a post-natal acceleration
(Chatterjee *et al.* 2005). In this paper, we propose an alternative explanation for the
origin of hyperfast NSs (cf. Gvaramadze 2007) based on the hypothesis that they could
be the remnants of *symmetric* supernova (SN) explosions of hypervelocity stars [HVSs;
the ordinary stars moving with extremely high ($\sim 1\,000\,\mathrm{km\,s^{-1}}$) peculiar velocities; e.g.
Brown *et al.* 2005]. A strong argument in support of this hypothesis comes from the fact
that the mass of one of the HVSs is $\geqslant 8\,M_\odot$ (Edelmann *et al.* 2005) so that this star
ends its evolution in a type II SN leading to the production of a hyperfast NS.

2. Hypervelocity stars and young massive star clusters

It is believed that the origin of HVSs could be connected to scattering processes involving the supermassive black hole (BH) in the Galactic centre (Hills 1988; Yu & Tremaine
2003; Gualandris *et al.* 2005). It is therefore possible that the progenitors of some hyperfast NSs were also ejected from the Galactic centre. The proper motion and parallax

---

† Other possible examples of hyperfast NSs are PSR B2224+65 (Chatterjee & Cordes 2004)
and RX J0822-4300 (Hui & Becker 2006).

measured for PSR B1508+55, however, indicate that this NS was born in the Galactic disk (Chatterjee *et al.* 2005). The kinematic characteristics of some high-velocity early B stars also suggest that these objects originated in the disk (e.g. Ramspeck *et al.* 2001). We consider the possibility that the HVSs (including the progenitors of hyperfast NSs) could be ejected not only from the Galactic centre but also from the cores of young ($< 10^7$ yr) massive ($\sim 10^4 - 10^5 \, M_\odot$) star clusters (YMSCs), located either in the Galactic disk or near the Galactic centre (cf. Gualandris & Portegies Zwart 2007).

3. Origin of hyperfast neutron stars

To check the hypothesis that the hyperfast NSs could be the descendants of HVSs which were ejected from the cores of YMSCs, we calculated (see Gvaramadze *et al.* 2007) the maximum possible ejection speed produced by dynamical processes involving close encounters between: *i*) two hard (Heggie 1975) massive binaries (e.g. Leonard 1991), *ii*) a hard binary and an intermediate-mass ($\sim 100 - 1\,000\,M_\odot$) BH (e.g. Portegies Zwart & McMillan 2002), and *iii*) a single star and a hard binary intermediate-mass BH (e.g. Gürkan *et al.* 2006). We find that main-sequence O-type stars cannot be ejected from YMSCs with peculiar velocities high enough to explain the origin of hyperfast NSs, but lower mass main-sequence stars or the stripped helium cores of massive stars could be accelerated to hypervelocities. We find also that the dynamical processes in the cores of YMSCs can produce stars moving with velocities of $\sim 200 - 400\,\mathrm{km\,s^{-1}}$ which therefore contribute to the origin of high-velocity NSs as well as to the origin of the bound population of halo stars (Ramspeck *et al.* 2001; Brown *et al.* 2007).

Acknowledgements

V. V. G. acknowledges the International Astronomical Union and the Russian Foundation for Basic Research for travel grants. A. G. and S. P. Z. acknowledge support from the Netherlands Organization for Scientific Research (NWO under grant No. 635.000.001 and 643.200.503), the Royal Netherlands Academy of Arts and Sciences (KNAW) and the Netherlands Research School for Astronomy (NOVA).

References

Brown, W. R., Geller, M. J., Kenyon, S. J., & Kurtz, M. J. 2005, *ApJ* 622, L33
Brown, W. R., Geller, M. J., Kenyon, S. J., Kurtz, M. J., & Bromley, B. C. 2007, *ApJ* 660, 311
Chatterjee, S., & Cordes, J. M. 2004, *ApJ* 600, L51
Chatterjee, S., *et al.* 2005, *ApJ* 630, L61
Edelmann, H., Napiwotzki, R., Heber, U., Christlieb, N., & Reimers, D. 2005, *ApJ* 634, L181
Gualandris, A., & Portegies Zwart, S. 2007, *MNRAS* 376, L29
Gualandris, A., Portegies Zwart, S., & Sipior, M. S. 2005, *MNRAS* 363, 223
Gürkan, M. A., Fregeau, J. M., & Rasio, F. A. 2006, *ApJ* 640, L39
Gvaramadze, V. V. 2007, *A&A* 470, L9
Gvaramadze, V. V., Gualandris, A., & Portegies Zwart, S. 2007, preprint: astro-ph/0702735
Heggie, D. C. 1975, *MNRAS* 173, 729
Hills, J. G. 1988, *Nat* 331, 687
Hui, C. Y., & Becker, W. 2006, *A&A* 457, L33
Leonard, P. J. T. 1991, *AJ* 101, 562
Portegies Zwart, S. F., & McMillan, S. L. W. 2002 *ApJ* 576, 899
Ramspeck, M., Heber, U., & Moehler, S. 2001, *A&A* 378, 907
Yu, Q., & Tremaine, S. 2003, *ApJ* 599, 1129

Dynamical Evolution of Dense Stellar Systems
Proceedings IAU Symposium No. 246, 2007
E. Vesperini, M. Giersz & A. Sills, eds.

© 2008 International Astronomical Union
doi:10.1017/S1743921308015974

Tracing Intermediate-Mass Black Holes in the Galactic Centre

Ulf Löckmann[1] and Holger Baumgardt[2]

Argelander Institut für Astronomie, Rheinische Friedrich-Wilhelms-Universität Bonn, Auf dem Hügel 71, 53121 Bonn, Germany

[1] email: uloeck@astro.uni-bonn.de

[2] email: holger@astro.uni-bonn.de

Abstract. We have developed a new method for post-Newtonian, high-precision integration of stellar systems containing a super-massive black hole (SMBH), splitting the forces on a particle between a dominant central force and perturbations. We used this method to perform fully collisional N-body simulations of inspiralling intermediate-mass black holes (IMBHs) in the centre of the Milky Way.

Keywords. Galaxy: centre, methods: n-body simulations, black hole physics, stellar dynamics

Most massive galaxies are now believed to host SMBHs. While such SMBHs have been directly observed as radio sources and compact massive objects, the existence of IMBHs with masses $10^2 \, M_\odot < M_{\mathrm{IMBH}} < 10^4 \, M_\odot$ is still a matter of debate.

In this work, we present new results on the inspiral of IMBHs, using a novel method for studying the dynamics of Galactic-centre-like systems where stars orbit a central SMBH on weakly perturbed Keplerian orbits.

1. Integration Method

The basic idea for our new method is to split the force calculation between the dominating central force (exerted by the SMBH) and the perturbing forces (due to the cluster stars). It is comparable to the so-called *mixed variable symplectic* methods (*MVS*, Wisdom & Holman 1991), as it makes use of Kepler's equation to integrate along the orbit. However, our method is not symplectic, as it is based on the Hermite scheme to allow for large N, close encounters, and use of the *GRAPE* special-purpose hardware.

For a given average number of steps per orbit, our method is almost a factor of 100 more accurate than the standard Hermite method, as it does not accumulate an error in the orbital motion around the SMBH. To account for relativistic effects, we extend our integration method by post-Newtonian correction terms.

2. IMBH inspiral

Theoretical arguments and N-body simulations have shown that a stellar system around a SMBH evolves into a cusp with an $\alpha = 1.75$ power-law density distribution (Bahcall & Wolf 1976; Baumgardt et al. 2004a,b).

Following the results of Baumgardt et al. (2006), we started our calculations with a $10^3 \, M_\odot$ IMBH on a circular orbit around a $3 \cdot 10^6 \, M_\odot$ SMBH with semi-major axis $a_{\mathrm{IMBH}} = 0.1 \, \mathrm{pc}$ (run A). Fig. 1 shows that the inspiral process follows the theoretical description very well until the IMBH has depleted the inner cusp and the density profile is flattened (like it has recently been observed for the Galactic centre).

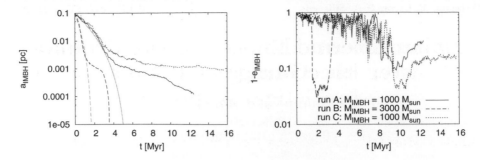

Figure 1. Evolution of the semi-major axes a_{IMBH} and eccentricities e_{IMBH} of three subsequent inspiralling IMBHs. The theoretical curve predicting the inspiral is left once the IMBH has depleted the inner cusp. As an IMBH acquires a highly eccentric orbit due to interactions with passing stars, emission of gravitational waves becomes important, eventually leading to coalescence with the SMBH.

If an additional IMBH spirals into such a flat cusp, the inspiral process due to two-body encounters may terminate at a central distance of 10^{-3} pc (see run C in Fig. 1). At this point, the estimated time until coalescence due to gravitational wave emission may be of the order of 1 Gyr, thus allowing for direct observation in the near future.

3. Ejection of hyper-velocity stars

During the inspiral process, a number of hyper-velocity stars (HVSs) are ejected from the system by close encounters with the IMBH.

Our simulations show that there should be a population of HVSs with very high velocities ($v > 1000$ km/s) which has not been observed so far. However, from the small amount of observational data currently available, the IMBH HVS ejection model can neither be excluded nor concluded for our Galaxy.

If a star's encounter with a massive black hole that leads to HVS ejection is close enough, tidal effects become important. One effect would be that, as a result of the encounter, rotation is induced in the star. We find that fast rotation of a HVS would be evidence for an IMBH in the Galactic centre, but can only be expected among the very fast moving HVSs ($v > 1000$ km/s) (see Löckmann & Baumgardt 2007, for details).

Acknowledgements

We are grateful to Douglas C. Heggie, Ulrich Heber, and Ingo Berentzen for useful discussions. We also thank Naohito Nakasato for the help with his SPH code. This work was supported by the DFG Priority Program 1177 'Witnesses of Cosmic History: Formation and Evolution of Black Holes, Galaxies and Their Environment'.

References

Bahcall J. N. & Wolf R. A., 1976, *ApJ*, 209, 214
Baumgardt H., Gualandris A., & Portegies Zwart S., 2006, *MNRAS*, 372, 174
Baumgardt H., Makino J., & Ebisuzaki T., 2004a, *ApJ*, 613, 1133
Baumgardt H., Makino J., & Ebisuzaki T., 2004b, *ApJ*, 613, 1143
Löckmann U. & Baumgardt H., 2007, *MNRAS*, submitted
Wisdom J. & Holman M., 1991, *AJ*, 102, 1528

Dynamical Evolution of Dense Stellar Systems
Proceedings IAU Symposium No. 246, 2007   © 2008 International Astronomical Union
E. Vesperini, M. Giersz & A. Sills, eds.   doi:10.1017/S1743921308015986

Environmental Effects on the Globular Cluster Blue Straggler Population: a Statistical Approach

Alessia Moretti[1],[2], Francesca De Angeli[3] and Giampaolo Piotto[1]

[1]Dipartimento di Astronomia - Università di Padova,
Vicolo dell'Osservatorio 3, I-35122 Padova
email: alessia.moretti@unipd.it

[2]INAF - Osservatorio Astronomico di Padova,
Vicolo dell'Osservatorio 5, I-35122 Padova

[3]Institute of Astronomy,
Madingley Road, CB3 0HA Cambridge, UK

Abstract. Blue stragglers stars (BSS) constitute an ubiquitous population of objects whose origin involves both dynamical and stellar evolution. We took advantage of the homogeneous sample of 56 Galactic globular clusters observed with WFPC2/HST by Piotto *et al.* (2002) to investigate the environmental dependence of the BSS formation mechanisms. We explore possible monovariate relations between the frequency of BSS (divided in different subsamples according to their location with respect to the parent cluster core radius and half mass radius) and the main parameters of their host GC. We also performed a Principal Component Analysis to extract the main parent cluster parameters which characterise the BSS family.

Keywords. stars:blue stragglers-luminosity function–Hertzsprung-Russell diagram- globular clusters:general

1. Introduction

BSS are located blue-ward and at brighter magnitudes than the turnoff (TO) in the color-magnitude diagram (CMD). The simple presence of such stars in a CMD poses serious challenges to the stellar evolution theory, since cluster stars with masses higher than the TO mass should have already been evolved off the main sequence (MS) toward the red giant branch (RGB). Explanations of the existence of these objects must take into account both the dynamical processes happening during the cluster lifetime and the stellar evolution itself. Two different mechanisms have been proposed so far to account for the existence of these peculiar objects: one describes BSS as the by–product of primordial binaries that simply evolve transferring their masses up to a complete coalescence (McCrea 1964, Carney *et al.* 2001), while the other one predicts that BSS are formed from the merger (collision) of two main sequence stars during the dynamical evolution of the cluster (Baylin 1995). We will refer to BSS of the first and second types *primordial* and *collisional* BSS, respectively. Recent results show that both formation mechanisms could be at work in a given cluster (Ferraro *et al.* 1997, Piotto *et al.* 2004, Davies *et al.* 2004, Mapelli *et al.* 2004, Mapelli *et al.* 2006).

369

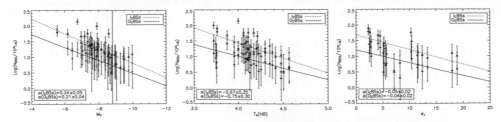

Figure 1. From left to right: normalized number of BSS as a function of the host cluster integrated absolute magnitude, the maximum temperature along the Horizontal Branch and the central velocity dispersion. Diamonds refer to the number of BSS outside the core radius of the host cluster, triangles to the number of BSS inside the core.

2. Analysis and Results

BSS selection was performed in two steps: first, BSS candidates were identified by eye in the CMDs (a visual inspection of each diagram is particularly important to disentangle BSS candidate from HB stars); later on, the BSS candidates were further selected by drawing a line approximately parallel to the sub-giant branch and at 0.7 magnitudes from the turnoff locations published by De Angeli *et al.* 2005. Most importantly, our selection has been made on the basis of our accurate error analysis, which stems from appropriate artificial star experiments (see Piotto *et al.* 2002 for more details).

After having selected BSS from the CMD diagrams we divided them according to their location with respect to the cluster center. In particular we analyzed BSS inside the core radius and outside it, as well as BSS inside the half mass radius and outside it. We studied monovariate relations with most of the characterizing cluster parameters and derived the possible significance of the fitted relations through the related errors coming from a bootstrap analysis. We used two main quantities, i.e. the total number of BSS and the normalised number of BSS (number of BSS in a given region divided by the total luminosity in the same region in unit of $10^4 L_\odot$).

We find that any subpopulation of BSS strongly depends on the luminosity of the cluster, on the extension of the cluster horizontal branch and on the central velocity dispersion (see Fig.1): more luminous clusters, clusters with a smaller central density, and a smaller central velocity dispersion have a higher BSS frequency. Moreover, we find that clusters having higher mass, higher central densities, and smaller core relaxation timescales possess on average more luminous BSS. Finally, different dependencies seem to hold for clusters with different integrated luminosity: brighter clusters show a BSS population that depends on the collisional parameter, while BSS in fainter clusters are mostly influenced by the cluster luminosity and the dynamical time–scales.

References

Bailyn, C. D. 1995, *ARAA*, 33, 133

Carney, B. W., Latham, D. W., Laird, J. B., Grant, C. E., & Morse, J. A. 2001, *AJ*, 122, 3419

Davies, M. B., Piotto, G., & De Angeli, F. 2004, *MNRAS*, 349, 129

De Angeli, F., Piotto, G., Cassisi, S., *et al.* 2005, *AJ*, 130, 116

Ferraro, F. R., Paltrinieri, B., Fusi Pecci, F., *et al.* 1997, *A&A*, 324, 915

Mapelli, M., Sigurdsson, S., Colpi, M., *et al.* 2004, *ApJL*, 605, L29

Mapelli, M., Sigurdsson, S., Ferraro, F. R., *et al.* 2006, *MNRAS*, 373, 361

McCrea, W. H. 1964, *MNRAS*, 128, 147

Piotto, G., King, I. R., Djorgovski, S. G., *et al.* 2002, *A&A*, 391, 945

Piotto, G., De Angeli, F., King, I. R., *et al.* 2004, *ApJL*, 604, L109

Dynamical Evolution of Dense Stellar Systems
Proceedings IAU Symposium No. 246, 2007
E. Vesperini, M. Giersz & A. Sills, eds.

© 2008 International Astronomical Union
doi:10.1017/S1743921308015998

Paucity of Dwarf Novae in Globular Clusters

P. Pietrukowicz and J. Kaluzny

Nicolaus Copernicus Astronomical Center, Bartycka 18, 00-716 Warsaw
email: pietruk@camk.edu.pl

Abstract. We have conducted an extensive photometric search for dwarf nova (DN) outbursts in 16 Galactic globular clusters (GCs). The survey was based on the rich photometric data collected by the Cluster AgeS Experiment (CASE) team. We have identified two new DNe. Together with previously known systems this gives the total number of 12 known DNe in 7 Galactic GCs. Inserting artificial light curves of "DNe" into frames of investigated clusters allowed us to assess completeness of the search. Our results clearly show that outbursting cataclysmic variables (CVs) are very rare in GCs in comparison to field CVs where half of the systems belongs to DNe. Recent X-ray observations of GCs lead to identification of hundreds of compact binaries. Many of them are promising candidates for CVs. The theory also predicts that dozens of white/red dwarf binaries should form in the cores of GCs via dynamical processes or internal evolution of the binaries. Our results rises the question about possible causes of paucity of outbursts in GCs.

Keywords. stars: dwarf novae, novae, cataclysmic variables, globular clusters: general

1. Observations and reductions

The observations (see Table 1) are made with the 1.0-m Swope telescope at Las Campanas Observatory, Chile. In our search we have used about 20,000 long exposures (80-600 sec) taken in Johnson V filter. The reductions were made using difference image analysis package DIAPL (http://www.camk.edu.pl/~pych/DIAPL) and DAOPHOT package (Stetson 1987) for profile photometry.

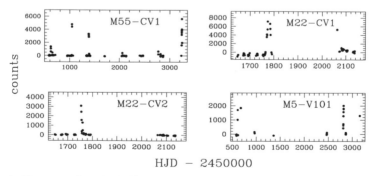

Figure 1. Two new (on the left) and two previously known (on the right) cluster DNe detected in our data

2. Results of the search for DNe

Our search for DNe in 16 GCs yielded two new objects, namely cataclysmic variable CV1 in the globular cluster M55 (Kaluzny *et al.* 2005) and CV2 in M22 (Pietrukowicz *et al.* 2005). In the remaining 14 GCs we found no new certain DNe. We easily recovered two well-known cluster DNe: M5-V101 and M22-CV1. Fig. 1 presents light curves in units of differential counts for all four objects. We note that we detected no light variations near positions of 27 CV candidates located in the fields of analysed clusters.

3. Simulations

We performed simulations to assess completeness of our search. We inserted into frames of three clusters, M22, M30, and NGC 2808, artificial images of erupting DNe and checked whether they would be detected in our search. The simulations were modelled using light curves of two prototype DNe, SS Cygni and U Geminorum, taken from the American Association of Variable Star Observers (AAVSO) International Database. We found that most of such stars would be detected in our data.

Table 1. Data on analyzed globular clusters

Cluster	Seasons	Total number of long exposures in V band	Total number of analyzed nights	Expected number of CVs in the core
M 4	1998-2000, 2002-2005	1981	61	4.8
M 5	1997-1999, 2002-2004	1043	34	36
M 10	1998, 2002	847	28	13
M 12	1999-2001	1236	41	4.6
M 22	2000-2001	2006	71	30
M 30	2000	340	23	39
M 55	1997-2004	3795	151	1.7
NGC 288	2004-2005	297	9	0.3
NGC 362	1997-1998, 2000-2005	1424	90	101
NGC 2808	1998-1999	312	33	208
NGC 3201	2001-2005	751	22	2.6
NGC 4372	2004-2005	601	19	0.7
NGC 6362	1999-2005	2585	104	0.9
NGC 6752	1996-1997	395	7	51
ω Cen	1999-2002	1571	76	30
47 Tuc	1993	551	34	200

4. Conclusions

Our search for DNe in 16 GCs resulted in identification of two new objects. Together with previously known systems this yields the total number of 12 known DNe in 7 Galactic GCs. We have confirmed that ordinary DNe are indeed very rare in GCs. Such a small number of DNe among expected cluster CVs is extremely low in comparison to field CVs of which half are DNe.

Possible explanations of rare outbursts in cluster CVs:

• absence of accretion discs due to strong magnetic fields of the white dwarfs (Ivanova *et al.* 2006)

• a combination of a low mass transfer rates and moderately strong white dwarf magnetic fields (Dobrotka *et al.* 2006)

• frequent stellar encounters affect stability of binary orbits thus affecting their ability to sustain accretion rate suitable for formation of an accretion disc (Shara & Hurley 2006)

References

Dobrotka, A. *et al.* 2006, *ApJ* 640, 288
Ivanova, N. *et al.* 2006, *MNRAS* 372, 1043
Kaluzny, J. *et al.* 2005, *MNRAS* 359, 677
Pietrukowicz, P. *et al.* 2005, *AcA* 55, 261
Shara, M. M. & Hurley, J. R. 2006, *ApJ* 646, 464
Stetson, M. M. 1987, *PASP* 99, 191

Dynamical Evolution of Dense Stellar Systems
Proceedings IAU Symposium No. 246, 2007
E. Vesperini, M. Giersz & A. Sills, eds.

XMM-*Newton* and *Chandra* Observations of Neutron Stars and Cataclysmic Variables in the Globular Cluster NGC 2808

M. Servillat[1], N. A. Webb[1], D. Barret[1], R. Cornelisse[2], A. Dieball[3], C. Knigge[3], K. S. Long[4], M. M. Shara[5] and D. R. Zurek[5]

[1]CESR, CNRS, Université Paul Sabatier, Toulouse, France
email: mathieu.servillat@cesr.fr

[2]Instituto de Astrofisica de Canarias, Santa Cruz de Tenerife, Spain

[3]School of Physics and Astronomy, University of Southampton, UK

[4]Space Telescope Science Institute, Baltimore, USA

[5]Department of Astrophysics, American Museum of Natural History, New York, USA

Abstract.
We report on XMM-*Newton* and *Chandra* observations of the globular cluster NGC 2808. We detect one quiescent low mass X-ray binary of the 3 ± 1 expected, if these systems are formed through encounters, and we show evidence for the presence of 20 ± 10 *bright* cataclysmic variables in the core with a luminosity above 4×10^{31} erg s^{-1}. We also review the specific nature of cataclysmic variables in globular clusters with reference to recent VLT/FORS1 observations of a cataclysmic variable in M 22.

Keywords. globular clusters: individual (NGC 2808, M 22), stars: neutron, cataclysmic variables, binaries: close

---

Globular clusters (GCs) are thought to harbour a large number of interacting binaries which are believed to delay their core collapse. With XMM-*Newton* and *Chandra*, it is possible to detect faint X-ray sources such as active binaries (ABs) and interacting binaries and their products, e.g. neutron star low mass X-ray binaries often in quiescence (qLMXBs), cataclysmic variables (CVs) and millisecond pulsars (MSPs). We discuss here only the core X-ray sources of NGC 2808 (for a discussion of XMM-*Newton* X-ray sources outside NGC2808's core, see Servillat, Webb & Barret 2008).

Based on the colour-flux diagram of all the detected sources (Fig. 1), we identify some of the core sources:

1 qLMXB. 3 ± 1 qLMXBs are expected if they are formed through encounters (Gendre *et al.* 2003). C2 is the only qLMXB candidate discovered with XMM-*Newton* (Servillat, Webb & Barret 2008). The spectrum of C2, after subtracting the contribution of source C1, is well fitted with a hydrogen atmosphere and has parameters typical for neutron stars. *Chandra* observations confirm the XMM-*Newton* identification.

~9 *bright* CVs. From observations of 23 GCs (Pooley & Hut 2006), we expect to detect 20 ± 10 CVs with luminosities above 4×10^{31} erg s^{-1} in NGC 2808, which is consistent with the global luminosity of the core (without C2). C1 is resolved into 4 Chandra sources, all consistent with CV emission. The XMM-*Newton* spectrum of C1 is well fitted with a high temperature bremsstrahlung model, typical of CVs (Baskill *et al.* 2005). C3 has faded by almost a factor 10 in the *Chandra* observation. C5 is undetected with *Chandra*, and one *Chandra* source was not detected with XMM-Newton, implying a flux variation of at least a factor 10.

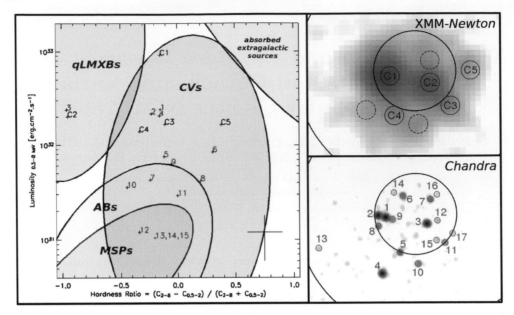

Figure 1. *Left*: colour-flux diagram of detected sources. It is divided into very approximate regions extrapolated from Pooley & Hut 2006. *Right*: images of the core of NGC 2808 from XMM-*Newton* and *Chandra* observations, where the core and half-mass radii are plotted.

- We note that the XMM-*Newton* and *Chandra* observations are complementary and well correlated.
- We find evidence for CV variability and maybe as many as three outbursts.
- The spectrum of C2 is helpful for constraining the neutron star equation of state as described in Webb & Barret (2007).
- Few outbursts have been detected in GC CVs compared to field CVs, leading to the idea that the two populations may harbour different properties. It was proposed that GC CVs have moderately strong magnetic fields that stabilize the accretion flow (Dobrotka *et al.* 2006). The mass of the white dwarf is also possibly higher in GC CVs (Ivanova *et al.* 2006). Another possibility is that because of crowding in GCs, many CV outbursts were not detected. From our two sets of X-ray observations of NGC 2808, we detect high flux variations for three CV candidates, possibly three outbursts. Two CVs in M22 have also been observed in outburst optically. From recent optical spectroscopy of these faint CVs with VLT/FORS1, we will determine the mass of the white dwarf and the magnetic field to determine whether these play a role in GC CVs.
- Ultra-violet data (Dieball *et al.* 2005) and other multi-wavelength observations will help in identifying the faintest sources (Servillat *et al. in preparation*).

References

Baskill, D. S., Wheatley, P. J. & Osborne, J. P. 2005, *MNRAS* 357, 626
Dieball, A., Knigge, C., Zurek, D. R., Shara, M. M. & Long, K. S. 2005, *ApJ* 625, 156
Dobrotka, A., Lasota, J-P. & Menou, K. 2006, *ApJ* 640, 288
Gendre, B., Barret, D., & Webb, N. 2003, *A&A*, 403, L11
Ivanova, N., Heinke, C. O., Rasio, F. A., *et al.* 2006, *MNRAS* 372, 1043
Pooley, D. & Hut, P. 2006, *ApJ* 646, L143
Servillat, M., Webb, N. A. & Barret, B., 2008, *A&A* submitted
Webb, N. A. & Barret, D. 2007, *ApJ* in press, arXiv:0708.3816

Part 7

Globular Cluster Systems

Dynamical Evolution of Dense Stellar Systems
Proceedings IAU Symposium No. 246, 2007
E. Vesperini, M. Giersz & A. Sills, eds.

© 2008 International Astronomical Union
doi:10.1017/S1743921308016013

An Update on the ACS Virgo and Fornax Cluster Surveys

Patrick Côté[1], Laura Ferrarese[1], Andrés Jordán[2], John P. Blakeslee[3], Chin-Wei Chen[1,4], Leopoldo Infante[5], Simona Mei[6], Eric W. Peng[1], John L. Tonry[7] and Michael J. West[8]

[1] Herzberg Institute of Astrophysics, Victoria, Canada; [2] Harvard-Smithsonian Center for Astrophysics, Cambridge, MA; [3] Washington State University, Pullman, WA; [4] National Central University, Taiwan; [5] Pontificia Universidad Católica de Chile, Chile; [6] Observatoire de Paris, France; [7] Institute for Astronomy, Honolulu, USA; [8] European Southern Observatory, Santiago, Chile

Abstract. We present a brief update on the ACS Virgo and Fornax Cluster Surveys — *Hubble Space Telescope* programs to obtain *ACS* imaging for 143 early-type galaxies in the two galaxy clusters nearest to the Milky Way. We summarize a selection of science highlights from the surveys as including new results on the central structure of early-type galaxies, the apparent continuity of photometric and structural parameters between dwarf and giant galaxies, and the properties of globular clusters, diffuse star clusters and ultra-compact dwarf galaxies.

Keywords. galaxies: elliptical and lenticular, cD galaxies: fundamental parameters – galaxies: nuclei – galaxies: photometry – galaxies: star clusters – galaxies: structure

1. The ACS Virgo and Fornax Cluster Surveys: An Introduction

With the installation of the *Advanced Camera for Surveys* (ACS) in 2002, the already outstanding imaging capabilities of the *Hubble Space Telescope* (HST) improved dramatically. To capitalize on the roughly factor of ten improvement in "discovery efficiency" (Ford *et al.* 1998), our team initiated — in Cycles 11 and 13 — ACS surveys of the early-type galaxy populations in the Virgo and Fornax Clusters: the two rich clusters nearest to the Milky Way. These *ACS Virgo and Fornax Cluster Surveys* (ACSVCS, Côté *et al.* 2004; ACSFCS, Jordán *et al.* 2007a) consist of HST imaging for 143 members of the Virgo and Fornax Clusters, all of which were previously classified on the basis of ground-based photographic imaging to have early-type morphologies (e.g., types E, S0, dE, dE,N or dS0; Binggeli, Sandage & Tammann 1985; Ferguson 1989). All ACS images were taken in WFC mode using a filter combination (F475W and F850LP) equivalent to the g and z bandpasses in the SDSS photometric system. The images cover a $200'' \times 200''$ field with $\approx 0.1''$ resolution, corresponding to a physical resolution of ≈ 8–9 pc at the distances of Virgo and Fornax (16.5 and 19.5 Mpc, respectively). The Virgo sample is complete for early-type galaxies brighter than $M_B \approx -19.2$ and 44% complete down to its limiting magnitude of $M_B \approx -15.2$, while the Fornax sample is complete down to $M_B \approx -16.1$. In addition to the ACS images that comprise the core datasets for both surveys, our group has obtained supplemental imaging and spectroscopy from WFPC2, Chandra, Spitzer, Keck, VLT, Magellan, KPNO, and CTIO.

The core science goals of the surveys include the measurement of luminosities, colors, metallicities, ages, and structural parameters for the many thousands of globular clusters associated with the early-type galaxies in these clusters, a high-resolution isophotal analysis of galaxies spanning a factor of ≈ 720 in luminosity, and the measurement of accurate distances for the full sample of galaxies using the method of surface brightness fluctuations (SBF). In this article, we give a brief summary of selected scientific highlights as of September 2007.

2. The Transition from Central Luminosity Deficit to Excess: No Core/Power-law Dichotomy

Pioneering HST imaging studies of the centers of early-type galaxies suggested an apparently abrupt transition in central stellar density at $M_B \sim -20.3$ mag — the so-called "core/power-law dichotomy" (e.g., Ferrarese *et al.* 1994; Lauer *et al.* 1995). These findings prompted and supported the widely held view that the bright ("core") and faint ("power-law") galaxies follow distinct evolutionary routes (e.g., Ebisuzaki, Makino & Okumura 1991; Faber *et al.* 1997). The evidence for such a dichotomy has lessened — although not entirely disappeared — following more recent studies that identified a population of galaxies with intermediate properties (Rest *et al.* 2001; Ravindranath *et al.* 2001).

By analyzing the ACS brightness profiles for the 100 ACSVCS galaxies, Ferrarese *et al.* (2006a) showed that Sérsic models provide more accurate parameterizations of the global brightness profiles than the so-called "Nuker models" (Lauer *et al.* 1995; Lauer *et al.* 2007) used in most previous analyses of HST brightness profiles (see also Graham *et al.* 2003; Graham 2004; Ferrarese *et al.* 2006c) and argued that the core/power-law dichotomy is an artifact introduced, in part, by the use of an inappropriate (i.e., power-law) parameterization of the outer profiles, combined with a tendency in previous work (which often relied on HST brightness profiles of limited radial extent) to not properly account for the compact stellar nuclei found in low- and intermediate luminosity galaxies (e.g., Graham & Guzmán 2003; Grant, Kuipers & Phillipps 2005; Côté *et al.* 2006).

For each ACSVCS and ACSFCS galaxy, azimuthally-averaged gz surface brightness profiles were derived as described in Ferrarese *et al.* (2006a). As we have noted elsewhere (e.g., Ferrarese *et al.* 2006abc; Côté *et al.* 2006, 2007; Jordán *et al.* 2007a), there is systematic behavior of the brightness profiles on small scales (see the left panels of Fig. 1): galaxies brighter than $M_B \sim -20$ typically show central light *deficits* with respect to the inward extrapolation of the Sérsic (1968) model that best fits the profile at radii larger than $\sim 0.1''-1''$, while the great majority of low- and intermediate-luminosity galaxies ($-19.5 \lesssim M_B \lesssim -15$) show central light *excesses*. Galaxies occupying a narrow range of intermediate luminosities ($-20 \lesssim M_B \lesssim -19.5$) are reasonably well fitted by Sérsic models over all radii (see also Graham *et al.* 2003).

In light of this systematic behavior, and to provide a more robust characterization of the inner regions of galaxies, Côté *et al.* (2007) introduced a parameter, $\Delta_{0.02} = \log{(\mathcal{L}_g/\mathcal{L}_s)}$ — where $\mathcal{L}_g$ and $\mathcal{L}_s$ are the integrated luminosities inside $0.02R_e$ of the observed profile and of the inward extrapolation of the outer Sérsic (1968) model — to describe the central luminosity deficit ($\Delta_{0.02} < 0$) or excess ($\Delta_{0.02} > 0$). We find that $\Delta_{0.02}$ varies smoothly over the range of ≈ 720 in blue luminosity spanned by the Virgo and Fornax sample galaxies, with no evidence for a dichotomy (see the right panel of Fig. 1). A review of the possible formation models for these central excesses was presented in Côté *et al.* (2006), including our favored interpretation (Côté *et al.* 2007) that they are the analogs of the "dense central cores" predicted by some numerical simulations to form as a result of gas inflows (see, e.g., Mihos & Hernquist 1994).

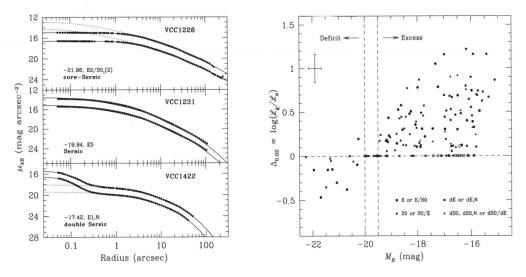

Figure 1. *(Left Panels)* Surface brightness profiles in g (F475W) and z (F850LP) for three galaxies from the ACS Virgo Cluster Survey, illustrating the transition from central "deficit" to "excess" relative to the inward extrapolation of Sérsic models that fit the outer profiles (dotted curves). See §2 and §3 for details on the parameterization of the profiles. *(Right)* Dependence of $\Delta_{0.02} = \log(\mathcal{L}_g/\mathcal{L}_s)$ on galaxy magnitude for ACSVCS and ACSFCS galaxies. Galaxies with central luminosity deficits have $\Delta_{0.02} < 0$ while those with central excesses have $\Delta_{0.02} > 0$. Note the smooth transition from central deficit to excess with decreasing galaxy luminosity. A typical errorbar is shown on the left side of the panel. Open symbols in this plot denote galaxies with "dE/dIrr transition" morphologies, dust, young stellar clusters and/or evidence of young stellar populations from blue integrated colors.

3. A Continuity of Structural Properties: No Dwarf/Giant Dichotomy

The commonly held view that early-type "giant" ($M_B \lesssim -18$) and "dwarf" ($M_B \gtrsim -18$) galaxies obey completely different photometric and structural scaling relations was introduced by Kormendy (1985) who argued that "there is a large discontinuity in the parameter correlations for bright [giant] ellipticals, including M32, and dwarf spheroidals [dwarfs]". However, it was noted by Jerjen & Binggeli (1997) and Graham & Guzmán (2003) that the Kormendy (1985) sample had a marked absence of galaxies with $M_B \sim -18$: the precise dividing point between the so-called dwarf and giant populations. In recent years, as high-quality photometric data have become available for more and more early-type galaxies, the reality of this dwarf/giant dichotomy has been questioned in a number of studies (see also early contributions from Binggeli, Sandage & Terenghi 1984 and Caldwell 1983, 1987): e.g., Jerjen & Binggeli (1997), Jerjen, Binggeli & Freeman (2000), Graham & Guzmán (2003), Graham *et al.* (2003), Gavazzi *et al.* (2005), Aguerri *et al.* (2005), Zibetti *et al.* (2005), Côté *et al.* (2006), Ferrarese *et al.* (2006a). In this section, we present new evidence from the ACS Virgo and Fornax Cluster Surveys — supplementing these data with structural parameters for additional faint galaxies in Virgo, Fornax and the Local Group — that there is a continuity in the photometric and structural properties of early-type galaxies for $M_B \gtrsim -20$.

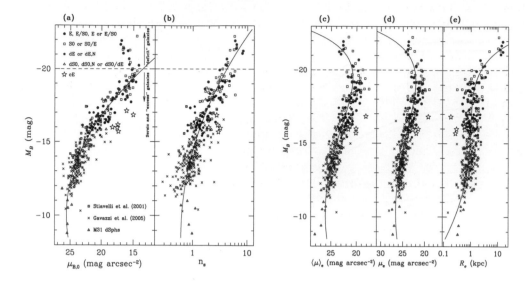

Figure 2. *(Panel a)* Relationship between absolute blue magnitude, M_B, and the intrinsic B-band central surface brightness for early-type galaxies from the ACS Virgo and Fornax Cluster surveys. Also shown are dE and E galaxies in the Virgo Cluster from Gavazzi *et al.* (2005), early–type Virgo dwarfs from Stiavelli *et al.* (2001) and M31 dSphs from McConnachie & Irwin (2006). The smooth curve shows the best-fit quadratic relation for galaxies fainter than $M_B = -20$ (indicated by the dashed line). *(Panel b)* Absolute blue magnitude plotted against Sérsic index, n_s. The best-fit quadratic relation is again shown by the smooth curve. *(Panel c)* Dependence on M_B of the average B-band surface brightness within the effective radius. *(Panel d)* Dependence on M_B of the B-band surface brightness measured at the effective radius. *(Panel e)* Dependence on M_B of the effective radius, R_e. The solid curves in panels *(c)*-*(e)* show the expected relations for Sérsic models, beginning with the M_B-μ_0 and M_B-n_s relations shown in panels *(a)* and *(b)*.

Panel *(a)* of Fig. 2 shows the relationship between absolute blue magnitude, M_B, and central surface brightness, $\mu_{B,0}$, for the Virgo and Fornax galaxies.† Absolute magnitudes have been derived using the SBF distances from Mei *et al.* (2007) and Blakeslee *et al.* (2008, in preparation). For bright galaxies ($M_B \lesssim -20$), core-Sérsic models (Graham *et al.* 2003) have been used to fit the profiles. For galaxies fainter than $M_B \sim -20$, a Sérsic law is a good representation of the entire profile when a nucleus is not present, while the profiles of nucleated galaxies are fitted by adding a second Sérsic component for the nucleus. Thus, for nucleated galaxies, the central surface brightnesses shown in Fig. 2a were measured using the inward extrapolation of the outer Sérsic component (i.e., excluding the central excess). Fig. 2b shows the corresponding relationship between M_B and Sérsic index, n_s.

Also included in Fig. 2ab are 23 early-type dwarf galaxies in Virgo and Fornax from Stiavelli *et al.* (2001). For consistency, we have re-fitted the WFPC2 profiles for these galaxies using the same family of models used for the ACS samples: i.e., single Sérsic models for non-nucleated galaxies and double Sérsic models for galaxies with nuclear excesses. At still lower luminosities, we plot the results of fitting Sérsic models to the brightness profiles of six M31 dSph galaxies from McConnachie & Irwin (2006). In these cases, the surface brightness at 0.05″ is measured after shifting the profiles to the distance of Virgo. Finally, we include 201 of the 226 early-type (i.e., dE or E) Virgo galaxies

---

† Measured at $R = 0.05''$, equivalent to one ACS/WFC pixel, or $\approx$ 4–5 pc at the distance of Virgo/Fornax.

observed by Gavazzi *et al.* (2005) with the 2.5m Isaac Newton Telescope, (i.e., omitting 25 galaxies that were already included in the ACS or WFPC2 samples).

The smooth curves in Fig. 2a and 2b show the best-fit quadratic relations for the combined sample. Since the "deficit" galaxies are thought to have central cores depleted by the dynamical evolution of supermassive black hole binaries, we have excluded these galaxies from the analysis. We have also excluded six Virgo or Fornax galaxies that have, at various times, been classified as members of the rare class of "compact ellipticals" (cEs): VCC1297, VCC1199, VCC1192, VCC1440, VCC1627 and FCC143 (indicated by the open stars in Fig. 2). Such galaxies are thought to have had their internal structure heavily modified by tidal interactions with nearby companions; indeed, all of these galaxies, except for VCC1440, are observed to have a bright, nearby companion with a similar velocity and SBF distance. These objects aside, we see from Fig. 2ab that there is a clear, continuous sequence in both panels extending from $M_B \sim -20$ down to the level of the M31 dwarfs – a range of more than 11 mag.

Fig. 2cbd shows three additional scaling relations for this same sample: the dependence on absolute blue magnitude of (1) mean surface brightness within the effective radius, $\langle \mu \rangle_e$; (2) surface brightness measured at the effective radius, μ_e; and (3) the effective radius, R_c. Once again, fainter than $M_B \sim -20$, there is no discontinuity in any of these relations. Moreover, as noted by Graham & Guzman (2003) the expected parameter correlations for Sérsic models can be calculated directly from the M_B-μ_0 and M_B-n$_s$ relations shown in Fig. 2ab. These relations are shown as the smooth curves in each panel. We conclude that there is no evidence for a "dwarf/giant dichotomy".

4. The Extragalactic Distance Scale

It is well known that the Virgo Cluster has a complex and irregular structure on the plane of the sky, spanning an area of ≈ 140 deg^2 (see Fig. 2 of Binggeli, Sandage & Tammann 1985). The cluster's structure *along the line of sight* is, however, a matter of longstanding debate with estimates of the back-to-front depth differing wildly. At one extreme, Young & Currie (1995) report a depth of $\pm$ 6–8 Mpc by employing the shape of the brightness profiles for dwarf elliptical galaxies as a distance indicator; a similar estimate based on Virgo spirals has been reported by Yasuda, Fukugita, & Okamura (1997) using the Tully-Fisher relation. On the other hand, measurements from the SBF method usually point to a much smaller depth (e.g., Neilsen & Tsvetanov 2000; Jerjen, Binggeli & Barazza 2004).

The ACSVCS was designed to provide accurate distances for our program galaxies using the z-band SBF method (see Fig. 7 of Côté *et al.* 2004). Details of our implementation of this technique were given in Mei *et al.* (2005ab) and, in Mei *et al.* (2007), Virgo's three-dimensional structure was studied using distances with a mean random uncertainty of 0.07 mag (0.5 Mpc) for 84 galaxies. Apart from five galaxies located at a distance of $d \sim 23$ Mpc — which are thus members of the background W' Cloud — we find the ACSVCS galaxies to have a narrow distribution around our adopted distance of $\langle d \rangle = 16.5 \pm 0.1$ (random mean error) ± 1.1 Mpc (systematic). The back-to-front depth of the cluster measured from our sample is 2.4 ± 0.4 Mpc (i.e., $\pm 2\sigma$ of the intrinsic distance distribution). M87 and M49 — the dominant galaxies of the A and B subclusters that comprise the Virgo Cluster — were found to lie at distances of 16.7 ± 0.2 and 16.4 ± 0.2 Mpc, respectively. Interestingly, there is some evidence for a weak correlation between velocity and line-of-sight distance that may be a faint echo of the cluster velocity distribution not having yet completely virialized. In three dimensions, the cluster has a

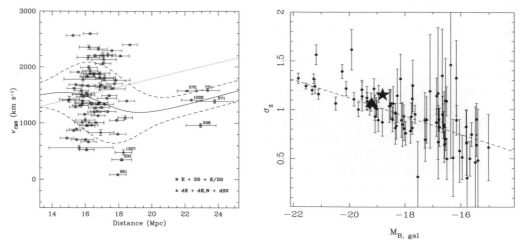

Figure 3. *(Left)* Velocity-distance relation for galaxies from the ACSVCS. Results for "giants" and "dwarfs" in the ACSVCS (as classified by Binggeli, Sandage & Tammann 1985) are shown as the red circles and blue triangles, respectively. The dotted lines shows the undisturbed Hubble Flow in the direction of Virgo for an assumed Hubble Constant of $H_0 = 73$ km s^{-1} Mpc^{-1}. The predicted distance-velocity relation for a line of sight passing through the cluster, based on the model of Tonry *et al.* (2000) for large-scale flows in the local universe, is shown by the solid (mean velocity) and dashed curves ($\pm 1\sigma$ limits). Note the presence of five galaxies associated with the W' Cloud at $d \approx 23$ Mpc. See Mei *et al.* (2007) for details. *(Right)* Gaussian dispersion, σ_z, vs. galaxy absolute blue magnitude for the z-band globular cluster luminosity functions (GCLFs) of 89 ACSVCS galaxies. The GCLF width varies systematically with host luminosity, being narrower in fainter galaxies. The two outliers at $M_B = -21.2$ and -19.9 correspond to the galaxies VCC798 and VCC2095, both of which have large excesses of diffuse star clusters (see §6 and Peng *et al.* 2006b). The large star is plotted at the spheroid luminosity and GCLF dispersion of the Milky Way, while the large triangle marks the bulge luminosity and GCLF dispersion of M31. See Jordán *et al.* (2006, 2007b) for details.

slightly triaxial distribution, with axis ratios of (1:0.7:0.5) and a principal axis inclined by $\sim 20°$–$40°$ from the line of sight.

In Jordán *et al.* (2006, 2007b), the SBF distances were used to examine the universality of the globular cluster luminonsity function (GCLF) — long thought to be a standard candle useful in distance estimation (see Harris 2001 and references therein). Using a Gaussian parameterization of the GCLF, a highly significant correlation was found between GCLF dispersion, σ_z, and galaxy luminosity, in the sense that the globular cluster systems of fainter galaxies have narrower luminosity functions (see the right panel of Fig. 3). This behavior was found to be mirrored by a steepening of the globular cluster mass function for $\mathcal{M} \gtrsim 3 \times 10^5 \mathcal{M}_\odot$, a mass regime in which the shape of the GCLF has probably not been strongly affected by dynamical evolution. In bright galaxies, the GCLF "turns over" at the canonical mass scale of $\mathcal{M}_{TO} \approx 2 \times 10^5 \mathcal{M}_\odot$. However, the turnover was also shown to scatter to significantly lower values, $\mathcal{M}_{TO} \sim (1-2) \times 10^5 \mathcal{M}_\odot$, in galaxies fainter than $M_B \gtrsim -18.5$ — an important consideration if the GCLF is to be applied as a distance indicator in "dwarf" galaxies.

Finally, we note that at the distance of the Virgo Cluster, $1''$ corresponds to a projected distance of ≈ 80 pc. Thus, Galactic globular clusters, with typical sizes of $r_h \approx 3$ pc, would be marginally resolved in our ACS images. As part of a customized data reduction pipeline (Jordán *et al.* 2004a), we measured photometric and structural parameters for candidate globular clusters by fitting the two-dimensional ACS surface brightness profiles with

PSF-convolved isotropic, single-mass King (1966) models; i.e., for each object classified probabilistically (Peng *et al.* 2006a) as a globular cluster candidate we measure, in both the g and z bandpasses, the total magnitude, King concentration index, c, and half-light radius, r_h. Jordán *et al.* (2005) showed that the r_h distribution has a characteristic form in the ACSVCS galaxies, with a peak at $\approx \langle r_h \rangle = 2.7 \pm 0.35$ pc (after correcting for weak dependencies on galaxy and globular cluster color, and underlying galaxy surface brightness). Thus, it appears that the mean cluster half-light radius in early-type galaxies can be successfully used as a standard ruler for distance estimation, with an overall precision of $\sim 15\%$.

5. The Globular Cluster Systems of Early-Type Galaxies

As the above discussion makes clear, a key objective of both the ACSVCS and ACSVCS is the study of thousands of globular clusters associated with the program galaxies. Indeed, the images are sufficiently deep that $\sim 90\%$ of the globular clusters falling within the ACS fields can be detected at a high level of completeness (Côté *et al.* 2004). While the analysis of the globular cluster systems in Fornax is still in progress, we have reported our findings for Virgo in several papers that discuss the cluster colors/metallicities (Peng *et al.* 2006a), sizes (Jordán *et al.* 2005), luminosities/masses (Jordán *et al.* 2006, 2007b), formation efficiencies (Peng *et al.* 2007), their distribution in the color-magnitude plane (Mieske *et al.* 2006), and the connection to low-mass X-ray binaries (Jordán *et al.* 2004b; Sivakoff *et al.* 2007). The reader is referred to these papers for complete details on these issues. Here we focus on a single topic: the color/metallicity distributions of globular clusters in the ACSVCS (e.g., Peng *et al.* 2006a).

Somewhat surprisingly, the ACSVCS galaxies were found to have bimodal, or asymmetric, globular cluster color distributions *at all luminosities* (see the left panel of Fig. 4). Almost all galaxies in the survey were found to possess a population of metal-poor clusters, whereas the fraction of metal-rich clusters was found to increase steadily with galaxy luminosity, from $\sim 15\%$ to 60% over the range $-22 \lesssim M_B \lesssim -15$. The colors of both sub-populations were found to correlate with host galaxy luminosity, with the red population having a steeper slope.

A key reason that the ACVCS and ACSFCS surveys were carried out in the F475W and F850LP filters is the excellent metallicity sensitivity afforded by the $(g - z)$ color index (see Fig. 4-6 of Côté *et al.* 2004). However, converting $(g - z)$ colors to metallicities presented something of a challenge since there was no empirical $(g - z)$–[Fe/H] relation for old stellar populations available at the outset of the survey. Peng *et al.* (2006a) therefore presented a preliminary (and non-linear) $(g - z)$–[Fe/H] relation based on CTIO 0.9m telescope imaging of Galactic globular clusters, and supplemented with data for several dozen clusters in the Virgo ellipticals M49 and M87. This relation was used to convert the observed color distributions into metallicity distributions, with the interesting result that the metallicities of the metal-poor and metal-rich cluster were found to vary similarly with respect to the stellar mass, $\mathcal{M}_*$, of the host galaxy: [Fe/H] $\propto \mathcal{M}_*^{0.2}$. Taken at face value, this would suggest a commonality in the conditions governing the formation of metal-poor and metal-rich globular clusters.

6. New Families of Hot Stellar Systems

As discussed in §1, the Virgo and Fornax surveys were planned with very specific science goals in mind, most notably a better understanding of the photometric and structural properties of the program galaxies and their surrounding globular clusters. Yet the

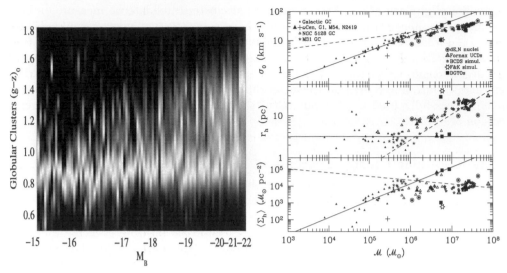

Figure 4. *(Left)* Globular cluster color distributions of ACSVCS galaxies ordered by absolute blue magnitude of the host galaxy. Each column in this image is a kernel density estimation of a galaxy's globular cluster color distribution, with white normalized to the peak density. The galaxies are ordered by luminosity, increasing from left to right. This image illustrates that all galaxies possess a population of blue clusters with similar color, while the color and relative fractions of red clusters is a strong function of galaxy luminosity. See Peng *et al.* (2006a) for details. *(Right)* Dynamical scaling relations for low-mass hot stellar systems: i.e., central velocity dispersion, half-light radius and mass surface density averaged within the half-light radius, are plotted against dynamical mass. The dashed line in each panel shows the downward extrapolation of the appropriate scaling relation for luminous elliptical galaxies. The solid lines in the top and bottom panels show the least-square fits to globular clusters in the Milky Way; in the middle panel, the solid line shows the median half-light radius of $r_h = 3.2$ pc for Galactic globular clusters. Note the apparent transition between the two types of stellar systems (i.e., globular clusters and ultra-compact dwarf galaxies) at $\mathcal{M} \approx 2 \times 10^6 \mathcal{M}_\odot$. See Haşegan *et al.* (2005) for details.

ACS images that form the basis of the surveys have proven rich datasets for studying new families of hot stellar systems that appeared serendipitously in the surveys: i.e., diffuse star clusters (DSCs, Peng *et al.* 2006b) and ultra-compact dwarf galaxies (UCDs) or, in our preferred nomenclature, dwarf-globular transition objects (DGTOs, Haşegan *et al.* 2005).

(1) *Diffuse Star Clusters.* Peng *et al.* (2006b) reported the detection of DSCs in twelve of the 100 ACSVCS program galaxies. Compared to globular clusters, DSCs have low luminosities ($M_V \gtrsim -8$) and a broad distribution of sizes ($3 < r_h < 30$ pc). They are, however, principally characterized by their low mean surface brightnesses which can be more than 3 mag fainter than a typical globular cluster ($\mu_g \gtrsim 20$ mag arcsec^{-2}). Moreover, while the sizes of globular clusters are constant with luminosity, DSCs are bounded at the bright end by an envelope of nearly constant surface brightness. The median DSC colors ($1.1 < g - z < 1.6$) are redder than metal-rich GCs and often as red as the galaxy itself. This suggests that most DSC systems thus have mean ages older than 5 Gyr or else have supersolar metallicities, implying that at least some DSCs are likely to be long-lived. In many respects, the DSCs therefore resemble the "faint fuzzies" identified in other nearby galaxies (e.g., Larsen & Brodie 2000). The closest Galactic

analogs to the DSCs appear to be the old open clusters, and if a DSC population did exist in the disk of the Milky Way, it would be very difficult to find.

(2) *Ultra-Compact Dwarfs and/or Dwarf-Globular Transition Objects.* UCDs are a potentially new class of stellar system first identified in spectroscopic surveys of the Fornax Cluster (Hilker *et al.* 1999; Drinkwater *et al.* 2000). In Haşegan *et al.* (2005), we studied the possible connection between globular clusters and UCDs by examining the photometric, structural and dynamical properties of compact objects having luminosities in the range $10^{6.25} \lesssim \mathcal{L}_V/\mathcal{L}_{V,\odot} \lesssim 10^{6.65}$ (DGTOs). Our principal finding from this analysis, which was based on an examination of the mass-based scaling relations for a sample of globular clusters and UCDs in the Virgo Cluster, the Local Group and NGC5128, was the detection of an apparent transition between these two types of stellar systems at a mass of $\mathcal{M} \approx 2 \times 10^6$ solar masses (see the right panel of Fig. 4). Furthermore, there is some evidence to suggest that either the presence of dark matter, or lengthy relaxation timescales, may distinguish UCDs from globular clusters. Five of the 13 DGTOs in Virgo were found to be associated with a single galaxy, M87, suggesting that proximity to the cluster center may be of critical importance for the formation of these rare objects. A detailed analysis of an expanded sample of DGTOs in Virgo and Fornax will be presented in Haşegan *et al.* (2008).

Acknowledgements

We thank Massimo Stiavelli and Alan McConnachie for kindly providing the brightness profiles for the low-luminosity galaxies in Virgo, Fornax and M31 discussed in §3.

References

Aguerri, *et al.* 2005, *AJ*, 130, 475
Binggeli, B., Sandage, A., & Terenghi, M. 1984, *AJ*, 89, 64
Binggeli, B., Sandage, A., & Tammann, G.A. 1985, *AJ*, 90, 1681
Caldwell, N. 1983, *AJ*, 88, 804
Caldwell, N. 1987, *AJ*, 94, 1116
Côté, P., *et al.* 2004, *ApJS*, 153, 223 (ACSVCS Paper I)
Côté, P., *et al.* 2006, *ApJS*, 165, 57 (ACSVCS Paper VIII)
Côté, P., *et al.* 2007, *ApJ*, in press (ACSFCS Paper II)
Drinkwater, M. J., *et al.* 2000, *PASA*, 17, 227
Ebisuzaki, T., Makino, J., & Okumura, S.K. 1991, *Nature*, 354, 212
Faber, S. M., *et al.*1997, *AJ*, 114, 1771
Ferguson, H. C. 1989, *AJ*, 98, 367
Ferrarese, L., *et al.* 1994, *AJ*, 108, 1598
Ferrarese, L., *et al.* 2006a, *ApJ*, 164, 334 (ACSVCS Paper VI)
Ferrarese, L., *et al.* 2006b, *ApJ*, 644, L21
Ferrarese, L., *et al.* 2006c, arXiv:astro-ph/0612139
Ford, H. C., *et al.* 1998, *Proc. SPIE*, 3356, 234
Gavazzi, G., *et al.* 2005, *A&A*, 430, 411
Graham, A. W., *et al.* 2003, *AJ*, 125, 2951
Graham, A. W., & Guzmán, R. 2003, *AJ*, 125, 2936
Grant, N. I., Kuipers, J. A., & Phillipps, S. 2005, *MNRAS*, 363, 1019
Harris, W. E. 2001, Saas-Fee Advanced Course 28: Star Clusters, 223
Haşegan *et al.* 2005, *ApJ*, 627, 203 (ACSVCS Paper VII)
Hilker, M., *et al.* 1999, *A&AS*, 134, 75
Jerjen, H., & Binggeli, B. 1997, in ASP Conference Series, Vol. 116, p. 239
Jerjen, H., Binggeli, B., & Freeman, K. C. 2000, *AJ*, 119, 593
Jerjen, H., Binggeli, B., & Barazza, F. D. 2004, *AJ*, 127, 771

Jordán, A., *et al.* 2004a, *ApJS*, 154, 509 (ACSVCS Paper II; J04)

Jordán, A., *et al.* 2004b, *ApJ*, 613, 279 (ACSVCS Paper III)

Jordán, A., *et al.* 2005, *ApJ*, 634, 1002 (ACSVCS Paper X)

Jordán, A., *et al.* 2006, *ApJ*, 651, L25.

Jordán, A., et al. 2007a, *ApJS*, 169, 213 (ACSFCS Paper I)

Jordán, A., *et al.* 2007b, *ApJS*, 171, 101 (ACSVCS Paper XII)

King, I. R. 1966, *AJ*, 71, 64

Kormendy, J. 1985, *ApJ*, 29573

Larsen, S. S., & Brodie, J. P. 2000, *AJ*, 120, 2938

Lauer, T. R., *et al.* 1995, *AJ*, 110, 2622

Lauer, T. R., *et al.* 2007, *ApJ*, 664, 226

McConnachie, A. W., & Irwin, M. J. 2006, *MNRAS*, 365, 1263

Mei, S., *et al.* 2005a, *ApJS*, 156, 113 (ACSVCS Paper IV)

Mei, S., *et al.* 2005b, *ApJ*, 625, 121 (ACSVCS Paper V)

Mei, S., *et al.* 2007, *ApJ*, 655, 144 (ACSVCS Paper XIII)

Mihos, J. C., & Hernquist, L. 1994, *ApJ*, 437, L47

Neilsen, E. H., & Tsvetanov, Z. I. 2000, *ApJ*, 536, 255

Peng, E. W. *et al.* 2006a, *ApJ*, 639, 95 (ACSVCS Paper IX)

Peng, E. W. *et al.* 2006b, *ApJ*, 639, 838 (ACSVCS Paper XI)

Peng, E. W. *et al.* 2007, *ApJ*, submitted (ACSVCS Paper XV)

Ravindranath, S., Ho, L. C., & Filippenko, A. V. 2002, *ApJ*, 566, 801

Rest, A., *et al.* 2001, *AJ*, 121, 2431

Sérsic, J. -L. 1968, *Atlas de Galaxias Australes* (Córdoba: Obs. Astron., Univ. Nac. Córdova)

Sivakoff, G. R., *et al.* 2007, *ApJ*, 660, 1246

Stiavelli, M., *et al.* 2001, *AJ*, 121, 1385

Tonry, J. L., *et al.* 2000, *ApJ*, 530, 625

Yasuda, N., Fukugita, M., & Okamura, S. 1997, *ApJS*, 108, 417

Young, C. K., & Currie, M.J. 1995, *MNRAS*, 273, 1141

Zibetti, S., *et al.* 2005, in *IAU Colloq. 198*, p. 380

Dynamical Evolution of Dense Stellar Systems
Proceedings IAU Symposium No. 246, 2007
E. Vesperini, M. Giersz & A. Sills eds

© 2008 International Astronomical Union
doi:10.1017/S1743921308016025

Giant Elliptical Galaxies: Globular Clusters and UCDs

William E. Harris[1]

[1]Physics & Astronomy, McMaster University, Hamilton, ON L8S 4M1 Canada
email: harris@physics.mcmaster.ca

Abstract. Explorations of the globular cluster populations in many nearby galaxies are revealing increasing connections to other dense stellar systems such as UCDs, DGTOs, and nuclear star clusters in dwarf galaxies. The nearest giant elliptical, NGC 5128, is now giving us a much-improved delineation of the GC Fundamental Plane of structural parameters, and indicates as well that the known correlation between GC scale size and metallicity is likely to be at least partly a projection effect coupled with the different spatial distributions of the metal-poor and metal-rich clusters. New photometry of the huge cluster populations around the giant Brightest Cluster Ellipticals, which allows us to study samples of many thousands of GCs at once, are now beginning to turn up surprising examples of "sequences" of high-mass GCs leading up to the UCD regime. Lastly, new modelling of cluster formation through a specially tuned semi-analytic galaxy formation code strongly suggests that the mass-metallicity relation now known to affect the blue GC sequence can arise fairly naturally out of such models, if significant numbers of the massive GCs actually represent the remnant nuclei of stripped dwarf satellites.

Keywords. galaxies: star clusters

1. Introduction

Our understanding of the range of properties that dense stellar systems can display in nature has experienced several healthy extensions in the past few years. Classic globular clusters, for long regarded as sitting in splendid isolation in the "fundamental plane" of structural properties of stellar systems, are now seen as having intriguing potential connections to other systems, particularly dE galaxies, Ultra-Compact Dwarfs, and the dense nuclei of dwarf galaxies. In this paper I will discuss a set of recent observations that relate to this central issue. The overarching theme is that the links between GCs and UCD-like objects are, indeed, likely to be more intricate than we had realized.

2. NGC 5128 and its Cluster System

NGC 5128 (Centaurus A) is the centrally dominant giant in the Centaurus galaxy group; at a distance of just (3.8 ± 0.2) Mpc, it is by far the nearest easily accessible giant elliptical. For comparison, the Virgo and Fornax clusters are at least three magnitudes farther away. With NGC 5128, we have a *unique* opportunity to study a globular cluster system in a gE galaxy at close range; and, since this galaxy contains $\simeq 1500$ GCs (Harris *et al.* 2006), it gives us a far larger sample of GCs to work on than all the Local Group galaxies combined.

One immediate advantage of this system is simply that we can test, quite straightforwardly, whether or not GCs in a giant elliptical are structurally similar to those in spirals and dwarfs. At $d = 3.8$ Mpc, a typical GC half-light diameter of $\simeq 6$ pc corresponds to an angular size of $0.33''$, which is easily resolved by HST imaging and even by the best-quality imaging from the ground. With HST, it is possible to resolve core radii too for many clusters, and to perform extensive tests of the basic question whether standard King-type models apply to them as well as in the Milky Way, M31, and the other Local Group galaxies. A second major advantage is that, simply because of sheer numbers, it is possible to find many "high-end" GCs with luminosities in the range of ω Cen, M31-G1, and even above. These ultramassive GCs are especially interesting because they might give clear links to other more recently discovered types of dense stellar systems – Ultra-Compact Dwarfs (UCDs), Dwarf-Globular Transition Objects (DGTOs), and nuclear star clusters in dwarf galaxies (e.g. Phillipps et al. 2001; Hasegan et al. 2005; Evstigneeva et al. 2007, among many recent studies).

The major barrier to carrying all this out in practice is a purely observational one: the GCs in NGC 5128 need to be found one by one, and since the galaxy is at moderate galactic latitude $b = 19^o$, field contamination from both foreground Milky Way stars and faint, small background galaxies is extensive. High-resolution imaging and radial velocity measurement are the main tools to use in picking out GCs from everything else. The complete catalog of the ~ 450 GCs known to date is summarized in Woodley et al. (2007), along with a kinematic and dynamic analysis of the entire system based on the most complete available set of radial velocities. Of these known clusters, about 150 have published structural parameters based on HST ACS and STIS imaging McLaughlin et al. (2007) and more will soon be available through a similar imaging program by Jordan et al. (2007).

A sample of the results so far is shown in Fig. 1, from McLaughlin et al. (2007). This graph displays two of the most commonly used forms of the structural "fundamental plane" for GCs (see, e.g., McLaughlin 2000 and McLaughlin & van der Marel 2005 for the most detailed recent descriptions). The two panels plot the cluster half-light radius R_h and central velocity dispersion σ against cluster mass. Notice particularly the change in slope of both relations above $\simeq 2 \times 10^6 M_\odot$: we are traditionally used to thinking of GCs as having a nearly uniform scale size $R_h \sim 3$ pc with some scatter, but at very high masses they become progressively more extended, approaching the downward extrapolation of the FP line for E galaxies. This change in the trend is also reflected in their central velocity dispersions, which continue to increase with M but not as rapidly as for fainter GCs. Other objects that have been isolated recently, including dwarf-globular transition objects (DGTOs; see Hasegan et al. 2006; Evstigneeva et al. 2006) and the nuclear star clusters in dE's and bulges (pointed pentagons in Fig. 1) contribute to the scatter at the high end but fall in just the same general region.

What is perhaps equally important is that the GCs in all the major nearby galaxies (NGC 5128, M31, the Milky Way) overlap completely with each other and, as far as we can tell, accurately define just the same GC fundamental plane (see especially Barmby et al. 2007 for more comparisons). The more we learn in detail about the structures, ages, and compositions of GCs in different galaxies, the more they continue to look like universal objects that are a common thread in the early history of galaxies.

3. Cluster Size versus Location

We are just beginning to exploit the full potential of the rich, nearby NGC 5128 system. A new ground-based imaging survey of the system has been completed with the Magellan IMACS camera, covering a 1.2×1.2—degree field centered on NGC 5128 and with a remarkable $0.45''$ resolution (Gomez et al. 2007, in preparation). This image quality makes it almost exactly equivalent to working on the Virgo and Fornax galaxies with HST. At this resolution, 90% of the GCs in NGC 5128 are resolved as nonstellar and so the heavy

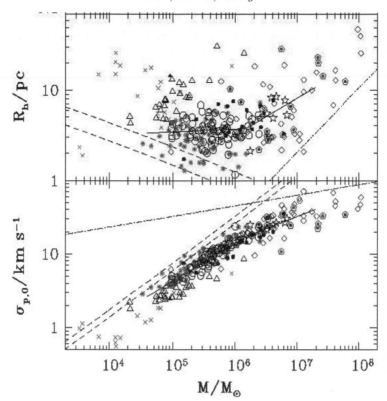

Figure 1. Two versions of the Fundamental Plane for globular clusters. The upper panel shows half-light radius versus cluster mass, the lower panel central velocity dispersion. Small crosses and stars are from the Milky Way and larger open symbols from NGC 5128. Other objects such as DGTOs and nuclear star clusters are the open pentagons with points. The pair of dashed lines in each graph show boundaries set by internal dynamical timescales: any cluster older than 20 relaxation times τ_{Rh} would fall below the upper line; any older than $40\tau_{Rh}$ would fall below the lower dashed line. Most such objects would now be completely evaporated. The dot-dashed line in each graph represents a line parallel to the E-galaxy Fundamental Plane relation: E galaxies lie just above that line in the R_h graph or just below it in the σ_p graph. See Barmby *et al.* (2007) and McLaughlin *et al.* (2007) for extensive discussion.

field contamination by foreground stars is immediately removed. Furthermore, the field coverage (extending out to projected radii of 50 kpc, equivalent to about $10R_{eff}$) will allow us to explore GC properties with radius more extensively than has been done in most other systems. After further selecting objects by color and morphology, we have isolated a sample of about 1000 GC candidates not previously known. Follow-up radial velocity programs are now underway to do the final rejection of contaminants, and will also form the basis for a much more comprehensive kinematic and dynamic study of the entire GC system.

A first scientific result from this new imaging database is a comprehensive set of measurements of structural parameters for 359 already-known clusters (Gomez & Woodley 2007). Though the cluster core radii are too small to be resolved with anything else except HST, the half-light radii and central concentrations can be well estimated with King-type models convolved with the point spread function. Both metal-poor and metal-rich clusters display very similar global distributions in r_e, but an intriguing trend emerges when the distributions are broken down by galactocentric distance. This is shown in Fig. 2. A small but by-now well established correlation is for the metal-richer GCs ($[\text{Fe/H}] > -1$) to have systematically smaller scale radii r_e than the metal-poorer ones ($[\text{Fe/H}] < -1$) in the same spatial region. This trend was first noted by Kundu & Whitmore (1998) and Kundu et al. (1999) for NGC 3115 and M87 and was since revealed in many other systems (see particularly Larsen et al. 2001; Jordan et al. 2005 for the largest such databases). One interpretation, explored by Larsen & Brodie (2003), is that it is a projection effect resulting from the fact that the low-metallicity clusters form a more spatially extended subsystem and thus fewer of them are deep in the central bulge region where tidal effects are strongest. This interpretation also assumes that there is a basic relation between GC size and galactocentric distance, such as the $r_e \sim R_{gc}^{0.5}$ scaling that applies to the Milky Way, and that both types of clusters obey at least approximately. A competing interpretation (Jordan 2004) is that it is a result strictly internal to the clusters, whereby mass segregation of their stars, coupled with shorter lifetimes on the red-giant branch for higher-metallicity stars, leads to the metal-richer clusters having smaller half-*light* radii.

One effective test of these two scenarios is suggested by Jordan and Larsen & Brodie: if it is due to a projection effect within the galaxy halo, then the size differences between the metallicity groups should die away at larger R_{gc}, far out from the central bulge. A hint that the differences do indeed die away at large R_{gc} was found by Spitler et al. (2006) for NGC 4594, where the GC measurements extend out to $\simeq 5R_{eff}$ (see their Fig.20). However, an even stronger test can already be made within NGC 5128. As Fig. 2 indicates, a clear trend for r_e to shrink with increasing metallicity is present for clusters projected within $1.0R_{eff}$; that is, within the bulge. The same trend, but weaker, is present for the next zone $1-2R_{eff}$. Beyond $2R_{eff}$, however, the correlation disappears. The only remaining effect is for increasing cluster-to-cluster scatter in their scale sizes in the outermost zones, out to $8R_{eff}$ and the limits of the current data (Gomez & Woodley 2007).

This overall picture is at least roughly consistent with the Larsen/Brodie model. The reason why the previously available data did not give a clearer picture before this is simply that most of it relied on HST WFPC2 and ACS imaging of galaxies in Virgo and comparable distances, and the small field sizes of these cameras restricted the measurements to clusters within $R_{gc} \lesssim 2R_{eff}$ – a region clearly too restricted to see the full correlation. Once the full sample of IMAC data has been assessed, we will be able to more than double the GC sample size and confront them with more detailed modelling. One puzzle, for example, is the basic question why this trend with R_{gc} should affect the *half-mass* radii of the clusters, which are relatively immune to dynamical evolution within the cluster, quite unlike the tidal radius. How does the cluster's location within the bulge or halo affect its intrinsic scale size? If it is not easily driven by external tidal influences during its lifetime, then once again, we may be forced to invoke the conditions at time of formation: deeper within the tidal field, the protocluster gas clouds should have already been subject to higher external pressure and thus take on smaller scale radii right from the start (Harris & Pudritz 1994). But these descriptive ideas need to be developed much more quantitatively.

4. Extensions to the UCD Regime: More Connections

Connections between the "top end" of the normal GC sequence and the still higher mass range typically occupied by structures like nuclear star clusters and Ultra-Compact Dwarfs may be far more direct than even the combined data from nearby galaxies has suggested. Fig. 3 shows the luminosities and colors for a population of several thousand GCs around the supergiant elliptical NGC 3311 (the central cD galaxy in A1060). This

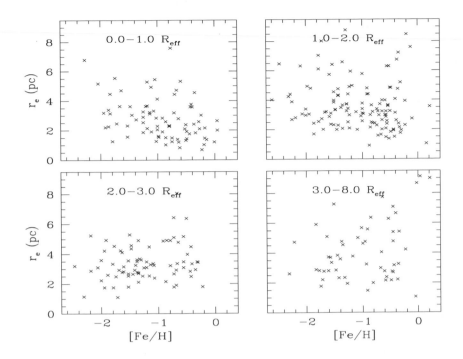

Figure 2. Effective radius r_e (equivalent to the half-light radius) for globular clusters in NGC 5128, plotted against cluster metallicity. Four zones are shown, in steps of increasing projected galactocentric distance (in units of the galaxy effective radius). For the innermost zones, GC size decreases with increasing metallicity; for the halo beyond $\simeq 2R_{eff}$, no trend with metallicity is evident.

photometry (Wehner & Harris 2007) reveals the normal bimodal color/metallicity distribution found almost everywhere, but it also shows a distinct upward extension of the red, metal-richer sequence consisting of about two dozen GCs. Archival HST/WFPC2 imaging available for about half of these shows that they are *just* barely more extended than the PSF width, which at the 54-Mpc distance of NGC 3311, suggests that they have typical radii $r_e \sim 10$ pc, rather similar to UCDs and the very biggest known GCs. Their masses, assuming $M/L \simeq 3$, range from $6 \times 10^6 M_\odot$ up to $3 \times 10^7 M_\odot$ and thus reach into the UCD regime as well.

It is noteworthy that this type of clearly delineated sequence extension has not been seen even in composite samples of thousands of GCs resulting from combining many galaxies, either Brightest Cluster Galaxies (Harris *et al.* 2006) or the entire Virgo Cluster Survey (Mieske *et al.* 2006). Clearly there are still individual site-to-site differences in the events forming these dense, luminous stellar systems that we do not yet understand.

The blue, metal-poor sequence of GCs has gained new attention recently because it has been found to exhibit a *mass-metallicity* relation (MMR) whereby the more massive GCs become systematically more metal-rich, following a scaling of heavy-element abundance with mass $Z \sim M_{GC}^{0.5}$. The red GC sequence displays no such trend, staying resolutely vertical in the color-magnitude diagram. The effect was first found through metallicity-sensitive color indices in several giant cD-type galaxies (Harris *et al.* 2006; Strader *et al.* 2006) and then in other galaxies of a wider range of sizes and types (Spitler *et al.* 2006;

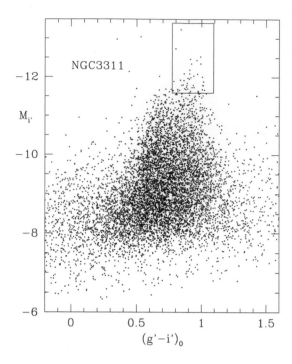

Figure 3. Luminosity versus color index for the globular cluster population around NGC 3311, the central giant elliptical in A1060 (the Hydra cluster). Note the normal "blue" and "red" GC sequences partially overlapping, but the top of the red sequence (boxed) continues upward to higher luminosity connecting with the UCD and DGTO regime.

Mieske *et al.* 2006). What is the origin of the MMR? Harris *et al.* (2006) adopt the view that the blue GCs formed at a very early stage of hierarchical merging, when they were still located within reasonably distinct pregalactic dwarfs of near-primordial gas. We then propose that on average, the more massive GCs should have formed within proportionately more massive host dwarfs (see Harris & Pudritz 1994) that would have been more able to hold onto SNe and stellar wind ejecta and thus self-enrich, leaving behind more enriched GCs. Mieske *et al.* (2006) discuss several other options but favor a similar route.

This intriguing new correlation may well be giving us an entirely new window onto the properties of the pregalactic dwarfs. Quantitative models are now starting to emerge. Rothberg *et al.* (2007) apply the Somerville semi-analytic galaxy formation code with modifications to track the properties of the remaining dwarf galaxies within the halo of a giant galaxy, and in particular the stripped nuclei of these dwarfs. An encouraging result is that, with the same "standard" set of input model parameters found to reproduce the global properties of large galaxies, these residual nuclei fall along the upper end of the blue GC sequence and have a MMR with a realistic slope: the more massive nuclei came from systematically larger dwarfs and experienced more self-enrichment. Increases in the parameter representing the supernova heating efficiency can reduce or eliminate the MMR, while decreases can, surprisingly, turn the residual nuclei into UCD-like objects.

Models like this clearly represent only a first tentative step into realistic interpretation. Nevertheless, the general direction looks promising, and is further reinforcing the links between these various types of compact, massive stellar systems.

Acknowledgements

The author gratefully acknowledges financial support from NSERC of Canada.

References

Barmby, P., McLaughlin, D. E., Harris, W. E., Harris, G. L. H., & Forbes, D. A. 2007, *AJ*, 133, 2764

Evstigneeva, E. A., Gregg, M. D., Drinkwater, M. J., & Hilker, M. 2007, *AJ*, 133, 1722

Gomez, M., & Woodley, K. A. 2007, *ApJL*, submitted

Harris, W. E., Harris, G. L. H., Barmby, P., McLaughlin, D. E., & Forbes, D. A. 2006, *AJ*, 132, 2187

Harris, W. E., & Pudritz, R. E. 1994, *ApJ*, 429, 177

Hasegan, M. *et al.* 2005, *ApJ*, 627, 203

Jordan, A. 2004, *ApJL*, 613, 117

Jordan, A. *et al.* 2005, *ApJ*, 634, 1002

Jordan, A. *et al.* 2007, *ApJL*, submitted

Kundu, A., & Whitmore, B. C. 1998, *AJ*, 116, 2841

Kundu, A., Whitmore, B. C., Sparks, W. B., Macchetto, F. D., Zepf, S. E., & Ashman, K. M. 1999, *ApJ*, 513, 733

Larsen, S. S., & Brodie, J. P. 2003, *ApJ*, 593, 618

Larsen, S. S., Brodie, J. P., Huchra, J. P., Forbes, D. A., & Grillmair, C. J. 2001, *AJ*, 121, 2974

McLaughlin, D. E. 2000, *ApJ*, 539, 618

McLaughlin, D. E., Barmby, P., Harris, W. E., Harris, G. L. H., & Forbes, D. A. 2007, *MNRAS*, submitted

McLaughlin, D. E., & van der Marel, R. P. 2005, *ApJS*, 161, 304

Mieske, S. *et al.* 2006, *ApJ*, 653, 193

Phillipps, S., Drinkwater, M. J., Gregg, M. D., & Jones, J B. 2001, *ApJ*, 560, 201

Rothberg, B., Harris, W. E., Somerville, R. S., Whitmore, B. C., & Woodley, K. A. 2007, *ApJL*, submitted

Spitler, L. R., Larsen, S. S., Strader, J., Brodie, J. P., Forbes, D. A., & Beasley, M. A. 2006, *AJ*, 132, 1593

Strader, J., Brodie, J. P., Spitler, L., & Beasley, M. A. 2006, *AJ*, 132, 2333

Wehner, E. M. H., & Harris, W. E. 2007, *ApJL*, in press

Woodley, K. A., Harris, W. E., Beasley, M. A., Peng, E. W., Bridges, T. J., Forbes, D. A., & Harris, G. L. H. 2007, *AJ*, 134, 494

Dynamical Evolution of Dense Stellar Systems
Proceedings IAU Symposium No. 246, 2007
E. Vesperini, M. Giersz & A. Sills

Observational Constraints on the Formation and Evolution of Globular Cluster Systems

Stephen E. Zepf

Department of Physics and Astronomy, Michigan State University, East Lansing, MI 48824,
USA
e-mail: zepf@pa.msu.edu

Abstract. This paper reviews some of the observational properties of globular cluster systems, with a particular focus on those that constrain and inform models of the formation and dynamical evolution of globular cluster systems. I first discuss the observational determination of the globular cluster luminosity and mass function. I show results from new very deep HST data on the M87 globular cluster system, and discuss how these constrain models of evaporation and the dynamical evolution of globular clusters. The second subject of this review is the question of how to account for the observed constancy of the globular cluster mass function with distance from the center of the host galaxy. The problem is that a radial trend is expected for isotropic cluster orbits, and while the orbits are observed to be roughly isotropic, no radial trend in the globular cluster system is observed. I review three extant proposals to account for this, and discuss observations and calculations that might determine which of these is most correct. The final subject is the origin of the very weak mass-radius relation observed for globular clusters. I discuss how this strongly constrains how globular clusters form and evolve. I also note that the only viable current proposal to account for the observed weak mass-radius relation naturally effects the globular cluster mass function, and that these two problems may be closely related.

Keywords. globular clusters:general, galaxies:star clusters

1. What is the Shape of the Globular Cluster Luminosity Function?

A natural starting point for determining the key physical processes that make the globular cluster luminosity function (GCLF) is to consider the best available observational constraints on the GCLF. The ideal observational sample for determining the GCLF would both very deep to constrain the faint, low-mass end of the cluster population, and have large numbers of clusters to provide adequate statistics, particularly for the rare very bright and very faint globular clusters. The Milky Way globular cluster system meets the first criterion of probing to very faint globular clusters, and has thus provided some of the basic data indicating the roughly lognormal shape of the GCLF, and the location of the turnover magnitude and estimated mass of the distribution. However, the Milky Way has a relatively poor globular cluster system, numbering only about 150 objects. As a result the Galactic GCLF provides very limited statistical power for any test of the behavior of the GCLF, particularly at the low and high mass ends, each of which are only expected to have a few clusters given the total number of Galactic globulars.

One way to overcome the problem of small numbers of globular clusters in the Milky Way is to study a galaxy with a much richer system of globular clusters. M87 is an obvious choice as it has a combination of an exceedingly rich globular cluster system (with almost two orders of magnitude more globular clusters than the Milky Way), and a moderate distance with its location in the Virgo cluster. However, because the distance to M87 is not negligible, to realize its potential for the study of the low-mass end of the GCLF

394

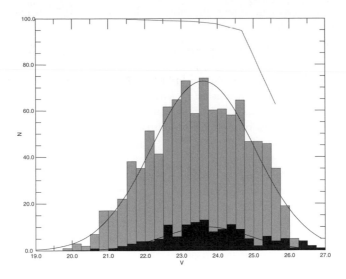

Figure 1. A plot of the globular cluster luminosity function for M87 (upper histogram in green) and the Milky Way (lower histogram in black), with the Milky Way data shifted to the 16 Mpc distance of Virgo. The M87 data is from Waters *et al.* (2006) and has been completeness corrected, using the completeness function shown as the upper (red) line. Gaussian fits to the M87 and Milky Way data are overplotted on the histograms. The plot makes immediately apparent the dramatic advantage of the much larger numbers of globular clusters in M87 for studying the effects of dynamical evolution on globular cluster systems. As detailed in Waters *et al.* (2006). the turnover magnitudes of the M87 and Milky Way GCLFs are identical to within the uncertainties, but the M87 GCLF is significantly broader. These conclusions are confirmed by the even deeper 50-orbit ACS dataset (Waters & Zepf 2008).

requires data that reach to faint apparent magnitudes. Moreover, because the number of background galaxies rises steeply to faint magnitudes, the data must have excellent spatial resolution to distinguish between faint compact galaxies and M87 globular clusters. At brighter magnitudes, ground-based imaging with various image classification and color cuts can create useful globular cluster samples (e.g. Rhode & Zepf 2001 for a detailed example). However, given the depths required to reach the faint end of the M87 GCLF, and the declining numbers of globular clusters and rising number of galaxies at these faint magnitudes, deep HST imaging is required to reliably probe the shape of the faint end of the GCLF in M87 and other similar galaxies.

A start in this direction was made by various multi-orbit HST WFPC2 studies (e.g. Kundu *et al.* 1999). These were able to establish that the peaks of the turnover of the GCLF were very similar in M87 and the Milky Way, and also suggested that the width of the M87 GCLF might be larger than that of the Milky Way. However, to accurately test the shape of the GCLF to very faint and low-mass globular clusters required yet deeper data. Until recently, such very deep HST data to probe the faint end of the GCLF were not available. This situation has changed due to extraordinarily deep imaging of M87 obtained originally for studying microlesning in the Virgo cluster. Two such datasets now exist. The first is a 30 orbit WFPC2 dataset, for which we have published an analysis of the faint end of the GCLF in Waters *et al.* (2006). The second is a 50 orbit ACS dataset which we are now analyzing. In Fig. 1, we show the GCLF resulting from our analysis of the 30-orbit WFPC2 dataset, presented in Waters *et al.* (2006). The plot makes immediately apparent the dramatic advantage of the combination of very deep

HST imaging and the large numbers of globular clusters in M87 compared to the Milky Way for studying the GCLF.

The very deep HST data on the M87 GCLF such as shown in Fig. 1 from Waters $et\ al.$ (2006) provide two key constraints on models of the dynamical evolution of globular clusters through evaporation. This can be readily seen when writing the mass loss of a globular cluster due to evaporation as $\dot{M} = kM^{\gamma}$. k is then the mass loss rate for a cluster of given mass M, and the exponent γ accounts for any dependence of this mass loss rate on cluster mass. In terms of comparison to the data, the mass loss rate is primarily constrained by the location of the turnover of the globular cluster mass function, and the dependence of the mass loss rate on M is constrained by the slope of the globular cluster mass function at low masses. These constraints are mostly insensitive to the initial globular cluster mass function for all extant models of two-body relaxation and evaporation. Clusters more massive than the turnover have experienced very little mass loss as a fraction of their total mass, and the low-mass end of the GCLF is composed of globular clusters which have lost an increasing fraction of their initial mass as one goes to fainter globular clusters, Thus the massive end of the observed GCLF is completely determined by the initial GCLF (and possibly dynamical friction for very centrally located clusters) and the low mass end of the observed GCLF is set almost entirely by mass loss from dynamical evolution. As an aside, we note that the shape around the turnover can be affected in detail by the initial mass function if this function is flat or decreases to lower masses.

We can then compare different theoretical models of mass loss due to two-body relaxation in globular clusters to our extraordinarily deep HST data on the M87 globular cluster system. Fig. 2 plots the results of the comparison of our M87 GCLF from the 30-orbit WFPC2 dataset to several theoretical models as shown in Waters $et\ al.$ (2006). In detail, three theoretical models are shown, each having a different dependence of the mass loss rate on mass as advocated in recent papers. One of these possibilities is that $\dot{M} = $ a constant (i.e. $\gamma = 0$), and that the mass loss rate is independent of mass (e.g. Fall & Zhang 2001 and references therein). Another possibility comes from the detailed N-body simulations of Baumgardt & Makino (2003, hereafter BM03) for which the mass loss rate is approximately $\dot{M} \propto M^{0.25}$. Finally, Lamers, Gieles, & Portegies Zwart (2005) found $\dot{M} \propto M^{0.38}$ from a combination of fits of the Baumgardt & Makino (2003) simulations and analytic arguments. Additionally, Fig. 6 in BM03 indicates that the correct mass to use in this calculation is the current mass of the globular cluster, and that in the BM03 simulations the mass loss rate of a globular cluster is not constant in time, and is described well by the $\dot{M} \propto M^{0.25}$ expression calculated for the current cluster mass. There was some discussion of this point at the meeting, but Fig. 6 from BM03 gives a mass loss rate due to evaporation that changes as the globular cluster mass evolves.

The comparison of the deep M87 data to models in Fig. 2 favors mass-loss which is independent of cluster mass. This may be somewhat surprising in that the most state of the art simulations would seem to suggest some mass dependence. In Waters $et\ al.$ (2006), we discuss some of the possible resolutions of these differences. One is simply the statistical difference, even in the very deep 30-orbit WFPC2 data is not overwhelming. A second issue noted in Waters $et\ al.$ (2006) is that the mass-to-light ratio (M/L) of the globular clusters may decrease as low-mass stars are preferentially lost from the cluster as it begins evaporating away. We are carrying out new studies that address these points. Our upcoming work uses a 50-orbit ACS dataset to both reach fainter magnitudes and lower masses and to increase the overall number. We also use the mass-to-light ratio evolution of a globular cluster as a function of its disruption time given in BM03 to

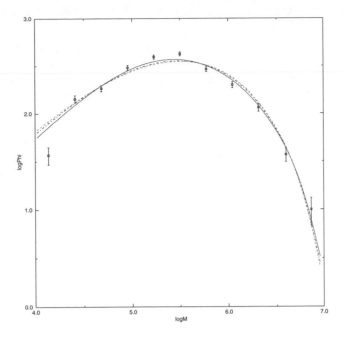

Figure 2. The mass function of the M87 globular cluster system from Waters *et al.* (2006) analysis of a 30-orbit HST dataset. The three lines are models with different dependencies of the evaporation mass loss rate on cluster mass. The solid line is $\dot{M} \propto M^0$, the dot dashed line is $\dot{M} \propto M^{0.25}$, and the dashed line is $\dot{M} \propto M^{0.38}$. The data plotted favor a mass loss rate that is independent of mass, as do new, even deeper data (Waters & Zepf 2008). This result is driven the by the slope from the turnover to the low-mass end of the mass function, and is not dependent on the initial globular cluster mass function. However, for completeness we adopted a Burkert & Smith (2000) initial mass function for this plot in order to match the bright end of the globular cluster system.

specifically account for the the changing M/L. The final results from this new work will be published in Waters & Zepf (2008).

2. The Unsolved Problem of the Constancy of the GCLF

Globular clusters closer to the center of a galaxy will have smaller tidal radii for a given cluster mass. This will cause globular clusters closer to the center of a galaxy to lose mass more quickly, and therefore a radial gradient in the globular cluster mass function and GCLF is expected. However, the Galactic GCLF shows no strong changes with radius. This discrepancy between basic theoretical expectation and observation has been noted for some time (e.g. Baumgardt 1998, Vesperini 1997 and references therein). One way around this problem is to have the globular cluster orbits become increasingly radial with distance from the center of the host galaxy, so that all globular clusters have very similar pericenters. Coupled with the assumption that the tidal radii are set at pericenter, one can then recover a constant GCLF with distance from the center of the host galaxy. This scenario was investigated in detail by Fall & Zhang (2001) who suggested it could account for the lack of an observed radial gradient in GCLFs.

The question is whether the existing globular cluster population has such extremely radial orbits. M87 provides an ideal place to test proposals for the physical origin of the radial constancy of the GCLF. As shown in Vesperini *et al.* (2003), HST data over

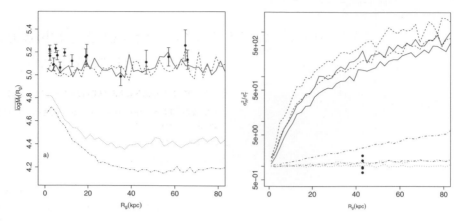

Figure 3. The plot on the left-hand side shows the final globular cluster mean mass vs. projected galactocentric distance for given an initial power-law globular cluster mass function and an initial anisotropy radius equal to 2 (solid line), 3 (dashed line), 15 (dotted line), and 150 kpc (dot-dashed line), from detailed calculations in Vesperini et al. (2003). Large anisotropy radii models are close to isotropic and small anisotropy radii models have orbits which become strongly radial at large galactocentric distances. The constancy of the GCLF with radius is clearly a problem for isotropic orbits. The plot on the right compares the anisotropy of the globular cluster orbits, using the same convention for the models in the plot on the left, to the observed anisotropy of the M87 globular cluster system, which is shown as the points and taken from Romanowsky & Kochanek 2001). This shows that strongly radial orbits are clearly inconsistent with the observed velocities. Thus, the dilemma that for a power-law initial mass function, evaporation-driven dissolution only produces a constant GCLF with galactocentric distance if the orbits becomes very radial with distance from the galaxy center, but the observed radial velocities rule out such radial orbits (see Vesperini et al. 2003 for details).

a very wide range of radii, from 1 to 75 kpc, indicate a generally constant GCLF with galactocentric distance. Other available HST data are consistent with the result, but none yet match this depth and very large radial range. The other key component of the M87 globular cluster system is that several radial velocities studies of large numbers of M87 globular clusters indicate that the current M87 globular cluster population does not have strongly radial orbits (e.g. Côté et al. 2001, Romanowsky & Kochanek 2001). Vesperini et al. (2003) show directly that their observations of a generally constant GCLF with distance from the galaxy center combined with the results from the radial velocity surveys rule out any model which appeals to strongly radial orbits in the outer regions of galaxies to explain the constancy of the GCLF with galaxy radius (see also Fig. 3).

What is the explanation for the generally constant GCLF seen across the Milky Way and M87? There are three extant proposals, each with its own problems and promise. In order discussed below, these proposals are - 1) that hierarchical merging continually mixes the cluster orbits so thoroughly that clusters now located at a wide range of distances from the center of a galaxy, i.e. 2 to 100 kpc, have on average had the same local galaxy potential during their lifetime as mentioned in Fall & Zhang (2001), 2) that early gas expulsion from forming clusters preferentially disrupts lower-mass proto-globular clusters, producing a turnover in the GCLF everywhere at early times (e.g. Baumgardt, Kroupa, & Parmentier 2008, and references therein), and 3) that globular clusters are formed with an initially power-law mass function, and have a cluster mass-concentration relation such that lower mass clusters have lower concentration and thus are much more likely to be disrupted by stellar mass loss (Vesperini & Zepf 2003). The difference between 2) and 3) is a matter of physical origin and timescale, as 3) occurs on a later timescale in a purely stellar cluster due to stellar mass loss and smaller concentrations for lower-mass clusters, while 2) happens more quickly in a proto-cluster which is still gas-rich.

The appeal of the hierarchical model is that it can work from a power-law initial cluster mass function, as likely found in the Antennae young cluster system (Zhang & Fall 1999) and consistent with many other young cluster system data. The problem with this proposal is that the complete mixing required to make the GCLF today the same at 2 kpc as 100 kpc has never been demonstrated in any hierarchical model. Mergers are generally found to *not* mix completely (e.g. White, S.D.M. 1980, Barnes 1988), and a wide variety of observational evidence, such as the presence of metallicity gradients in elliptical galaxies, supports the idea that the most bound material in the progenitors tends to remain the most bound in the merger product. It is also important to note that this thorough mixing must continue until recent epochs. If a population was thoroughly mixed in galactic radius at early times, but was mostly in place at a redshift of one, a radial dependence of the GCLF would be created, as much of the evolution of the cluster would be dominated by its tidal radius at its current location (see Fig. 5 in Vesperini *et al.* 2003).

Models 2 and 3 which introduce physical processes other than two-body relaxation to help set the scale of the turnover of the globular cluster mass function naturally avoid this problem of the radial constancy of the GCLF. They do not require strongly radial orbits to achieve a constant GCLF, and thus do not violate the observed anisotropy constraints. By far the biggest challenge to these models is the evidence that the mass function for young globular cluster systems is a power-law extending to masses below the current turnover mass observed in old systems. This power-law to lower masses is best established for the Antennae (Zhang & Fall 1999), but is also consistent with a wide range of other data on young cluster systems. The key question is whether the age of the bulk of the Antennae clusters is older than the age by which these early processes will have disrupted low-mass clusters. As Parmentier & Gilmore (2007) and others point out, disruption by gas expulsion is not absolutely immediate, but would be expected to take of order of tens of Myr for an unbound cluster to disperse itself into the field. The bulk of the Anntenae cluster sample ranges up to about 100 Myr, so this is somewhat older, although it can be argued whether this difference is large enough to be fatal for the model.

The hypothesis that the low-mass clusters are disrupted somewhat later, due to stellar mass loss after they are purely stellar systems (Vesperini & Zepf 2003), has an important advantage in comparison with the Antennae data that it happens later than the proposed gas expulsion. One key here is establishing whether the globular cluster mass-concentration relation observed for the Milky Way globular clusters is primordial. If it is, it seems unavoidable that stellar mass loss will cause some preferential dissolution of low-mass globular clusters, and cause the globular cluster mass function to flatten or possibly turnover at lower masses (Vesperini & Zepf 2003 and discussion therein). Another key test of the mechanism by which most low-mass globular clusters are destroyed is to study intermediate-aged globular cluster systems, as there is no doubt these are old enough any early destruction mechanism should have changed their mass functions. To date this has proved challenging. Although deep optical data alone may not be able to distinguish intermediate and old globular cluster systems, results suggestive of a power-law mass function below the turnoff mass in intermediate-age systems have been found by Paul Goudfrooij and collaborators (e.g. Goudfrooij *et al.* 2007). One way forward is deep HST near-infrared imaging like that obtained for the intermediate-age system in

NGC 4365 (Kundu *et al.* 2005) which allows for a clean enough separation of globular cluster sub-populations to test the dependence of the mass function on age.

3. The Implications of the Weak Mass-Radius Relationship for Globular

The masses and radii of globular clusters are for the most part uncorrelated, with only a very shallow relation between the two. This has been established both in the Milky Way (e.g. van den Bergh *et al.* 1991, Djorgovski & Meylan 1994, Ashamn & Zepf 1998), and for extragalactic cluster systems (e.g. Waters *et al.* 2006, Jordán *et al.* 2005). This absent or very shallow mass-radius relation stands in contrast to nearby every other type of astronomical object to which globular clusters might be compared. For example, galaxies have a clear mass-radius relation, and clusters of galaxies do as well. Perhaps most importantly scaling relations for molecular clouds, the natural progenitors of star clusters, give $R \propto M^{0.5}$. Studies of globular clusters, both Galactic and extragalactic, give a dramatically shallower relation, with the best current constraint probably being the $R \propto M^{0.04}$ found for M87 globulars by Waters *et al.* (2006).

Furthermore, studies of young globular cluster systems have also found a very weak mass-radius relation, first in NGC 3256 (Zepf *et al.* 1999), and now in a number of galaxies (Larsen 2004, Scheepmaker *et al.* 2007). The lack of a mass-radius correlation in both young and old globular cluster systems strongly suggests the reason for its surprising absence is *not* due to a long-term evolutionary process, but must be closely related to the formation and early evolution of globular clusters.

The question then is how the formation and early evolution of globular cluster produces such a weak mass-radius relation, particularly when the progenitor clouds seem to have a typically strong relation. Ashman & Zepf (2001) considered many possibilities for physical mechanisms to account for the weak mass-radius relation of globular clusters. Most of these failed, including such standard ideas as a Schmidt law relating star formation efficiency to density. The one solution that works is to adopt a star formation efficiency proportional to the binding energy of the molecular cloud. Because lower mass clusters have less binding energy per unit mass than higher mass clusters, in this case, low mass clusters have a lower star formation efficiency. As a result of this lower star formation efficiency, when the remaining gas is lost from the clusters, low mass clusters will expand more in response to this mass loss.

Note that this is generally true of any proposal in which lower-mass young clusters lose a greater fraction of their mass than higher mass clusters. Lower mass clusters then respond to this greater mass loss by expanding more than higher mass clusters, assuming the mass loss happens adiabatically. Thus, the radii of low and high-mass clusters become more similar. The same effect might occur in cases in which the cluster loses mass because of stellar mass loss, if more massive globular clusters lose a smaller fraction of their mass.

An invariable outcome of models that successfully produce a weak mass-radius relation is that they flatten the globular cluster mass function (Ashman & Zepf 2001). Exactly how much they do so depends on the specifics of the model, both how the expansion of a cluster is related to the star formation efficiency, and what the final mass-radius relation is. For example, adopting the final radius of the cluster r_f is the initial radius r_i divided by the efficiency ϵ, that is $r_f = r_i/\epsilon$, and the efficiency is proportional to the binding energy per unit mass of the initial system to the power n, that is $\epsilon \propto (M_i/r_i)^n$, then the final slope of the globular cluster system mass function, α is $\alpha = (2\beta + n)/(n + 2)$, where β to be the initial slope (see Section 4 of Ashman & Zepf 2001). If n is around 0.5, corresponding to a significant but modest mass-radius relation, then the difference between the initial and final mass slopes is small, about 10%. If n is large, such as $n = 1$ which gives no mass-radius relation, then the difference between the initial and final mass slopes is larger.

Real cluster evolution is undoubtedly more complicated, but these calculations provide an effective framework for showing the connections between the question of the weak mass-radius relation and of the globular cluster mass function. Specifically, the most viable current proposal for the origin of the weakness of the mass-radius relation for globular clusters generically produces a change a flattening of the globular cluster mass function (see Ashman & Zepf 2001). Whether this flattening is substantial or modest depends on the exact mass-radius relation and to some extent on whether the mass loss and expansion occurs adiabatically. Therefore, one of the obvious observational challenges is to determine the mass-radius relation for globular cluster systems as accurately as possible. A second challenge is to determine the mass function in globular cluster systems that are old enough to have experienced the bulk of their stellar mass loss. If these still have steep power law mass functions like those of molecular clouds, then either the mass-radius relation must be pushed to the limits of the current constraints, or another solution for the weakness of the mass radius relation that does not effect the cluster mass function must be found. The underlying key points are that the data for both young and old globular cluster systems indicate a very weak mass-radius relation, and that extant explanations to produce this have implications for the globular cluster mass function.

Acknowledgements

Much of the work described above has been carried out in collaboration with Chris Waters, Enrico Vesperini, and Keith Ashman. I gratefully acknowledge support for this work from NSF award AST-0406891 and grants GO-8592 and GO-10543 from the Space Telescope Science Institute.

References

Ashman, K. M. & Zepf, S. E. 1998, *Globular Custer Systems*, (Cambridge University Press)
Ashman, K. M., & Zepf, S. E. 2001, *AJ*, 122, 1888
Barnes, J. E. 1988, *ApJ* 331, 699
Baumgardt, H. 1998, *A&A* 330, 480
Baumgardt, H., Kroupa, P. & Parmentier, G 2008, *MNRAS*, in press
Baumgardt, H. & Makino, J. 2003, *MNRAS* 340, 227
Burkert, A. & Smith, G.H. 2000, *ApJ* 542, L95
Côté, P., *et al.* 2001, *ApJ* 559, 828
Djorgovski, S. & Meylan, G. 1994, *AJ* 108, 1292
Fall, S. M. & Zhang, Q. 2001, *ApJ* 561, 751
Goudfrooij, P., Schweizer, F., Gilmore, D., & Whitmore, B. C. 2007, *AJ* 133, 2737
Jordán, A. *et al.* 2005 *ApJ* 634, 1002
Kundu, A., Whitmore, B. C., Sparks, W. B., Macchetto, F. D., Zepf, S. E., & Ashman, K. M. 1999, *ApJ* 513, 733
Kundu, A., *et al.* 2005, *ApJ (Letters)* 634, 41
Lamers, H. J. G. L. M., Geiles, M., & Portegies Zwart, S. F. 2005, *A&A* 429, 173
Larsen, S. S. 2004, *A&A* 416, 537
Parmentier, G., & Gilmore, G. 2007, *MNRAS* 377, 352
Rhode, K. L., & Zepf, S. E. 2001, *AJ* 121, 210
Romanowsky, A. J. & Kochanek, C. S. 2001, *ApJ* 553, 722
Scheepmaker, R. A., *et al.* 2007, *A&A* 469, 925
van den Bergh, S., Morbey, C., & Pazder, J. 1991, *ApJ* 375, 594
Vesperini, E. 1997, *MNRAS* 287, 915

Vesperini, E., & Zepf, S. E. 2003, *ApJ (Letters)* 587, L97

Vesperini, E., Zepf, S. E., Kundu, A., & Ashman, K. M. 2003, *ApJ* 593, 760

Waters, C. Z., Zepf, S. E., Lauer, T. R., Baltz, E. A., Silk, J. 2006, *ApJ* 650, 885

Waters, C. Z., & Zepf, S. E. 2008, *ApJ in preparation*

White, S. D. M. 1980, *MNRAS* 191, 1P

Zhang, Q. & Fall, S. M. 1999, *ApJ* 527, L81

Dynamical Evolution of Dense Stellar Systems
Proceedings IAU Symposium No. 246, 2007
E. Vesperini, M. Giersz & A. Sills, eds.

Dynamical Evolution of Globular Clusters in Hierarchical Cosmology

Oleg Y. Gnedin[2] and José L. Prieto[2]

[1] University of Michigan, Department of Astronomy, Ann Arbor, MI 48109-1042, USA
ognedin@umich.edu

[2] The Ohio State University, Department of Astronomy, Columbus, OH 43210, USA
prieto@astronomy.ohio-state.edu

Abstract. We probe the evolution of globular clusters that could form in giant molecular clouds within high-redshift galaxies. Numerical simulations demonstrate that the large and dense enough gas clouds assemble naturally in current hierarchical models of galaxy formation. These clouds are enriched with heavy elements from earlier stars and could produce star clusters in a similar way to nearby molecular clouds. The masses and sizes of the model clusters are in excellent agreement with the observations of young massive clusters. Do these model clusters evolve into globular clusters that we see in our and external galaxies? In order to study their dynamical evolution, we calculate the orbits of model clusters using the outputs of the cosmological simulation of a Milky Way-sized galaxy. We find that at present the orbits are isotropic in the inner 50 kpc of the Galaxy and preferentially radial at larger distances. All clusters located outside 10 kpc from the center formed in the now-disrupted satellite galaxies. The spatial distribution of model clusters is spheroidal, with a power-law density profile consistent with observations. The combination of two-body scattering, tidal shocks, and stellar evolution results in the evolution of the cluster mass function from an initial power law to the observed log-normal distribution.

Keywords. globular clusters: general

---

1. Giant Molecular Clouds at High Redshift

The outcomes of many proposed models of globular cluster formation depend largely on the assumed initial conditions. The collapse of the first cosmological 10^6 $M_\odot$ gas clouds, or the fragmentation of cold clouds in hot galactic corona gas, or the agglomeration of pressurized clouds in mergers of spiral galaxies could all, in principle, produce globular clusters, but only if those conditions realized in nature. Similarly, while observational evidence strongly suggests that all stars and star clusters form in molecular clouds, the initial conditions for cloud fragmentation are a major uncertainty of star formation models.

The only information that we actually have about the initial conditions comes from the early universe, when primordial density fluctuations set the seeds for structure formation. These fluctuations are probed directly by the anisotropies of the cosmic microwave background radiation. Cosmological numerical simulations study the growth of these fluctuations via gravitational instability, in order to understand the formation of galaxies and all other structures in the Universe. The simulations begin with tiny deviations from the Hubble flow, whose amplitudes are set by the measured power spectrum of the primordial fluctuations while the phases are assigned randomly. Therefore, each particular simulation provides only a statistical description of a representative part of the Universe, although current models successfully reproduce major features of observed galaxies.

Kravtsov & Gnedin (2005) attempted to construct a first self-consistent model of star cluster formation, using an ultrahigh-resolution gasdynamics cosmological simulation

403

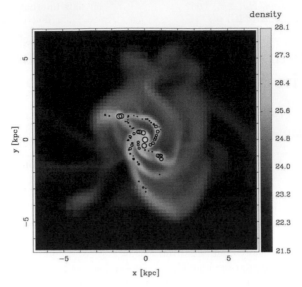

Figure 1. A massive gaseous disk with prominent spiral arms, seen face-on at redshift $z = 4$ in a process of active merging. The gas density is projected over a 3.5 kpc slice. In our model star clusters form in giant gas clouds, shown by circles with the sizes corresponding to the cluster masses. From Kravtsov & Gnedin (2005).

with the Adaptive Refinement Tree code. They identified supergiant molecular clouds in high-redshift galaxies as the likely formation sites of globular clusters. These clouds assemble during gas-rich mergers of progenitor galaxies, when the available gas forms a thin, cold, self-gravitating disk. The disk develops strong spiral arms, which further fragment into separate molecular clouds located along the arms as beads on a string (see Fig. 1).

In this model, clusters form in relatively massive galaxies, with the total mass $M_{\rm host} > 10^9$ $M_\odot$, beginning at redshift $z \approx 10$. The mass and density of the molecular clouds increase with cosmic time, but the rate of galaxy mergers declines steadily. Therefore, the cluster formation efficiency peaks at a certain extended epoch, around $z \approx 4$, when the Universe is only 1.5 Gyr old. The host galaxies are massive enough for their molecular clouds to be shielded from the extragalactic UV radiation, so that globular cluster formation is unaffected by the reionization of cosmic hydrogen. As a result of the mass-metallicity correlation of progenitor galaxies, clusters forming at the same epoch but in different-mass progenitors have different metallicities, ranging between 10^{-3} and 10^{-1} solar. The mass function of model clusters is consistent with a power law $dN/dM \propto M^{-\alpha}$, where $\alpha = 2.0 \pm 0.1$, similar to the observations of nearby young star clusters.

2. Orbits of Globular Clusters

We adopt this model to set up the initial positions, velocities, and masses for our globular clusters. We then calculate cluster orbits using a separate collisionless N-body simulation described in Kravtsov, Gnedin & Klypin (2004). This is necessary because the original gasdynamics simulation was stopped at $z \approx 3.3$, due to limited computational resources. By using the N-body simulation of a similar galactic system, but complete to $z = 0$, we are able to follow the full dynamical evolution of globular clusters until the present epoch. We use the evolving properties of all progenitor halos, from the outputs with a time resolution of $\sim 10^8$ yr, to derive the gravitational potential in the whole

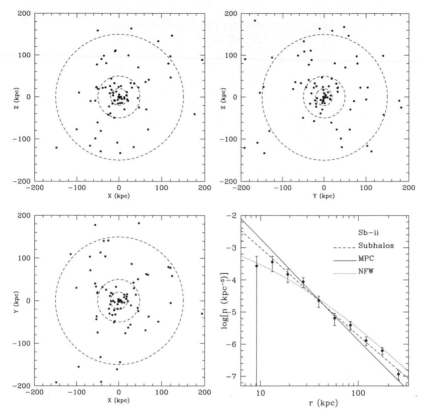

Figure 2. Spatial distribution of surviving model clusters in the Galactic frame. Dashed circles are to illustrate the projected radii of 20, 50, and 150 kpc. The number density profile (*bottom right*) can be fit by a power law, $n(r) \propto r^{-2.7}$. The distribution of model clusters is similar to that of surviving satellite halos (*dashed line*) and smooth dark matter (*dotted line*). It is also consistent with the observed distribution of metal-poor globular clusters in the Galaxy (*solid line*), plotted using data from the catalog of Harris (1996).

computational volume at all epochs. We convert a fraction of the dark matter mass into flattened disks, in order to model the effect of baryon cooling and star formation on the galactic potential. We calculate the orbits of globular clusters in this potential from the time when their host galaxies accrete onto the main (most massive) galaxy. Using these orbits, we calculate the dynamical evolution of model clusters, including the effects of stellar mass loss, two-body relaxation, tidal truncation, and tidal shocks.

We consider several possible scenarios, some with all clusters forming in a short interval of time around redshift $z = 4$, and others with a continuous formation of clusters between $z = 9$ and $z = 3$. Below we discuss the spatial and kinematic distributions of globular clusters in the best-fit model with the synchronous formation at $z = 4$.

In our model, all clusters form on nearly circular orbits within the disks of progenitor host galaxies. Depending on the subsequent trajectories of the hosts, clusters form three main subsystems at present time. *Disk clusters* formed in the most massive progenitor that eventually hosts the present Galactic disk. These clusters, found within the inner 10 kpc, do not actually stay on circular orbits but instead are scattered to eccentric orbits by perturbations from accreted galactic satellites. *Inner halo clusters*, found between 10 and 60 kpc, came from the now-disrupted satellite galaxies. Their orbits are inclined with respect to the Galactic disk and are fairly isotropic. *Outer halo clusters*, beyond 60

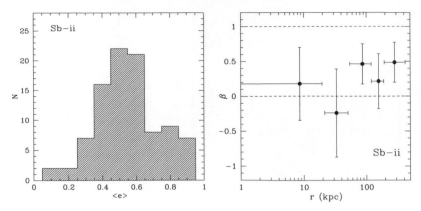

Figure 3. *Left:* Average eccentricity distribution of the surviving model clusters. *Right:* Anisotropy parameter β as a function of radius. Vertical error bars represent the error of the mean for each radial bin, while horizontal error bars show the range of the bin. Horizontal dashed lines illustrate an isotropic ($\beta = 0$) and a purely radial ($\beta = 1$) orbital distributions.

kpc from the center, are either still associated with the surviving satellite galaxies, or were scattered away from their hosts during close encounters with other satellites and consequently appear isolated.

Mergers of progenitor galaxies ensure the present spheroidal distribution of the globular cluster system (Fig. 2). Most clusters are now within 50 kpc from the center, but some are located as far as 200 kpc. The azimuthally-averaged space density of globular clusters is consistent with a power law, $n(r) \propto r^{-\gamma}$, with the slope $\gamma \approx 2.7$. Since all of the distant clusters originate in progenitor galaxies and share similar orbits with their hosts, the distribution of the clusters is almost identical to that of the surviving satellite halos. This power law is similar to the observed distribution of the metal-poor ([Fe/H] < -0.8) globular clusters in the Galaxy. Such comparison is appropriate, for our model of cluster formation at high redshift currently includes only low metallicity clusters ([Fe/H] $\leqslant$ -1). Thus the formation of globular clusters in progenitor galaxies with subsequent merging is fully consistent with the observed spatial distribution of the Galactic metal-poor globulars.

Fig. 3 shows the kinematics of model clusters. Most orbits have moderate average eccentricity, $0.4 < \langle e \rangle < 0.7$, expected for an isotropic distribution. The anisotropy parameter, $\beta = 1 - v_t^2/2v_r^2$, is indeed close to zero in the inner 50 kpc from the Galactic center. At larger distances, cluster orbits tend to be more radial. There, in the outer halo, host galaxies have had only a few passages through the Galaxy or even fall in for the first time.

3. Evolution of the Globular Cluster Mass Function

Using these orbits, we now calculate the cluster disruption rates. Sophisticated models of the dynamical evolution have been developed using N-body simulations as well as orbit-averaged Fokker-Planck and Monte Carlo models. Several processes combine and reinforce each other in removing stars from globular clusters: stellar mass loss, two-body scattering, external tidal shocks, and dynamical friction of cluster orbits. The last three are sensitive to the external tidal field and therefore, to cluster orbits. While a general framework for all these processes has been worked out in the literature, the knowledge of realistic cluster orbits is essential for accurate calculations of the disruption.

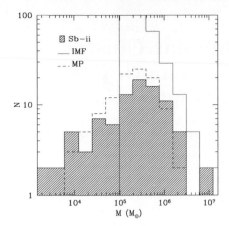

Figure 4. Evolution of the mass function of clusters in our best-fit model from an initial power law (*solid line*) to a peaked distribution at present (*histogram*), including mass loss due to stellar evolution, two-body relaxation, and tidal shocks. For comparison, dashed histogram shows the mass function of metal-poor globular clusters in the Galaxy.

Fig. 4 shows the transformation of the cluster mass function from an initial power law, $dN/dM \propto M^{-2}$, into a final bell-shaped distribution. In this model all globular clusters form at the same redshift, $z = 4$, or about 12 Gyr ago. The half-mass radii, R_h, are set by the condition that the median density, M/R_h^3, is initially the same for all clusters and remains constant as a function of time. Over the course of their evolution, numerous low-mass clusters are disrupted by two-body relaxation while the high-mass clusters are truncated by tidal shocks. The present mass function is in excellent agreement with the observed mass function of the Galactic metal-poor clusters.

This result by itself is not new. Previous studies of the evolution of the cluster mass function have found that almost any initial function can be turned into a peaked distribution by the combination of two-body relaxation and tidal shocks. However, the efficiency of these processes depends on the cluster mass and size, $M(t)$ and $R_h(t)$. *The new result is that we find that not all initial relations $R_h(0) - M(0)$ and not all evolutionary scenarios $R_h(t) - M(t)$ are consistent with the observed mass function.*

Consider two examples. (i) If the half-mass radius R_h is kept fixed for clusters of all masses and at all times, the median density $M(t)/R_h^3$ decreases as the clusters lose mass. Two-body scattering becomes less efficient and spares many low-mass clusters, while tidal shocks become more efficient and disrupt most high-mass clusters. The final distribution is severely skewed towards small clusters. (ii) If the size is assumed to evolve in proportion to the mass, $R_h(t) \propto M(t)$, the cluster density increases with time. As a result, all of the low-mass clusters are disrupted by the enhanced two-body relaxation, while the high-mass clusters are unaffected by the weakened tidal shocks. The final distribution is skewed towards massive clusters.

Only our best-fit model with $M(t)/R_h^3(t) = const$ successfully reproduces the mass function and spatial distribution of metal-poor globular clusters in Galaxy. We are now investigating the formation of metal-rich clusters in galactic mergers at lower redshifts.

References

Harris W., 1996, *AJ*, 112, 1487
Kravtsov A., Gnedin O., & Klypin A., 2004, *ApJ*, 609, 482
Kravtsov A. & Gnedin O., 2005, *ApJ*, 623, 650

Dynamical Evolution of Dense Stellar Systems
Proceedings IAU Symposium No. 246, 2007
E. Vesperini, M. Giersz & A. Sills, eds.

© 2008 International Astronomical Union
doi:10.1017/S1743921308016050

Clues to Globular Cluster Evolution from Multiwavelength Observations of Extragalactic Systems

Arunav Kundu[1] **Thomas J. Maccarone**[2] **and Stephen E. Zepf**[1]

[1]Department of Physics & Astronomy, Michigan State University, East Lansing, MI 48824,
USA
email: akundu@pa.msu.edu, zepf@pa.msu.edu

[2]School of Physics & Astronomy, University of Southampton, Southampton, UK SO17 1BJ
email: tjm@phys.soton.ac.uk

Abstract. We present a study of the globular cluster (GC) systems of nearby elliptical and S0 galaxies at a variety of wavelengths from the X-ray to the infrared. Our analysis shows that roughly half of the low mass X-ray binaries (LMXBs), that are the luminous tracers of accreting neutron star or black hole systems, are in clusters. There is a surprisingly strong correlation between the LMXB frequency and the metallicity of the GCs, with metal-rich GCs hosting three times as many LMXBs as metal-poor ones, and no convincing evidence of a correlation with GC age so far. In some of the galaxies the LMXB formation rate varies with GC color even within the red peak of the typical bimodal cluster color distribution, providing some of the strongest evidence to date that there are metallicity variations within the metal-rich GC peak as is expected in a hierarchical galaxy formation scenario. We also note that any analysis of subtler variations in GC color distributions must carefully account for both statistical and systematic errors. We caution that some published GC correlations, such as the apparent 'blue-tilt' or mass-metallicity effect might not have a physical origin and may be caused by systematic observational biases.

Keywords. globular cluster systems, low mass X-ray binaries, blue-tilt

1. Introduction

High resolution Chandra X-ray images of nearby ellipticals and S0s have resolved large numbers of point sources, confirming a long-standing suggestion that the hard X-ray emission in many of these galaxies is predominantly from X-ray binaries. In early type galaxies most of the bright, $L_X \gtrsim 10^{37}$ erg s^{-1} sources seen in typical Chandra observations must be low mass X-ray binaries, binary systems comprising a neutron star or black hole accreting via Roche lobe overflow from a low mass companion, since they generally have stellar populations that are at least a few Gyrs old.

An important characteristic of LMXBs is that they are disproportionately abundant in globular clusters. Even though GCs account for $\lesssim 0.1\%$ of the stellar mass in the Galaxy, they harbor about 10% of the $L_X \gtrsim 10^{36}$ erg s^{-1} LMXBs , indicating a probability of hosting a LMXB that is at least two orders of magnitude larger than for field stars. This has long been attributed to efficient formation of LMXBs in clusters due to dynamical interactions in the core. Early type galaxies are ideal for studies of the LMXB-GC link as they are particularly abundant in globular clusters. The identification of LMXBs with these simple stellar systems that have well defined properties such as metallicity and age provides a unique opportunity to probe the effects of these parameters on LMXB formation and evolution.

2. The Effect of GC Environment on LMXB Formation & Evolution

Fig. 1 plots the colors, magnitudes, half-light radii, and galactocentric distances of globular clusters identified in HST-WFPC2 images of NGC 4472 (Kundu, Maccarone & Zepf 2002 [KMZ02]), the brightest elliptical in the Virgo cluster. The well known Gaussian globular cluster luminosity function and the bimodal globular cluster color (metallicity) distribution are apparent. The large symbols indicate the GCs that host LMXBs. LMXBs are preferentially found in the most luminous and red (metal-rich) GCs. Statistical tests (KMZ02) reveal a marginal tendency of LMXBs to favor GCs with smaller half light radius, and no convincing correlation with galactocentric distance. These correlations have subsequently been confirmed in other galaxies (Kim *et al.* 2006; Kundu, Maccarone, & Zepf 2007 [KMZ07]; Sivakoff *et al.* 2007). The presence of bright LMXBs in ≈4% of GCs and the association of ≈40% of LMXBs with GCs in NGC 4472 is also typical of the values in other early type galaxies.

The underlying reasons for these correlations provide a window into both the dynamics of GCs and the physics of LMXBs. Since luminous clusters are known to be denser than less luminous ones and obviously have more stars, the higher rate of LMXB formation in these clusters due to the consequently higher dynamical interaction rate is not surprising. One the other hand there are no obvious dynamical reasons for the three times larger rate of LMXBs in the red, metal-rich, globular clusters as compared to the metal-poor ones.

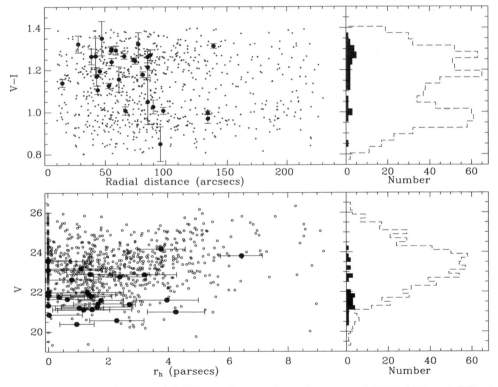

Figure 1. Top: The V-I colors of GCs vs. distance from the center of NGC 4472 and GC color distribution. LMXB-GC matches are indicated by filled circles/bins. Bottom: V magnitude of globular clusters vs. half light radius and the globular cluster luminosity function. LMXBs are preferentially located in the brightest, most metal-rich GCs. There is a weak anti-correlation with GC half-light radius and no obvious correlation with galactocentric distance. Each of these broad correlations (or lack thereof) have been confirmed in other galaxies.

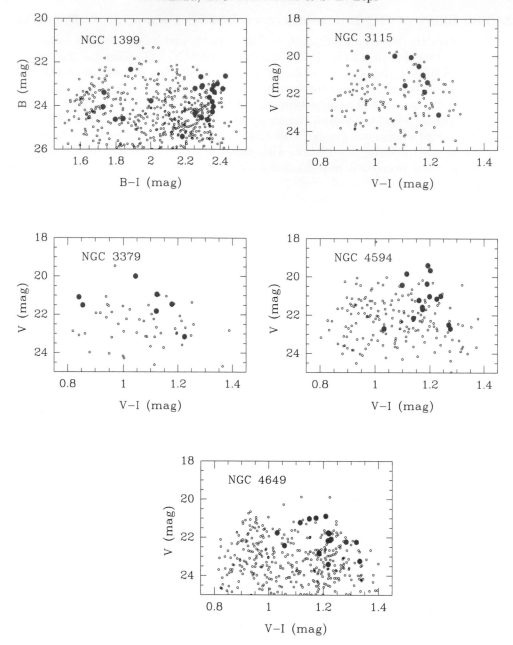

Figure 2. Color-magnitude diagrams for globular cluster candidates in five elliptical galaxies with known bimodal cluster metallicity distributions. Filled points represent clusters with LMXB counterparts. LMXBs are clearly found preferentially in luminous (high mass) and red (metal-rich) globular clusters. There is a clear enhancement of LMXBs in the reddest GCs in the metal-rich GCs of NGC 1399 providing some of the strongest evidence to date that there is metallicity substructure within the red GCs in this giant elliptical galaxy.

LMXBs are similarly found preferentially in the metal-rich Galactic GCs, but in the past this was often dismissed as the consequence of the location of these clusters in the bulge, where the dynamical evolution of GCs may be accelerated. The lack of correlation of the LMXBs with galactocentric distance in the large GC samples of ellipticals (KMZ02;

KMZ07) argues against this possibility and establishes a primary correlation with GC color.

It has also been suggested that enhanced rate of LMXBs in metal-rich GCs may be because of the younger ages of these clusters. However infrared imaging, which yields more accurate constraints on the metallicities of GCs and helps disentangle the age-metallicity degeneracy, has established that the optical colors of globular cluster systems that have clearly bimodal distributions are primarily a tracer of metallicity (Kundu & Zepf 2007) and LMXBs are indeed preferentially located in metal-rich GCs (Kundu *et al.* 2003; Hempel *et al.* 2007). Fig. 2 shows the color magnitude distributions of five galaxies with confirmed bimodal color distributions and the factor of three enhancement with metallicity.

A very interesting feature of the NGC 1399 distribution is that not only is there a larger fraction of LMXBs in the metal-rich sub-population of clusters, but in fact LMXBs are preferentially located in the most metal-rich GCs. This is the most convincing evidence to date that there is metallicity structure within the metal-rich peak of GCs, as is expected from hierarchical models of galaxy and globular cluster system formation. It is not clear if the reason that this feature is obvious only in NGC 1399 is because of the location of this galaxy at the center of the Fornax cluster which leads to a particularly efficient enrichment history, or because of the larger color baseline of this data set. Although the Sivakoff *et al.* (2007) analysis is broadly in agreement about the metallicity trend and finds a linear increase in the fraction of LMXBs in GCs with metallicity, it suggests that the rate of LMXBs in the reddest, and consequently most metal-rich GCs actually drops. We note that LMXBs are found preferentially in the brightest globular clusters, and if the GC sample in Sivakoff *et al.* (2007) were to be restricted to similar luminosities this apparent discrepancy would disappear. In other words the reddest GCs in the Sivakoff sample are the faint ones which have scattered to these colors due to observational uncertainties and are not representative of the true underlying color/metallicity of the GCs.

There have been some theoretical attempts to explain the metallicity effect. Maccarone, Kundu, & Zepf (2004) suggest that irradiation of the donor star by the LMXB is key. Higher metallicity objects can dissipate this energy through line cooling while a wind is generated in low metallicity star, thus lowering the LMXB lifetime. Ivanova (2006) on the other hand suggests that the absence of an outer convective layer in solar mass metal-poor stars limits magnetic braking and the formation of mass transferring LMXBs.

Various groups have attempted to derive the dependence of the observed LMXB rate on the globular cluster metallicity and mass by assuming a specific M/L ratio and color-metallicity correlation (Jordan *et al.* 2004; Smits *et al.* 2006; Sivakoff *et al.* 2007). These studies generally agree that the mass dependence is roughly linear and the metallicity dependence is approximately $Z^{0.25}$. Jordan *et al.* (2004) and Sivakoff (2007) further attempt to link the probability of finding a LMXB in a GC to the rate of stellar interactions in the core of a GC. However, this requires measuring the core radii of GCs which is a small fraction of even a HST ACS pixel for these galaxies. Thus the core radius is derived by extrapolation of other measured GC properties. Smits *et al.* (2006) show that the Jordan *et al.* (2004) results are as expected if there is no information about the core radius of the clusters in their sample. While the core radii of extragalactic GCs are a challenge for the present data sets, the half-light radii can indeed be measured and its effect on other measured GC parameters must be accounted for carefully. Some of the consequences of not doing so are outlined next.

3. 'Blue-Tilts', Mass-Metallicity Trends and other Correlations in the Properties of Halo Globular Clusters

Many recent HST-based studies of globular cluster systems have reported the discovery of color-metallicity trends in the blue, metal-poor sub-population of globular clusters (Strader *et al.* 2006; Mieske *et al.* 2006; Harris *et al.* 2006). This small trend of brighter blue clusters appearing a few hundredths of a magnitude redder per magnitude of brightness has been dubbed the 'blue-tilt', and is viewed as evidence of a mass-metallicity trend. This has been interpreted as evidence of self-enrichment of the more massive clusters either because they started out with dark matter halos that have since been stripped, or because they formed in large gas clouds. This effect is larger in more massive galaxies, and more apparent when the color magnitude diagram is plotted using the redder of the two filters under consideration for the magnitude axis (Mieske *et al.* 2006).

We have reanalyzed some of these data sets and find that there is a small, but measurable, mass-radius relationship for GCs in these galaxies (Kundu & Zepf 2007b in preparation). Since the blue GCs are on average larger, this trend is better defined for the metal-poor clusters. Studies of extragalactic cluster systems typically assume either a uniform aperture correction or attempt to fit the profile within a small aperture in order to minimize the uncertainties introduced by the galaxy background. We show in Kundu & Zepf (2007b) that the blue tilt is likely the consequence of photometric bias introduced by the effect of the mass-radius relationship on such photometric techniques.

It is also important to note that the uncertainties in color and magnitude are not independent, and hence not orthogonal on a color magnitude diagram. In fact when the redder of the two filters is used for the magnitude axis in a color-magnitude diagram the uncertainties are parallel to the direction of the 'blue-tilt' and hence amplify the effect. Conversely the uncertainties tend to negate the 'blue-tilt' when the blue filter is plotted on the magnitude axis of a color magnitude plot, thus explaining the filter effect. Moreover, the underlying galaxy light is known to be redder for more massive galaxy. This explains the apparent variation of the strength of the 'blue-tilt' with galaxy mass.

Finally we note that various studies have suggested that there is a small color metallicity trend in the mean colors of the metal-poor globular clusters with host galaxy mass (e.g. Strader *et al.* 2006). We find a trend of larger mean sizes of blue GCs with larger galaxy mass, which may cause this effect. Thus, within the observational uncertainties the properties of metal-poor halo clusters appear to be remarkably uniform everywhere.

References

Harris, W. E., *et al.* 2006, *ApJ*, 636, 90
Hempel, M., Zepf, S., Kundu, A., Geisler, D., & Maccarone, T. J. 2007, *ApJ*, 661, 768
Ivanova, N. 2006, *ApJ*, 636, 979
Jordán, A., *et al.* 2004, *ApJ*, 613, 279
Kim, E., *et al.* 2006, *ApJ*, 647, 276
Kundu, A., Maccarone, T. J., & Zepf, S. E. 2002, *ApJ* (Letters), 574, L5
Kundu, A., Maccarone, T. J., & Zepf, S. E. 2007, *ApJ*, 662, 525
Kundu, A., Maccarone, T. J., Zepf, S. E., & Puzia, T. H. 2003, *ApJ* (Letters), 589, L81
Kundu, A., & Zepf, S. E. 2007, *ApJ* (Letters), 660, L109
Maccarone, T. J., Kundu, A., & Zepf, S. E. 2004, *ApJ*, 606, 430
Mieske, S., *et al.* 2006, *ApJ*, 653, 193
Sivakoff, G. R., et al. 2007, *ApJ*, 660, 1246
Smits, M., Maccarone, T. J., Kundu, A., & Zepf, S. E. 2006, *A&A*, 458, 477
Strader, J., Brodie, J. P., Spitler, L., & Beasley, M. A. 2006, *AJ*, 132, 2333

Dynamical Evolution of Dense Stellar Systems
Proceedings IAU Symposium No. 246, 2007
E. Vesperini, M. Giersz & A. Sills, eds.

The Origin of the Universal Globular Cluster Mass Function

G. Parmentier[1] and G. Gilmore[2]

[1] Argelander Institut fuer Astronomie, University of Bonn, Auf dem Huegel 71, D-53121 Bonn, Germany
Scientific Research Worker of Fonds National de la Recherche Scientifique, Belgium
Humboldt Fellow
email: gparm@astro.uni-bonn.de

[2] Institute of Astronomy, University of Cambridge, Madingley Road, Cambridge CB3 0HA, UK
email: gil@ast.cam.ac.uk

Abstract. Evidence favouring a Gaussian initial mass function for systems of old globular clusters has accumulated over recent years. We show that a bell-shaped mass function may be the imprint of expulsion from protoclusters of the leftover star forming gas due to supernova activity. Owing to the corresponding weakening of its gravitational potential, a protocluster retains a fraction only of its newly formed stars. The mass fraction of bound stars extends from zero to unity depending on the star formation efficiency achieved by the protoglobular cloud. We investigate how such wide variations affect the mapping of the protoglobular cloud mass function to the initial globular cluster mass function. We conclusively demonstrate that the universality of the globular cluster mass function originates from a common protoglobular cloud mass-scale of about 10^6 $M_\odot$ among galaxies. Moreover, gas removal during star formation in massive gas clouds is highlighted as the likely prime cause of the predominance of field stars in the Galactic Halo.

Keywords. globular clusters: general, Galaxy: halo, Galaxy: formation, galaxies: star clusters

1. Introduction

The cluster mass function (i.e. the number of clusters per logarithmic mass interval $dN/d\log m$) is one of the primary characteristics of any globular cluster system hosted by a massive galaxy. Intriguingly, it proves to be almost independent of the size, the morphological type or the environment of the host galaxy. This universal globular cluster mass function is well fitted by a Gaussian with a mean of $\log (m/M_\odot) \simeq 5.2$ and a standard deviation of ~ 0.6. Globular clusters having evolved over a Hubble-time in their galactic environment, their initial mass function has remained controversial, with two competing hypotheses. It may have been a featureless power-law with a slope of ~ -1, the Gaussian function characteristic of old globular cluster populations then resulting from a purely evolutionary effect, namely, the preferential removal of the more vulnerable low-mass clusters (Fall & Zhang 2001). Yet, the present-day mass function represents an equilibrium state so that the initial one may also have been a Gaussian similar to that today (Vesperini 1998). *If so, the Gaussian shape is the fossil imprint of the cluster formation process, this holding the clue to the universality of the observed globular cluster mass function.* Parmentier & Gilmore (2005) and Vesperini *et al.* (2003) provide evidence for a Gaussian initial globular cluster mass function in the Galactic halo and in the giant elliptical M87, respectively. Theoretical support for a bell-shaped initial cluster mass function has been missing so far however (although see Kroupa & Boily 2002).

As a result of the high star formation efficiency (SFE) required to form a bound cluster, the initial cluster mass function has often been postulated to mirror that of their gaseous progenitors. Actually, numerous studies (e.g. Hills 1980; Lada; Margulis & Dearborn 1984; Fellhauer & Kroupa 2005) point that star forming clouds must be better than 30-50 % efficient in converting gas into stars to produce bound stellar clusters. The limited variations in the SFE (i.e., less than a factor of 3) may then guarantee that the initial mass function of the clusters is that of their parent clouds. However, following the dispersal of the residual star forming gas by the combined actions of stellar winds and supernova explosions, the newly formed stars suddenly find themselves in a shallower gravitational potential, resulting into either the escape of some of them or even the complete disruption of the protocluster. Therefore, the initial mass of a stellar cluster is not determined by the SFE ϵ only. It depends on the mass fraction of the cluster parent cloud which is turned into stars *remaining bound after the dispersal of the gaseous component*. Specifically, the cluster initial mass m_{init} obeys

$$m_{init} = F_{bound} \times \epsilon \times m_{cloud} \,, \qquad (1.1)$$

where m_{cloud} is the mass of the gaseous progenitor and F_{bound} is the mass fraction of stars remaining bound after gas removal. The bound fraction F_{bound} ranges from 0 (when the ϵ is smaller than a threshold value $\epsilon_{th} \simeq 0.35$) up to 1 (see the solid line in the left panel of Fig. 1). As a result, the assumed mirroring effect between the mass function of the cluster forming clouds on the one hand and the initial cluster mass function on the other hand can no longer be taken for granted, the latter depending on the former *and* on gas removal through the variations in the quantity $F_{bound} \times \epsilon$. We now investigate how the initial cluster mass function differs with respect to the protoglobular cloud mass function as a result of gas removal.

2. From a universal protoglobular cloud mass-scale to the universal Gaussian globular cluster mass function

As a first step, we assume a power-law protoglobular cloud mass spectrum

$$\mathrm{d}N \propto m^{\alpha} \mathrm{d}m \,, \qquad (2.1)$$

with $\alpha \simeq -1.7$, as is observed for giant molecular clouds and their star forming cores in the Local Group of galaxies (e.g. Rosolowski 2005). Star forming regions are characterized by a range in their respective star formation efficiency ϵ, so that the protoglobular cloud mass spectrum is convolved with an ϵ probability distribution function, which we describe by a decreasing power-law of slope δ and core r_c, that is:

$$\wp(\epsilon) = \frac{\mathrm{d}N}{\mathrm{d}\epsilon} = c_1 \left(1 + \frac{\epsilon}{r_c}\right)^{\delta} + c_4 \,. \qquad (2.2)$$

The two parameters c_1 and c_4 are determined so as to satisfy the two following constraints: (1) the integration of the probability distribution over the range $\epsilon = [0,1]$ is unity, and the probability $\wp(\epsilon)$ is zero when $\epsilon = 1$. The formation of a bound star cluster requires its gaseous progenitor to achieve $\epsilon > \epsilon_{th}$, i.e., the *local* star formation efficiency must be greater than $\simeq 0.3 - 0.4$. On the scale of a galaxy, star formation proceeds inefficiently, so the *global* star formation efficiency may be of order a few per cent only. The core r_c and the slope δ of the efficiency distribution $\wp(\epsilon)$ are thus bounded so that the mean star formation efficiency, namely, the mass fraction of gas converted into stars for an entire system of protoglobular clouds, is one per cent.

Following the onset of supernova activity, the gas-embedded cluster gets exposed as its residual gas is removed. Not only does the protocluster lose its gaseous component, it also loses a fraction of its initial *stellar* mass. We account for this phase by matching each efficiency value ϵ to the corresponding fraction F_{bound} of bound stars. The F_{bound} vs. ϵ relation we are using is shown as the solid line in the left panel of Fig. 1. The initial mass m_{init} of globular clusters is then derived following equation 1.1.

Finally, the initial globular cluster mass function is evolved up to an age of 13 Gyr with Baumgardt & Makino 's (2003) equation 12, which they derived by fitting the results of a large set of N-body simulations taking into account the effects of stellar evolution, of two-body relaxation and of cluster tidal truncation. At that stage, we obtain the model goodness of fit by comparing the evolved globular cluster mass function and the Old Halo globular cluster mass function.

If the protoglobular cloud mass function is a featureless power-law, then the initial cluster mass function is a power-law with the same slope, by virtue of the mass-independence of the star formation efficiency ϵ and of the bound mass fraction F_{bound}. If the protoglobular cloud mass function shows a lower mass-limit in the form of a truncation, however, it evolves into a bell-shaped initial cluster mass function. We have discussed how the cluster initial mass function responds to variations in the input parameters of our model (see Parmentier & Gilmore 2007). We have successively varied the slope α of the cloud mass spectrum, its lower and upper limits, the slope δ and the scale-length r_c of the star formation efficiency distribution $\wp(\epsilon)$, and the efficiency threshold ϵ_{th} required to retain a bound core of stars (equivalently the gas removal time-scale τ_{gr} measured in units of the protocluster crossing-time τ_{cross}; see the solid/dashed curves in the left panel of Fig. 1). The turnover location is mostly sensitive to the lower limit m_{low} of the protoglobular cloud mass spectrum. The observed universality of the turnover of the globular cluster mass function would therefore originate from a common value among galaxies for the lower mass limit of protoglobular clouds, possibly with second-order variations driven by differences in the slopes of the cloud mass spectrum, that of the efficiency distribution $\wp(\epsilon)$, as well as by differences in the gas removal time-scale. The right panel of Fig. 1 shows the specific case of a protoglobular cloud mass function truncated at $m_{low} \simeq 6 \times 10^5$ M$_\odot$ (solid line with plus signs), along with the corresponding initial and 13-Gyr old cluster mass functions (solid line with plain/open symbols, respectively). The latter matches the observed Old Halo globular cluster mass function, with an incomplete gamma function of $Q \simeq 0.1$.

The bottom panel of Parmentier & Gilmore 's (2007) figure 7 illustrates the goodness of fit of the evolved globular cluster mass function for various model parameters. The probability distribution function for the star formation efficiency $\wp(\epsilon)$ (i.e. δ and r_c) is adjusted so that the present globular cluster mass fraction in the Galactic halo is 2%, as observed. That panel implies that a power-law protoglobular cloud mass spectrum with a narrow mass range, e.g. 10^6 M$_\odot \lesssim m_{cloud} \lesssim 2 \times 10^6$ M$_\odot$, leads to a good fit ($Q \simeq 0.1$) of the modelled cluster mass function onto the observed one. This suggests that the present-day halo cluster mass distribution, which covers two decades in mass, may equally-well arise from a characteristic mass for the protoglobular clouds. In order to investigate this point more closely, we now assume that the protoglobular cloud mass function obeys a Gaussian. The best solution under that assumption is shown in the right panel of Fig. 1, along with the power-law case which we have just discussed (dashed curves with plus signs, filled symbols and open symbols represent the protoglobular cloud mass function, the initial and evolved globular cluster mass functions). A Gaussian protoglobular cloud mass function with a mean of 2×10^6 M$_\odot$ and a logarithmic standard deviation smaller than 0.4 provides a good fit to the Old Halo cluster mass function.

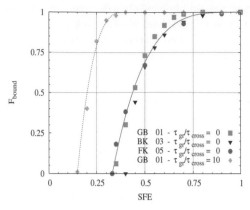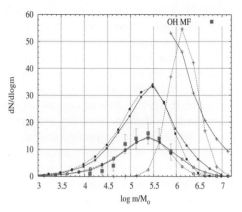

Figure 1. *Left panel:* Relations between the fraction F_{bound} of stars remaining bound to the protocluster after gas removal and the star formation efficiency ϵ achieved by the gaseous progenitor. The solid/dashed lines correspond to the case of rapid/slow gas removal (i.e, $\tau_{gr} \ll \tau_{cross}$ or $\tau_{gr} \gg \tau_{cross}$). Data are from Geyer & Burkert (2001), Boily & Kroupa (2003) and Fellhauer & Kroupa (2005) (respectively quoted as GB01, BK03 and FK05). *Right panel:* Initial/evolved (plain/open circles) cluster mass functions corresponding to two distinct mass functions for their gaseous progenitors: a power-law with a spectral index $\alpha = -1.7$ and truncated at a mass of $6 \times 10^5 M_\odot$ and a Gaussian mass function with a mean logarithmic mass of 6.15 and a standard deviation of 0.3 (solid/dashed curves with plus signs, respectively). The Old Halo globular cluster mass function is depicted by the full squares.

In order for our model to explain the universal Gaussian globular cluster mass function, the protoglobular clouds must thus be characterized by an almost invariant high-mass scale, either in the form of a lower truncation of the power-law cloud mass spectrum ($m_{low} \simeq 6 \times 10^5 M_\odot$) or in the form of a high mean logarithmic mass of $\simeq 6.2$ if the clouds are distributed following a Gaussian in log-mass. Although that issue remains debated, various models for the formation of the first bound objects in the Universe actually suggest that globular cluster gaseous progenitors are characterized by that high a mass-scale (see Parmentier & Gilmore 2007, their section 3.5, for an in-depth discussion). That narrow mass range is then broadened and turned into a bell-shaped cluster initial mass function by gas removal, while the $\simeq 10^6 M_\odot$ protoglobular cloud mass-scale guarantees that the turnover settles at the observed value.

3. The origin of halo field stars

In the present-day Galaxy, the formation of unbound stellar groups is the rule and not the exception. Most field stars in the Galactic disc likely originate from embedded clusters which either lost a fraction of their original members or were disrupted while emerging out of their natal clouds (Lada & Lada 2003). It is likely that that paradigm also characterizes the formation of the stellar halo. Actually, 98 % of its mass consists of field stars and so large a mass fraction cannot be accounted for by the secular evaporation and disruption of globular clusters over a Hubble-time, regardless of the shape of the cluster initial mass function (Parmentier & Gilmore 2005).

The violent relaxation phase affecting protoclusters following the expulsion of their residual star forming gas constitutes a prime candidate to explain the origin of field stars in the Galactic halo, without conflicting with the well-accepted paradigm following which most stars form in clusters. In our fiducial cases of the right panel of Fig. 1, at an age of 13 Gyr, the total mass fraction of field stars $f_{FS} = 98\%$ in the halo arises mostly

($f_{FS} = 91\%$) from cluster infant mortality, i.e. most halo field stars are given off by star forming regions whose efficiency ϵ is less than the star formation efficiency threshold ϵ_{th}.

It is worth noting that this may help understand why field stars and globular cluster stars sometimes show different patterns in their light element abundances. Chemical abundance anomalies observed in globular cluster stars are usually ascribed to the accretion onto their surface of stellar winds of intermediate mass stars ascending the asymptotic giant branch, a process made feasible by the dense stellar environment characteristic of globular clusters. In contrast, to the possible exception of binary stars, *ab initio* field stars remain unaffected by this external pollution.

4. Conclusions

In this paper, we have presented detailed simulations highlighting how the mass function of protoglobular clouds evolves into that of gas-free bound star clusters as a result of the expulsion of the residual star forming gas due to supernova activity. Combining our model generating cluster initial mass functions with the cluster evolutionary model of Baumgardt & Makino (2003), we have subsequently investigated which input parameters reproduce both the present-day mass function of the Old Halo clusters and their present-day mass fraction in the stellar halo. Our model naturally explains the universality of the globular cluster mass function among galaxies *if the protogalactic era sets a characteristic mass of about 10^6 M$_\odot$ for the protoglobular clouds independent of the host galaxy*. We point however that our model still lacks a crucial ingredient, namely, the tidal field of the host galaxy since the F_{bound} vs. SFE relation we use has been derived in the case of isolated globular clusters. The tidal radius of a cluster depending on its mass, the tidal field may also contribute to the shape of the initial globular cluster mass function. Whether the turnover location is affected as well remains to be investigated.

Acknowledgements

This research was supported by a Marie Curie Intra-European Fellowship within the 6th European Community Framework Programme. GP also acknowledges support from the Belgian Science Policy Office in the form of a Return Grant and the Alexander von Humboldt Foundation in the form of a Research Fellowship.

References

Baumgardt, H. & Makino, J. 2003, *MNRAS* 340, 227
Boily, C. M. & Kroupa, P. 2003, *MNRAS* 338, 665
Fall, S. M. & Zhang, Q. 2001, *ApJ* 561, 751
Fellhauer, M. & Kroupa, P. 2005, *MNRAS* 630, 879
Geyer, M. P. & Burkert, A. 2001, *MNRAS* 323, 988
Hills, J. G. 1980, *ApJ* 225, 986
Kroupa, P. & Boily, C. M. 2002, *MNRAS* 336, 1188
Lada, C. J., Margulis, M., & Dearborn, D. 1984, *ApJ* 285, 141
Lada, C. J. & Lada, E. A. 2003, *ARA&A* 41, 57
Parmentier, G. & Gilmore, G. 2005, *MNRAS* 363, 326
Parmentier, G. & Gilmore, G. 2007, *MNRAS* 377, 352
Rosolowski, E. 2005, *PASP* 117, 1403
Vesperini, E. 1998, *MNRAS*, 299, 1019
Vesperini, E., Zepf, S. E., Kundu, A., & Ashman, K.M. 2003, *ApJ* 593, 760

Dynamical Evolution of Dense Stellar Systems
Proceedings IAU Symposium No. 246, 2007
E. Vesperini, M. Giersz & A. Sills, eds.

Masses and M/L Ratios of Bright Globular Clusters in NGC 5128

M. Rejkuba[1], P. Dubath[2], D. Minniti[3] and G. Meylan[4]

[1] ESO, Karl-Schwarzschild-Strasse 2, D-85748 Garching, Germany
email: mrejkuba@eso.org

[2] Observatoire de Geneve, ch. des Maillettes 51, 1290 Sauverny, Switzerland

[3] Pontificia Univ. Católica de Chile, Vicuña Mackenna 4860, Santiago 22, Chile

[4] Ecole Polytechnique Fédérale de Lausanne, Observatoire, 1290 Sauverny, Switzerland

Abstract. We present an analysis of the radial velocities and velocity dispersions for 27 bright globular clusters in the nearby elliptical galaxy NGC 5128 (Centaurus A). For 22 clusters we combine our new velocity dispersion measurements with the information on the structural parameters, either from the literature when available or from our own data, in order to derive the cluster masses and mass-to-light (M/L) ratios. The masses range from $1.2 \times 10^5 M_\odot$, typical of Galactic globular clusters, to $1.4 \times 10^7 M_\odot$, similar to more massive dwarf globular transition objects (DGTOs) or ultra compact dwarfs (UCDs) and to nuclei of nucleated dE galaxies. The average M/L_V is 3 ± 1, larger than the average M/L_V of globular clusters in the Local Group galaxies. The correlations of structural parameters, velocity dispersion, masses and M/L_V for the bright globular clusters extend the properties established for the most massive Local Group clusters towards those characteristic of dwarf elliptical galaxy nuclei and DGTOs/UCDs. The detection of the mass-radius and the mass-M/L_V relations for the globular clusters with masses greater than $\sim 2 \times 10^6 M_\odot$ provides the link between "normal" old globular clusters, young massive clusters, and evolved massive objects.

Keywords. galaxies: star clusters, galaxies: elliptical and lenticular, cD, galaxies: individual (NGC 5128)

1. Introduction

The properties of globular clusters and the observed correlations between their various internal structural and dynamical parameters offer empirical constraints for the formation of globular clusters and for the star formation history of the host galaxy. A large number of empirical relations have been found between various properties (core and half-light radii, surface brightnesses, velocity dispersions, concentrations, luminosities, metallicities, etc.) of the globular clusters in the Milky Way and its satellite galaxies (Djorgovski & Meylan 1994; McLaughlin & van der Marel 2005). Many of them are mutually dependent due to the fact that globular clusters have very simple structures that can be well approximated by isotropic, single-mass King (1966) models.

Ultra compact dwarf galaxies (UCDs) or dwarf globular transition objects (DGTOs) are compact massive objects discovered first in the Fornax cluster (Hilker *et al.* 1999), and later also in Virgo, Centaurus and Hydra clusters (Haşegan *et al.* 2005; Mieske *et al.* 2007; Wehner & Harris 2007). They are more luminous, more massive, and more extended than typical Galactic globular clusters. Their ages are similar to those of old globular clusters (see review by Hilker 2006). Are UCDs/DGTOs related to massive globular clusters, or are they more similar to compact dwarf galaxies?

Young massive clusters (YMCs) form in starbursts and mergers. Their masses are in the range of those of the most massive globular clusters observed around massive elliptical galaxies ($> 10^6$ M$_\odot$), or even comparable to UCDs ($> 10^7$ M$_\odot$). Question is whether they will evolve to become normal massive globular clusters or even more extreme UCDs.

Given the bright magnitudes and large masses of UCDs/DGTOs and YMCs it is appropriate to compare their properties to those of the most massive "normal" globular clusters, which are typically found around elliptical galaxies. While our Galaxy contains only a handful of clusters with masses in excess of 10^6 M$_\odot$, many more are expected to be present in giant elliptical galaxy halos. At the distance of 3.8 Mpc (Rejkuba 2004), the closest giant elliptical galaxy and easiest to observe in detail is NGC 5128 (=Centaurus A). Martini & Ho (2004) presented first measurements of masses and mass-to-light (M/L) ratios of 14 bright clusters in this galaxy. This sample was extended by adding new mass and M/L$_V$ measurements of bright globular clusters observed with UVES echelle spectrograph at ESO VLT, and with EMMI at ESO NTT (Rejkuba *et al.* 2007). We summarize here the main results of that work.

2. Velocity dispersion measurements

We have observed 23 bright clusters in NGC 5128 with UVES Echelle spectrograph at Kueyen VLT, and 10 clusters with EMMI at NTT. Velocity dispersions of these clusters were measured from the observed integrated light spectra using the well established cross-correlation technique. The templates for cross-correlation were bright stellar spectra, obtained during the same observing runs as our bright clusters. The cross-correlation of star-star spectra was used to determine instrumental profiles, and extensive simulations were done to assess the accuracy of the measurements and systematics. According to these simulations our velocity dispersion measurements were accurate to ~ 1 km/s for almost all our targets.

Our target's apparent V-band magnitudes ranged from 17.1 to 19.4, corresponding to absolute magnitudes of $M_V = -11.1$ to -8.9. The measured velocity dispersions were between 5 and 30.5 km/s. For comparison, the central velocity dispersion of the most massive Galactic globular cluster ω Cen is 22 km/s, while G1 in M31 has 27.8 km/s. Due to finite slit width (1") and the seeing of mostly ~ 0.8" during the observations, aperture corrections were small. The measured velocity dispersions were on average $\sim 6\%$ smaller than the central velocity dispersion.

In total we derived velocity dispersions for 22 clusters observed with UVES and 10 clusters observed with EMMI. Five clusters were in common to the two samples, hence we measured velocity dispersions for 27 different clusters. Our measurements are in very good agreement with those of Martini & Ho (2004), for 10 clusters in common, and the measurements for clusters observed with both UVES and EMMI agree within the errors. Nowadays there are in total 31 different globular clusters in NGC 5128 with accurate velocity dispersion measurements.

3. Masses and M/L ratios

Structural parameters, in particular the cluster half-mass radius r_h, are necessary to derive the masses. We have taken the structural parameters from the literature, where available, but for 7 clusters we derived their structural parameters by fitting King profiles to the high resolution FORS1 images (Rejkuba *et al.* 2007). In total we had structural parameters for 22 clusters in our sample.

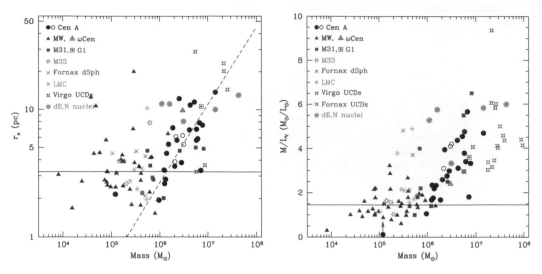

Figure 1. Left: Mass vs. effective (half-light) radius for massive star clusters in NGC 5128 is compared to DGTOs/UCDs and dE,N nuclei, as well as to the less massive globular clusters in the Local Group galaxies: the Milky Way, M31, M33, LMC, and Fornax dSph. A relation consistent with Faber-Jackson relation for hot galaxy sized systems emerges for objects more massive than $1 - 2 \times 10^6$ M$_\odot$. Dashed line shows the dependence for bright elliptical galaxies (Eq. 12 from Haşegan *et al.* 2005), while the solid line represents the median $r_e = 3.2$ pc for Galactic globular clusters. Right: Mass vs. M/L for massive clusters and UCDs. The solid line is the average dynamically determined $M/L = 1.45$ for Galactic globular clusters.

Masses were calculated using the Virial Theorem:

$$M_{vir} \simeq 2.5 \frac{3\sigma^2 r_h}{G} \tag{3.1}$$

where $3\sigma^2$ is the mean square velocity of the stars (assuming an isotropic velocity distribution), and r_h is the cluster half-mass radius, which is related to the half-light (effective) radius r_e through $r_e \approx 3r_h/4$. All the observed targets had radial velocities and structural parameters within the ranges expected for globular clusters belonging to NGC 5128, hence confirming their membership.

The least massive cluster, R122 (Rejkuba 2001), in our sample had 1.2×10^5 M$_\odot$. This mass is typical for Galactic globular clusters, but its magnitude ($M_V = -10.2$) indicates brighter than average cluster. However, its magnitude, structural parameters, and mass have larger errors due to the presence of stellar light, probably from a foreground source, which was visible in the integrated light spectra. The masses of all the other targets ranged from 1×10^6 to 1.4×10^7 M$_\odot$. Only the most luminous Galactic globular clusters have masses in excess of 10^6 M$_\odot$, and the most massive cluster in our Galaxy, ω Cen, has 3×10^6 M$_\odot$. UCDs/DGTOs and some YMCs have masses in the range of $10^6 - 10^8$ M$_\odot$ (Kissler-Patig, Jordán, & Bastian 2006).

Dividing the derived masses with V-band luminosities we derived M/L ratios for our targets. Leaving aside R122, M/L_V ratios range from 1.1 to 5.7, with an average of 3 ± 1, which is larger than average $M/L_V = 1.45$, determined dynamically for the Milky Way globular clusters (McLaughlin 2000).

4. Correlation relations

We examined relations of central velocity dispersion (σ_0) with luminosity and with mass for the bright clusters in NGC 5128. They are both within the errors similar to those obeyed by the Galactic globular clusters, but the relation between σ_0 and mass shows a small deviation for the most massive objects (for details see Rejkuba *et al.* 2007).

The most remarkable results are shown in Fig. 1. In both panels we compare the properties of massive globular clusters in Cen A with globular clusters and other massive evolved compact objects. The references to the literature data are as follows: filled dots are used for our Cen A sample (Rejkuba *et al.* 2007); open symbols are clusters from Martini & Ho (2004) that are not in common with our sample; old globular clusters of the Milky Way, Fornax dSph, and LMC are from McLaughlin & van der Marel (2005); M33 clusters from Larsen *et al.* (2002); M31 clusters from Dubath & Grillmair (1997); ω Cen and G1 data from Meylan *et al.* (1995), and Meylan *et al.* (2001), respectively; Virgo and Fornax UCDs are from Haşegan *et al.* (2005), and Hilker *et al.* (2007); and dE,N nuclei data are from Geha, Guhathakurta & van der Marel (2002).

The left panel shows that the massive globular clusters follow similar mass-effective radius relation as other compact evolved objects. The dashed line (which is not a fit to the data) indicates that this relation is essentially the same one obeyed by the bright ellipticals. The YMCs follow the same mass-radius relation (Kissler-Patig *et al.* 2006).

In the right panel Fig. 1 displays the relation between the mass and M/L ratio for bright globular clusters and UCDs/DGTOs. The ordinary globular clusters in the Milky Way, M31, M33 and other smaller galaxies do not show any dependence of mass on M/L, and the average dynamically determined $M/L_V = 1.45$ for the Galactic globular clusters is shown with the horizontal solid line (McLaughlin 2000). However, the clusters more massive than $1 - 2 \times 10^6$ $M_\odot$, including ω Cen and G1 start obeying a relation where the larger the mass the higher the M/L.

5. Discussion and conclusions

We have explored the possible causes for the increase in M/L_V for the bright massive clusters: (i) the stellar and dynamical evolution of clusters in the tidal field of giant elliptical galaxy; (ii) the metallicity; and (iii) the initial mass function (IMF).

Baumgardt & Makino (2003) have made extensive numerical N-body simulations of the dynamical evolution of star clusters in the tidal field of the Galaxy. They find that the clusters in the last 10% of their lifetime, before being dissolved, increase their M/L ratios very fast. These results were also reproduced by a simple analytical description of the evolution of star clusters in the tidal field by Lamers *et al.* (2005). Therefore, if all the most massive clusters were caught during the last stage of their lifetime, an increase in the M/L ratio could be expected. However, this interpretation has difficulties to reproduce the constant M/L ratios for lower mass clusters, at the same time as an increase for the most massive clusters. It is expected that preferentially first the low-mass clusters are dissolved, but the details depend somewhat on the orbits and the galactic potential. Moreover, the relaxation times for the clusters more massive than 10^6 $M_\odot$ are longer than the Hubble time, and orbits passing extremely closely to the galactic center would be necessary for their disruption in less than 13 Gyr. Given the wide range of galactocentric distances of our targets, ranging up to 23.4 kpc, dynamical evolution cannot reproduce the observed trend of mass vs. M/L_V.

The more metal-rich stellar population is expected to have higher M/L. While on average the observed clusters are more metal-rich than the Milky Way sample, the colors of Cen A clusters are not consistently redder for the more massive or for clusters with increasing M/L_V. Instead there is a scatter in the color-M/L_V diagram, showing that some blue, metal-poor clusters have rather high M/L_V ratios of 4–5.

It is possible to select a combination of stellar evolutionary model parameters (IMF, age, metallicity) in order to fit the M/L ratios for Cen A clusters with single stellar population. However, in that case the models do not fit the Galactic globular clusters, indicating that other parameters need to be taken into account when interpreting the M/L and r_{eff} relations with mass.

It is quite probable that these massive objects do not have the simple single-age stellar populations, but that due to higher masses and deeper potential wells, they could retain the gas in the early evolution and have a self-enriched second population. In that case the simple stellar population models that are typically assumed for globular clusters, would not be valid any more, and a more sophisticated modelling would be required. Unfortunately at the distance of NGC 5128, we do not have the possibility to obtain accurate photometry for many stars necessary to detect multiple stellar populations. An observational fact in favour of this possibility is that the globular clusters that are known to have multiple stellar populations, ωCen in our Galaxy, and G1 in the M31, lie along the same mass vs. r_{eff} and mass vs. M/L_V relations as our targets.

Finally we point out that the bright globular clusters in NGC 5128 connect the regions of diagrams occupied by "normal" globular clusters and more massive evolved UCDs/DGTOs indicating perhaps a connection in formation and evolution history between these classes of objects. Also the mass-radius relation in YMCs could survive up to much later times, if these young massive clusters evolve into some of the most massive globular clusters.

References

Baumgardt, H. & Makino, J. 2003 *MNRAS* 340, 227

Djorgovski, S. & Meylan, G. 1994, *AJ* 108, 1292

Dubath, P. & Grillmair, C. J. 1997, *A&A* 321, 379

Geha, M., Guhathakurta, P. & van der Marel, R. P. 2002, *AJ* 124, 3073

Haşegan, M., Jordán, A., Côté, P., *et al.* 2005, *ApJ* 627, 203

Hilker, M., Infante, L., Vieira, G., *et al.* 1999, *A&AS*, 134, 75

Hilker, M. 2006, in: T. Richtler & S. Larsen (eds.) *"Globular Clusters - Guides to Galaxies"*, *Concepcion, Chile* (Springer), astro-ph/0605447

Hilker, M., Baumgardt, H., Infante, L., *et al.* 2007, *A&A* 463, 119

King, I. R. 1966, *AJ* 71, 64

Kissler-Patig, M., Jordán, A. & Bastian, N. 2006, *A&A* 448, 1031

Lamers, H. J. G. L. M., Gieles, M., Bastian, N., *et al.* 2005 *A&A* 441, 117

Larsen, S. S., Brodie, J. P, Sarajedini, A. & Huchra, J. P. 2002, *AJ* 124, 2615

Martini, P, & Ho, L. C. 2004, *ApJ* 610, 233

Meylan, G., Mayor, M., Duquennoy, A. & Dubath, P. 1995, *A&A* 303, 761

Meylan, G., Sarajedini, A., Jablonka, P., *et al.* 2001, *AJ* 122, 830

Mieske, S., Hilker, M., Jordán, A., *et al.* 2007 *A&A* 472, 111

McLaughlin, D. E. 2000, *ApJ* 539, 618

McLaughlin, D. E. & van der Marel, R. P. 2005, *ApJ* 539, 618

Rejkuba, M. 2001, *A&A* 369, 812

Rejkuba, M. 2004, *A&A*, 413, 903

Rejkuba, M., Dubath, P., Minniti, D. & Meylan, G. 2007, *A&A* 469, 147

Wehner, E. & Harris, W. E. 2007, *ApJ Lett* accepted, arXiv:0708.1514v1

Dynamical Evolution of Dense Stellar Systems
Proceedings IAU Symposium No. 246, 2007
E. Vesperini, M. Giersz & A. Sills, eds.
© 2008 International Astronomical Union
doi:10.1017/S1743921308016086

Slow Evolution of a System of Satellites Induced by Dynamical Friction

Serena E. Arena and Giuseppe Bertin

Dipartimento di Fisica, Università degli Studi di Milano,
via Celoria 16, I-20133 Milano, Italy
E-mail: serena.arena@unimi.it, giuseppe.bertin@unimi.it

Abstract. The evolution induced by dynamical friction on a spherical shell of rigid satellites interacting directly with the particles sampling the host elliptical galaxy is followed by means of N-body simulations for a variety of shell–galaxy configurations.

Keywords. stellar dynamics, methods: n-body simulations, galaxies: kinematics and dynamics

1. Introduction

Elliptical galaxies often host significant systems of globular clusters. The properties of the clusters change with time because of stellar evolution, internal stellar dynamical effects, and tidal interactions with the host galaxy. In addition, the distribution and kinematics of such systems can evolve because of the dynamical friction felt by the individual globular clusters while moving through the host galaxy.

Here, by means of N-body simulations, we focus on the evolution induced by dynamical friction by considering a system of rigid satellites initially placed on a quasi-spherical shell. With respect to the study of friction on a single heavy object, this configuration represents an intermediate step in the direction of studying a realistic three-dimensional distribution of globular clusters, with the advantage of defining a dynamical problem in which quasi-spherical symmetry is preserved in the course of evolution and of testing to what extent dynamical friction can be thought of as a basically local process. We consider a variety of such shell configurations, characterized by different mass and phase space properties, and investigate their evolution in realistic galaxy models.

The realistic galaxy models adopted are a one-parameter family, derived from the $f^{(\nu)}$ distribution function, with a central isotropic core and a radially anisotropic envelope (Trenti & Bertin (2005)). The satellites are rigid Plummer spheres of radius R_f, initially distributed in a spherical shell of thickness $2R_{shell}$, positioned at a distance $r_{shell}(0)$ from the galaxy center. We have considered two configurations of the shell. The first has a density profile given by $\rho_{shell}(r) = \rho_{shell_0} e^{-4[(r-r_{shell}(0))/R_{shell}]^2}$, with satellites on circular orbits; in the second configuration, satellites move on eccentric orbits with positions and velocities extracted from the distribution function of the galaxy.

The evolution of the system is followed by means of the N–body collisionless (mean–field) particle–mesh code described in Trenti (2005). The galaxy is made of 2.5×10^5 particles and the shell of $N_f = 20$ and 100 satellites. For further details, see Arena & Bertin (2007) and Arena (2007).

2. The evolution induced in the spherical shell of satellites

Each satellite loses energy and angular momentum because of dynamical friction with the particles composing the galaxy (that represent stars and/or dark matter). As a re-

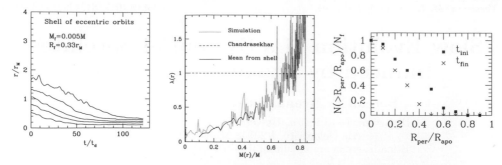

Figure 1. Left frame: evolution of the Lagrangian radii for the mass distribution of a shell made of satellites on eccentric orbits. Middle frame: ratio λ of the Coulomb logarithm measured in the simulations to that predicted by the Chandrasekhar theory; the dotted line refers to a single satellite. Right frame: evolution of the eccentricity of the satellite orbits, see text for details. The scales r_M, M, and t_d are the host galaxy half–mass radius, mass, and dynamical time.

sult, the shell moves slowly towards the center of the galaxy, while its thickness decreases (left frame of Fig. 1). In the final stage of evolution, the shell settles down in a quasi–equilibrium configuration without actually reaching the galaxy center. Shells made of more massive or less extended satellites remain at a larger distance from the galaxy center. The shell thickness decreases more for shells made of less extended satellites. Shells starting at different distances from the galaxy center reach the same final quasi–equilibrium configuration. The fall time observed in the simulations is longer than that predicted by the Chandrasekhar theory of dynamical friction; indeed, the measured Coulomb logarithm is smaller (middle frame of Fig. 1).

In the simulation of shells with satellites on eccentric orbits, we have studied the process of circularization. The right frame of Fig. 1 illustrates the distribution of the ratio of the pericenter to the apocenter R_{per}/R_{apo} for the satellite orbits. The quantity $N(> R_{per}/R_{apo})$ is the number of satellites with ratio R_{per}/R_{apo} greater than the value shown on the x–axis. By comparing the initial distribution of points (filled squares) with that of the final configuration (crosses), we conclude that the number of satellites with eccentric orbits *increases* during evolution. The host galaxy responds non–adiabatically to the infalling shell with a decrease of its density concentration and of its pressure anisotropy in the radial direction (Arena *et al.*(2006)).

The characteristics of the entire process depend on the properties of the host galaxy, especially on its density concentration. Indeed, in low-density concentration galaxy models (e.g. polytropes; see Bertin, Liseikina & Pegoraro (2003)) the final configuration of the shell is characterized by a larger thickness and a larger distance from the galaxy center with respect to the evolution in models with higher density concentration (e.g. the $f^{(\nu)}$ models). The study of the evolution induced by dynamical friction of a shell of satellites is a first step towards the study of the interaction with the host galaxy of a more realistic three–dimensional distribution of satellites.

References

Arena, S. E., Bertin, G., Liseikina, T., & Pegoraro, F. 2006, *A&A*, 453, 9
Arena, S.E., & Bertin, G. 2007, *A&A*, 463, 921
Arena, S. E. 2007, Ph.D. Thesis, Università degli Studi di Milano, Milano
Bertin, G., Liseikina, T., & Pegoraro, F. 2003, *A&A*, 405, 73
Trenti, M. 2005, Ph.D. Thesis, Scuola Normale Superiore, Pisa
Trenti, M., & Bertin, G. 2005, *A&A*, 429, 161

Dynamical Evolution of Dense Stellar Systems
Proceedings IAU Symposium No. 246, 2007
E. Vesperini, M. Giersz & A. Sills, eds.

© 2008 International Astronomical Union
doi:10.1017/S1743921308016098

Sizes of Confirmed NGC 5128 Globular Clusters

Doug Geisler[1], M. Gómez[1]
K. A. Woodley[2], W. E. Harris[2]
and G. L. H. Harris[3]

[1] Departamento de Fisica, Universidad de Concepción, Concepción, Chile
email: dgeisler@astro-udec.cl

[2] Department of Astronomy, McMaster University, Toronto, Ontario, Canada

[3] Department of Astronomy, University of Waterloo, Toronto, Ontario, Canada

Abstract. We present results from a new wide-field study of the NGC 5128 globular cluster system. We have obtained new high resolution images with the Magellan 6.4m + IMACS camera. Our images cover an area of 1.2x1.2 sq. degrees and have a seeing of 0.45". This allows us to not only resolve most of the globular clusters (GCs) but also derive their structural parameters. These are combined with existing Washington photometry in order to select by metallicity. We present here results for a subsample of 359 GCs which includes all currently confirmed GC members of the system. Our derived sizes are in very good agreement with those derived from ACS data. We find, as expected, that the metal-rich GCs in the inner regions ($r < 10'$) are 26% smaller than their metal-poor components, but in the outer region this normal trend is reversed. We compare our GCs to previous results for GCs, UCDs, etc. in the luminosity - size plane and find substantial overlap between different types of objects, indicating more of a continuum in these properties.

Keywords. galaxies: star clusters

1. Introduction

At a distance of ~ 3.5 Mpc, NGC 5128 is the nearest giant elliptical. It contains several thousand globular clusters (GCs). They are ideal targets for addressing a number of important questions regarding GC and galaxy formation.

2. Observations and Analysis

We have recently obtained superb new imaging data for this system with the Magellan 6.5m + IMACS. Our images cover an area of 1.2x1.2 sq. degrees in a set of 25 fields 15.4' on a side. Pixels are 0.11" and the seeing was 0.45". This allows us to cover essentially the entire GC system.

This data allows us to not only resolve most of the GCs but also derive their structural parameters, using the ISHAPE program from S. Larsen. We have first applied this technique to a total of 359 objects which includes ALL currently confirmed GC members from radial velocity data.

The sizes we derive show excellent agreement with those found from HST/ACS data using an entirely independent technique. This gives us increased confidence in our sizes. We then combine our structural parameters with our existing Washington photometry which provides an excellent metallicity indicator.

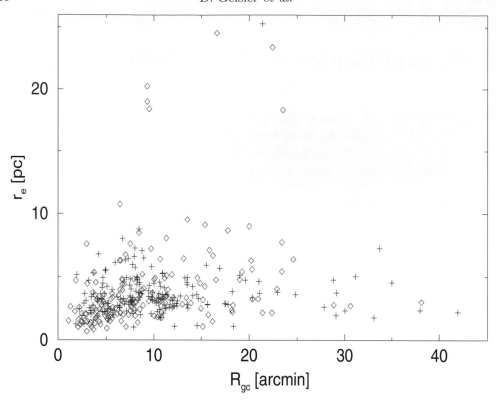

Figure 1. GC size (in pc) vs. galactocentric distance (in ') for 359 confirmed GCs in NGC 5128. (Red) squares are GCs more metal-rich than [Fe/H]=-1 and (blue) crosses are GCs more metal-poor than this limit.

3. Results

We divide our sample into metal-rich (RGCs) and metal-poor (BGCs) clusters and plot their sizes vs. galactocentric radius in Fig. 1. In the inner 10', we find the usual result: the 107 BGCs are 26% larger than the 107 RGCs. BUT in the outer region, the 79 BGCs are 23% SMALLER than the 58 RGCs! Note the substantial fraction of RGCs with sizes > 8pc in this region. This is the first time such a clear trend has been found. Perhaps projection effects are responsible for this difference.

We finally compare our clusters to previous results for star clusters, UCDs, etc . in the M_V :size plane. Although one speaks of different types of objects, implying different behavior in such a diagram, we instead find substantial overlap between different types of objects, indicating more of a continuum in these properties.

Acknowledgements

D.G. gratefully acknowledges support from the Chilean *Centro de Astrofísica* FONDAP No. 15010003.

Dynamical Evolution of Dense Stellar Systems
Proceedings IAU Symposium No. 246, 2007
E. Vesperini, M. Giersz & A. Sills, eds.

© 2008 International Astronomical Union
doi:10.1017/S1743921308016104

Ultra-Compact Dwarf Galaxies – More Massive than Allowed?

Michael Hilker[1], S. Mieske[1], H. Baumgardt[2] and J. Dabringhausen[2]

[1] ESO, Karl-Schwarzschild-Str. 2, 85748 Garching bei München, Germany
email: mhilker@eso.org, smieske@eso.org

[2] AIfA, Universität Bonn, Auf dem Hügel 71, 53121 Bonn, Germany
email: holger@astro.uni-bonn.de, joedab@astro.uni-bonn.de

Abstract. Dynamical mass estimates of ultra-compact dwarfs galaxies and massive globular clusters in the Fornax and Virgo clusters and around the giant elliptical Cen A have revealed some surprising results: 1) above $\sim 10^6 M_\odot$ the mass-to-light (M/L) ratio increases with the objects' mass; 2) some UCDs/massive GCs show high M/L values (4 to 6) that are not compatible with standard stellar population models; and 3) in the luminosity-velocity dispersion diagram, UCDs deviate from the well-defined relation of "normal" GCs, being more in line with the Faber-Jackson relation of early-type galaxies. In this contribution, we present the observational evidences for high mass-to-light ratios of UCDs and discuss possible explanations for them.

Keywords. galaxies: star clusters, galaxies: dwarf, galaxies: kinematics and dynamics

1. Introduction

The so-called ultra-compact dwarf galaxies (UCDs) are very massive ($10^6 M_\odot < M < 10^8 M_\odot$), old, compact stellar systems that were discovered in nearby galaxy clusters about a decade ago (Hilker *et al.* 1999, Drinkwater *et al.* 2000). Their nature is unknown yet. Maybe they are remnant nuclei of disrupted galaxies, or maybe they are merged stellar super-clusters formed in interacting galaxies. Regardless of what UCDs actually are, some properties divide them from "ordinary" globular clusters (GCs). The half-light radii of UCDs scale with luminosity reaching ~ 90 pc for the most massive UCDs. Unlike GCs, their densities within the half-light radii are not increasing with mass but stay at a constant level or even decrease. Thus UCDs are not that compact at all when compared to $10^6 M_\odot$ GCs, but certainly much denser than dwarf ellipticals of comparable mass.

2. Mass determinations and results

To estimate the masses of UCDs a new modeling program has been developed that allows a choice of different representations of the surface brightness profile (i.e. Nuker, Sersic or King laws) and corrects the observed velocity dispersions for observational parameters (i.e. seeing, slit size). The derived dynamical masses are compared to those expected from stellar population models. For more details, see Hilker *et al.* (2007).

The masses, central densities and mass-to-light (M/L) ratios of different hot stellar systems (GCs, UCDs, dEs, bulges and ellipticals) were compared with each other (Dabringhausen *et al.* 2007, in prep.). The findings are as follows: 1) In the central density vs. mass plane, there seems to be an upper limit of about $10^4 M_\odot/pc^3$ for GCs of $\sim 10^6 M_\odot$. UCDs scatter towards lower densities with increasing mass. 2) In the M/L vs. mass plane, the M/L ratio increases with mass above $\sim 10^6 M_\odot$ (see Fig. 1, right panel), reaching values typical for bulges and ellipticals. 3) When plotting a normalised M/L

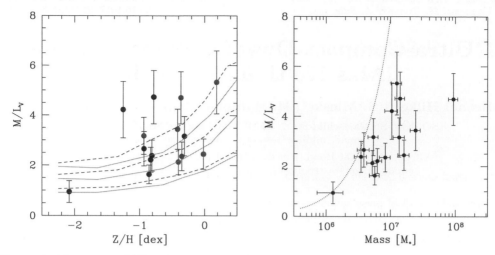

Figure 1. Most recent M/L_V determinations for Fornax UCDs from FLAMES/UVES observations (Mieske *et al.* 2008). Left: SSP models (5, 9 and 13 Gyr) from Bruzual & Charlot (2003, solid) and Maraston (2005, dashed). Right: The dotted line indicates the observational limit.

ratio (taking out the metallicity dependence on M/L) vs. mass, the objects more massive than a few times $10^6 M_\odot$ show systematically higher M/L values than the lower mass 'normal' GCs. These high values cannot easily be explained with standard single stellar population models (see Fig. 1, right panel). Interestingly, the transition from low-M/L to high-M/L objects corresponds to the mass regime (10^6-$10^7 M_\odot$) where the relaxation time at the half-light radius exceeds a Hubble-time.

3. Possible explanations for unusually high M/L ratios

1) Dark matter: This would imply a very high DM density within the core radius of UCDs. A cuspy NFW halo with 10^8-$10^{12} M_\odot$ would be needed. Might UCDs be surviving dense low mass DM sub-structures?

2) Tidal heating: UCDs might be out of dynamical equilibrium (Fellhauer & Kroupa 2006). However, very eccentric orbits would be needed to observe high M/L-UCDs.

3) Top-heavy IMF: Remnants of massive stars (stellar BHs, neutron stars and white dwarfs) might contribute to the unseen mass and increase the M/L value. An IMF slope of $\alpha = -1$ to -1.5 for $M > 1 M_\odot$ would be needed.

4) Bottom-heavy IMF: Many low mass stars might contribute to the high M/L value ($\alpha = -2.35$ low mass slope might explain it).

For a detailed discussion of these points, see Dabringhausen *et al.* (2007, in prep.) and Mieske *et al.* (2007, in prep.).

References

Bruzual, G. A., Charlot, S. 2003, *MNRAS* 344, 1000
Drinkwater, M. J., Jones, J. B., Gregg, M. D., Phillipps, S. 2000, *PASA* 17, 227
Fellhauer, M., Kroupa, P. 2006, *MNRAS* 367, 1577
Hilker, M., Baumgardt, H., Infante, L., *et al.* 2007, *A&A* 463, 777
Hilker, M., Infante, L., Vieira, G., Kissler-Patig, M., Richtler, T. 1999, *A&AS* 134, 75
Maraston, C. 2005, *MNRAS* 362, 799

Dynamical Evolution of Dense Stellar Systems
Proceedings IAU Symposium No. 246, 2007
E. Vesperini, M. Giersz & A. Sills, eds.

© 2008 International Astronomical Union
doi:10.1017/S1743921308016116

GMOS Spectroscopy of Globular Clusters in Dwarf Elliptical Galaxies

Bryan W. Miller[1], Jennifer Lotz[2], Michael Hilker[3], Markus Kissler-Patig[3] and Thomas Puzia[4]

[1] Gemini Observatory, Casilla 603, La Serena, Chile
email: bmiller@gemini.edu

[2] National Optical Astronomy Observatory, 950 N. Cherry Ave., Tucson, AZ 85719, USA
email: lotz@noao.edu

[3] European Southern Observatory, Karl-Schwarzschild-Str.2 85748 Garching, Germany
email: mhilker@eso.org, mkissler@eso.org

[4] Herzberg Institute of Astrophysics, 5071 West Saanich Road, Victoria, BC V9E 2E7, Canada
email: thomas.puzia@nrc.ca

Abstract. We present a Gemini/GMOS program to measure spectroscopic metallicities and ages of globular clusters (GCs) and nuclei in dwarf elliptical galaxies in the Virgo and Fornax Clusters. Preliminary results indicate that the globular clusters are old and metal-poor, very similar to the GCs in the Milky Way halo. The nuclei tend to be more metal-rich than the globular clusters but more metal-poor and older, on average, than the stars in the bodies of the galaxies. The [α/Fe] ratio appears to be solar for the GCs, nuclei, and dEs, but the uncertainties do not exclude some globular clusters from being enhanced in alpha elements.

Keywords. galaxies: star clusters, galaxies: dwarf, galaxies: nuclei

---

Dwarf elliptical (dE) galaxies are the dominant type of galaxy in galaxy clusters and they may be related to pre-galactic fragments in hierarchical formation scenarios. They also contain relatively large numbers of globular clusters (GCs; Miller & Lotz 2007). Photometry has shown that the GCs have $(V - I)$ colors similar to those of old, metal-poor Galactic clusters but that the mean color increases with galaxy luminosity (Lotz *et al.* 2004). The galaxies themselves have redder colors and the nuclei fall in between. However, from photometry alone it is not possible to distinguish the effects of age and metallicity on the colors.

Therefore, multi-object spectroscopy of the dE GC photometric candidates and nuclei were carried out with both GMOS instruments on the Gemini telescopes during four semesters between 2002A through 2004B. In all cases the B600 grating was used with 0.75 arcsec slits, giving a wavelength coverage from approximately 3500Å to 6500Å at $R \sim 1400$. Between one and three masks were observed for each galaxy depending on the number of candidates. The typical exposure time per mask is about five hours (some objects are observed in more than one mask) and the image quality varied between 0.7 and 0.85 arcsec. The data were reduced with the Gemini IRAF package with additional steps for handling bad pixel masks, relative quantum efficiency corrections between the three CCDs, and correcting the spectral shapes for slit losses due to the differences between the parallactic angles and the PAs of the slits. The spectra of many of the faint GCs have signal-to-noise too faint for line-strength analysis. Therefore, the spectra of all GC candidates for a given galaxy with radial velocities within about 200 km/sec of the velocity of the nucleus were combined to produce a "mean" GC spectrum. These mean spectra have $S/N \sim 30 - 40$ and are used for measuring line indices.

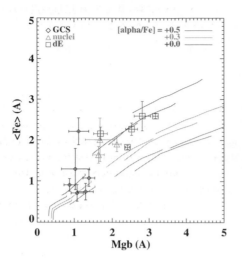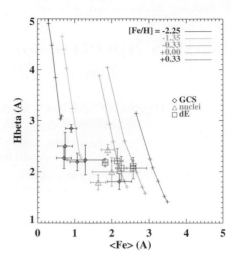

Figure 1. Line indices for GCs and nuclei from GMOS spectroscopy are compared with dE background light (Geha *et al.* 2003) and stellar evolutionary models (Thomas *et al.* 2003). The left plot shows ⟨Fe⟩ versus Mgb with models for different [α/Fe]. The right plot uses models with [α/Fe] = 0.0 and shows that the GCs are old and metal-poor while the dEs are more metal-rich and somewhat younger. The nuclei appear to have ages and metallicities between those of the GCs and dE field stars.

A preliminary comparison of the line indices of the GCs, nuclei, and background light in dEs is given in Fig. 1. Stellar evolutionary model of Thomas *et al.* (2003) are overplotted. For the nuclei and field stars [α/Fe] ≈ 0, indicating a star formation timescale longer than 1 Gyr. The uncertainties in the GC data make it difficult to determine [α/Fe]. It is consistent with the solar value but some GCs could be α enhanced. The right plot in Fig. 1 give models of different ages and metallicities in the ⟨Fe⟩ − Hβ diagram. In general the GCs are old and metal-poor, the nuclei are more metal-rich, and the galaxy field stars are even more metal-rich, −0.5 ≲ [Fe/H] ≲ 0, and a few Gigayears younger. Thus, the color-trends are mostly a function of metallicity but there are also important age differences that the photometry can not discern. Future work includes a thorough analysis of the line indices and M/L estimates from the velocity dispersions.

Acknowledgements

Based on observations obtained at the Gemini Observatory, which is operated by the Association of Universities for Research in Astronomy, Inc., under a cooperative agreement with the NSF on behalf of the Gemini partnership: the National Science Foundation (United States), the Particle Physics and Astronomy Research Council (United Kingdom), the National Research Council (Canada), CONICYT (Chile), the Australian Research Council (Australia), CNPq (Brazil), and CONICET (Argentina).

References

Geha, M., *et al. AJ*, 126, 1794
Lotz, J. M., Miller, B. W., & Ferguson, H. C. 2004, *ApJ*, 613, 262
Miller, B. W. & Lotz, J. M. 2007, *ApJ*, in press (astro-ph/0708.2511)
Thomas, D., *et al. MNRAS*, 339, 897

Dynamical Evolution of Dense Stellar Systems
Proceedings IAU Symposium No. 246, 2007
E. Vesperini, M. Giersz & A. Sills, eds.

Formation of Galactic Nuclei by Globular Cluster Merging

P. Miocchi and R. Capuzzo Dolcetta

Department of Physics, Universitá di Roma "La Sapienza", P.le A. Moro, 5, I-00185, Italy
email: miocchi,roberto.capuzzodolcetta@uniroma1.it

Abstract. Recent HST observations have revealed that compact sources exist at the centers of many galaxies across the Hubble sequence. These sources are called "nuclear star clusters" (NCs), because their structural properties and scaling relationships are similar to those of globular clusters (GCs). It has been also found that the relationship between the masses of NCs and that of the host galaxies is similar to that obeyed by supermassive black holes (SBHs). In this observational frame, the hypothesis that galactic nuclei may be the remains of GCs driven inward to the galactic center by dynamical friction and there merged, finds an exciting possible confirm. In this short paper we report of our recent results on GC mergers obtained by mean of detailed N-body simulations.

Keywords. stellar dynamics, methods: n-body simulations, globular clusters: general, galaxies: nuclei, galaxies: kinematics and dynamics, galaxies: star clusters

1. Introduction

In many early-type galaxies (Côté *et al.* 2006) and late-type spirals (Böker *et al.* 2002, Rossa *et al.* 2006) evident compact nuclei have been resolved, showing a luminosity distribution much better fitted by an extended (King's) profile rather than by a point source (Fig. 1). Such compact nuclei are likely the low-mass counterparts of nuclei hosting SBHs detected in bright galaxies (see also Wehner & Harris 2006). It is clear that these characteristics of galaxies nuclei well fit into a "dissipationless" scenario of multiple GCs merging in the inner galactic regions. So far, not many N-body simulations have been presented in the literature that study the dissipationless formation of NCs. Among these, we remind those by Fellhauer & Kroupa (2005) finding that super-massive star clusters, like W3 in NGC 7252, are very likely the merging remnants of smaller systems. Also Bekki *et al.* (2004) examined the merger formation scenario of NCs, though with simulations done in a simplified way, neglecting the role of the galactic external potential and with a relatively low resolution.

2. The Simulations

Here we report briefly of some of the results presented in Capuzzo-Dolcetta & Miocchi (2007b, *in preparation*), where we studied whether and how the merging of various massive GCs decayed by dynamical friction in the inner galactic region may occur. These results give substance to the interpretation of the formation of galaxies nuclei via merging of decayed GCs (see Capuzzo Dolcetta & Vicari (2005) and references therein). We consider GCs as N-body systems moving within a triaxial galaxy represented by an analytical potential, subjected also to dynamical friction. We studied the merging process occurring among four GCs already decayed within 100 pc from the galactic center. The galaxy where the GCs move is represented by a self-consistent triaxial potential. Simulations are done with the parallel 'ATD' N-body code (Miocchi & Capuzzo Dolcetta,

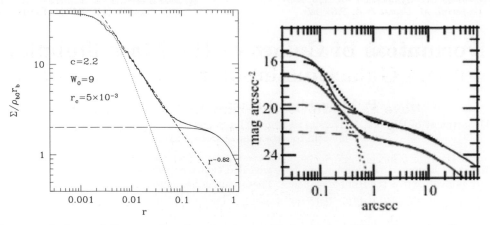

Figure 1. Left panel: Projected surface density profile for the last NC configuration (in proper units), overlapped to the galaxy background profile (long dashed line) so to give the total surface density (solid line). The short dashed line indicates the $r^{-0.82}$ behaviour, while the dotted one is the core best-fit King profile whose parameters are also reported. Right panel: surface brightness profile of VCC 1661 (black squares); the lower solid line is the superposition of a fitting Sérsic model for the galaxy (long dashed) and a King model for the core region (dotted); the upper solid line is the fitting with a point source instead of the King model. The two curves are translated for display convenience. From Côté *et al.* 2006.

2002), with a total of 10^6 particles. See Capuzzo-Dolcetta & Miocchi (2007a) for more details.

3. Results

After ~ 15 Myr, corresponding to ~ 20 galactic core crossing times, the merging is completed and the resulting system attains a quasi-equilibrium configuration. This gives a total projected density profile in the nuclear region that is remarkably similar to those recently observed in the central regions of nucleated early-type galaxies in the Virgo cluster (Fig. 1), as well as in nearby late-type spirals. The final NC morphology is that of an axisymmetric ellipsoid (axial ratios 1.4:1.4:1, ellipticity ~ 0.3) without figure rotation. In the velocity dispersion-mass plane, the NC is located closer to the scaling relation followed by GCs than to that of elliptical galaxies.

References

Bekki, K., Couch, W. J., Drinkwater, M. J. & Shioya, Y. 2004, *ApJL* 610, L13

Böker, T., Laine, S., van der Marel, R. P., Sarzi, M., Rix, H. -W., Ho, L. C. & Shields, J. C. 2002, *AJ*, 123, 1389

Capuzzo Dolcetta, R. & Miocchi, P. 2007a, in: J. Knapen, T. Mahoney & A. Vazdekis (eds.), *Pathways Through an Eclectic Universe* (San Francisco: ASP), in press (astro-ph/0709.0455)

Capuzzo-Dolcetta, R. & Vicari, A. 2005, *MNRAS* 356, 899

Côté, P. *et al.* 2006, *ApJS* 165, 57

Fellhauer, M. & Kroupa, P. 2005, *MNRAS* 359, 223

Miocchi, P. & Capuzzo Dolcetta, R. 2002, *A&A* 382, 758

Rossa, J., van der Marel, R.P., Boeker, T., Gerssen, J., Ho, L.C., *et al.* 2006, *AJ* 132, 1074

Wehner, E. H. & Harris, W. 2006, *ApJL* 644, L17

Dynamical Evolution of Dense Stellar Systems
Proceedings IAU Symposium No. 246, 2007
E. Vesperini, M. Giersz & A. Sills, eds.

Dynamical Evolution of the Mass Function of the Galactic Globular Cluster System

Jihye Shin[1], Sungsoo S. Kim[1] & Koji Takahashi[2]

[1]Dept. of Astronomy and Space Science, Kyung Hee University, Korea
email: jhshin@ap1.khu.ac.kr

[2]Dept. of Informational Society Studies, Saitama Institute of Technology, Japan

Abstract. Using the most advanced anisotropic (2D) Fokker-Planck (FP) models, we calculate the evolution of the mass functions of the Galactic globular cluster system (GCMF). Our models include two-body relaxation, binary heating, tidal shocks, dynamical friction, stellar evolution, and realistic cluster orbits. We perform 2D-FP simulations for a large number of virtual globular clusters and synthesize these results to study the relation between the initial and present GCMFs. We found two probable IGCMFs that eventually evolve into the Milky Way GCMF : truncated power-law, and log-normal model with higher initial low mass limit and peak mass than the earlier studies.

Keywords. Globular Clusters, Fokker-Planck method.

1. Introduction

There have been many studies on the GCMF with various simulation methods. Vesperini (1997), Vesperini & Heggie (1997), Vesperini (1998), Baumgardt (1998), and Baumgardt & Makino (2003) used an N-body method, Okazaki & Tosa (1995), Ostriker & Gnedin (1997), Vesperini (1997), and Fall & Zhang (2001) used simple analytical models, and Murali & Weinberg (1997a), Murali & Weinberg (1997b), and Murali & Weinberg (1997c) used an isotropic FP model. However, not all of the the related physics (i.e., tidal shocks, realistic orbits, dynamical friction, stellar evolution, and a wide range of cluster parameters) were considered by any of the previous studies.

To consider all of the related physics for the evolution of the star cluster, we use the most advanced 2D-FP developed by Takahashi & Lee (2000). The main difference of our calculation from the past studies is the realistic cluster orbit moving in the axisymmetric Galactic potential with eccentricities. We adjust the magnitude of the tidal shocks and dynamical friction with time according to the realistic velocity and galactocentric distance. Thanks to relatively short computing times of 2D-FP, we perform a total of 578 simulations with various initial conditions.

2. Results

The two-body relaxation causes the low mass part of the GCMF to decrease faster, and this changes the shape of GCMF to be log-normal even if the initial GCMF (IGCMF) is a power-low function. The contribution of the external effects on the GCMF is found to be more important than in earlier studies, and this is mainly because the realistic orbits considered by us result in faster cluster evolution. This generally causes our peak cluster masses smaller than earlier studies.

The IGCMFs that best fit the Milky Way GCMF at 12Gyr are a power-law IGCMF with $M_{low} = 10^6 M_\odot$ and a log-normal IGCMF with $M_{peak} = 10^6 M_\odot$ (M_{low} is the lower

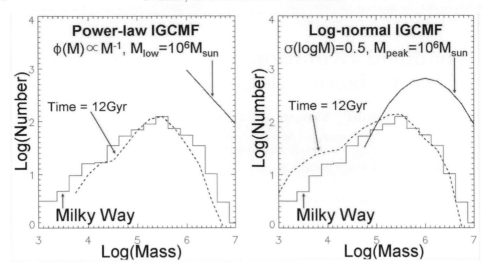

Figure 1. Two probable initial globular cluster mass functions that eventually evolve into the Milky Way GCMF.

cluster mass cutoff and M_{peak} is the peak cluster mass). These IGCMFs have higher initial M_{low} or M_{peak} than the earlier studies. This is caused by the contribution of external effects that effectively shift the overall location of GCMF to the lower mass side. We find that the shape of the GCMF is mainly determined by the massive clusters with $10^6 M_\odot \leqslant M \leqslant 10^7 M_\odot$. In spite of the good fit to the Milky Way GCMF, the feasibility of the power-law IGCMF truncated at $10^6 M_\odot$ is somewhat questionable because there is no proper theoretical base for such a high M_{low}. Therefore, we propose the log-normal model with $M_{peak} = 10^6 M_\odot$ as a best-fit IGCMF for the Milky Way. Log-normal IGCMFs could be naturally generated by expulsion from the proto-cluster of star forming gas due to supernova activity (Parmentier & Gilmore 2007).

References

Baumgardt, H. 1998, *A&A*, 330, 480
Baumgardt, H. & Makino, J. 2003, *MNRAS*, 340, 227
Fall, S. & Zhang, Q. 2001, *ApJ*, 561, 751
Gnedin, O. Y., Hernquist, L., & Ostriker, J. P. 1999, *ApJ*, 514, 109
Gnedin, O. Y., Lee, H. M., & Ostriker, J. P. 1999, *ApJ*, 522, 935
Murali, C. & Weinberg, M. D. 1997a, *MNRAS*, 288, 749
Murali, C. & Weinberg, M. D. 1997b, *MNRAS*, 288, 767
Murali, C. & Weinberg, M. D. 1997c, *MNRAS*, 291, 717
Okazaki, T. & Tosa, M. 1995, *MNRAS*, 274, 48
Ostriker, J. P. & Gnedin, O. Y. 1997, *ApJ*, 474, 223
Parmentier, G. & Gilmore, G. 2007, *MNRAS*, 377, 352
Takahashi, K. & Lee, H. M. 2000 *MNRAS*, 316, 671
Vesperini, E. 1997 *MNRAS*, 287, 915
Vesperini, E. & Heggie, D. C. 1997 *MNRAS*, 289, 898
Vesperini, E 1998 *MNRAS*, 299, 1019

Part 8

Computational Aspects of Simulations of Dense Stellar Systems

Dynamical Evolution of Dense Stellar Systems
Proceedings IAU Symposium No. 246, 2007
E. Vesperini, M. Giersz & A. Sills, eds.

Dancing with Black Holes

S. J. Aarseth

Institute of Astronomy, Madingley Road, Cambridge CB3 0HA, UK
email: sverre@ast.cam.ac.uk

Abstract. We describe efforts over the last six years to implement regularization methods suitable for studying one or more interacting black holes by direct N-body simulations. Three different methods have been adapted to large-N systems: (i) Time-Transformed Leapfrog, (ii) Wheel-Spoke, and (iii) Algorithmic Regularization. These methods have been tried out with some success on GRAPE-type computers. Special emphasis has also been devoted to including post-Newtonian terms, with application to moderately massive black holes in stellar clusters. Some examples of simulations leading to coalescence by gravitational radiation will be presented to illustrate the practical usefulness of such methods.

Keywords. celestial mechanics, methods: n-body simulations

1. The chain concept

In the study of strong gravitational interactions, utilization of the chain data structure can be very beneficial. Over many years, the original chain regularization method (Mikkola & Aarseth 1993) has proved to be effective in star cluster simulations containing binaries. As we shall see in the following, it is also a useful tool in connection with time transformations which do not employ the usual coordinates. By introducing one or more dominant masses these advantages become more apparent. Such problems fall naturally into three classes according to the number of massive objects and each class requires special attention. At the simplest level we have the case of one central massive body which dominates the motion of other members within a certain distance. The role of the reference body is readily seen in the case of three interacting particles which can be studied by three-body regularization (Aarseth & Zare 1974). This idea was extended to an arbitrary membership (Zare 1974). However, a natural application was lacking until the problem of black holes (BHs) became a challenge for simulators in recent years. The aptly named wheel-spoke regularization (Aarseth 2003a) has now been adapted to study compact subsystems containing a single massive object.

Historically speaking, a special method for a BH binary was implemented in an *N*-body code first. Here the main idea is based on a time-transformed leapfrog scheme (TTL) suitable for dealing with large mass ratios (Mikkola & Aarseth 2002). Remarkably, this method yields machine precision for unperturbed two-body motion. Although regularized chain coordinates are not employed directly, the accuracy is improved by using relative quantities with respect to the nearest massive body.

Alternative methods may be needed for problems involving more than two massive objects. The recent algorithmic regularization (Mikkola & Tanikawa 1999; Preto & Tremaine 1999; Mikkola & Merritt 2006) appears to be a promising way of studying such systems. Indeed, the masses only play a kinematical role in one formulation, suggesting that it may be applicable to systems with large mass ratios. However, it should be emphasized that for practical reasons any of the above methods are of necessity limited to relatively small particle numbers, or in other words, compact subsystems.

2. Post-Newtonian formulation

Relativistic effects in stellar systems are usually associated with high densities or very massive objects. However, quite large values of the eccentricity may also occur for certain types of initial conditions (Aarseth 2003b, Berczik *et al.* 2006; Iwasawa, Funato & Makino 2006) such that the orbital shrinkage by gravitational wave emission becomes significant. This stage is characterized by velocities reaching an appreciable fraction of the speed of light, c. The original expressions for modelling post-Newtonian effects (Soffel 1989) were subsequently replaced by an equivalent scheme which facilitates the evaluation of consistent two-body elements (Blanchet & Iyer 2003; Mora & Will 2004). This development enables the equation of motion to be written in the concise form

$$\frac{d^2\mathbf{r}}{dt^2} = \frac{M}{r^2}\left[(-1+A)\frac{\mathbf{r}}{r} + B\mathbf{v}\right], \tag{2.1}$$

where the scaled quantities A, B represent the post-Newtonian terms. An expansion of increasing complexity up to $1/c^6$ is available for implementation.

Given the perturbing force $\mathbf{F}_{\mathrm{GR}}$, the corresponding energy loss can be obtained by integrating $\mathbf{F}_{\mathrm{GR}} \cdot \mathbf{v}dt$ in regularized form. Here the secular change is due to gravitational wave radiation represented by the terms $A_{5/2}, B_{5/2}$, while the two first orders are connected with precession. The radiation time-scale in N-body units is (Peters 1964)

$$t_{\mathrm{GR}} = \frac{5c^5\,a^4}{64m_im_0(m_i+m_0)}\frac{(1-e^2)^{7/2}}{g(e)}, \tag{2.2}$$

with m_0 usually the dominant mass and $g(e)$ a known function of the eccentricity (about 4 for large values). Even a typical hard binary with $a \simeq 1 \times 10^{-4}$ would only yield $t_{\mathrm{GR}} \simeq 1000$ for $e = 0.999$ and the present model parameters ($c \simeq 15\,000$). Rather extreme dynamical evolution is therefore required to reach the relativistic regime.

3. Multiple regularization schemes

In the following we summarize the main ideas involved in the three multiple regularization schemes outlined above. Although the TTL method (Mikkola & Aarseth 2002) does not deal specifically with the removal of singularities, it allows arbitrarily close encounters (including collisions) to be studied. Here the time transformation $t' = 1/W$ is combined with leapfrog integration of physical coordinates and velocities. The key feature is to replace an explicit evaluation of the auxiliary variable $W = \Omega(r)$ by the differential equation

$$\dot{W} = \mathbf{v} \cdot \frac{\partial\Omega}{\partial\mathbf{r}}, \tag{3.1}$$

which is usually a slowly varying quantity and integrated by $W' = \dot{W}/\Omega$. The function Ω may be taken as the sum of all inverse two-body distances when large mass ratios are present. This leads to equations of motion for the relative coordinates and velocities, $\mathbf{r}, \mathbf{v}$ of the form

$$\mathbf{r}'_i = \frac{\mathbf{v}'_i}{W}, \quad \mathbf{v}'_i = \frac{\mathbf{F}_i}{\Omega}, \tag{3.2}$$

with $\mathbf{F}_i$ the usual N-body acceleration.

The solutions are formulated as a set of leapfrog equations, with the quantity W integrated over the time-step h by

$$W_1 = W_0 + h\frac{\mathbf{v}_0+\mathbf{v}_1}{2\Omega(\mathbf{r}_{\frac{1}{2}})} \cdot \frac{\partial\Omega(\mathbf{r}_{\frac{1}{2}})}{\partial\mathbf{r}_{\frac{1}{2}}}. \tag{3.3}$$

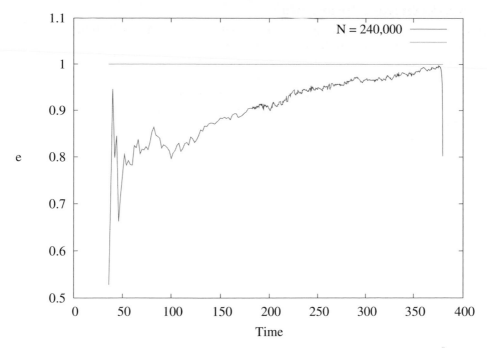

Figure 1. Eccentricity evolution of a massive BH binary for $N = 2.4 \times 10^5$.

The addition of an external perturbation $\mathbf{f}_i$ gives rise to the energy equation

$$E' = \sum m_i \mathbf{v}_i \cdot \frac{\mathbf{f}_i}{\Omega} \tag{3.4}$$

which can be integrated in the same way as the other equations. As in the post-Newtonian formulation, velocity-dependent terms can be included by using the implicit mid-point method and solved by iteration.

In the case of one dominant body, it is natural to treat the system as a wheel-spoke with all members connected to the central hub (Zare 1974). This representation can now be seen as an alternative to the chain geometry. Hence the introduction of regularized coordinates and momenta, $\mathbf{Q}$ and $\mathbf{P}$, as well as their transformations can be taken over directly from chain regularization. After multiplying by the time transformation function $g(\mathbf{q}, \mathbf{p})$ the resulting Hamiltonian can be written as

$$\Gamma^* = g(\mathbf{q}, \mathbf{p}) \left[H(\mathbf{Q}_i, \mathbf{P}_i) - E \right]. \tag{3.5}$$

Here the Hamiltonian function itself should be equal to the energy, E, along the solution path, with deviations due to numerical errors. The choice of time transformation as the inverse Lagrangian (in scaled physical units) has proved highly effective.

In contrast to standard two-body regularization where $g = r$ and the singularities are removed explicitly, the equations of motion derived from (3.5) must be differentiated term by term without employing the simplifying condition $\Gamma^* = 0$. Again the effect of perturbers in changing the internal energy must be taken into account and added to any post-Newtonian contributions.

The third method, algorithmic regularization (Mikkola & Merritt 2006), makes use of the implicit mid-point rule to achieve a time-symmetric leapfrog scheme. In this way, velocity-dependent terms of the post-Newtonian expansion can be handled by the extrapolation method. This elegant algorithm produces exact solutions for unperturbed

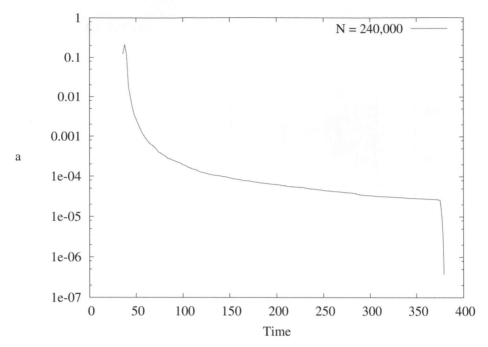

Figure 2. Semi-major axis of a massive BH binary for $N = 2.4 \times 10^5$.

two-body motion as well as accurate results in strongly interacting few-body systems. An attractive feature is that the method works for arbitrary mass ratios. The new formulation employs two equivalent time transformations for coordinates and velocities, respectively, given by

$$t'_q = \frac{1}{\alpha T + B}, \quad t'_v = \frac{1}{\alpha U + \beta \Omega + \gamma}, \qquad (3.6)$$

where α, β and γ are dimensionless constants and T, U and B are the kinetic, potential and positive binding energy. As above, Ω can be taken to be the inverse sum of all separations. Moreover, the slowly varying quantities Ω and B are obtained by integration. These time transformations are qualitatively similar to the inverse Lagrangian used in wheel-spoke regularization. We also note that the chain data structure is utilized in order to prevent loss of accuracy.

4. *N*-body implementations

Each method requires a suitable subsystem to be present which is ideally of a long-lived nature. In the first instance, this usually occurs after the emergence of a hard binary containing one massive object near the centre. Other nearby members are then added to the subsystem using standard selection criteria as for chain regularization. Conversely, any particles moving away from the BH system are included with the perturbers. These procedures entail updating the internal energy consistently without explicit evaluation of the dominant terms. The membership change also requires initialization of force polynomials for the new centre of mass and any ejected particle. Since the simulations are made on a GRAPE-6A (or GRAPE-6 initially), this procedure requires making differential force corrections on the host computer for interactions between subsystem members

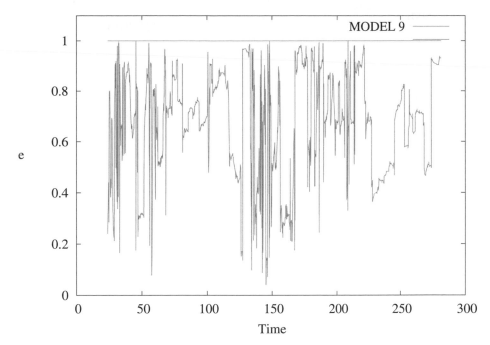

Figure 3. Eccentricity of the innermost binary, model M9.

and perturbers. Finally, a check of total energy conservation is facilitated by separate integration of the perturbations and relativistic contributions.

The special treatment for adapting the TTL method to a large-N simulation is analogous to that of chain regularization (Aarseth 2003a). The equations of motion are integrated to high accuracy and include the effect of perturbers selected in the usual way. Conversely, the evaluation of the perturber forces take into account the structure of the subsystem. In order to study a BH binary with massive components, we choose two identical spherical systems in an elliptic orbit with eccentricity $e_{\rm orb} = 0.8$ and a massive object at the centre of a cusp-like density distribution. In such a case of two merging clusters, the massive binary formed on a short time-scale (about 36 N-body units) and reached the hard binary stage ($a = 4 \times 10^{-4}$) at twice this time (see Fig. 2).

The algorithms for the wheel-spoke regularization have been described in considerable detail elsewhere (Zare 1974; Aarseth 2003a; Aarseth 2007). Here we emphasize that each subsystem interaction with the massive reference body is treated as a standard KS regularization (Kustaanheimo & Stiefel 1965), while the other internal interactions are subject to a small softening to smooth near-collisions of point-masses. Again the first-order equations of motion are advanced by a high-order integrator (Bulirsch & Stoer 1966). In view of the large mass ratio employed (around 300), only a few perturbers are usually selected for 3–4 members of the subsystem in this preliminary investigation.

A sequential decision-making strategy has been implemented for including relativistic effects, initially up to 2.5PN (Aarseth 2003b). The main idea is to treat only the most dominant two-body motion. Moreover, the relevant terms are included progressively according to the value of $t_{\rm GR}$. For convenience we choose $c = 3 \times 10^5 / V^*$, where V^* is the rms velocity in $\rm km\,s^{-1}$. Hence for a specified total mass, relativistic effects can be examined via different values of the half-mass radius and the equilibrium velocity dispersion.

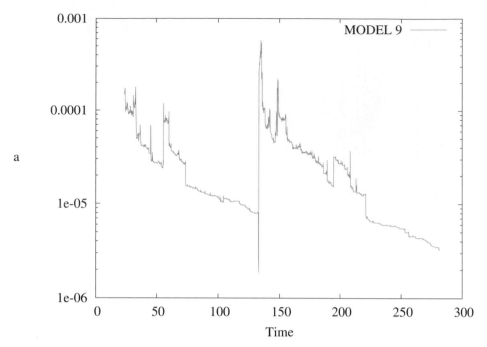

Figure 4. Semi-major axis of the innermost binary, model M9. GR coalescence occurred at
$t = 32, 33, 56, 133, 148, 195, 260$.

If the time-scale falls below a certain value (say $t_{\mathrm{GR}} \simeq 1000$) or the hyperbolic velocity exceeds $0.001c$, we first activate the radiation term by itself. The terms 1PN, 2PN, 3PN are then included for progressively shorter values chosen experimentally (say 100, 10, 1). This procedure is based on the assumption that precession only plays an important role in detuning the eccentricity growth in the later stages, which has been tested for some idealized cases of nearly isolated two-body motion. In order to reach the relativistic regime, it is necessary to achieve considerable shrinkage by dynamical evolution and/or a large eccentricity. Unless stellar disruption occurs, gravitational coalescence is adopted if the distance falls inside three Schwarzschild radii, $6m_0/c^2$. Note that the two-body elements a and e needed for decision-making are evaluated consistently with the corresponding post-Newtonian expansion (Mora & Will 2004) in order to avoid spurious effects.

An early N-body simulation of relativistic effects in compact systems (Lee 1993) examined binary formation by gravitational radiation capture using KS regularization. More recently, this approach was extended up to 2.5PN for a similar study of collisional runaway (Kupi, Amaro–Seoane & Spurzem 2006). A fully consistent scheme was employed, leading to a shorter merging time-scale for the central object.

5. Some results

The special methods described above have been employed to study centrally concentrated systems containing one or more massive bodies. First we display the behaviour of a massive binary which formed as the result of two merging clusters, each with $N_0 = 1.2 \times 10^5$ equal-mass particles and a single body with mass ratio $m_0/m_i = (2N_0)^{1/2}$. Experience shows that a square root relation for the mass is still sufficient for such a binary to dominate the central region. Here the eccentricity increased steadily until

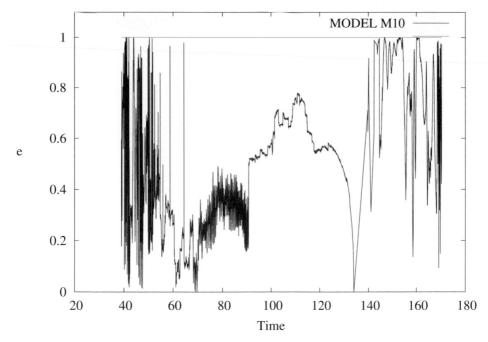

Figure 5. Eccentricity of the innermost binary, model M10.

relativistic effects became important ($V^* \simeq 16\,\mathrm{km\,s}^{-1}$). At present there is no good theoretical explanation for this behaviour. Likewise, a significant shrinkage of the semi-major axis took place. Most of this was due to dynamical evolution by sling-shot interactions until the final stage of rapid energy loss which led to gravitational coalescence.

The onset of the rather steep decrease in Fig. 2 is due to the slightly delayed experimental activation of the relativistic terms, which were included up to order $1/c^5$ (Aarseth 2003b). However, the eccentricity increase to a large value ($e = 0.996$) by dynamical effects shown in Fig. 1 would also speed up the shrinkage. Although various refinements have been added over the years, this early application of the TTL method demonstrated its usefulness in dealing with a difficult problem.

Simulations with the wheel-spoke method have been made for point-mass systems as well as cluster models containing white dwarfs of mass $1\,M_\odot$ and radius $r^* = 5 \times 10^{-5}$ au. With only one BH present, the qualitative behaviour is now quite different. The evolution of a point-mass model is illustrated in Fig. 3 and 4. In this model, a rather small value of the half-mass radius was used ($r_\mathrm{h} \simeq 0.1\,\mathrm{pc}$ or $c \simeq 5000$) and hence the coalescence distance is larger than in similar models with $r_\mathrm{h} \simeq 1\,\mathrm{pc}$ which also produced 5–6 such events. A number of large amplitude excursions in eccentricity can be seen. This behaviour is the hallmark of Kozai oscillations (Kozai 1962) which in many cases are induced by the second innermost stable orbit (Aarseth 2007).

The corresponding evolution of semi-major axis in Fig. 4 is characterized by several episodes of orbital shrinkage, followed by coalescence. Each coalescence is controlled by the minimum pericentre distance, $a(1 - e)$, where the semi-major axis has a tendency to decrease as a result of dynamical evolution (although reversed near the middle).

Several white dwarf models were also investigated using similar initial conditions (Aarseth 2007). In such cases, disruption takes place when the pericentre distance falls below about $7r^*$. Even so, this is sufficiently small for significant energy loss by

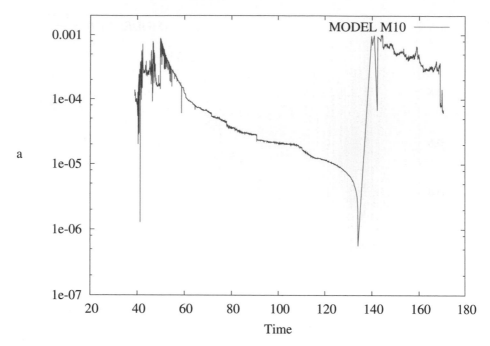

Figure 6. Semi-major axis of the innermost binary, model M10.

gravitational radiation, and hence would act to increase the number of disruptions which may be observable by LISA technology.

Recent efforts have been concerned with evaluating the algorithmic regularization code (Mikkola & Merritt 2006) in an N-body environment, again with the above realistic conditions of a white dwarf population. Extensive tests with a single massive object confirmed the qualitative results obtained by the wheel-spoke code. At the same time, the decision-making for efficient usage was improved in response to various technical problems. Very recently, a more general mass distribution was studied. In order to bypass the generation of appropriate initial conditions, several simulations were started from $t = 39$ of a wheel-spoke model by inserting artificially two massive objects with eccentric orbits in the inner region. The code is sufficiently robust to tolerate such a discontinuity and the evolution progressed normally.

As expected, the presence of two extra heavy objects leads to orbital decay towards the centre by dynamical friction. In the first such attempt (model $M10$), the extra masses were quite modest (4×10^{-4} and 3×10^{-4}) with a favourable value for the rms velocity, $V^* \simeq 300 \, \mathrm{km \, s^{-1}}$ (or $c = 1000$ in N-body units). Consequently the early evolution in Figs. 5 and 6 was qualitatively similar to some previous models, with six coalescence events due to the Kozai mechanism. At a later stage, all three heavy objects formed a compact subsystem, after which the lightest was ejected but not with sufficient energy to escape. There followed a long period with the dominant binary shrinking slowly by GR energy loss until coalescence. The third BH also returned to the centre before this event and underwent coalescence some 10 time units later. Consequently, the later stage was again dominated by a single massive body and three more mergers took place.

We also report on two models ($M11$ and $M12$) with more realistic parameters. Two additional BHs were added with masses 2×10^{-3} and 1×10^{-3} in a less dense system with $V^* \simeq 20 \, \mathrm{km \, s^{-1}}$. Consequently, $c = 15\,000$ was adopted. Starting from $t = 39$ as

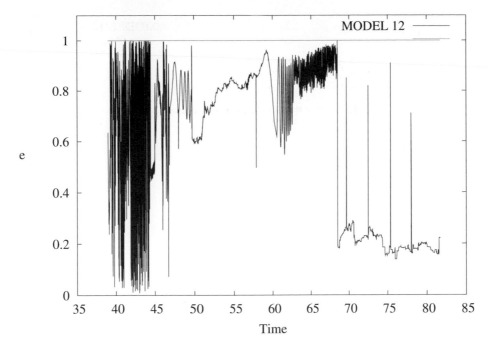

Figure 7. Eccentricity of the innermost binary, model M12.

above, the second and third BHs already formed part of the compact subsystem at $t \simeq 44$ and $t \simeq 48$, respectively. The latter was ejected but returned for an energetic sling-shot interaction (Saslaw, Valtonen & Aarseth 1974) leading to escape from the cluster. The subsequent semi-major axis declined to $a_{BH} \simeq 3.5 \times 10^{-5}$ at $t = 112$ when the calculation was halted, with the GR time-scale still quite large ($e \simeq 0.66$). It is noteworthy that the final binding energy exceeded 33 % of the total energy, all due to dynamical effects.

Model $M12$ exhibited another type of behaviour, illustrated in Figs. 7 and 8. The early stage ($t < 46$) gave rise to Kozai oscillations up to $e = 0.9999$ involving the primary BH but only sufficient to touch the relativistic regime. The second BH formed a binary with the first soon after ($t \simeq 50$), whereupon the innermost binary became wider. In view of the considerably smaller rms velocity, relativistic effects only played a minor role. However, near $t = 68$, the eccentricity of the dominant binary ($M = 5 \times 10^{-3}$) did reach $e_{max} = 0.99998$, induced by the third BH with small eccentricity and large inclination, which resulted in GR coalescence. This model therefore ended up with a low-eccentricity BH binary, decreasing slowly in size by dynamical means only.

6. The future

In this contribution, we have demonstrated the practical usefulness of three special regularization methods for treating one or more black holes in a dense stellar system. Moreover, each method is capable of dealing with considerable dynamical shrinkage and the calculation can be extended to the relativistic regime if necessary. Such problems offer a rich complexity of outcomes for future studies. In particular, it will be desirable to extend the simulations to larger systems. Based on present experience, a careful treatment is needed to deal with short time-scales which are a feature of compact subsystems. Observational imprints of such interactions will undoubtedly reveal new secrets.

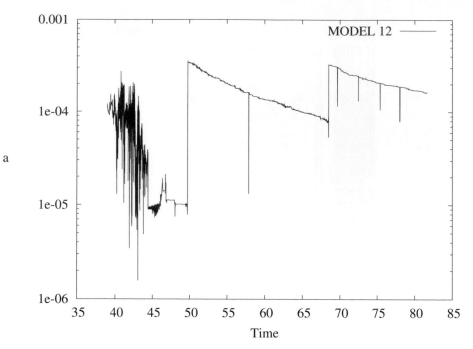

Figure 8. Semi-major axis of the innermost binary, model M12.

Acknowledgements

I am greatly indebted to Dr Seppo Mikkola for providing the stand-alone algorithmic regularization code as well as much technical advice.

References

Aarseth, S. J. 2003a, *Gravitational N-Body Simulations* (Cambridge Univ. Press)
Aarseth, S. J. 2003b, *AP&SS* 285, 367
Aarseth, S. J. 2007, *MNRAS* 378, 285
Aarseth, S. J. & Zare, K. 1974, *Celes. Mech.* 10, 185
Berczik, P., Merritt, D., Spurzem, R., & Bischof, H. 2006, *ApJ* (Letters) 642, L21
Blanchet, L. & Iyer, B. 2003, *Class. Quant. Grav.* 20, 755
Bulirsch, R. & Stoer, J. 1966, *Num. Math.* 8, 1
Iwasawa, M., Funato, Y., & Makino, J. 2006, *ApJ* 651, 1059
Kozai, Y. 1962, *AJ* 67, 591
Kupi, G., Amaro–Seoane, P., & Spurzem, R. 2006, *MNRAS* 371, L45
Kustaanheimo, P. & Stiefel, E. 1965, *J. Reine Angew. Math.* 218, 204
Lee, M. H. 1993, *ApJ* 418, 147
Mikkola, S. & Aarseth, S. J. 1993, *Celes. Mech. Dyn. Astron.* 57, 439
Mikkola, S. & Aarseth, S. J. 2002, *Celes. Mech. Dyn. Astron.* 84, 343
Mikkola, S. & Merritt, D. 2006, *MNRAS* 372, 219
Mikkola, S. & Tanikawa, K. 1999, *MNRAS* 310, 745
Mora, T. & Will, C. 2004, *Phys. Rev. D* 69, 104021
Peters, P. C. 1964, *Phys. Rev.* 136, B1222
Preto, M. & Tremaine, S. 1999, *AJ* 118, 2532
Saslaw, W. C., Valtonen, M. & Aarseth, S. J. 1974, *ApJ* 190, 253
Soffel, M. H. 1989, *Relativity in Astrometry, Celestial Mechanics and Geodesy* (Springer)
Zare, K. 1974, *Celes. Mech.* 10, 207

Dynamical Evolution of Dense Stellar Systems
Proceedings IAU Symposium No. 246, 2007
E. Vesperini, M. Giersz & A. Sills, eds.

Virtual Laboratories and Virtual Worlds

Piet Hut

Institute for Advanced Study, Princeton, NJ 08540, USA
email: piet@ias.edu

Abstract. Since we cannot put stars in a laboratory, astrophysicists had to wait till the invention of computers before becoming laboratory scientists. For half a century now, we have been conducting experiments in our virtual laboratories. However, we ourselves have remained behind the keyboard, with the screen of the monitor separating us from the world we are simulating. Recently, 3D on-line technology, developed first for games but now deployed in virtual worlds like Second Life, is beginning to make it possible for astrophysicists to enter their virtual labs themselves, in virtual form as avatars. This has several advantages, from new possibilities to explore the results of the simulations to a shared presence in a virtual lab with remote collaborators on different continents. I will report my experiences with the use of Qwaq Forums, a virtual world developed by a new company (see http://www.qwaq.com).

Keywords. sociology of astronomy, methods: numerical, methods: laboratory

1. Introduction

A year ago, I was vaguely familiar with the notion of virtual worlds. I had read some newspaper articles about Second Life,† which seemed mildly interesting, but I had no clear idea about what it would be like to enter such a world. All that changed when I was invited to give a popular talk on astronomy, in Videoranch‡, another much smaller virtual world. I realized how different this type of medium of communication is from anything I had tried before, whether telephone or email or instant messaging or shared screens. There was a sense of presence together with others that was far more powerful and engaging than I had expected. I quickly realized the great potential of these worlds for remote collaboration on research projects.

Since then, I have explored several virtual environments, with the aim of using them as collaboration tools in astrophysics as well as in some interdisciplinary projects in which I play a leading role. By and large my experiences have been encouraging, and I expect these virtual worlds to become the medium of choice for remote collaboration in due time, eventually removing the need for most, but not all, long-distance travel. The main question seems to be not so much whether, but rather when this will happen. My tentative guess would be five to ten years from now, but I may be wrong: the technology is evolving rapidly, and things may change even sooner.

In any case, I predict that ten years from now we will wonder how life was before our use of virtual worlds, just like we are now wondering about life before the world wide web, and the way we were wondering ten years ago about life before email.

2. What is a Virtual World

Twenty years ago, there was a lot of hype about virtual reality, with demonstrations of people wearing goggles for three-dimensional vision and gloves that gave a sense of

---

† http://secondlife.com
‡ http://www.videoranch.com

touch. These applications have been slow to find their way into the main stream, partly because of technical difficulties, partly because it is neither convenient nor cheap to have to wear all that extra gear.

In contrast, a very different form of virtual reality has rapidly attracted millions of people: game-based technology developed for ordinary computers, without any need for special equipment. About ten years ago, on-line 3D games, shared by many users, made their debut. In such a game, each player is represented by a simple animated figure, called an avatar, that the player can move around through the three-dimensional world. What appears on the screen is a view of the virtual world as seen through the eyes of your avatar, or as seen from a point of view several feet behind and a bit above your avatar, as you prefer.

In this way, a virtual world is a form of interactive animation movie, in which each participant plays one of the characters. Currently, many millions of players take part in these games, the most popular of which is World of Warcraft†. In addition, other virtual worlds have sprouted up that have nothing to do with games, or with killing dragons or other characters. Players enter these worlds for social reasons, to meet people to communicate with, or to find entertainment of various forms. Currently the most popular one is Second Life (SL).

A lot has been written about SL, as a quick Google search will show you. Businesses have branches in SL, various universities including Harvard and MIT have taught classes, and political parties in the French elections earlier this year have been represented there. SL has its own currency, the Linden dollar, convertible into real dollars through a fluctuating exchange rate, as if it were a foreign currency. In many ways, SL functions like a nation with its own economic, social and political structure.

3. Virtual Spaces as Information Tools

The world wide web has revolutionized global exchange of information. The notion of global connectivity has been novel, but the arrangement of content has not proceeded much beyond that of the printed press, with an element of tv or movies added. The dominant model is a bunch of loose-leaved pages, which are connected through a tree of pointers, allowing the user to travel in an abstract way through the information structure. As a result, it is often difficult to retrace your steps, to remember where you've been, or to take in the whole layout of a site.

In contrast to the abstract nature of the two-dimensional web, virtual worlds offer a very concrete three-dimensional information structure, modeled after the real world. While these worlds are virtual in being made up out of pixels on a screen, the experience of the users in navigating through such a world is very concrete. Virtual worlds call upon our abilities of perception and locomotion in the same way as the real world does. This means that we do not need a manual to interpret a three-dimensional information structure modeled on the world around us: our whole nervous system has evolved precisely to interact with such a three-dimensional environment.

Remembering where you have seen something, storing information in a particular location, getting an overview of a situation, all those functions are far more natural in a 3D environment than in an abstract 2D tree of web pages. One might argue that the technological evolution of computers, beyond being simply 'computing devices', has moved in this direction from the beginning. The only reason that it has taken so long is the large demand on information processing needed to match our sensory input.

† http://www.worldofwarcraft.com

Fortunately, the steady increase in processing power of personal computers is now beginning to make it possible for everyone to be embedded in a virtual world, whenever they choose to do so, from the comfort of their own home or office. As long as you have a relatively new computer with a good graphics card and broadband internet access, there are many virtual worlds waiting for you to explore. Some of them, like Second Life, offer a free entry-level membership, only requiring payment when you upgrade to more advanced levels of activity. Getting started only requires you to download the client program to your own computer; after only several minutes you are then ready to enter and survey that virtual world.

4. Virtual Spaces as Collaboration Tools

Email and telephone have given us the means to collaborate with colleagues anywhere on earth, in a near-instantaneous way. Yet both of them have severe limitations, compared to face-to-face meetings. In neither medium can you simply point to a graph as an illustration of a point you want to make, nor can you use a blackboard to scribble some equations or sketch a diagram. Three new types of tools have appeared that attempt to remedy these shortcomings.

One approach is to use video conferencing. Each person can see one or more others, in a video window on his or her own computer. While this gives more of a sense of immediate contact, compared to a voice-only teleconference call, it is not easy to use this type of communication to share any but the simplest types of documents.

Another approach has been to give each participant within an on-line collaboration access to a window on his or her computer that is shared between all of them. Whatever one person types will be visible by all others, and in many cases everybody is connected through voice as well, as in a conference call.

The third approach, the use of a virtual world, not only combines some of the advantages of both, it also adds extra features. Unlike a video conference, where participants have rather limited freedom of movement, virtual worlds offer the possibility of exploring the whole space. And in some worlds at least, everybody present in a room can gather in front of a shared screen that is embedded in the virtual world, in order to discuss its contents.

5. Qwaq Forums

After exploring a few different virtual worlds, I settled on Qwaq† as the company of choice for my initial experiment in using virtual spaces as collaboration tools. Qwaq is a new start-up company that provides the user with ready-made virtual offices and other rooms, called *forums*. There you can easily put up the contents of various files on wall panels. Whether they are pdf files, jpeg figures, powerpoint or openoffice presentations, or even movies, you can simply drag them with your mouse from your desktop onto the Qwaq screen and position them on a wall within the virtual world shown on your screen. As soon as you do that, the file becomes visible for all other users present in the same virtual room. The rooms persist between sessions: when users later visit the same room, your files are still there to be seen.

In addition to such useful files, that can be watched and discussed by a group of users, Qwaq also allows web browsers to be opened in a wall panel. In that way, any piece of information on the web becomes instantly available for perusing by the participants in

† http://www.qwaq.com

a Qwaq forum. This is not only convenient, it helps give the people present a sense of embedding and actual presence in the room, given that their whole discussion takes place in the same virtual space, without a need to jump out of Qwaq into other applications. And watching movies together, avatars can even enjoy meta-virtual presentations within their virtual environments!

One of the most interesting features is the presence of blackboards and editors in wall panels. In this way, users can illustrate their discussions with drawings and they can type their main points directly into a file that can be jointly edited by those present. Later, each user can easily download a copy of that file onto their own computer.

Almost all of our discussions are held through direct voice communication. While there is an option for text chatting, the advantage of using a headset with a microphone to directly talk to each other is so large that we hardly ever use text. The main exception is to exchange a few words while someone is giving a lecture, in order not to interrupt the speaker, or to ask a question to the speaker which can then be answered in due time.

The underlying software environment used in Qwaq is Croquet, derived from Squeak, a language based originally on Smalltalk. Unlike the more traditional server-centered virtual world architectures, Croquet is based on peer-to-peer communication, with potentially far better scaling properties. Alan Kay, a pioneer of the 2-D windowing system for personal computers, was the primary visionary behind the Croquet system, which now has accrued a thriving community of open source contributors.

6. Two Experiments

After I learned about Qwaq, at the MediaX† conference at Stanford in April 2007, I started two independent experiments by launching two independent initiatives, or in Qwaq terminology, two 'organizations'. The first Qwaq organization, later called MICA, was aimed at my astrophysics colleagues. The second organization, called WoK Forums, was aimed at a widely interdisciplinary group of scholars.

† http://mediax.stanford.edu/

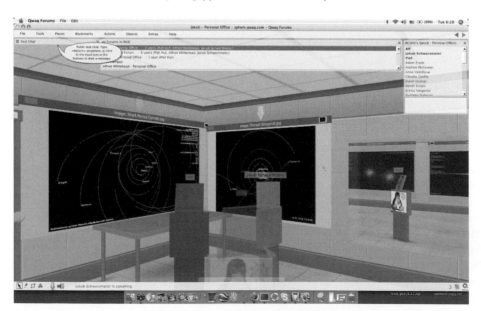

Figure 1. An early MICA meeting.

6.1. *MICA*

MICA stands for Meta-Institute for Computational Astrophysics, with *meta* derived from the term *metaverse* which is sometimes used to describe virtual worlds. During a couple months in the summer of 2007, we started to explore the use of Qwaq Forums. One function was to simply provide a meeting place for people to talk informally, a place that can play a role similar to that of a drinking fountain or a tea room in an academic department. Other activities were the organization of seminars and meetings focused on particular topics of research.

An example of the latter was the MUSE initiative, which stands for MUlti-scale MUlti-physics Scientific Environment for simulating dense stellar systems. During the MODEST-7a† meeting in Split, Croatia, all participants of the workshop were given an account in MICA, to give them a chance to follow up their discussions and collaborations after the end of the workshop.

6.2. *WoK Forums*

WoK stands for Ways of Knowing‡, a broadly interdisciplinary initiative that was started in 2006, with the aim of comparing the scientific approach to knowledge with other approaches such as those of art, spirituality, philosophy and every-day life. For half a year now, starting in the spring of 2007, we have had daily meetings in WoK Forums, with many in-depth discussions about notions such a using your own life as a lab.

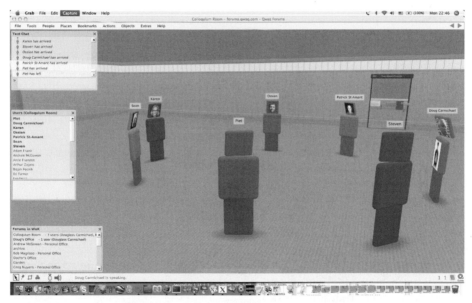

Figure 2. A recent WoK Forums meeting.

Currently we have about two dozen active participants, mostly from Europe and North America, attending on average one or more meetings a week. They range from leading figures in fields such as cognitive science, psychology, medicine, physics and finance, to graduate students and postdocs as well as independent scholars and other professionals.

† http://www.manybody.org/modest.html
‡ http://www.waysofknowing.net

7. A Tale of Three Surprises

When I started the two Qwaq organizations, MICA and WoK Forums, in May 2007, I did not know what to expect in any detail, given the novelty of the medium of virtual worlds as a collaborative tool for academic investigations. However, I had some rough picture of what I thought was likely to happen:

- a quick start for my astronomy group, a slow start for my interdisciplinary group;
- virtual worlds as a way to facilitate existing collaborations;
- an emphasis on using tools: web browsers, 3D objects, etc.

To my great surprise, all three expectations turned out to be wrong. What happened instead was:

- my interdisciplinary group took off right away;
- I found myself and others creating new collaborations;
- 3D presence was far more important than specific tools.

7.1. *First surprise*

I had expected that the computational astronomers whom I had invited to MICA would quickly take to the new environment. After all, most of them had many years of experience working with rather advanced computer tools, and many had designed and written their own code and toolboxes. In contrast, many of the broadly interdisciplinary researchers that I had gathered were not particularly computer savvy. I was wondering whether they would have any interest at all in getting into a new kind of product that they first would have to download, and then would have to learn to navigate in.

I was wrong on both counts. The latter group showed an immediate interest. Even though I had started slowly with weekly meetings, there was strong interest in more frequent gatherings, and soon we began to meet on a daily basis. In contrast, the former group, for whom I had started off with a daily 'astro tea time' showed little interest initially, and most meetings found me being in the tea room all by myself.

It took a while before it dawned upon me what was happening. The main reason was that widely interdisciplinary activities do not have any traditional infrastructure, in terms of journals, workshops, societies and other channels to fall back on. Those people interested in transcending the borders of their own discipline, not only into the immediately adjacent discipline but into a range of other disciplines, have very little to lean on. By offering a forum for discussions, I was effectively creating an oasis in a desert, attracting many thirsty fellow travelers.

In contrast, many of my astrophysics colleagues complain that nowadays there are already too many meetings and joint activities, and that it has become increasingly harder to find time to sit down and do one's own original research, amidst the continuing barrage of email, faxes, and cell phone conversations. For them I had created yet one more fountain in a fountain-filled park.

However, once my astro colleagues started to trickle in, many of them did find the new venue to be of great interest. And I had a trick to increase the trickle: threatening to close MICA sufficed to catch people's attention, and to increase attendance. Switching from the initial daily meetings to weekly meetings also helped considerably. Having a dozen people in a room discussing the latest news in computational astrophysics clearly is a lot more fun that being by yourself or with just one other random person during a daily tea time.

Meanwhile, the daily WoK Forums meetings continue to attract between half a dozen to a dozen participants on a daily basis, and the attendance continues to grow.

7.2. *Second surprise*

I had expected to kick start my virtual world activities by bringing in existing teams of collaborators, offering them the chance to continue what they were doing already, but in a new medium. Perhaps this new approach would later attract other individuals, who might be interested in joining or in starting their own projects, but that was not my initial objective.

Rarely in my life have I so completely misjudged a situation. Getting an existing group to make the transition to a totally new mode of communication turned out to be effectively impossible. Trying to change given ways of doing things provoked far more resistance than I had expected, in both my astrophysics and my interdisciplinary collaborations. Simply put, that just didn't work, period.

This became so obvious, very early on, that I had no choice but try a completely different tag. I went through my address book, and gathered names of people who just might have some interest in trying out a new medium, providing them with some bait, at the off chance that they might bite. I had no idea what criterion to use, in order to attract potential players, given the novelty of the new setup, so I just threw my net widely, waiting to see what would happen.

Roughly half the people I contacted did not reply. Of the half that did reply, roughly half told me that it all sounded fascinating but that they had no time in the foreseeable future to engage in new fun and games. Of the people who did want to give it a try, more than half quickly got discouraged after trying once or twice, and not getting immediate gratification one way or another. But many of those who remained at the end of this severe selection process were wildly enthusiastic, considering themselves to be pioneers in a whole new world.

Even in retrospect, I could never have predicted whom of my colleagues would fall in the ten percent group of early adopters. I still do not see any clear pattern or set of characteristics separating those who rushed in right away from those choosing to remain sitting on the fence. Many of those of whom I had been convinced that they would embrace virtual worlds did not, and quite a few whom I had contacted without much expectation turned out to jump in right away. In fact, for some of the early players I had not anticipated their interest at all. I had contacted them mainly so as not to make them feel left out when they would hear that I had contacted their seemingly more promising friends!

Given this randomly hit-or-miss way of collecting early players, any notion of starting with existing teams rapidly went out the window. What I wound up with was a bunch of enthusiastic tourists, eager to look around in the new virtual world that opened up unexpected horizons, with doing any kind of real work seemingly far from their mind. They were lured into a new adventure, with new toys.

After a while, though, many of the tourists began to settle down, and they started to behave more like neighbors. They began to get to know each other, although many of them had never met in real life. Among the MICA participants, there were some old hands in computational astrophysics, but there also was a freshly minted PhD in the field of education, Jakub Schwarzmeier† from Pilsen, in the Czech Republic, who happened to have written some astrophysical simulations as part of his educational research. The MICA snapshot above shows the room that Jakub created, with me visiting him together with Alf Whitehead who is a graduate student in astrophysics in a remote study course in Australia while making a living as a manager of a team of Ruby programmers in Toronto. Both Jakub and Alf had independently contacted me by email, without having

† http://home.zcu.cz/~schwarz1/index.html

met me in person, less than half a year before I started MICA, offering their help with my ACS‡ project, so it was natural to invite them into MICA.

Finally, some of the tourists that had turned into neighbors finally began to turn into collaborators. Seeing each other regularly, and becoming familiar with each others' interests, they began to spawn new ideas, some of which led to new projects, with little connection to the original motivation for them to enter the virtual world where they had met. This has happened repeatedly in my interdisciplinary organization, even though there the discrepancy between people's background and interests was the largest. In my astrophysics organization, the first mile stone was reached when Evghenii Gaburov and James Lombardi started to write a paper together within MICA, Evghenii in Amsterdam, Holland, and Jamie in Meadville, Pennsylvania, USA, which led to a preprint in July 2007 (Gaburov *et al.* 2007). As far as I know, this is the first astrophysics paper that has an explicit acknowledgment to a virtual world as the medium in which it was created.

7.3. *Third surprise*

I had expected that the main attraction of a virtual world would surely be the lure of toys: being able to design and build 3D objects, to use web browsers in-world, to travel through output of simulations, all that good stuff. The Qwaq software designers had already put an attractive example of a simulation output in their world, in the form of a simple model of an NaCl crystal. I had expected my fellow astronomers to quickly come in with their galaxy models, following in the footsteps of the Qwaq folks.

I also had thought they would quickly start playing directly with the software offered by Qwaq. In addition to existing applications, Qwaq offers possibilities for scripting new ones, using Python, and the underlying Croquet offers even more ways to get into the nuts and bolts of the whole setup. I had expected my colleagues, especially students with more time on their hands, to come in to play like kids in a candy store.

Once more I was wrong. In a place full of toys, it was the place itself, not the collection of toys, that formed a magnet. The main attraction for coming into Qwaq Forums was presence. Presence in a persistent space, a watering hole that quickly became a familiar meeting ground, this is what was felt to be the single most important aspect of the whole enterprise. Everything else was clearly secondary.

It goes back to the difference between the abstract nature of the two-dimensional world wide web, versus the concrete sense of 'being there' that we get when we enter a virtual world. Hundreds of millions of years of evolution of our nervous system, in all its perceptive, motor, and processing aspects, have prepared us for being at home in a three-dimensional life-like spatial environment.

Sharing such an environment with others turned out to be a factor that was far more important than I could have guessed. I, too, was amazed to experience the difference between a meeting in MICA or WoK Forums, on the one hand, and being part of a traditional phone conference with the same number of individuals, on the other. Teleconference calls are among the least pleasant chores to be part of, in my work. It is not always clear who is talking, there is often little real engagement, and the whole thing just feels uninspired, leading the participants to doodling or reading their email or being otherwise distracted.

In contrast, a meeting of half an hour in a virtual world feels totally different. There is a palpable sense of presence. You can see where everybody is located, people can move around and gather in front of a blackboard or poster or powerpoint presentation, and you can even hear where people are, through the stereo nature of the sound communication.

‡ http://www.ArtCompSci.org

8. Conclusion

8.1. *Lessons learned*

Of the two groups that I have invited into virtual spaces, interdisciplinary researchers were the most eager early adopters. Astrophysicists were much slower to get started, but once they were in and saw the potential of this new medium, they could quickly use the infrastructure they already had in their own field to produce new results, such as writing a preprint within a virtual space.

Individual early adopters in both groups did not come in as teams. Instead, they met whoever else was there, behaving first as tourists, then as neighbors, and only later as potential collaborators, spontaneously creating new research projects. In this way, everything that happened in virtual spaces was serendipitous; trying to get existing projects moved into virtual spaces encountered too much resistance. But even these serendipitous activities took place only after significant encouragement. To get a group of people to adapt to a new medium seems to take a considerable and ongoing amount of prodding, using whatever carrots and sticks one can find. Trying to organize any type of new activity in academia resembles the proverbial challenge of 'herding cats.'

The main attraction of meeting in a virtual space has turned out to be the shared presence in a persisting space that the participants sense and get hooked to. After a number of meetings with various stimulating conversations, the regulars want to keep coming back to the familiar setting, where they know they can meet other interesting people, old friends as well as new acquaintances. Being able to visit such a space at the click of a button is a great asset. Whether at home or at work, or briefly logged in at an airport, the virtual space is always there, and with enough participants, chances are that you will meet people whenever you log in. It can function like a tea room in an academic department, but then in a portable form, always and everywhere within reach, a curious mix of attributes.

One major obstacle that I have encountered is the fact that the earth is round. Never before have I been so conscious of the fact that we all live in different time zones. Spatial distances may drop away, when people meet in virtual spaces, but temporal zone changes don't. In my interdisciplinary group, where we have experimented for several months now with daily meetings, I was forced to introduced meetings twice every day, in order to accommodate the fact that the participants live on different continents. In addition to time zone restrictions, some participants prefer to log in from home in the evening, others from work during the day. Scheduling a weekly colloquium has been rather difficult, with some people forced to get up very early and others having to stay up till late at night.

As a result of all this, the critical mass needed to sustain a 'tea time' where enough people show up spontaneously is much larger in a virtual space than it is in an academic department. With ten people in a building, and a fixed tea time at 3 pm, chances are that at least five people show up at any given tea. With twice-daily meetings in a virtual space, and many participants showing only once a week, you need more like a hundred people in total, to guarantee the presence of five people per meeting. And if the attendance often falls below five, there may not be enough diversity to attract regular attendance.

Trying to organize people to attend events in a virtual space has something in common both with running a department and with organizing a workshop. Like the former, it requires persistent management, unlike putting together a workshop that is a one-shot event. And even though it is much easier to establish a virtual space, compared to getting the funding and spending the time to build a physical building, it is also easy to underestimate the time it takes to establish an attractive infrastructure. Try to image what it would be like to run a never-ending workshop, and you get the idea.

In the short run, there is no ideal solution to the management problem. Trying to run things purely by committee is unlikely to work, nor will it be easy to find a single individual willing to do the brunt of the work needed to set up and maintain the infrastructure of a purely virtual organization for academic research. Progress is likely to come from some kind of middle ground, with a small core group of enthusiasts willing to spend significant amounts of time getting things going, in a typical 'open source' kind of atmosphere, setting the tone by their personal example.

8.2. *Next steps*

So far, the two organizations that I have founded, MICA and WoK Forums, are still very much in their initial phase where people are getting to know each other and are getting to know the virtual environment and its possibilities. What will happen next is difficult to predict. As always in a new medium, the most interesting developments will be those that nobody expected. Even so, there are a few obvious next steps.

One thing-to-do is to create some form of library or archive, containing a chronicle of what has happened in a given virtual space. After people give lectures, it will be good to keep at least their powerpoint presentations. When people hold discussions, it would be great to catch their conclusions in a type of wiki or other structure for text that is easy to enter. It would be great if a whole session in a virtual space could be captured on video, and stored for later viewing within a room in that same virtual space.

For computational science applications, such as large-scale simulations in astrophysics, virtual spaces can be at the same time places for people to meet, and places where those people can run their experiments. With individuals represented as avatars, it is natural for them to enter virtual laboratories where they are running their simulations. Instead of the scientist sitting in front of the computer and the simulation taking place at the other side of the screen, there are many advantages in letting the scientist enter the screen and the simulated world directly. By traveling through a simulation, one can become much more intimately familiar with the details of a simulation.

Finally, here is one more intriguing possibility. If researchers who are geographically remote start writing code together within a virtual space, we can literally capture all that is said and done while writing the code. By keeping the full digital record of a coding session, and indexing it to the lines of code that were written during that session, future users of that code will always have the option to travel back in time to get full disclosure of all that happened during the writing. Many of us, struggling with legacy code that was written decades ago, would be happy to give a minor fortune for the possibility of making such a trip back in time. This approach to massively overwhelming documentation is in the spirit of what Jun Makino and I have suggested on our Art of Computational Science website†, as a move from *open source* to *open knowledge* (Hut 2007).

Acknowledgments

I thank Sukanya Chakrabarti, Derek Groen, Andrew McGowan, Sean Murphy, Greg Nuyens, Rod Rees and Patrick St-Amant for their helpful comments on the manuscript.

Reference

Gaburov, E., Lombardi, J. C., & Portegies Zwart, S. 2007, *preprint, http://arxiv.org/ abs/0707.3021*

Hut, P. 2007, *Prog. Theor. Phys. Suppl.*, 164, 38 *(http://arxiv.org/abs/astro-ph/0610222)*.

† http://www.ArtCompSci.org

Dynamical Evolution of Dense Stellar Systems
Proceedings IAU Symposium No. 246, 2007
E. Vesperini, M. Giersz & A. Sills, eds.

Current Status of GRAPE Project

J. Makino

Division of Theoretical Astronomy, National Astronomical Observatory of Japan, 2-21-1
Osawa, Mitaka, Tokyo, 181-8588
email: makino@cfca.jp

Abstract. I'll summarize the current status of GRAPE project. GRAPE-6, completed in 2002, has been used by a number of people, for a wide variety of problems such as planet formation, star cluster dynamics, galactic nuclei, and cosmology. In 2004, we started the development of the next-generation machine, GRAPE-DR. GRAPE-DR has a architecture radically different from that of previous GRAPEs. It does not have hardwired pipeline for gravitational force calculation but a large number of small and simple programmable processors. This change made it possible to apply GRAPE-DR to a wide range of problems to which GRAPE was not efficient, and at the same time it helps us to explore new algorithms for N-body simulations. The GRAPE-DR chip was completed in 2006, and second prototype board was completed in May 2007. We hope to have full production-level board commercially available by the end of year 2007. A single board will offer the theoretical peak speed of 2 Tflops, about 20 times as that of a single PCI card version of GRAPE-6.

Keywords. methods: n-body simulations, globular clusters: general

1. Introduction

The N-body simulation technique, in which the equations of motion of N particles are integrated numerically, has been one of the most powerful tools for the study of astronomical objects, such as the solar system, star clusters, galaxies, clusters of galaxies, and large-scale structures of the universe.

In particular, for the study of the dynamical evolution of star clusters, N-body simulation is now an essential tool. There are certainly faster methods like Monte-Carlo or direct integration of the Fokker-Planck equation. However, they need a number of assumptions and simplifications, which need to be tested and justified through the comparison with the result of N-body simulations. On the other hand, the main limitation of N-body simulations is their high computational cost. Calculation cost scales as $O(N^{3.3})$, where N is the number of particles. The calculation cost per time step is $O(N^2)$, if we use simple direct summation. We have additional power of $1/3$, since the average size of the time steps must be small enough so that one particle would not move the distance larger than the distance to its nearest neighbor. The final power of 1 comes from the ratio between the relaxation timescale and dynamical timescale.

In the last two decades, we have developed a series of special-purpose computers, GRAPE (GRAvity piPE). The basic idea behind the GRAPE system is to build specialized hardwares which calculate the gravitational interaction between particles, and connect then to usual general-purpose computers. All calculations other than the calculation of interactions are done on the side of the general-purpose computers. Since most of the computing time is spent on the calculation of interactions, we can accelerate the overall calculation just by accelerating the interaction calculation, and the high-performance hardware which calculates only the gravitational interaction is relatively easy to design.

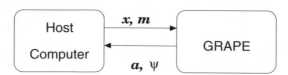

Figure 1. Basic concept of a GRAPE system.

In the rest of this paper, we briefly overview the GRAPE project and the ongoing GRAPE-DR project.

2. GRAPE project

2.1. *Basic Concept*

In our GRAPE project, we accelerate the calculation of gravitational interaction between particles by developing a computer specialized for that operation. Fig. 1 shows the basic structure of a GRAPE system. The calculation of the interaction between stars is handled by the special-purpose computer, while all other calculations, such as the time integration of stars, I/O, analysis and diagnostics are handled by a host computer. For the host computer, we used either a UNIX-running workstation or a PC (usually with Linux).

This hybrid architecture has several very important advantages. First of all, since the special-purpose part is dedicated to a single function, we can use a highly optimized architecture for that part. For GRAPE designs from GRAPE-1 (Ito *et al.* 1991) to GRAPE-6 (Makino *et al.* 2003), we adopted a fully pipelined processor designed specifically for the calculation of gravitational interaction between particles.

This fully-pipelined architecture means almost all transistors on a chip are used to implement arithmetic units, and each arithmetic unit can be optimized to a specific function assigned to it. Latest microprocessors such as Intel Core 2 Quad or Quad-core AMD Opteron has around 10^9 transistors, while a floating point arithmetic unit requires around 10^5 transistors. These microprocessors typically have eight arithmetic units. Thus, around 99.9% of all transistors are used for something other than the arithmetic units. On the other hand, a GRAPE-6 chip, consisting of only 10^7 transistors, have around 400 arithmetic units. Thus, the peak performance of a GRAPE chip is much higher than that of a general-purpose microprocessor made with the same technology, or the technology 5–10 years more advanced.

2.2. *Project history*

We started the development of GRAPE-type machines back in 1989. The first machine, GRAPE-1, was an experimental hardware with a very short word format (relative force accuracy of 5% or so), and was not really suited for simulations of collisional systems. However, its exceptionally good cost-performance ratio made it useful for simulations of collisionless systems. Also, we developed an algorithm to accelerate the Barnes–Hut tree algorithm using GRAPE hardware (Makino 1991), and developed GRAPE-1A (Fukushige *et al.* 1991), which was designed to achieve good performance with the treecode. Thus, the GRAPE approach turned out to be quite effective, not only for collisional simulations, but also for collisionless simulations as well as SPH simulations (Umemura *et al.* 1993; Steinmetz 1996). GRAPE-1A and its successors, GRAPE-3 (Okumura *et al.* 1993) and GRAPE-5 (Kawai *et al.* 2000), have been used by researchers worldwide for many different problems.

GRAPE-4 (Makino *et al.* 1997) was a single-LSI implementation of GRAPE-2, or actually that of HARP-1 (Makino *et al.* 1993), which was designed to calculate force and its time derivative. A single GRAPE-4 chip calculated one interaction in every three clock cycles, performing 19 operations. Its clock frequency was 32 MHz and peak speed of a chip was 640 Mflops.

A major difference between GRAPE-4 and previous machines was its size. GRAPE-4 integrated 1728 pipeline chips, for a peak speed of 1.08 Tflops. The machine was composed of 4 clusters, each with 9 processor boards. A single processor board housed 48 processor chips, all of which shared a single memory unit through another custom chip to handle predictor polynomials. GRAPE-4 chip used two-way virtual multiple pipeline, so that one chip looked like two chips with half the clock speed. Thus, one GRAPE-4 board calculated the forces on 96 processors in parallel. Different boards calculated the forces from different particles, but to the same 96 particles. Forces calculated in a single cluster were summed up by a special hardware within the cluster.

In 2002, we completed GRAPE-6 (Makino *et al.* 2003). It is a direct successor of GRAPE-4. The processor chip of GRAPE-6 integrates six force-calculation pipelines, each of which can calculate one interaction per clock cycle. The clock frequency of GRAPE-6 is 90 MHz. Thus, a single GRAPE-6 chip is around 50 times faster than a single GRAPE-4 chip. The total system with 2,048 chip offers the theoretical peak speed of 64 Tflops.

The concept of special-purpose computer for the long-range interaction between particle can be applied to other particle-based simulations. In fact, there were a number of attempts to develop special-purpose computers for molecular dynamics, and some of them used the pipeline architecture rather similar to GRAPE pipeline (Bakker & Bruin 1988; Fine, Dimmler & Levinthal 1991). We also applied the GRAPE architecture to molecular dynamics, starting with GRAPE-2A (Ito *et al.* 1994). It was followed by the custom-chip version, MD-GRAPE (Fukushige *et al.* 1996), and then by massively parallel MDM (Narumi *et al.* 1999). An even faster Protein Explorer was completed in 2006 (Narumi *et al.* 2006).

3. GRAPE-DR

3.1. *Problem with special-purpose architecture*

In July 2004, we were awarded the grant to develop the next-generation GRAPE system, which we call GRAPE-DR. This grant was, however, not for a special-purpose computer for astrophysical *N*-body simulation, but for a programmable massively-parallel processor. In the following we overview what is GRAPE-DR and why we chose to develop such a system.

As we summarized in the previous section, as far as the achieved speed is concerned, GRAPE hardwares have been pretty successful. Moreover, in the field of the dynamical evolution of star clusters, most of recent *N*-body simulations were performed on GRAPE hardwares. Thus, at the time of the completion of GRAPE-6, it was clearly desirable to develop next-generation GRAPE hardware. However, there was one quite practical limitation. The initial cost to design and fabricate a custom LSI chip has been increasing exponentially. For the processor of GRAPE-4 (year 1992), we paid around 200K USD as the initial cost. For GRAPE-6 (1997), we paid around 1.5M USD. A new design in 2004 would have costed at least 5M USD, and that in 2008 nearly 10M USD. The total amount of grant for GRAPE-4 was 2.5M USD, and that for GRAPE-6 was 5M. We need the total grant at least three times as much as that for the initial cost of the chip, in order to make a machine with reasonable price-performance ratio. Thus, the grant of around 15M USD was necessary to start the development of the next-generation GRAPE in 2003-4.

This amount of money was way too much for the project to develop a computer which can be used in a relatively narrow field within the theoretical astrophysics.

There were several possible approaches for this problem. One would be just to forget about building a custom processor and relax. I sometimes think this might be the best approach, but there are still several other options. The second one is to use FPGAs, or field-programmable gate-array chips. An FPGA chip is a mass-produced LSI, in which a user can "program" arbitrary logic circuits. An FPGA chip consists of a number of logic elements and a programmable network which connects them. A logic element is essentially a small SRAM table with 4-5 address bits. Each logic elements can realize any combinatorial logic circuits with 4-5 inputs, and by programming the connection network one can implement more complex circuits.

The advantage of FPGA chips is that we do not have to pay the initial development cost of the chip. The disadvantage is that the size of the logic circuit which can fit into an FPGA chip is much smaller than that can fit into a custom LSI chip made using the same manufacturing technology. A logic element requires at least a few hundred transistors, while a logic gate in a custom chip requires around 10. This difference also results in the difference in the speed. Thus, using an FPGA chip for high-accuracy gravitational force calculation has been difficult.

The third approach is to try to get larger grant, for example through international collaborations. This is the way followed by many big projects in basic science, and observational astronomy is no exception. However, this is not the way to develop a special-purpose computer, since a very important requirement for the project to develop a special-purpose computer is that the development timescale must be short, in order to be able to outperform general-purpose computers. In the case of big projects in particle physics or observational astronomy, the long development time is not a fatal issue, since there are no other facilities which can do the same experiment or observation. However, in the case of a computer, the difference is essentially in just the speed. If the machine is not faster than general-purpose computers at the time of the completion, it has no value.

The fourth approach is to design a machine which can be used for applications other than the calculation of gravitational interaction between particles. There are again several approaches to achieve this goal, but with any approach, it is clear that the performance that can be achieved is significantly less that what is possible with traditional GRAPE design specialized to just one function. Even so, this approach can still be better than other approaches, in particular that of using FPGAs.

3.2. *The GRAPE-DR architecture*

Fig. 2 shows the basic structure of the new programmable GRAPE. It consists of a number of processing elements (PEs), each of which consists of an FPU and a register file. They all receive the same instruction from outside the chip, and perform the same operation.

Compared to the classic SIMD architecture such as that of TMC CM-2, the main difference are the followings.

a) PEs do not have large local memories.

b) There is no communication network between PEs.

We introduce these two simplifications so that a large number of PEs can be integrated into a single chip. If we want to have a large memory connected to each PE, the only economical way is to attach DRAM chips. However, once we decide to use external memory chips, it becomes very difficult to integrate large number of processors into a chip, since an external memory with sufficient bandwidth is practically impossible to add.

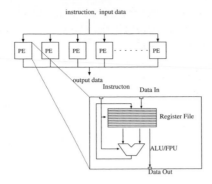

Figure 2. Basic structure of an SIMD processor.

A communication network is not very expensive, as far as it is limited into a single chip. A two-dimensional mesh network would be quite natural, for physically two-dimensional array of PEs on a single silicon chip. However, such a two-dimensional network poses a very hard problem, if we try to extend it to multi-chip systems. Again, external wires would be too costly.

If we eliminate the inter-PE communication network right from the beginning, we have no problem in constructing multi-chip systems, since PEs in different chips need not be connected.

Thus, this simple architecture has two advantages. First, we can integrate a very large number of PEs into a single chip. Second, it is easy to construct a system with multiple chips. As a result, we can construct a system with very high peak performance.

One problem with this architecture, when used as GRAPE, is that the number of processors is too large. A single chip can integrate several hundred PEs, and in order to use one PE efficiently, it is desirable to calculate the forces on several particles in one PE. Thus, the number of particles on which the forces are calculated in parallel becomes more than one thousand, which is generally too large for the numerical simulation of star clusters with individual timestep.

Of course, this problem was already there with GRAPE-4 or GRAPE-6, and we solved this problem by adding a reduction network, which takes the summation of partial forces calculated on many pipelines.

In the case of GRAPE-4, one processor board houses 48 pipeline chips and calculates forces on 96 particles in parallel. An additional summation circuit on another board takes the summation of forces from up to nine processor boards. In the case of GRAPE-6, one processor chip calculates the forces on 48 particles, and one board houses 32 processor chips. We added a reduction tree on each processor board. It takes the summation of forces calculated on 32 processor chips.

In the case of GRAPE-DR, each chip has 512 PEs, which is already a large number. So we added a reduction tree to each processor chip. We divided 512 processors to 16 groups each with 32 PEs, and added a reduction tree which takes the summation of results from these 16 groups. This reduction tree must be programmable. One node of the reduction tree of GRAPE-DR chip consists of the floating-point addition unit and integer ALU, which have the same logic design as those used in PEs, and the instructions are given from outside the chip essentially in the same way as the instructions to the PEs are supplied.

Thus, the processor has the architecture shown in Fig. 3. We added a buffer memory to each processor group, so that it can store the data sent from the external memory or the host computer.

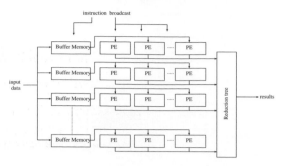

Figure 3. Modified SIMD architecture.

In this way, PEs in different blocks can calculate the forces from different particles. In addition, the reduction network allows multiple PEs in different blocks to calculate the force on the same particle from different particles. Thus, the efficiency for small-N systems or short-range force is greatly improved. In the following, we call these blocks of PE as broadcast blocks (BBs) and the buffer memory as broadcast memory (BM).

Note that the hardware cost of the buffer memory and reduction network is very small, since their cost is proportional to the number of blocks, which is a small fraction of the number of PEs.

We call this architecture GRAPE-DR, or Greatly Reduced Array of Processor Elements with Data Reduction.

4. Design of the GRAPE-DR chip

In this section, we overview the design of the GRAPE-DR chip. It integrates 512 simple processing elements (PEs), organized into 16 broadcast blocks. Each PE can do one floating-point addition and one multiplication in single precision per clock cycle, or one addition and one multiplication in double precision in every two clock cycles. The clock frequency is 500MHz and the theoretical peak performance is 512Gflops in single precision and 256 Gflops in double precision.

4.1. *PE architecture*

We designed the PE architecture so that the hardware is simple and yet can achieve high performance. Fig. 4 shows the architecture of a PE of the GRAPE-DR chip. A PE consists of a floating-point adder, a floating-point multiplier, an integer ALU, a three-port general-purpose register file (GP reg), a single-port local memory, and an additional dual-port working register (T register). The T register is used to store temporary values. The local memory augments the size of the register file. The address generator for the local memory supports the indirect addressing, by arrowing the content of the T register to be used as the address of the local memory. It also supports constant-stride access during vector operations. Storing of the results to memory units (GP reg and local memory) can be controlled by mask registers. Mask registers can store the flag output of the integer ALU and the floating-point adder.

Each PE has two fixed inputs, PEID and BBID. PEID gives the index of PE within its broadcast block, while BBID gives the index of the broadcast block itself. Using these fixed-number inputs and mask registers, we can control individual PEs independently.

The broadcast memory is dual-ported. With the current GRAPE-DR design, the data in the broadcast memory can be written directly to all of GP register, T register and

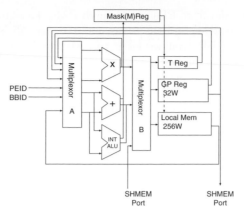

Figure 4. Structure of a Processor Element.

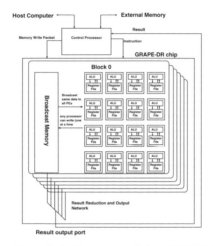

Figure 5. Overall Architecture of the GRAPE-DR chip.

the local memory, while only the data in the GP register can be transferred to to the broadcast memory.

The basic data format is a 72-bit floating-point format, with 1-bit sign, 11-bit exponent and 60-bit mantissa. We call this format double-precision. It also supports single-precision format with 24-bit mantissa. The integer ALU operates on 72-bit integers. The floating-point adder unit also work in 72-bit double-precision data, but it has the flag to round the output to single-precision format. Also, it has the flag to handle unnormalized numbers, for both the input and output.

The integer ALU can perform most of basic integer arithmetic and logical operations, including shift operations.

4.2. *Chip architecture*

Fig. 5 shows the overall architecture. We so far showed PEs in a two-dimensional grid, but from hardware point of view it is more appropriate to regard the structure as a two-level hierarchy. The chip consists of multiple broadcast blocks (BBs) of PEs. Each block consists of PEs and a broadcast memory (BM). All BBs receive the same data and instruction from outside the chip. The outputs of BBs are reduced by the reduction network.

All communication to and from PEs are through BMs. To write a data to one PE, first we write that data to the BM, and then transfer it to the PE's memory (or registers). To read out the data in a PE, first we let it to transfer the data to the BM, and then use the reduction network to output the data.

The reduction network has the binary tree structure, and each tree node has the floating-point adder and integer ALU of the same design as those of PEs. Thus, we can apply many different reduction operations, such as summation, multiplication, max, min, and, or etc.

4.3. *Programming GRAPE-DR*

In the case where we use this new GRAPE-DR processor as the replacement of traditional special-purpose GRAPE chip for particle-particle interaction calculation, programming is not a very large issue. We can simply write down the microcode for the gravitational force calculation, and communication library routines etc. This is not much different from the softwares necessary for traditional GRAPE hardwares. The only difference is that we also need to write the microcode, which is just several tens of lines.

For other kind of particle-particle interactions, the development process is quite similar to that for gravitational force calculation, and much of the communication library codes can be recycled. Thus, it is more efficient to let some software generate the communication library from higher-level specifications, in the way similar to the PGPG system (Hamada, Fukushige & Makino 2005). Also, writing the horizontal microcode by hand is hard, even for just a few lines. We have developed a simple symbolic assembly language, in which the program is written in a more or less human-readable way. Compiler languages are also under development.

4.4. *Parallel GRAPE-DR system*

So far, we have discussed the design of a single chip. In practice, in order to achieve a reasonable performance, it is necessary to use many of these chips for one application. In other words, we need to discuss how to construct a large parallel system.

We continued the approach we used for previous GRAPE hardwares (Fukushige, Makino & Kawai 2005). The GRAPE-DR hardware will be designed as a relatively small attached processor for UNIX/Linux running workstations or PCs, and large parallel systems will be constructed just by assembling large PC clusters in which each node is connected to small GRAPE-DR hardware.

One GRAPE-DR card will house 4 processor chips, each with its own off-chip memory. The data transfer speed between the host and GRAPE-DR card can be the bottleneck, but current fast interface standards like 8-lane PCI-Express would offer reasonable bandwidth, at least for the current GRAPE-DR chip.

We plan to complete a 4096-chip system by early 2009. It will have the theoretical peak performance of 2 Pflops for single precision and 1 Pflops for double precision. Most likely, it will be a 512-node system each with two GRAPE-DR cards.

5. Development status

We finished the physical design of the GRAPE-DR chip by the end of 2005, and received the first sample chips in May 2006. Left panel of Fig. 6 shows the top-level layout image of the chip. Each white square is one PE. The die size is 18mm by 18mm.

We have developed the GRAPE-DR test board (see the right photo of Fig. 6), which houses one GRAPE-DR chip around the same time and confirmed the operation of the chip with both the test vectors and for real applications. The test board consists of one

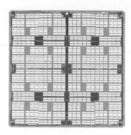

Figure 6. GRAPE-DR chip layout (left) and GRAPE-DR test board (right).

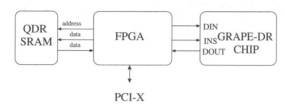

Figure 7. GRAPE-DR chip test board block diagram.

GRAPE-DR chip, one FPG chip (Altera Stratix II), and one memory chip. The interface to the host is PCI-X, and we used the IP core from PLDA. Fig. 7 shows the block diagram of the GRAPE-DR test board.

We have finished the development of the second board with PCI-Express interface and large on-board memory with DDR2 DRAM.

The measured maximum power consumption of the GRAPE-DR chip was 65W.

6. Discussion

6.1. *Comparison with related projects*

The design of ClearSpeed CX600 is quite similar to GRAPE-DR. It consists of 96 PEs, each with integer ALU, FMUL, FADD, integer MAC, 5-port register file and 6KB of memory. Compared to GRAPE-DR, the main difference is the lack of the support for the hierarchical structure (broadcast memory and reduction network). Since the number of PEs in the CX chip is still relatively small, the reduction network might not be crucial for the application performance. Its peak speed for matrix multiplication is 25 Gflops, which is about 1/10 of that of a GRAPE-DR chip.

Recent GPUs with the so-called "Unified Shader" architecture, in particular nVidia GeForce 8800, can be used as GPGPU (General-Purpose GPU), in the way rather similar to a GRAPE-DR processor chip. The peak performance numbers of GeForce 8800 and GRAPE-DR chip are rather similar. The former can perform 128 single-precision multiplications and 128 multiply-and-add operations also in single precision, at the clock speed of 1.35GHz. Thus, the theoretical peak performance is 518 Gflops. The peak performance of a GRAPE-DR chip is 512 Gflops. The transistor count of GeForce 8800 is 681M, while that of GRAPE-DR is 450M. Both are manufactured using TSMC 90nm process. Compared to GPUs, a GRAPE-DR chip lacks the fast external memory. In practice, it is not easy to use the large external memory of GPUs efficiently, since the communication bandwidth of a GPU with its host is rather slow. Compared to a GRAPE-DR chip, a GPU chip lacks the reduction tree, hardware support for double-precision operation, and few other minor things which help to achieve reasonable performance for the used as

GRAPE. At this point, it is not clear how a GRAPE-DR board compares with GPGPU for real applications.

Acknowledgments

The authors thank Toshiyuki Fukushige, Yoko Funato, Piet Hut, Toshikazu Ebisuzaki, and Makoto Taiji for discussions related to this work. The GRAPE-DR chip design was done in collaboration with IBM Japan and Alchip company. We thank Ken Namura, Mitsuru Sugimoto, and many others from these two companies. The design of the control processor on the prototype board was done by Takeshi Fujino. This research is partially supported by the Special Coordination Fund for Promoting Science and Technology (GRAPE-DR project), Ministry of Education, Culture, Sports, Science and Technology, Japan.

References

Bakker, A. F. & Bruin C., 1988, in: B. J. Alder (ed.), *Special Purpose Computers*, (San Diego: Academic Press), p. 183

Fine, R., Dimmler, G., & Levinthal, C. 1991 *PROTEINS: Structure, Function, and Genetics*, 11, 242

Fukushige, T., Ito, T., Makino, J., Ebisuzaki, T., Sugimoto, D., & Umemura, M. 1991, *PASJ*, 43, 841

Fukushige, T., Makino, J., & Kawai, A. 2005 *PASJ*, 57, 1009

Fukushige, T., Taiji, M., Makino, J., Ebisuzaki, T., & Sugimoto, D. 1996 *ApJ* , 468, 51

Hamada, T., Fukushige, T., & Makino, J. 2005 *PASJ*, 57, 799

Ito, T., Makino, J., Ebisuzaki, T., & Sugimoto, D. 1990 *Computer Physics Communications*, 60, 187

Ito, T., Fukushige, T., Makino, J., Ebisuzaki, T., Okumura, S. K., Sugimoto, D., Miyagawa, H., & Kitamura, K. 1994 *PROTEINS: Structure, Function, and Genetics*, 20, 139

Kawai, A., Fukushige, T., Makino, J., & Taiji, M. 2000 *PASJ*, 52, 659

Makino, J. *PASJ*, 43, 621

Makino, J., Fukushige, T., Koga, M., & Namura, K. 2003 *PASJ*, 1991, 55, 1163

Makino, J., Fukushige, T.K. Okumura, S., & Ebisuzaki, T. 1993 *PASJ*, 45, 303

Makino, J., Taiji, M., Ebisuzaki, T., & Sugimoto, D. 1997 *ApJ* , 480, 432

Narumi, T., Susukita, R., Ebisuzaki, T., McNiven, G., & Elmegreen, B. 1999 *Molecular Simulation*, 21, 401

Narumi, T., Ohno, Y., Okimoto, N., Koishi, T., Suenaga, A., Futatsugi, N., Yanai, R., Himeno, R., Fujikawa, S., Taiji, M., & Ikei M. 2006 in: *Proceedings of SC06*, (ACM Press), CD-ROM

Okumura, S. K., Makino, J., Ebisuzaki, T., Fukushige, T., Ito, T., Sugimoto, D., Hashimoto, E., Tomida, K., & Miyakawa, N. 1993 *PASJ*, 45. 329

Steinmetz, M. *MNRAS*, 278, 1005

Umemura, M., Fukushige, T., Makino, J., Ebisuzaki, T., Sugimoto, D., Turner, E. L., & Loeb, A. 1993 *PASJ*, 45, 311

Dynamical Evolution of Dense Stellar Systems
Proceedings IAU Symposium No. 246, 2007
E. Vesperini, M. Giersz & A. Sills, eds.

© 2008 International Astronomical Union
doi:10.1017/S1743921308016177

Fully Self-Consistent N-body Simulation of Star Cluster in the Galactic Center

M. Fujii[1,3], M. Iwasawa[2,3], Y. Funato[2] and J. Makino[3]

[1]Department of Astronomy, Graduate School of Science, The University of Tokyo,
7-3-1 Hongo, Bunkyo, Tokyo 113-0033
email: fujii@cfca.jp

[2]Department of General System Studies, Graduate School of Arts and Sciences,
The University of Tokyo, 3-8-1 Komaba, Meguro, Tokyo 153-8902
email: iwasawa@cfca.jp, funato@artcompsci.org

[3]Division of Theoretical Astronomy, National Astronomical Observatory of Japan,
2-21-1 Osawa, Mitaka, Tokyo, 181-8588
email: makino@cfca.jp

Abstract. We have developed a new tree-direct hybrid algorithm, "Bridge". It can simulate small scale systems embedded within large-N systems fully self-consistently. Using this algorithm, we have performed full N-body simulations of star clusters near the Galactic center (GC) and compared the orbital evolutions of the star cluster with those obtained by "traditional" simulations, in which the orbital evolution of the star clusters is calculated from the dynamical friction formula. We found that the inspiral timescale of the star cluster is shorter than that obtained with traditional simulations. Moreover, we investigated the eccentricities of particles escaped from the star cluster. Eccentric orbit of the star cluster can naturally explain the high eccentricities of the observed stars.

Keywords. galaxies: star clusters, methods: n-body simulations, Galaxy: center, kinematics and dynamics, stellar dynamics

---

A few dozens of very young and massive stars have been found in the central parsec of the Galaxy (Krabbe *et al.* 1995; Paumard *et al.* 2006). Some of these stars have high eccentricities (Paumard *et al.* 2006; Lu *et al.* 2006). In the central parsec, in situ formation of these stars seems difficult because of the strong tidal field of the central black hole (BH). Star cluster inspiral scenario (Gerhard 2001) can overcome this difficulty, but numerical simulations have shown that it would take too long time for inspiraling (Portegies Zwart *et al.* 2003; Gürkan & Rasio 2005). In these works the orbit of the cluster was calculated using the dynamical friction formula (Chandrasekhar 1943). This "traditional" analytic approach might have overestimated the inspiral timescale. Fujii, Funato & Makino (2006) showed that the orbital decay of the satellite is much faster than those calculated analytically. This difference was caused by particles escaped from the satellite. It should also occur in the case of star clusters. Therefore, a fully self-consistent N-body simulation is necessary.

However, such a fully self-consistent N-body simulation has been impossible with conventional numerical methods. While star clusters need a very accurate scheme, galaxies contain too many particles to use it. To overcome this problem, we have developed a new tree-direct hybrid scheme, the "Bridge" scheme (Fujii *et al.* 2007a). In this scheme, the internal interactions of star clusters are calculated accurately using the direct Hermite scheme, while all other interactions are calculated with the tree algorithm (Barnes & Hut 1986). We combined these two methods by extending the idea of the MVS scheme (Wisdom & Holman 1991; Kinoshita, Yoshida, & Nakai 1991).

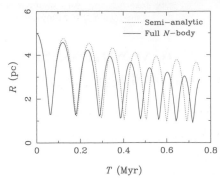 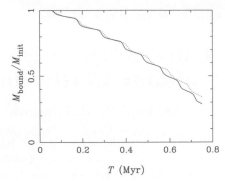

Figure 1. The distance of the star cluster from the GC (left) and the bound mass of the star cluster (right) plotted as a function of time. Solid and dotted curves show the result of the full N-body simulation and that of the "traditional" simulations, respectively. The initial orbit of the star cluster is eccentric

Using the Bridge scheme, we performed fully self-consistent N-body simulations of star clusters within their parent galaxies and compared the results with those obtained using the "traditional" simulations (Fujii *et al.* 2007b). We adopted a King model as a model of the star cluster and assigned each star a mass randomly drawn from a Salpeter (1955) initial mass function. We calculated two orbits; one is circular with the initial position from the GC, r_{init}, is 2 pc and the other is eccentric with $r_{\mathrm{init}} = 5$ pc. We used GRAPE-6 (Makino *et al.* 2003) for force calculation.

Fig. 1 shows the orbit and bound-mass of the star cluster with eccentric orbit. The orbital decay in the full N-body simulation is faster by 30–40% than that in traditional simulation. On the other hand, the evolution of the bound mass of the star cluster is almost the same between the two. This result suggests that previous studies underestimated the inspiral timescale of star clusters. In addition, eccentric orbits are preferable,because star clusters with eccentric orbits can approach to the GC much faster than those with circular orbits. We also investigated the eccentricities, e, of stars escaped from the star cluster (i.e. unbound stars). In the case of the eccentric orbit, the eccentricities of the escapers are higher than those in the case of the circular orbit. Thus, if the star cluster is initially in an eccentric orbit, it naturally explains the observations.

References

Barnes, J. & Hut, P. 1986, *Nature* 324, 446
Chandrasekhar, S. 1943, *ApJ* 97, 255
Fujii, M., Funato, Y., & Makino, J. 2006. *PASJ* 58, 743
Fujii, M., Iwasawa, M., Funato, Y., & Makino, J. 2007a, *PASJ* accepted, arXiv: 0706.2059
Fujii, M., Iwasawa, M., Funato, Y., & Makino, J. 2007b, arXiv: 0708.3719
Gerhard, O. 2001, *ApJ* 546, L39
Gürkan, M. A. & Rasio, F. A. 2005, *ApJ* 628, 236
Kinoshita, H., Yoshida, H. & Nakai, H. 1991, *Cel. Mech. and Dyn. Astr.* 50, 59
Krabbe, A. *et al.* 1995, *ApJ* 447, L95
Lu, J.R. *et al.* 2006, *JPhCS* 54, 279
Makino, J., Fukushige, T., Koga, M., & Namura, K. 2003, *PASJ* 55, 1163
Paumard, T. *et al.* 2006 *ApJ* 643, 1011
Portegies Zwart, S. F., McMillan, S. L. W., & Gerhard, O. 2003, *ApJ* 593, 352
Salpeter, E. E. 1955, *ApJ* 121, 161
Wisdom, J. & Holman, M. 1991, *AJ* 102, 152

Dynamical Evolution of Dense Stellar Systems
Proceedings IAU Symposium No. 246, 2007
E. Vesperini, M. Giersz & A. Sills, eds.

Test of the Accuracy of Approximate Methods to Handle Distant Binary-Single Star Encounters

Yoko Funato[1], D.C. Heggie[2], P. Hut[3] and Jun Makino[4]

[1]Department of Arts and Sciences, University of Tokyo,
Present address: General System Studies, Dept. Arts and Sciences, University of Tokyo, Komaba, Meguro-ku, Tokyo, 153 Japan
email: funato@artcompsci.org

[2]School of Mathematics, University of Edinburgh, UK
email:d.c.heggie@ed.ac.uk

[3]Institute for Advanced Study, USA
email:piet@ias.edu

[4]Division of Theoretical Astronomy, National Astronomical Observatory Japan,
email:makino@cfca.jp

Abstract. In the numerical simulations of evolution of star clusters, binary-single star interactions frequently take place. Since the direct integration of them is time consuming, distant interactions between binaries and field stars are often integrated by using some approximations. Traditionally the effect of the error caused by the approximated treatment is regarded as small enough to be ignored. However, if we have a binary-dominated core, the energy drift is large.

In this study, we perform numerical experiments to evaluate the effect of neglecting the weak perturbation from distant single particles. We developed an N-body integrator which can manipulate multiple precision floating point numbers.

Keywords. methods:n-body simulations

1. Introduction

1.1. *A Binary and a single star interactions*

Binary-single star interactions have an important role for the evolution of star clusters. Though the importance of binary-single star interactions has been well established, to evaluate their effect quantitatively is difficult and requires numerical simulations. However, the numerical simulation of the binary-single star interactions in star cluster evolution is also difficult even for the cases of weak interactions, such as distant encounters of binary-single stars and quasi-stable hierarchical triples. In this study, we investigate the correctness of the approximated techniques used in the numerical study of these weak interactions.

In order to numerically integrate accurately, the stepsize should be less than 0.01 times the period of the binary. The stepsize is much smaller than the orbital time of the encounter so that the number of steps becomes large. This causes two problems: too long computing and growth of accumulation of round-off errors.

To avoid these problems, usually approximated techniques are employed. One of which is an individual time step algorithm. The other is a single star approximation of the binary.

Though these techniques have been used, the adequacy of these approximations is not well understood. There are several mechanisms which may not be treated correctly in

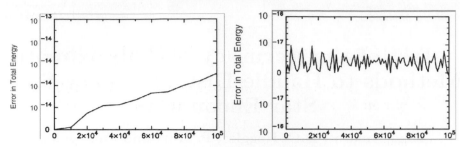

Figure 1. Growth of errors in total energy. *Left:* double precision *Right:* octuple precision

previous numerical studies. First, it is not well known when a binary can be treated as a single mass and when it should be treated as two stars. Second, though the energy can be regarded as an adiabatic invariant, the evolution of eccentricity of inner binary is not negligible (e.g., Heggie & Rasio, 1996). Third, the interactions between stars with large different time scale, such as stars in the core and those in the halo, are also important for the entire cluster evolution (Makino 2007; Nitadori 2007).

We plan to do numerical experiments to check the correctness of the approximate treatments of binaries. In order to avoid the accumulation of round-off errors, a high precision integrator is required. Using this integrator, we will carry out numerical experiments to estimate effect of weak distant encounters to cluster evolution.

2. Numerical experiment

For our experiment, we wrote an integrator which can manipulate any precision floating number. In our integrator, the precision is a free parameter. For the integrator we implemented the time-symmetric 4th order Hermite scheme to avoid an error growth due to truncation errors.

In Fig. 1 the result of the integration of the evolution of a hierarchical triple is shown. In Fig. 1, the left and right panels correspond to a double and octuple precision calculation, respectively. In this calculation the stepsize is set so as to the truncation error per one step is $\sim 10^{-16}$. Here the errors in total energy is plotted against time. In the left panel, the growth of error takes place. This growth is due to the accumulation of round-off errors. The right panel shows the result using the octuple precision integrator. Contrary to the result shown on the left panel, no error growth is observed. These figures show that our integrator works as well as expected.

Using this integrator, we are going to perform experiments of binary-single star interactions in future.

References

Heggie, D. C. & Rasio, F. 1996, *MNRAS.* 282, 1064
Makino, J. 2007, this volume
Nitadori, K. 2007, this volume

Dynamical Evolution of Dense Stellar Systems
Proceedings IAU Symposium No. 246, 2007
E. Vesperini, M. Giersz & A. Sills, eds.

© 2008 International Astronomical Union
doi:10.1017/S1743921308016190

TKira—A Hybrid N-Body Code

Ernest N. Mamikonyan[1], Stephen L. W. McMillan[1], Simon F. Portegies Zwart[2,3] and Enrico Vesperini[1]

[1]Department of Physics, Drexel University, Philadelphia, PA 19104, USA

[2]Astronomical Institute "Anton Pannekoek", University of Amsterdam, Amsterdam, the Netherlands

[3]Section Computational Science, University of Amsterdam, Amsterdam, the Netherlands

Abstract. Accurately modeling the evolution of a star cluster in a strong tidal field poses unique computational challenges. We present a hybrid code that combines the strengths of two different approaches to computing gravitational forces. The internal, collisional, dynamics of the cluster is followed with a direct N-body integrator, KIRA, while the galactic tidal field is modeled with a cosmological code, GADGET, that uses a Barnes-Hut tree to evaluate gravitational forces in $O(N \log N)$ time. The quadrupole moment at the center of mass of the cluster is used to compute the external potential and provides a mechanism for mass loss. This forms a robust, bidirectional interaction. The advantages of combining two highly-developed and well-established software packages at such high level are obvious and many; not the least of the these is the ability to include other physical processes, e.g., stellar evolution.

One problem to which we applied this technique is the evolution of a dense star cluster near the Galactic Center. We are also using this code to explore the effects of the strong time variation in the tidal field of merging galaxies on the evolution of young star clusters forming during the merger.

Keywords. methods: n-body simulations, galaxies: star clusters, Galaxy: center, galaxies: interactions

---

Traditional methods of modeling dynamical friction on star clusters have been, principally, semianalytical approximations based on assumptions of smoothness of the galactic potential. While for a certain, limited, class of problems it is valid, this approach is clearly inappropriate for many interesting scenarios of cluster evolution, e.g., time variable tidal field.

We present a scheme that evolves the external system together with the cluster using a collisionless method. We decided to use existing software packages for the two parts of the system: the KIRA integrator from STARLAB (Portegies Zwart *et al.* (2001)) for detailed modeling of the internal dynamics of the cluster and GADGET (Springel *et al.* (2001)) for the external collisionless system. The simulation proceeds in the following way:

(*a*) The cluster and the galactic system are created and initialized separately, using appropriate tools.

(*b*) The particle that will represent the cluster in the external system is initialized with its mass and the desired position and velocity. At this point the initial conditions are fully specified; the next steps are shown in Fig. 1 and form the main loop of the simulation.

(*c*) GADGET integrates the external system for Δt.

(*d*) Quadrupole moment $Q_{ij} = \frac{\partial a_i}{\partial x_j}$ at the center of mass of the cluster is computed by numerical differentiation.

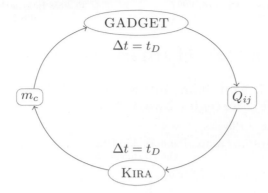

Figure 1. The high-level structure of TKIRA. After GADGET evolves the external system for Δt, the quadrupole moment $Q_{ij} = \frac{\partial a_i}{\partial x_j}$ is obtained at the cluster center of mass. KIRA uses this to approximate the external potential, evolving the cluster for another Δt; it then iteratively computes the new bound mass.

(*e*) KIRA integrates the cluster for Δt, using Q_{ij} as the estimate of the external field.

(*f*) The new bound mass of the cluster is estimated based on particle energies. A separate, distance criterion is used to permanently remove cluster particles that have strayed beyond two Jacobi radii.

While the original motivation for the development was to investigate the dynamical friction of dense nuclear clusters, it quickly became apparent that the same code can be applied to much more general problems involving strong, time-variable tidal field. To this end, we are considering the internal dynamics of a cluster whose parent galaxy is undergoing a merger. This is a considerably more difficult problem with a much larger parameter space. A somewhat related but considerably more general effort to simulate multiscale astrophysical systems is the *Multiscale Multiphysics Scientific Environment*; (see `http://muse.li`).

Even from the limited range of systems to which we have applied this code, several improvements have already become obvious. For problems where the changing mass of the cluster does not alter its orbit, the last step *(f)* can be safely omitted from the calculation. As there is no feedback of the cluster on the external potential, the quadrupole moment along the trajectory of the cluster particle can be now precomputed ahead of time, completely decoupling the two systems. Another refinement would be to include more than one cluster in the same simulation; this would open the doors to even more interesting systems. Unfortunately, this requires more substantial modification to the scheme.

Acknowledgments. This work was supported in part by NASA grants NNG04GL50G and NNX07AG95G, NWO grants 635.000.001 and 643.200.503, and the Netherlands Advanced School for Astronomy (NOVA).

References

Portegies Zwart, S. F., McMillan, S. L. W., Hut, P., & Makino, J. 2001, *MNRAS*, 321, 199
Springel, V., Yoshida, N., & White, S. D. M. 2001, *New Astron.*, 6, 79

Dynamical Evolution of Dense Stellar Systems
Proceedings IAU Symposium No. 246, 2007
E. Vesperini, M. Giersz & A. Sills, eds.

6th and 8th Order Hermite Integrator Using Snap and Crackle

Keigo Nitadori[1], Masaki Iwasawa[2] and Junichiro Makino[3]

[1]Department of Astronomy, Graduate School of Science, The University of Tokyo,
7-3-1 Hongo, Bunkyo-ku, Tokyo 113-0033, Japan
email: nitadori@cfca.jp

[2]Department of General System Studies, Graduate School of Arts and Sciences,
The University of Tokyo, 3-8-1 Komaba, Meguro-ku, Tokyo 153-8902
email: iwasawa@cfca.jp

[3]National Astronomical Observatory of Japan, Mitaka, Tokyo, 181-8588, Japan
email: makino@cfca.jp

Abstract. We present sixth- and eighth-order Hermite integrators for astrophysical N-body simulations, which use the derivatives of accelerations up to second order (*snap*) and third order (*crackle*). These schemes do not require previous values for the corrector, and require only one previous value to construct the predictor. Thus, they are fairly easy to be implemented. The additional cost of the calculation of the higher order derivatives is not very high. Even for the eighth-order scheme, the number of floating-point operations for force calculation is only about two times larger than that for traditional fourth-order Hermite scheme. The sixth order scheme is better than the traditional fourth order scheme for most cases. When the required accuracy is very high, the eighth-order one is the best.

Keywords. stellar dynamics, methods: n-body simulations, methods: numerical

1. Direct calculation of higher order derivatives

The gravitational acceleration from particle j to particle i and its first three time derivatives (we call them *jerk*, *snap* and *crackle*) are expressed as

$$\boldsymbol{A}_{ij} = m_j \boldsymbol{r}_{ij}/r_{ij}^3, \tag{1.1}$$

$$\boldsymbol{J}_{ij} = m_j \boldsymbol{v}_{ij}/r_{ij}^3 - 3\alpha \boldsymbol{A}_{ij}, \tag{1.2}$$

$$\boldsymbol{S}_{ij} = m_j \boldsymbol{a}_{ij}/r_{ij}^3 - 6\alpha \boldsymbol{J}_{ij} - 3\beta \boldsymbol{A}_{ij}, \tag{1.3}$$

$$\boldsymbol{C}_{ij} = m_j \boldsymbol{j}_{ij}/r_{ij}^3 - 9\alpha \boldsymbol{S}_{ij} - 9\beta \boldsymbol{J}_{ij} - 3\gamma \boldsymbol{A}_{ij}, \tag{1.4}$$

with

$$\alpha = (\boldsymbol{r}_{ij} \cdot \boldsymbol{v}_{ij})/r_{ij}^2, \tag{1.5}$$

$$\beta = (|\boldsymbol{v}_{ij}|^2 + \boldsymbol{r}_{ij} \cdot \boldsymbol{a}_{ij})/r_{ij}^2 + \alpha^2, \tag{1.6}$$

$$\gamma = (3\boldsymbol{v}_{ij} \cdot \boldsymbol{a}_{ij} + \boldsymbol{r}_{ij} \cdot \boldsymbol{j}_{ij})/r_{ij}^2 + \alpha(3\beta - 4\alpha^2), \tag{1.7}$$

where $\boldsymbol{r}_i$, $\boldsymbol{v}_i$, $\boldsymbol{a}_i$, $\boldsymbol{j}_i$ and m_i are the position, velocity, total acceleration, total jerk and mass of particle i, and $\boldsymbol{r}_{ij} = \boldsymbol{r}_j - \boldsymbol{r}_i$, $\boldsymbol{v}_{ij} = \boldsymbol{v}_j - \boldsymbol{v}_i$, $\boldsymbol{a}_{ij} = \boldsymbol{a}_j - \boldsymbol{a}_i$ and $\boldsymbol{j}_{ij} = \boldsymbol{j}_j - \boldsymbol{j}_i$ (Aarseth 2003).

The increase of the number of floating point operations required in the high-order schemes is small. In the fourth-order scheme (Makino 1991a; Makino & Aarseth 1992), we need 60 operations for one calculation of acceleration and jerk, if we count 10 operations

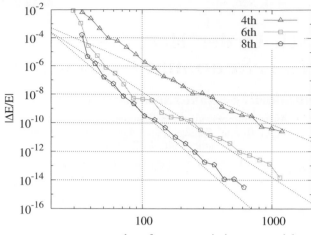

Figure 1. Maximum relative deviation of the total energy during the time integration for 10 time units, as a function of average number of timesteps per particle per unit time. Triangles, squares and pentagons represent the results of 4th-, 6th- and 8th-order schemes. The three dotted lines indicate the expected scaling relations for 4th-, 6th- and 8th-order algorithms.

for each division and square-root (Warren *et al.* 1997; Nitadori *et al.* 2006). In the sixth-order scheme, we need 97 operations for the acceleration, jerk and snap and even in the eighth-order scheme, we need only 144 operation for up to the crackle term (Nitadori & Makino 2007).

Fig. 1 shows the relation between the relative energy error after the integration for 10 time units of a 1024-body Plummer model and the average number of timesteps per particle per unit time. Here, we used the standard N-body unit (Heggie & Mathieu 1986), a softened potential with $\varepsilon = 4/N = 1/256$, and the block timestep algorithm (McMillan 1986; Makino 1991b) where all timesteps are restricted to be powers of two.

We can clearly see that the error of sixth- and eighth-order schemes are proportional to Δt^6 and Δt^8, as expected. For the relative accuracy of 10^{-8}, the sixth-order scheme allows the average timestep which is almost a factor of three larger than that necessary for the fourth-order scheme. For the relative accuracy of 10^{-10}, the eighth-order scheme allows the average timestep which is almost a factor of seven larger than that necessary for the fourth-order scheme. Even for the relatively low accuracy of 10^{-6}, the sixth-order scheme allows about a factor of two larger timestep than the fourth-order scheme does.

References

Aarseth, J. S., 2003, *Gravitational N-Body Simulations* [Cambridge Univ. Pr.]
Heggie, D. C. & Mathieu, R. D. 1986, The Use of Supercomputers in Stellar Dynamics, 267, 233
Makino, J., 1991, *ApJ* 369, 200
Makino, J. 1991b, *PASJ*, 43, 859
Makino, J. & Aarseth, S. 1992, *PASJ*, 44, 141
McMillan, S. L. W. 1986, in The Use of Supercomputer in Stellar Dynamics, ed. P. Hut & S. McMillan (New York: Springer), 156
Nitadori, K., Makino, J., & Hut, P. 2006, *New Astronomy*, 12, 169
Nitadori, K. & Makino, J. 2007, ArXiv e-prints, 708, arXiv:0708.0738
Warren, M. S., Salmon, J. K.,Becker, D. J., Goda, M. P., & Sterling, T. 1997, In *The SC97 Proceedings*, CD–ROM. IEEE, Los Alamitos, CA

Dynamical Evolution of Dense Stellar Systems
Proceedings IAU Symposium No. 246, 2007
E. Vesperini, M. Giersz & A. Sills, eds.

Embryo to Ashes
Complete Evolutionary Tracks, *Hands-off*

O. Yaron[1], A. Kovetz[1,2] and D. Prialnik[1]

[1] Department of Geophysics and Planetary Sciences, Sackler Faculty of Exact Sciences
[2] School of Physics & Astronomy, Sackler Faculty of Exact Sciences, Tel Aviv University, Israel

Abstract. We present a new stellar evolution code and a set of results, showing its capability to calculate full evolutionary tracks for a wide range of masses and metalicities. The code is meant to be used also in the context of modeling the evolution of dense stellar systems, for performing live evolutionary calculations both for 'normal' ZAMS/PRE-MS models, but mainly for 'non-canonical' (i.e. merger-products) stellar configurations. For such tasks, it has to be robust and efficient, capable to run through all phases of stellar evolution without interruption or intervention. Here we show a few examples of evolutionary calculations for stellar populations I and II, and for masses in the range 0.25–64 $M_\odot$.

Keywords. stars: evolution, Hertzsprung-Russell diagram

1. The Evolution Code

Basic Scheme The equations governing the structure and evolution of a star are those of continuity, hydrostatic equilibrium, energy transfer (radiative or convective), energy balance and composition balance. These differential equations are approximated by finite difference equations, to be solved by an iterative Newton-Raphson method. Following Eggleton (1971), the equations of structure and composition are solved simultaneously with a mass distribution function – implementing an adaptive mesh.

Input Physics Equations of state include Coulomb and quantum corrections; H and He ionization equilibria, H_2 creation, pair creation and pressure ionization (Pols *et al.* 1995) are taken into account. Using OPAL opacities (Iglesias & Rogers 1996) for high temperatures ($4.50 < logT < 8.70$) and Ferguson *et al.* (2005) for low temperatures, we create for a required metalicity our own set of opacity tables, covering both the Hydrogen mass fractions and Carbon/Oxygen excesses mass fractions from 0 to $[1 - Z]$. Electron scattering are incorporated for the highest temperatures according to Iben's fit. The radiative opacities – κ_{rad} are supplemented with conductive opacities – κ_{cond} (when applicable) as obtained from Cassisi *et al.* (2007). Nuclear reaction rates are from Caughlan & Fowler (1988) – following H, α and CO – burning. Neutrino losses are according to Itoh *et al.* (1996) – accounting for neutrino formation processes of pair annihilation, photo annihilation, plasma decay, bremsstrahlung and recombination. Several mass-loss recipes can be tested, variations to the original Reimers (1975) for advanced (POST-RGB) stages. The results displayed in this work were obtained using Bloecker (1995) $\dot{M}_{B1}$ expression (based on Bowen 1988).

Computational details Our automatically varying time steps, determined mainly by limits imposed on the changes allowed during time step, span a wide dynamic range – from $\lesssim 1\ sec$ (e.g. core He flash) to $\gtrsim 10^8\ years$ (MS). The grid mass shells span a range of $\sim 10^{-15} M_\odot$ (WD atmosphere) to $\gtrsim 10^{-1}\ M_\odot$ (inert stellar core). The typical number of grid points is kept in between 150 and 200, and a typical number of time steps for a complete evolution track is 1000.

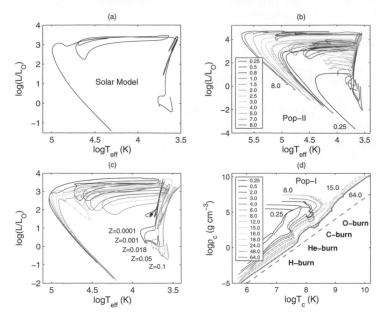

Figure 1. (a) Solar model – $1M_\odot$, $Y = 0.29$, $Z = 0.018$. (b) Complete tracks for Pop. II models ($Z = 0.001$), $0.25 - 8.0$ $M_\odot$. (c) Varying metalicities $Z = 0.0001$ to 0.1 for $1M_\odot$. MS T_{eff} and L decrease with increasing Z's; consequently – duration on the MS increases ($\tau_{MS} = 5.43$, 5.71, 9.01, 14.8 *and* 18.3 *Gyr* for the 5 Z's – 0.0001 *to* 0.1, respectively). (d) Evolution of the central stellar density and pressure of Pop. I models in the range $0.25 - 64$ $M_\odot$. Dotted line has a slope of 3 (as obtained for the $log\rho_c - logP_c$ relation of hydrostatic equilibrium under ideal gas law). Nuclear burning phases are marked along the tracks.

2. Results

In the panels of Fig. 1 we present selected preliminary calculations (serving mainly for proof of capability, specific applications are to follow). Panel (a) shows a calibration solar model, following through all evolutionary stages and ending as a cooling 0.55 $M_\odot$ CO WD. Features at 4.5 Gyr match the present sun to an accuracy of 1% or better. Panel (d) shows the evolution of the stellar central points in $log\rho_c - -logT_c$ plane, exhibiting the branching off between intermediate-mass stars that end their lives as WDs, and massive stars that go through advanced nuclear burning stages, ending their lives in dynamic core collapse (SN) – followed by the code until very high central pressures are attained and the adiabatic exponent approaches 4/3 throughout the core. Future work will extend the study and analysis of non-canonical evolution.

References

Bloecker, T. 1995 *A&A* 297, 727

Bowen, G. H. 1988 *ApJ* 329, 299

Cassisi, S., Potekhin, A. Y., Pietrinferni, A., Catelan, M., & Salaris M. 2007 *ApJ* 661, 1094

Caughlan, G. R. & Fowler, W. A. 1988 *Atomic Data and Nuclear Data Tables* 40, 283

Eggleton, P. P. 1971, *MNRAS* 151, 351

Ferguson, J. W., Alexander, D. R., *et al.* 2005 *ApJ* 623, 585

Iglesias, C. A. & Rogers, F. J. 1996, *ApJ* 464, 943

Itoh, N., Hayashi, H., Nishikawa, A., & Kohyama, Y. 1996 *ApJS* 102, 411

Pols, O. R., Tout, C. A., Eggleton, P. P. & Han, Z. 1995, *MNRAS* 274, 964

Reimers, D. 1975 *MSRSL* 8, 369

Author Index

Aarseth, S.J. – **437**
Alexander, T. – 275
Amaro-Seoane, P. 346
Anders, P. – **187**
Anderson, J. – 263
Apple, R. – **261**
Arena, S.E. – **423**

Banerjee, S. – **246**
Barret, D. – 373
Bassa, C. – 301
Bastian, N. – **32**
Baumgardt, H. – **36**, 171, 187, 195, 367, 427
Bedin, L.R. – 277
Belkus, H. – **357**
Belokurov, V. – 189
Berczik, P. – 346
Berentzen, I. – 346
Bergond, G. – 336
Bergmann, M. – 341
Bertin, G. – 423
Bhatt, B.C., – 73
Blakeslee, J.P. – 377
Boily, C.M. – **311**
Braden, E.K. – **105**
Brewer, J. – 263

Caloi, V. – 156
Capuzzo Dolcetta, R. – 431
Chambers, J.E. – 273
Charles, P.A. – 361
Chatterjee, S. – **151**
Chen, C.-W. – 377
Chen, W.P. – 73
Chumak, Y. – **107**
Church, R.P. – 273
Clark, P.C. – 3
Clarke, C.J. – **23**
Colpi, M. – 359
Cornelisse, R. – 373
Côté, P. – **377**

Dabringhausen, J. – 427
D'Antona, F. – **156**
Davies, M.B. – 176, 273
Davis, D.S. – **263**
De Angeli, F. – 273, 369
de Grijs, R. – 269
De Marchi, G. – **161**
Devecchi, B. – **359**
Dieball, A. – 321, **361**, 373

Downing, J.M.B. – **265**
Dubath, P. – 418

Eggleton, P.P. – **228**, **267**
Evans, N.W. – 189

Fellhauer, M. – **189**
Ferrarese, L. – 377
Ferraro, F.R. – **281**
Ferreira, B. – 46
Fiestas, J. – **166**
Fregeau, J.M. – 151, **239**, 351
Frinchaboy, P.M. – **109**
Froebrich, D. – 50
Fujii, M. – **467**
Fukushige, T. – **191**, 251
Funato, Y. – 467, **469**

Gaburov, E. – **193**
Gebhardt, K. – 341
Geisler, D. – **425**
Geller, A.M. – **111**
Ghosh, S.K. – 73
Ghosh, P. – 246
Gieles, M. – **171**, 193
Giersz, M. – **99**, 121
Gilmore, G.F. – 176, 189, 413
Glover, S.C.O – 3
Glebbeek, E. – **363**
Gnedin, O.Y.– **403**
Gómez, M. – 425
Gouliermis, D.A. – **61**
Gualandris, A. – 365
Gvaramadze, V.V. – **365**

Hannikainen, D. – 361
Harris, G.L.H. – 425
Harris, H.C. – 111
Harris, W.E. – **387**, 425
Heggie, D.C. – 99, **121**, 277, 469
Heinke, C.O. – 316
Hilker, M. – **427**, 429
Hurley, J.R. – **89**
Hut, P. – **447**, 469

Infante, L. – 377
Ivanova, N. – **316**
Iwasawa, M. – 467, 473

Jordán, A. – 377

Kaluzny, J. – 371

Kharchenko, N.V. – 115, 117
Kim, S.S.– 433
King, I.R. – **131**
King, N.L. – 65
Kisseleva-Eggleton, L. – 267
Kissler-Patig, M. – 429
Klessen, R.S. – **3**, 50
Knigge, C. – **321**, 331, 361, 373
Kouwenhoven, M.B.N. – **269**
Kovetz, A. – 475
Kroupa, P. – **13**, 36, 71
Kumar, M.S.N. – 50
Kundu, A. – 336, **408**
Kupi, G. – 275

Lada, E.A. – 46
Lajoie, C.-P. – **271**
Lamb, J.B. – **63**, 65
Lamers, H.J.G.L.M. – 171, 187
Lanzoni, B. – 281, **326**
Leigh, N. – **331**
Löckmann, U. – **367**
Long, K.S. – 321, 361, 373
Lotz, J. – 429

Maccarone, T.J. – **336**, 408
Mackey, A.D. – **176**, 273
Maheswar, G. – 73
Maíz-Apellániz, J. – 321
Makino, J. – **457**, 467, 469, 473
Malmberg, D. – **273**
Mamikonyan, E. – **471**
Mapelli, M. – 359
Mardling, R.A. – **199**
Mathieu, R.D. – **79**, 105, 111, 277
McClure, D. – 111
McMillan, S. – **41**, 181, 471
Mei, S. – 377
Meibom, S. – 105
Mengel, S. – **113**
Merritt, D. – 346
Meylan, G. – 418
Mieske, S. – **195**, 427
Mikkola, S. – **218**
Miller, B.W. – **429**
Milone, A. – 277
Minniti, D. – 418
Miocchi, P. – **431**
Moretti, A. – 277, **369**
Mylläri, A. – 209

Nielsen, D. – 109
Nitadori, K. – **473**
Noyola, E. – **341**

Oey, M.S. – 63, **65**
Ojha, D.K. – 73

Olczak, C. – **67**, 69
Orlov, V. – 209

Padmanabhan, T. – 311
Paiement, A. – 311
Pandey, A.K. – 73
Paresce, F. – 161
Parker, J.Wm. – 65
Parmentier, G. – **413**
Peng, E.W. – 377
Perets, H.B. – **275**
Pfalzner, S. – 67, **69**
Pflamm-Altenburg, J. – **71**
Pietrukowicz, P. – **371**
Piotto, G. – **141**, 277, 369
Piskunov, A.E. – 115, 117
Pols, O.R. – 363
Pooley, D. – 301
Portegies Zwart, S. – 41, 181, 365, 471
Porth, O. – 166
Possenti, A. – 359
Preto, M. – 346
Prialnik, D. – 475
Prieto, J. L. – 403
Pulone, L. – 161
Puzia, T. – 429

Ransom, S.M. – **291**
Rasio, F.A. – 151, 316, 351
Rastorguev, A. – 107
Rejkuba, M. – 277, **418**
Rhode, K.L. – 336
Richer, H.B. – 263
Román-Zúñiga, C.G. – **46**
Röser, S. – **115**, 117
Rubinov, A. – 209

Sagar, R. – 73
Salzer, J.J. – 336
Schilbach, E. – 115, **117**
Schmeja, S. – **50**
Scholz, R.-D. – 115, 117
Servillat, M. – **373**
Shara, M.M. – 321, 361, 373
Sharma, S. – **73**
Shih, I.C. – 336
Shin, J. – **433**
Sills, A. – 271, 331
Smith, L.J. – **55**
Sommariva, V. – **277**
Spurzem, R. – 166, 265, **346**
Sweatman, W.L. – **233**

Tacconi-Garman, L.E. – 113
Takahashi, K.– 433
Tanikawa, A. – 191, **251**

Tonry, J.L. – 377
Trenti, M. – **256**

Umbreit, S. – **351**
Urminsky, D. – **235**

Van Bever, J. – 357
Vanbeveren, D. – 357
Valtonen, M. – **209**
Verbunt, F. – **301**
Vesperini, E. – 41, **181**, 471

Webb, N.A. – 373
West, M.J. – 377
Wilkinson, M.I. – 176, 189, 273
Woodley, K.A. – 425

Yaron, O. – **475**

Zepf, S.E. – 336, **394**, 408
Zinnecker, H. – **75**
Zurek, D.R. – 321, 361, 373

Douglas Heggie

Piet Hut, Douglas Heggie and Linda Heggie

Jun Makino and Piet Hut

Christian Boily and Peter Eggleton

Ivan King

Mirek Giersz, Rainer Spurzem and Sverre Aarseth

Giampaolo Piotto

Steve Zepf

Stefan Umbreit and Marc Freitag

Eva Grebel, Andrea Dieball and Genevieve Parmentier

Steve McMillan

Frank Verbunt

Franca D'Antona

Francesco Ferraro

Doug Geisler and Bill Harris

A view of Capri at sunset

490

Capri
(Thanks to Bob Rood for this and most of the pictures included in this volume)